Risk, Systems and Decisions

Series Editors

Igor Linkov
U.S. Army ERDC, Vicksburg, MS, USA

Jeffrey Keisler
College of Management, University of Massachusetts
Boston, MA, USA

James H. Lambert
University of Virginia, Charlottesville, VA, USA

Jose Figueira
University of Lisbon, Lisbon, Portugal

Health, environment, security, energy, technology are problem areas where man-made and natural systems face increasing demands, giving rise to concerns which touch on a range of firms, industries and government agencies. Although a body of powerful background theory about risk, decision, and systems has been generated over the last several decades, the exploitation of this theory in the service of tackling these systemic problems presents a substantial intellectual challenge. This book series includes works dealing with integrated design and solutions for social, technological, environmental, and economic goals. It features research and innovation in cross-disciplinary and transdisciplinary methods of decision analysis, systems analysis, risk assessment, risk management, risk communication, policy analysis, economic analysis, engineering, and the social sciences. The series explores topics at the intersection of the professional and scholarly communities of risk analysis, systems engineering, and decision analysis. Contributions include methodological developments that are well-suited to application for decision makers and managers.

More information about this series at http://www.springer.com/series/13439

Cynthia H. Stahl • Alan J. Cimorelli

Environmental Public Policy Making Exposed

A Guide for Decision Makers and Interested Citizens

For dear Geoffrey, from that bright-eyed 9 yr old listening to his first MIRA presentation to the next generation of critical thinkers, you are my bright star. Love, Mom

🦅 Springer

Cynthia H. Stahl
US EPA
Philadelphia, PA, USA

Alan J. Cimorelli
US EPA (retired)
Philadelphia, PA, USA

ISSN 2626-6717   ISSN 2626-6725 (electronic)
Risk, Systems and Decisions
ISBN 978-3-030-32129-1   ISBN 978-3-030-32130-7 (eBook)
https://doi.org/10.1007/978-3-030-32130-7

This Springer imprint is published by the registered company Springer Nature Switzerland AG
The registered company address is: Gewerbestrasse 11, 6330 Cham, Switzerland

Foreword

I am pleased and honored to contribute this Foreword to this important book. I believe that post-normal science, which I developed with Silvio Funtowicz, has been a very useful insight into the workings of science in relation to policy. I know that it has helped many people resolve their perplexity at the difference between science as practiced and science as imagined by teachers, philosophers, and publicists. I also know that the theory as we have developed it is incomplete in several respects, both on the side of conceptual clarity and on the side of practical applicability. Contributions under the heading of PNS have mainly been applications of the basic insight, as expressed in the mantra "facts uncertain…" and the quadrant-rainbow diagram. To the question, "just how do you <u>do</u> PNS?", our answers have necessarily been a bit vague. For this, I make no apology. Silvio and I were resolved to express the basic insight as best as we could and to refrain from legislating for a practice that at that time lay totally in the future. Then as time went by and PNS gained in acceptance, we found ourselves fully occupied with studies at the more general and theoretical level. We did not have the opportunity to become "reflective practitioners" of the sort of science that we were advocating. All along, I have been keenly aware that until scientists have some clear guidance on post-normal practice, the insight will remain on the margins. Speaking somewhat paradoxically, we need to make post-normal science the new normal, even in Kuhn's sense of a semiskilled practice, in order for the basic insight to come to full maturity and effectiveness.

For this to be accomplished, we have needed a handbook, what used to be a called a *vade mecum*, offering help at all levels, from a simple rubric through to worked examples of real practice with all its complexities and difficulties. This is why I am delighted to welcome this book by Cynthia Stahl and Alan Cimorelli. It is solidly based on their extensive practice at the EPA, of involving stakeholders and communities in very real, complex issues of environmental protection. And it offers organizing concepts and techniques, whereby those complexities can be managed, to mutual benefit. I have a particular pleasure in seeing how they have adopted the "clumsy solutions" approach, as this was developed by my colleagues, Mike Thompson and Steve Rayner, and has been insufficiently appreciated up to now. Their list of requisite steps reminds me of the maxims of W. Edwards Deming, who brought quality assurance to the world of manufacture and then broader practice, animated by his idealistic vision.

This broader perspective motivates my other reason for appreciation of the book. It follows from another prophetic essay by Steve Rayner, "How fair is safe enough?". Published in 1987 with coauthor Robin Cantor, this was the first explicitly ethical critique of the technocratic approach to risk management. It anticipated what Silvio and I said about the "extended peer community" of post-normal science. Back then, such considerations were marginal; experts ruled the governance of risk and environment. But now, thanks to the populist revolt, it is all different. The perceived arrogance and hypocrisy of the economic/cultural elites affects science policy as much as politics. "Vaccination hesitation" may indeed be due as much to deficiencies in the delivery system as to anti-vaxx propaganda, but anti-GM and antinuclear campaigns have a force of their own. We have recently been reminded how the Chernobyl disaster destroyed public trust in the Soviet system. It is generally assumed that "it couldn't happen here," but for those who dare to know, that assumption is decidedly optimistic. It would not be an idle exercise to run scenarios on the social and political consequences of the collapse of one of the hundred American spent-fuel cooling pools.

The credo of the Declaration of Independence, that governments derive their just powers from the consent of the governed, is now understood to apply directly to science, risks, and the environment. Making this new constitution real will require the integration of a broad vision with a practical methodology. A good start has been made by the authors of *Environmental Policy Making Exposed*. I congratulate them, and I hope to assist in its use and development.

Oxford, UK   Jerome Ravetz
July 2019

Contents

Chapter 1
Environmental Policy: The Current Paradigm

Abstract Written for public policy practitioners and interested public policy stakeholders, this is a book about the challenges of public policy decision making that are not being met by current approaches. Applying the principles and lessons learned from the literature on decision analytic methods, stakeholder participation approaches, and public policy making, an innovative but feasible methodology for environmental decision and public policy making is unveiled. The Multi-criteria Integrated Resource Assessment (MIRA) approach provides a policy decision analytic interface between the social and physical sciences that was previously unavailable. Stakeholder inclusiveness, transparency, learning, scientific data usage, the construction of social and scientific indicators, and addressing uncertainty are overlaid across the common set of steps that are used in all multi-criteria decision making. This book challenges the current science-based decision making paradigm that the rational and objective application of scientific data alone constitutes the ideal decision making process. Effective environmental policy making requires meeting stakeholder goals by balancing science and values. This book examines current policy decision making processes and presents some dilemmas that can impede finding policy solutions. Proceeding from the conceptual to the practical, this book includes a final word on the policy implications of viewing commonly used indices through the MIRA lens.

Keywords Public policy process · Decision making · Environmental management · Stakeholder involvement · Uncertainty analysis

1.1 Introduction

Written for public policy practitioners and interested public policy stakeholders, this is a book about the challenges of public policy decision making that are not being met by current approaches. Applying the principles and lessons learned from the literature on decision analytic methods, stakeholder participation approaches, and public policy making, an innovative but feasible methodology for environmental

© Springer Nature Switzerland AG 2020
C. H. Stahl, A. J. Cimorelli, *Environmental Public Policy Making Exposed*,
Risk, Systems and Decisions, https://doi.org/10.1007/978-3-030-32130-7_1

1

decision and public policy making is unveiled. The Multi-criteria Integrated Resource Assessment (MIRA) approach provides a policy decision analytic interface between the social and physical sciences that was previously unavailable. Stakeholder inclusiveness, transparency, learning, scientific data usage, the construction of social and scientific indicators, and addressing uncertainty are overlaid across the common set of steps that are used in all multi-criteria decision making. This book challenges the current science-based decision making paradigm that the rational and objective application of scientific data alone constitutes the ideal decision making process. Effective environmental policy making requires meeting stakeholder goals by balancing science and values. This book examines current policy decision making processes and presents some dilemmas that can impede finding policy solutions. Proceeding from the conceptual to the practical, this book includes a final word on the policy implications of viewing commonly used indices through the MIRA lens.

There are a range of methods used in decision making because there are a range of circumstances under which decisions are being made. The simplest is when an urgent situation arises such as when there are no pedestrians or cars nearby and a dog runs out in front of your car. With little or no thinking, you slam on the brakes and avert the crisis. This type of decision making is relegated to a spinal cord response that is based on an unambiguous value to avoid killing an animal. In other decision making situations, the value judgments may not be as definitive and the consequences not as unconsciously directed or as dire, for example, what to eat for dinner or what to wear for work. However, along the decision making continuum, there is a point when the decision is neither an unconscious response nor inconsequential, e.g., when stakeholder values are either uncertain or disparate, the stakes are high, or the consequences are likely to be important. This book is written for those situations.

While the experience of the authors is in environmental public policy making, this is a book that uses a variety of illustrative examples to demonstrate concepts applicable to all public policy decision making. Public policy making is a multi-criteria assessment process that requires engaging diverse stakeholders, understanding perspectives, employing socio-economic and scientific data, engaging experts, and incorporating all of these to assess and choose policy alternatives. For public policy decision makers, practitioners, and citizens interested in engaging in the policy process, this book offers basic guidelines intended to prepare them to evaluate different tools and approaches while offering a step-by-step guide to MIRA. Each chapter in this book is designed to present and integrate the conceptual and methodological foundations for a new public policy making paradigm. As a result, interested citizens and practitioners should be in a better position to understand and participate in the public policy decision making process. Throughout this book, the term "stakeholders" refers to both decision makers and citizens who are both interested in and affected by public policy decisions. While public policy decision makers have the authority for making the final decision, an effective public policy process considers citizen stakeholders an integral part of the process leading up to the final decision.

Chapter 1 is intended for a policy-interested audience that has no experience with decision methodologies but may be curious about them. It is an overview of policy decision making that includes an outline of the common steps found in all forms of the decision making process (the Requisite Steps). It offers by way of four public policy stories, examples of how these steps are utilized. These stories raise questions about transparency, inclusivity, and successfulness in the current public policy decision making paradigm. Readers are introduced to Decision Uncertainty, a learning-oriented process that policy makers can use to improve the transparency, inclusivity, and effectiveness of the policy decision making process.

Chapter 2 defines and describes Decision Uncertainty. For those also interested in the conceptual foundation of this element, which is missing from current decision making processes, this chapter includes the historical highlights of science policy decision making that led to the development of MIRA. Chapter 3 synthesizes the concepts in Chaps. 1 and 2 to specifically describe the "open solution approach," a next-generation public policy framework and methodology derived from the primarily conceptual "clumsy solution approach." Chapter 3 is intended for policy practitioners who want to understand how to conduct effective policy making. All readers will gain the skills to critique approaches and tools for their utility in policy decision making. MIRA, a specific kind of public policy open solution approach, is introduced.

Chapters 4 and 5 together describe the process, analytical strategy, and methodology that comprise the MIRA approach. They also illustrate how to conduct a MIRA analysis and use the analytical tools within MIRA to increase understanding about what elements drive the ranking of decision alternatives. These chapters are intended for practitioners interested in evaluating MIRA for their own use or those interested in more specifically understanding the details of this open solution approach to become more engaged stakeholders. These chapters also provide a process to evaluate approaches used by others. Using a simple example, Chap. 4 describes the MIRA initialization process and defines terms, methods, and tools useful to that process. Chapter 5 describes the iteration of the Requisite Steps and evaluates an expansion of the Chap. 4 example using the MIRA analytical tools. With the additional knowledge gained in Chap. 5, practitioners will be prepared to analyze a wide variety of case studies.

Chapter 6 discusses the application of the MIRA open solution approach to another frequently used policy tool, the index. As a composite of multiple criteria, an index can be viewed as another type of multi-criteria problem, even when no decision is explicitly made. Indices used at the US EPA, the Organisation for Economic and Co-operative Development (OECD), *Newsweek*, the Dow Jones, and other organizations are described. Using the principles of the MIRA open solution approach, this chapter describes the benefits of understanding index construction for personal decision making.

1.2   Steps Used in All Decision Making (The Requisite Steps)

Whenever decisions are made, a similar methodology is employed. However, in most situations, that methodology is implicit and, hence, invisible. Even when unilateral decisions are made, there is a methodology involved. An implicit methodology is also in play when we jump to conclusions based on observations. For example, we frequently, and automatically, judge distance by how small and/or fuzzy an object appears to us. Typically, we fare well with this approach, such as when we avoid hitting people or other cars as we drive. We understand that the goal is to avoid hitting people or cars and employ judgments about distance and speed as well as our own capability to steer and stop. All these steps are employed without the formal recognition that a methodology was used. However, this is not always foolproof. For example, on a cloudy or foggy day, airline pilots' judgments are altered. The pilots' use of their own judgment (i.e., their implicit methodology) can cause them to jump to the wrong conclusions. This could be catastrophic if their own judgment were the only means by which they are able to judge distance. However, this disaster is avoided because when circumstances impair pilots' built-in methodology, they are trained to use instrumentation as a replacement. Similarly, when decision making becomes more complex, using more explicit approaches can be necessary to ensure that goals are met [1, 2].

Whether or not decision support tools are used, decision making utilizes an inherent set of steps, which are explicit, implicit, or some combination of the two. These are the Requisite Steps. Stakeholders who understand these steps are better able to participate in the process. The Requisite Steps can be generalized and synthesized into categorical actions that represent the components common to all decision making processes, including decision analytic tools.

These Requisite Steps are as follows:

1. Engage stakeholders.
2. Define the problem.
3. Identify possible solutions.
4. Determine and organize criteria.
5. Gather data and build metrics.
6. Judge the significance of the data.
7. Apply values.
8. Combine all the information to evaluate alternative actions.

The process ends when stakeholders have proceeded through these eight steps and have chosen among the alternatives or decided to remain with the status quo.

These steps are utilized without exception in decision making, whether making a split-second decision or a more considered decision. Consider the simple example of purchasing a family car, as described using the Requisite Steps. The process may start with the family getting together to begin a discussion (Requisite Step 1: Engage stakeholders). This discussion may center on why a car is needed; perhaps the present car is old, or it is too small for the growing family (Requisite Step 2: Define the problem). Based on the family's needs (i.e., how the problem is defined), a set of possible car options is identified (Requisite Step 3: Identify solutions). In this example, each solution is represented by a different car whereby the family intends to choose one car at the end.

Before the family can begin to decide among the possible choices, they will identify the factors to select a car. Some of these factors are cost, comfort, fuel economy, performance, and reliability (Requisite Step 4: Determine and organize criteria). Typically, in current formal and informal decision making, only some of these factors are explicitly identified. For each of the cars being considered, the family gathers data including recommendations from other family or friends, searching the Internet, checking vehicle reviews, and taking test drives of specific car models (Requisite Step 5: Gather data and build metrics).

Armed with the data they have gathered, the family assesses the significance of each car's data relative to their own personal circumstances and needs. For example, if members of a wealthy family compare the cost of a new car against its reliability, they may consider that a highly reliable car is as significant as a car that costs $45,000. In contrast, members of a less wealthy family may make the comparison between reliability and cost differently by determining that a highly reliable car is of equal significance to paying $25,000 for the car (Requisite Step 6: Judge the significance of the data). Although both families consider reliability and cost, how they assess the relative significance of each is based on their financial perspective. The family will also need to determine which criteria are more important and by how much. This is often a partially implicit process as individuals may share opinions about their major criterion but not others. One person may indicate a willingness to spend more money for comfort but not for gas economy (comfort more important than fuel economy) while another may indicate a greater willingness to trade off reliability for increased towing power (reliability less important than towing power) (Requisite Step 7: Apply values).

Once all this information is considered together, implicitly or explicitly, the family comes to an agreement and selects a car (Requisite Step 8: Combine all the information to evaluate alternative actions). Note that the choices made throughout the process (how to define the problem, what criteria are relevant, and how to weigh the relative importance of the criteria) are specific to one family. Another family choosing a car may decide to describe the problem differently, pick different criteria, and weigh them differently. There is no single right way to choose a new car.

Car buying is a simple example of decision making described using the Requisite Steps. However, car buying decisions are generally not the kinds of decisions worth spending a large investment of time or resources to evaluate. In fact, most people typically decide to buy a new car by relying primarily on their own intuition and experience, without realizing that they have implicitly followed the eight Requisite Steps in decision making.

For decision makers, a simple, short decision making process is best. If a decision is non-controversial or if stakeholders are not aware of the decision affecting them, a simple process makes sense to decision makers. However, in public policy situations where the stakes are high and people of conflicting interests are involved, a decision arrived at in this way becomes problematic. In these situations, stakeholders may object to their exclusion from the process, or it may be found that the approach taken does not resolve the problem, the problem is exacerbated by it, or a new problem is produced. Decision makers may be accused of being influenced by the last, the loudest, or the most powerful voice heard. Rancor and chaos may follow the announcement of the decision and decision makers may be accused of acting arbitrarily or capriciously. Legal action may follow or be threatened.

When decisions are more controversial, or the results are found to be unacceptable to some stakeholders, there are increased demands for more inclusive stakeholder participation and transparency in the decision making process. In the public sector, the need for inclusiveness, transparency, and the explicit justification of the final decision is particularly important. Public sector issues are complex and multidisciplinary and require the input, perspective, and agreement of a diverse stakeholder group.

In the following sections, four policy stories are shared to demonstrate how the Requisite Steps are embedded in policy decision making and how the judgments that were made were not objective, but value-laden. These stories represent a sample of real policy decision making that involves the US Environmental Protection Agency (US EPA), a corporation, and both historical and current societal issues.

1.2.1   DDT

When Rachel Carson published *Silent Spring* in 1962, she warned of industrial chemicals doing terrible and irreversible harm to the planet [3]. The pesticide dichlorodiphenyltrichloroethane (DDT) became the poster child of industrial chemicals, even though it was developed and used as a substitute for more toxic metal-based pesticides. The scientific evidence showed that while a metabolite of DDT, DDE (dichlorodiphenyldichloroethylene), was the cause for the thinning of bird eggshells, DDT also provided benefits by globally saving 100 million lives from malaria by killing the mosquitoes that bear this disease. It was also used in World War II to delouse US troops to protect against typhus [4]. The fact is that DDT has relatively low human toxicity compared with wildlife toxicity [5]. When considering only its human benefits, DDT's long half-life in the environment, which reduced the need for frequent applications, made DDT the simple pesticide of choice.

For a time, the use of DDT to combat diseases such as malaria, typhus, and dengue fever and provide protection for commercial agricultural crops against pest damage seemed an obvious choice. Without explicitly realizing it, decision makers likely considered the benefits of low human toxicity and long half-life to combat troublesome pest problems, a winning choice when considered alone. However, after Carson's book raised public awareness that chemicals damage the environment, the resulting public outcry, and a subsequent Congressional hearing, DDT was banned in the USA in 1972 by the US Environmental Protection Agency Administrator, William Ruckelshaus [6]. Currently, DDT is still manufactured in the USA and continues to be used outside the USA, as it is still an extremely effective strategy against the continuing public health threats of malaria and dengue fever. For example, in 2015 and 2016, there were outbreaks of dengue fever, a mosquito-borne disease, on the Big Island of Hawaii [7]. The DDT story suggests that data alone do not reveal what the policy choice must be. The same data could have been used in support of continuing DDT use. The US EPA decision to ban DDT was based on the emergence of strong human values in favor of wildlife concerns, as balanced by US public health concerns. The DDT decision was a values choice and could have been made decided and justified differently. There is no single right answer or solution.

Carson's *Silent Spring* was an exposé. As a warning about society's love affair with technological solutions, Carson used the example of synthetically produced chlorinated industrial chemicals to challenge people to think about their values regarding the natural environment. Her goal was to ultimately change the social values that allowed for the degradation of the natural environment by raising awareness and alarms about the chemicals adversely affecting it. Carson and her supporters focused exclusively on the risks from industrial chemicals to the natural environment rather than offering a more complete picture that would have also included the benefits of these chemicals to human health [8]. Advocacy for certain positions is often a common phenomenon when complex data are involved in a public policy issue. Rachel Carson's book changed public opinion and the country's values regarding the relationship between human activity and the environment.

When the US EPA deliberated about whether to ban DDT, it conducted Congressional hearings [6]. While the hearings were not public, the Agency engaged science experts in the toxicology and the effects of DDT on human health and wildlife and created a publicly accessible record for that deliberation. Evaluated against the Requisite Steps of policy decision making, after public outcry generated by Carson's book, the US EPA deliberations about DDT used science experts to define the DDT policy problem (Step 1, Engage stakeholders, and Step 2, Define the problem). During the Congressional hearings, there were a lot of data presented but not explicitly organized for decision making (Step 4, Determine and organize criteria). Determining which data were more important and by how much was solely a US EPA determination (Step 5, Gather data and build metrics, and Step 6, Judge the significance of the data). The US EPA also deliberated non-scientific questions such as whether low human toxicity or high bird reproductive toxicity is more important (Step 7, Apply values). The US EPA's final decision to ban DDT concluded the process (Step 8, Combine all the information to evaluate alternative actions). Based on the Congressional hearing record, the only policy solutions contemplated (Step 3, Identify possible solutions) were to ban DDT or to continue to allow its use. The US EPA decision makers were focused on whether the data supported a ban of DDT and did not solicit, consider, or evaluate other policy options. Would the public have been better served if other alternatives were considered such as limiting its seasonal use or limiting its use to indoor settings?

Today, public health policy decisions continue to be difficult and the public policy process is still mostly implicit. The threat of the Zika virus in 2016 raised new awareness about the dangers of mosquito-borne disease [9]. For the USA, the first cases of Zika discovered in Puerto Rico and Florida required those public health officials to consider strategies that avoided the use of DDT. The pesticides available included naled (an organophosphate neurotoxin) and pyriproxyfen (an insect larvae growth hormone). Puerto Rico and Florida public health officials deliberated about how to balance the efficacy, cost and environmental effects of their use [10–13]. Armed with the same data, public health officials in Puerto Rico and Florida came to different decisions about what pesticide to use. In Puerto Rico, they chose to use pyriproxyfen, while in Florida, naled was used. Without explicit recognition of the Requisite Steps in the public policy process, the public is left to wonder how the issues were defined, what data were used, and how those data were used.

1.2.2   The Ford Pinto

The building and marketing of the Ford Pinto is an example of a corporate decision that affected public policy on vehicle safety. In 1968, Ford Motor Company introduced a new subcompact car, the Pinto, which was designed to weigh no more than 2000 pounds and cost no more than $2500 at a time when cars cost, on average, $2800 and average personal income was about $6200 [14]. Many believed that in his rush to get a competitive car on the market, Lee Iacocca, Ford's president at that time, compressed the usual 3.5 years of research, design, and testing for a new car into 2 years. Like 90% of the cars on the market in 1968, the Pinto's gas tank was in the rear of the vehicle, which met all federal safety standards at the time. The National Highway Traffic Safety Administration (NHTSA) crash tests and safety standards, now widely known, were not yet approved. After many deaths from rear end collisions that resulted from the explosion of the Pinto's flawed gas tank design, subsequent investigations revealed that Ford, while not specifically determining this for the Pinto, estimated that technological safety fixes amounted to about $11 (equivalent to about $80 in 2018) per vehicle. An accompanying production delay would also be needed to reduce the estimated 180 car fire-related fatalities per year [15]. At the time, the significance of the cost and time-delay data for Ford outweighed the significance that was ascribed to the safety data. Their decision to build and sell the Pinto as-is was based on their expert judgment that the number of deaths represented a low probability of occurrence. In retrospect, many people, including perhaps those at Ford today, would argue that Ford officials made a poor policy decision.

While an internal company decision, Ford's process can still be aligned with the Requisite Steps of decision making. Iacocca, as Ford's president, had the final decision and it was unclear to what extent he engaged his designers, engineers, marketing and sales personnel, and customer perspectives in the process to bring the Pinto to market (Step 1). How the small car market and the need for a lightweight, low-price point vehicle were defined, what data were selected as relevant to analyze the problem, and what alternatives to the Pinto were considered are also unclear (Steps 2, 3, 4, and 5). The decision to produce and sell the Pinto showed that the fire risk data were not deemed significant enough to warrant concern (Step 6). Consumers are left to wonder to what degree costs and time delays were valued more than the risk to human life (Step 7). Although many of the Requisite Steps in this example were unclear, they were followed, even if unknown to the decision maker and stakeholders.

1.2.3   Incarceration in America

Eastern State Penitentiary, a prison facility located in Philadelphia, opened in 1829 in response to a new public policy for prison reform: solitary confinement. Prior to this, jails were crowded as men, women, and child prisoners were housed together in large, dirty, and disease-ridden pens [16]. Jailers sold food and alcohol to the prisoners. Rape and robbery were common. People were jailed for crimes that ranged from stealing a loaf of bread to murder. There were many more prisoners in jail for a wide range of crimes than the existing jails could handle and punishments for all crimes were similar. Philadelphia leaders wanted to balance justice, punishment, and rehabilitation in the treatment of these prisoners but did not know how. The Philadelphia Society for Alleviating the Miseries of Public Prisons advocated for a different model. The design and development of the Eastern State Penitentiary (ESP) sought to change the harsh treatment of prisoners by considering crime a "moral disease" where prisoners, when given the proper opportunity, could repent and be rehabilitated. With heating, flush toilets, and showers in every private cell, the ESP had luxuries that most law-abiding citizens of the time, including US President Andrew Jackson, did not have. However, solitary confinement meant that food was distributed through feeding holes in the cell doors and extraordinary measures were taken to keep prisoners isolated from each other with only the Bible for company.

The ESP model was copied around the world in at least 300 other prisons but there were doubters, such as Charles Dickens and William Roscoe, who warned that isolation tortured the minds of the prisoners [17]. History would prove that Dickens and Roscoe were right to raise concerns, as the implementation of solitude at the ESP was discovered to be more punishment than redemption for these prisoners. Prisoners went crazy. Ultimately, this public policy experiment was determined to be a failure, and in 1913, the ESP changed its policy practice to allow prisoners to eat and play sports together, until it closed in 1971, due to overcrowding.

To state what may be obvious to many today, all social policies reflect the power, race, religion, gender, and economic relationships of the time. What is less transparent is how the social context and values specifically affect the Requisite Steps in the decisions made regarding the ESP. With the advantage of historical perspective, the case of the ESP reflects the dominant social values of the time that many today consider racist, classist, and patriarchal. Elite white men used the penal system to validate and reinforce their concept of social order and hierarchy. Their penal system included training ESP prisoners as workers and taking advantage of free labor, by ensuring that inmates, especially poor citizens and immigrants, women, slaves, and freed black men, were taught their place through punishment, forced labor, and religious conversion. Whether pertaining to incarceration or any other kind of public policy making, social context and values play important roles in determining who participates, how the problem is defined, what and how information is used, and what kinds of solutions are considered.

Then, and now, it is possible to define the prison policy problem in more than one way. For example, the prison policy problem can be defined as one of rehabilitation or punishment, or a combination of the two. The lack of clarity about how the problem was (and is) defined creates additional problems downstream when deciding what data needs to be used to both assess the current condition and determine whether there is success. Did the data on crime make the ESP solitude and labor model seem an obvious and good policy choice? Perhaps so, if overcrowding and inmate-to-inmate criminal activities were paramount. But the data on the effects of solitary confinement might have given Philadelphia prison reformers pause if a prisoner's mental state and rehabilitation had been considered. Even today, with the lessons of ESP and knowing the impacts of solitary confinement on psyches, its use is still debated in the US prison system [18]. Some states currently prohibit the use of solitary confinement, but others do not. If the data on the effects of solitary confinement of inmates "speak for themselves," would it have pointed policy makers away from its use? Policy makers at the time focused on a set of data related to in-prison rape, abuse, and theft and believed that separation of the perpetrators and victims involved in these crimes via solitary confinement was the right solution. Clearly, there are many considerations beyond the data on solitary confinement and in-prison crimes that continue to plague decision makers about how to deal with incarceration.

The prison situation, whether in the nineteenth or the twenty-first century, is a multi-faceted problem of crowding, crime, disease, punishment, inmate rehabilitation, and justice. Furthermore, in the nineteenth century, the elite decision makers had little interest in engaging multiple stakeholders (Step 1) about alternative perspectives on the problem (Steps 2, 3, 5, and 6) or the data that is available to support a decision (Step 4). They preferred to describe the problem that reflected their own worldview and shortened the Requisite Steps in the decision making process. They implicitly used and judged the relative importance of the data (Step 7) on prison crowding, prison crime, solitary confinement on mental health, and other factors. The prison policy discussion today is still controversial and worrisome; however, the question for policy makers seeking more comprehensive and enduring solutions is whether a more transparent and explicit decision making process could be helpful.

1.2.4 Hunger in America

There are plenty of hungry people in the world and many of them live in one of the richest food-producing countries in the world, America. In 2014, Feeding America, the nation's largest domestic hunger-relief organization, estimated that more than 48 million Americans were food insecure. US food insecurity is defined by the US

Economic Research Service as reduced quality, variety, or desirability of diet, including disrupted eating patterns or reduced food intake [19]. Of these 48 million Americans, more than 15 million of them are children and ten million are seniors [20]. This means that more than 15% of the total US population, 20% of children, and 15% of seniors are food insecure. Food insecurity, defined by the US Department of Agriculture (USDA), means the lack of access to, at times, enough food for an active, healthy life for all household members, including limited or uncertain availability of nutritionally adequate foods [21]. Access to food, among other things, requires food availability as well as the ability to pay for the food. The price of an average meal can differ widely across the country, affecting the ability to acquire food, even if it is available.

There are many efforts intended to address the hunger in America problem. These include programs that address the immediate lack of food access such as the Federal Supplemental Nutrition Assistance Program (SNAP) (formerly, the Food Stamp Program) and nonprofit programs such as the What a Waste™ Program of the National Foundation to End Senior Hunger and the Backpack Program of Feeding America [20, 22]. At the risk of oversimplifying this complex policy problem to make a point about its multi-dimensionality, hunger can be considered a symptom, result, or cause of poverty, the lack of employment, or low minimum wages. In addition, gender inequalities such as domestic abuse, divorce, or other traumatic life events that unequally affect women (and their children) result in hunger disproportionately affecting women and children. Women are often less educated and employable and have less financial and other resources than their male counterparts. A network of social welfare programs, implemented by state and federal agencies, work on a longer-term basis to address a wide variety of the issues that affect, result from, and relate to hunger. These programs include housing, health care, disability, employment preparation and job assistance, unemployment benefits, and personal bankruptcy assistance.

The variety of ways that hunger can be defined as a problem, including who the affected stakeholders are and what data are relevant to describing, assessing, and valuing the problem, has produced a wide range of policy programs across different spatial and temporal scales. Many people have been well served by these programs, but hunger continues to be a difficult and persistent public policy problem. In one possible and reasonable conception of the hunger policy problem, advocates, assuming the role of stakeholders for the hunger problem (Step 1), focus on food insecurity and hungry school children (Step 2) who typically receive lunch while in school but who have little or no access to food over the weekends when school is not in session. The Feeding America Backpack Program identified the solution as the need to provide ready-to-eat food in backpacks that are to be distributed in school prior to the weekends (Step 3). The relevant data include the number of school children whose only regular meal during the week is the provided school lunch, the availability of food pantries in their neighborhood, and the work hours and health status of their parents or guardians (Steps 4 and 5). Based on their problem definition, data choices, and values, these hunger advocates use this information, and their judgment about the significance and importance of these factors, to assess the need for and the success of the Backpack Program (Steps 6, 7, and 8).

Another reasonable way to define the hunger problem is to consider food insecurity as a problem of families. Defined as such, one solution is to provide families with access to food banks or pantries with canned food and/or produce. When stakeholders are defined as families with the means to prepare food, the assumption is that these families have access to can openers, cooking and eating utensils, and cooking appliances to utilize the food available from food banks. Left out of this scenario are the hungry homeless families (and individuals) who require food to be ready to eat, as they lack the means to prepare and store food. All these definitions of the hunger problem are valid. However, how hunger is defined determines the possible solutions.

For most citizens not part of these hunger-alleviating organizations, it is difficult to know how these individual hunger policy program discussions are conducted, how the problem is defined, what data is used, which experts are consulted, and how value judgments are made in choosing who and how to help. As a public policy problem, is hunger a single problem, with multiple causes, or multiple public policy problems that each require their own unique course of action? Could applying the Requisite Steps of decision making with inclusiveness, transparency, and learning offer the promise to discover new ways to address this complex and multi-faceted problem?

1.3 The Decision Making Reality: People Make Decisions

In the four policy stories above, policy decisions were made using the Requisite Steps, even if applied to different degrees and mostly implicitly. Problems were defined, either on behalf of the stakeholders or, more often, without them. Without engaging the relevant stakeholders, the selection and use of information and data were biased by the perspectives of the decision makers. Experts may have been the decision makers or specific experts may have been engaged but, in either case, they were often the only stakeholders. Values were applied to determine priorities as well as which contributing factors were more important. No matter how the process was conducted, conclusions were drawn, and decisions were made. Although these four stories illustrate a small group of policy problems, they make it possible to think about the general applicability and implementation of the Requisite Steps across a variety of policy decision making. They raise important questions about whether more inclusiveness and transparency could have helped to define the public policy problem, raise important concerns, and offer additional solutions. Would a more transparent policy making process have allowed decision makers to more completely address the public policy problem, including garnering greater support for those final decisions?

1.3.1 Just the Facts

Knowingly or unknowingly, many people believe that disagreements related to environmental and other science-based societal problems are due to the lack of data or facts. They use this premise to guide how they participate in processes not just related to policy making but also in education and communication. A common strategy that assumes that disagreements can be resolved with more factual education often leads to the communication strategies and educational curricula focused on disseminating and explaining more facts. This strategy is supported by the erroneous premise that the Requisite Steps can be simplified to focus simply on fact gathering (Step 5). In this way of thinking, the lack of factual understanding rather than potential differences in perspectives and values is believed to be the barrier to policy agreements.

This strategy is consistent with that used in science research, where the discovery of facts is paramount to understanding. The goal is to know and better understand the physical, chemical, and biological mechanisms that drive our ecosystem and underpin our society. All too often the strategy used for tackling policy problems is based on the confidence that once enough data is amassed the solution will reveal itself. Scientific research, as contrasted from the analysis of science-based policy issues, seeks to answer questions such as how do cells replicate and pass on genetics, what conditions interfere with fish growth, and what happens to human blood when aspirin is taken and why? These and many more complex questions, which relate to the exclusive discovery of facts, are important for establishing a knowledge base on which all good decisions and policies should be based. However, this knowledge alone cannot determine what policies should be implemented. Knowing the answers to these questions does not tell us anything about whether we should conduct stem cell research to seek a cure for human disease, establish numeric standards for water quality, or treat all heart patients with aspirin. Many policy makers and scientists have challenged the reliance of science and facts alone to make good public policy [23–26]. They warn about oversimplification in policy decision making when values and other perspectives are not considered.

What is needed to make science-based policy decisions, and what distinguishes policy making from science research, is the deliberation and application of our values. For example, the public policy problem inherent with developing a human clone is not simply figuring out how to do it but whether we should. When the US EPA banned DDT or when the Philadelphia leaders chose solitary confinement, there were values employed. Stakeholder values determine what action to take in response to a societal issue. This is the exercise of choice. While it is true that policy problems are often complex, this complexity can also often be confounded by the misdirected focus on facts, without an understanding of the values that surround the problem, when policy decisions need to be made [27]. What is needed is the explicit performance of the Requisite Steps that allows stakeholders to deliberate and balance facts and values in a process of discovery that is not limited to the discovery of facts.

Therefore, the strategy of relying on just facts to work through a current environmental disagreement is problematic. Many government agencies and nonprofit organizations are among those that utilize such a strategy, which might be referred to as a type of environmental or community education program. Providing missing information is expected to eliminate, or sufficiently reduce, conflict in policy making. Sometimes it does, but often it does not. This is because this strategy assumes that all people are (or should be) rational thinkers. With the proper facts and good education, rational thinkers are expected to arrive at the same opinion.

1.3.2 Assumed Rationality

Reflecting successful practices from the mathematical and physical science traditions, the decision analytic field is dominated by economists and others that design tools and approaches with rules and procedures that are intended to make decision making more scientific and, hence, both rational and objective. Objectivity is defined as being factual. Rationality is generally defined as being reasonable and, in the economics discipline, rationality is further defined as making choices to optimize individual benefit. Under the rational choice paradigm, ideal decision making is achieved when objectivity is combined with rationality. But actual decision making conflicts with this paradigm. For example, when investigators compared how chemical toxicity was evaluated under the European Regulation, Evaluation, Authorisation and Restriction of CHemicals (REACH) program with other European frameworks to determine whether to ban a chemical, they found that there was no single rational or objective way to make this assessment [28]. The same chemicals evaluated under different programs produce different conclusions. Others have found that decision analytic tools designed for maximizing rationality and objectivity are often found to be unsatisfactory to decision makers because they neglect to consider how people really think about decision making.

Decision analysts generally presume that rationality and objectivity need to be rigorously applied to achieve good decision making. Based on that premise, many decision tools are developed with a primary focus on objectivizing the choices. Removal of stakeholder-introduced subjectivity, including value judgment, is a design feature of these kinds of tools. Inherent in this design strategy is the presumption that stakeholders are too irrational, inconsistent, and emotional to be trusted with making decisions on their own. Furthermore, generalizing the ready-to-use tools by removing the means for decision analysts to tailor tool rules based on specific policy decision problems is considered a desired tool feature. This is consistent with the premise that all policy problems can, and should, be reduced to a set of core elements. Therefore, independent of stakeholders and specific policy context, tool designers make judgments about what kinds of algorithms will be built into their tools to avoid introducing bias. In the language of the Requisite Steps, in this model of objective and rational decision making, Step 1, Engage stakeholders, is unnecessary and Step 2, Define the problem, is undesirable.

Another way to introduce rationality to decision making is to standardize decision making to monetary choices. Economists use rational choice theory to model social behavior, which assumes that individuals will make prudent and logical choices that maximize their own benefits while minimizing their disbenefits [29, 30]. The premise of rational choice economic theory is that it is possible to standardize based on monetary equivalents and have these monetary equivalents apply to all decisions, a practice that is considered objective. Under this theory, removing what is perceived as irrational stakeholder behavior, or values, is often accomplished by standardizing tools and/or processes that are modeled after scientific tools or methods. This standardization limits the possibilities to only those that are consistent with the algorithms and data contained in those tools. Because choice in public policy decision making is about how stakeholders value many kinds of facts (not just monetary equivalents), the rational choice theory applied to decision making is problematic.

Rationality and objectivity have two major problematic presumptions when they are part of the tools or approaches intended for decision making: (1) that the set of circumstances in which the decision is being made is understood without explicit discussion and can be generalized and (2) a set of a priori standardized rules to govern how to value data and varying perspectives is needed. Presuming that a set of circumstances can be generalized ignores the potentially important differences in stakeholder groups deliberating different policy issues, even as these issues may require the use of the same underlying information. Suppose a stakeholder group considering policy alternatives to increase access to national parks decides to conduct a survey to solicit stakeholders' willingness to pay (WTP) to enter national parks. The WTP survey incrementally increases entrance fees to determine the point at which stakeholders are unwilling to pay. These survey results produce a WTP utility curve for national park entrance fees. Now suppose another stakeholder group is considering policy alternatives to determine how tourism dollars contribute to a local economy. This group decides to utilize the WTP utility curves that were originally developed to determine how to set the fees for entrances to national parks. These two groups are likely to use the same information in different ways [31, 32]. However, the context is different for these two policy issues. Therefore, it is more likely that the WTP curves are more appropriate for the original context in which they were solicited, rather than for other subsequent uses where the circumstances are likely to be quite different.

The second problem with rationality and objectivity is that standardized rules typically presume how disparate pieces of data are to be combined. Typically, decision analytic tools have built-in standardized rules that include the presumptive use of data through its equations or algorithms that establish trade-offs among the different criteria. Examples include utility functions and mental models [33–35]. In utility functions, economists standardize human preferences relative to some good or service. There are many types of mental models but, typically, they try to parameterize human behavior to standardize how choices are made, or to predict what choices should, or will, be made. Some mental model tool developers seek to determine what kinds of incentives might be utilized to steer decision makers toward more rational choices.

Many analysts have studied and written about how people think about decisions and how decision making occurs [35]. A variety of methodologies and approaches have been proposed to help understand, communicate, and sometimes quantify these processes [36–39]. Although each of these methodologies has different strengths and weaknesses, as part of their design, they each seek to objectify or impose rationality in the process by minimizing stakeholder-exercised judgment and choice. Often unknowingly, users of these methodologies relinquish the opportunity to clarify their own problem definition by defaulting to a definition that is represented within those methodologies. Problem definition is represented through a variety of tools in their methodologies, which include (1) "dashboards" through which users either select a subset of the pre-built indicators already contained within the tool or import their own data into the tool as long as it complies with the tool's format requirements and (2) algorithms contained within the methodology that organize, expertly judge, and weigh the relative importance of the data.

People making decisions are confronted with conflicting facts and values. All decision making is an effort to reconcile these conflicts. The issue is how this reconciliation occurs. Rationality and objectivity shortcut and shortchange the Requisite Steps by screening who can participate and constraining the process to data and procedures deemed rational and objective. Particularly when there is conflict and the stakes are high, objectivity is often seen as the arbiter in a landscape of pure reason and logic. Opinions, values, and perspective are diminished as they are considered irrational, illogical, and irrelevant to the decision making process. The result is that most decision makers do not use decision support tools.

1.3.3   Instilling Values

Instilling values early in life is another commonly pursued strategy, particularly in environmental science. Many environmental education programs often have a component designed to focus on teaching certain environmental values. Typically, this is achieved by highlighting selected aspects of the science or facts. For example, children are typically taught the importance of clean water as a single-minded goal. This approach focuses on one step, Step 7 (Apply values), with the additional presumption that once taught, these values remain fixed. Choosing clean water is more realistically part of a values-laden trade-off, balancing other environmental, economic, and social goals. Therefore, although the need for clean water is important, it is only one part of the story, if the goal is to provide students with the future capability to critically think and engage in environmental public policy making. Instilling values is often employed as a hopeful and long-term environmental education strategy that is designed to reduce future environmental disagreements. Therefore, when there are environmental disagreements, many environmental professionals as well as activists typically pursue one of two strategies: (1) provide more facts (most communication strategies are these) or (2) provide classroom or field experiences introducing students to the wonders and beauty of nature, wildlife, and outdoor experiences (most environmental education schooling are these). None of these are bad activities per se but expecting instilled values to drive potential future decision making does not necessarily train good citizens.

Neither the strategy of providing just facts nor just values properly acknowledges or embraces the Requisite Steps. Neither of these presumptions alone fully explains why there are public policy disagreements. Even when the facts are presented, how those facts are used can easily differ among stakeholders. Stakeholders do not share the same values. To proceed as if they do is to seek to standardize values that are based on either mere facts or assumed common experiences. This diminishes the experiences and values of the stakeholders who hold different perspectives.

1.4 Current Tools and the Requisite Steps

Deliberating complex policy issues can easily overwhelm the most diligent and well-intentioned decision makers and stakeholders. The complaints about current decision making processes include those pertaining to the exclusion of stakeholders, disagreements about how to define the problem, disagreements or lack of under-standing about what data are relevant, and how to manage different stakeholder values [40–44]. Stakeholders, unhappy about the final policy decision, most fre-quently complain that the decision making process was poorly conducted because they cannot understand how the final decision was justified. In some cases, they believe that another decision was better justified. As a result, many practitioners have developed different approaches, methods, and tools intending to make the decision making process more manageable, quantitative, and objective. Unwilling to have their judgment usurped by a "black box" with data and algorithms that they did not choose or necessarily understand, many decision makers prefer to rely on their own methods, even if these are not explicit or transparent. A better way to help decision makers and engage stakeholders is needed.

1.4.1 Designing Without Success

Current approaches and methodologies for public policy analysis and decision mak-ing share a major failure, which is that they fail to explicitly recognize the Requisite Steps or to perform them well. For example, many approaches focus almost exclu-sively on gathering facts and data [45]. Other approaches value all criteria equally or constrain users to select from a pre-determined set of data [46–48]. The problem with current tools is that they contain or utilize (1) pre-determined data catalogues and built-in algorithms, (2) one-pass reductionism, and (3) incomplete implementa-tion of the Requisite Steps. Each of these issues is explained in more detail.

1.4.1.1 Defaults

Pre-determined data catalogues are those systems that contain data or indicators from which users select to conduct their analyses. These catalogues determine which data can be used. Built-in algorithms are those systems that contain default procedures, which determine how data can be used. Pre-determined data catalogues and built-in algorithms are most commonly found together in the same tool. The US EPA Recovery Potential Screening (RPS) tool is an example of such a combined tool. The RPS groups data into social, ecological and stressor categories [49]. Practitioners use RPS to calculate an index to determine the relative recovery poten-tial for a group of watersheds. RPS, as with other tools containing data catalogues, contains a pre-selected set of data within each of those catalogues. In the RPS, the data catalogue for social indicators contains 33 pieces of pre-formatted data such as proximity to universities or funding availability. Users are limited to evaluating their watersheds, counties, or other alternatives using those data. Such technical features, designed by experts, tend to eliminate or shortcut discussions, giving the appearance of providing efficient and effective decision making. In some cases, these data may be precisely what stakeholders want but it is likely that, in other cases, it is not. Therefore, data catalogue tools like the RPS usurp stakeholder choice. Stakeholders may be deceived into thinking that the only choices available are those provided in the tool. The RPS index calculation is a default algorithm.

Default algorithms pre-determine how the data are used in the analysis. Sometimes these default algorithms are knowingly built and perhaps even guarded as a proprietary business asset. At other times, tool designers are themselves unaware that they have utilized a default algorithm. Commonly, the default algorithm forces all criteria to be valued equally, i.e., deemed equally important in the analysis. Again, users are shortchanged because they are not offered the opportunity to explicitly contemplate and decide how they want to use the data.

Both data catalogues and default algorithm tools limit the kinds of problems that can be defined and tackled while presuming who the stakeholders are. Moreover, providing a group of facts accepted by experts, via the limited data catalogue or default algorithm, is thought by the tool designers to be one means of ensuring that stakeholders are presented only with the "right" facts and the "right" way to use them. Although the users of the US EPA's RPS are likely to analyze similar problems related to watershed recovery, other more generalized decision making tools also contain pre-loaded data or indicators, which simplifies the process for stakeholders evaluating similar topics. The use of Willingness to Pay (WTP) utility curves is a form of a data catalogue. Data or indicators constructed for one purpose and made available for other purposes without a clear understanding of their nature contribute to a poor understanding of the problem. Even when metadata are available, the relevance of the data, including an understanding of the kind(s) of problem(s) that the data or indicators were designed to address, is insufficiently described. Decision tools with built-in defaults (data or algorithms) use the data in a context that likely differs from the stakeholders' current context, which is problematic. Consequently, decision makers using these decision tools without a full understanding of the default context contained in these tools are making it more difficult, if not impossible, to find effective solutions for their own policy problem.

1.4.1.2 Reductionism

Reductionism is modeled on the scientific tradition of reducing a problem to its components, solving the component, and re-aggregating the whole. Reductionism is attractive to many scientists and decision makers who find the original problem unwieldy (many stakeholders, large amounts of information, and highly contentious) with no clear technical solutions. Applying reductionism is problematic because public policy problems are irreducible. Removing a cohort of stakeholders or ignoring certain data changes the problem. When the public policy problem is reduced to a science or technical problem, there is the threat that "we are doing the wrong things righter" [50]. Decision makers have ignored or abandoned these inflexible methods, most likely without understanding why. However, decisions makers' substitute for that decision making process is often no better. The use of alternative but non-transparent, exclusive, and top-down controlled processes also disenfranchises stakeholders and potentially inflames the environmental issues [51, 52].

Therefore, applying reductionism as a one-pass system in current tools is also problematic. When iteration is not part of the process, refinements to problem definition, data, and other decision-influencing factors based on stakeholder learning are unavailable or truncated. These are lost opportunities for the discovery needed to arrive at an acceptable action plan that is based on situation-specific decision making circumstances [53].

1.4.1.3 Incomplete Requisite Steps

There are other tools that are designed to implement one or two of the Requisite Steps without consideration for the others. One such example is the Analytic Hierarchy Process (AHP) developed by Saaty for decision making [54]. AHP is primarily designed to help perform Requisite Step 7 (Apply values), which it does well, but unless the user specifically ensures that AHP is part of a more inclusive decision making approach, the technique itself ignores the other Requisite Steps. AHP is a mathematical means to solicit stakeholder values via pair-wise comparisons between decision criteria. In this process, stakeholders are asked "Which criterion is more important, A or B, and by how much?" In pair-wise fashion, the same question is asked of criterion B and C and A and C and so on. The linear algebraic algorithms, used in AHP, offer a very useful mathematical check (the consistency ratio) to determine whether stakeholder responses are internally consistent. Therefore, although there is no right way to respond to these pair-wise solicitations, once a stakeholder indicates that A is more important than B and B is more important than C, it would be inconsistent to say that C was more important than A. The Analytic Hierarchy Process is an example of an approach that can provide substantial value in how Step 7 can be performed because it is transparent and explicit. AHP offers a mathematical algorithm to perform Steps 7 and 8 (Combine all the information to evaluate alternative actions). As a decision analytic methodology, AHP can be quite helpful, but alone it is insufficient as a comprehensive approach to good public policy decision making.

Stakeholders that use decision tools and approaches without evaluating or questioning them are potentially cheating themselves out of a means toward discovering workable alternative policy solutions. Designers of these tools usurp the decision making process.

1.5 Designing for Success

Designing for success in policy making requires tools and approaches that offer decision makers a combination of ease, utility, and flexibility to meet their analytical needs. By design, all tools contain operational elements to make the decision maker's task easier. However, good tools and approaches allow users to determine whether these presumptions work for them and, if necessary, to change them to better suit their circumstances and decision making needs. In this way, users have the flexibility to make suitable changes if the specific context differs from the presumptive context.

1.5.1 Requisite Steps: Necessary but Not Sufficient

Even when the Requisite Steps of decision making are recognized and performed well, the public policy process can still be frustrating, exclusive and unenlightening for stakeholders interested in discovering and examining policy alternatives. This is because, unlike learning about science or other facts, where a one-way communication from expert to student can be adequate, public policy problems require stakeholders to learn from each other. In doing so they develop a common understanding about the problem definition as well as the information important for tackling the problem. This kind of learning requires experimentation, examination, comparison and reflection at each Requisite Step. Approaching the decision making process in this way gives stakeholders the ability to determine how the choices made in combination across the Requisite Steps affect the attractiveness of the alternatives being considered. When stakeholders struggle with articulating the problem definition, they are learning from each other and building trust in each other and in the process. When stakeholders question what data are relevant to the problem definition, they are learning. Discussing and identifying potential solutions to consider, and challenging each other's values and priorities, is also part of this learning process.

Through the iterative process, stakeholders learn how differences in each of these aspects (what data is selected, how significant it is within the decision context, and what values are applied to define the importance among criteria) influence the relative attractiveness of the alternatives. Therefore, although clarity in recognizing and performing the Requisite Steps well is a necessary condition for good decision making, unless it is also possible to learn about the relationship between the data and the attractiveness of the alternatives, the Requisite Step process alone is insufficient for public policy decision making. Without such means, it will be difficult to converge on an accepted and effective policy decision.

1.5.2 Decision Uncertainty

Decision making ends when stakeholders decide on an alternative or choose to remain with the status quo. When stakeholders have a chance to consider many different alternatives and to gain a better understanding of the reasoning behind arriving at the final choice, there is greater satisfaction and it is likely that the decision will be easier to implement. At each of the eight Requisite Steps, the stakeholders become appropriately engaged in the discussion about their concerns, values, and what they think are the relevant data. Iteration avoids the reductionism, but stakeholders need a benchmark to know when the iteration is enough, and a decision can be made. This benchmark is Decision Uncertainty, which is the compounded uncertainty from all the Requisite Steps that influence the stakeholders' confidence in their decision [55] . When there is uncertainty about the problem definition or what data is relevant, these contribute to Decision Uncertainty. Stakeholders assess Decision Uncertainty each time they change their analysis such as when they clarify or refine their problem definition or agree on the selected data. When stakeholders are satisfied that Decision Uncertainty has been minimized, the iteration stops. By introducing the capability to discover how the facts, expert judgments, and different stakeholder perspectives color the attractiveness of available policy alternatives, it is possible to have increased clarity about where the real conflicts are. In general, these conflicts are not about the science but about how to use the science [56, 57].

1.6 Summary

From simple to complex, the methods in current decision practice, usually implicitly, utilize the Requisite Steps of decision making. These steps are as follows: (1) Engage stakeholders, (2) Define the problem, (3) Identify possible solutions, (4) Determine and organize criteria, (5) Gather data and build metrics, (6) Judge the significance of the data, (7) Apply values, and (8) Combine all the information to evaluate alternative actions. Although all decision making utilizes these Requisite Steps, stakeholders are shortchanged when they are not explicitly performed or performed to allow stakeholder learning. Default or standardized procedures or preselected data usurp stakeholder choice and truncate the policy making process. Stakeholders are robbed of the opportunity to learn from each other, discover new information, develop new understanding, or build trust with each other. All these opportunities are important for practicing better public policy making and the promise of finding more sustainable policy decisions.

References

1. Committee on Human Factors Commission on Behavioral Social Sciences Education, National Research Council (1983) Research needs for human factors. National Academy Press, Washington, DC
2. Fischhoff B, Harvey J (1981) No man is a discipline. In: Cognition, social behavior, and the environment. Erlbaum, Hillsdale, pp 579–583
3. Carson R (1962) Silent spring. Houghton Mifflin Company, Boston
4. National Pesticide Information Center (2000) DDT: technical fact sheet. NPIC. http://npic.orst.edu/factsheets/ddtgen.pdf. Accessed 11/23/18
5. Agency for Toxic Substances and Disease Registry, Centers for Disease Control and Prevention (2002) Public health statement: DDT, DDE, and DDD. ATSDR. https://www.atsdr.cdc.gov/phs/phs.asp?id=79&tid=20. Accessed 11/23/18
6. Ruckelshaus WD (1972) Consolidated DDT hearings: opinion and order of the administrator. Fed Regist 37(131):13369–13376
7. State of Hawaii DoH, Disease Outbreak Control Division (2018) Disease types. State of Hawaii, Department of Health. https://health.hawaii.gov/docd/disease-types/mosquito-trans-mitted/. Accessed 5/27/2019
8. Malkin M, Fumento M (1996) Rachel's Folly: the end of chlorine. Competitive Enterprise Institute.   http://cei.org/sites/default/files/Michael%20Fumento%20-%20Rachel's%20 Folly%20The%20End%20of%20Chlorine.pdf. Accessed 11/23/18
9. Centers for Disease Control and Prevention (2016) Zika virus: 2016 case counts in the US. U.S. Department of Health and Human Services. https://www.cdc.gov/zika/reporting/2016-case-counts.html. Accessed 8/6/2018
10. CBS News (2016) CDC struggling to wipe out mosquito carrying Zika virus. CBS Interactive, Inc.
11. National Pesticide Information Center (2016) Pyriproxyfen: General Factsheet. NPIC. http://npic.orst.edu/factsheets/pyriprogen.pdf. Accessed 11/23/18
12. National Pesticide Information Center (2018) Pesticides used in mosquito control. http://npic.orst.edu/pest/mosquito/mosqcides.html. Accessed 11/23/18
13. U.S. Environmental Protection Agency (2011) Naled technical: product registration notifica-tion. U.S. Environmental Protection Agency, Washington, DC
14. U.S. Department of Commerce, Bureau of the Census (1969) Consumer reports: household income in 1968 and selected social and economic characteristics of households, vol 65. U.S. Department of Commerce, Washington, DC
15. Wojdyla B (2011) The top automotive engineering failures: the Ford Pinto fuel tanks. In: Popular mechanics. Hearst Communications Inc., New York
16. Woodham C (2008) Eastern state penitentiary: a prison with a past. Smithsonian
17. Manion J (2015) Liberty's prisoners: Carceral culture in early America. University of Pennsylvania Press, Philadelphia
18. Melamed S (2018) Why is 1 in 10 Philly inmates still confined 'in the hole'? Philadelphia Inquirer, 5/17/2018
19. Feeding America (2014) Hunger in America 2014, national report. Feeding America, Chicago
20. National Foundation to End Senior Hunger (2018) What a waste. Paraclete Multimedia. http://nfesh.org/. Accessed 11/23/2018
21. U.S. Department of Agriculture, Economic Research Service (2018) Food security in the U.S. USDA. https://www.ers.usda.gov/topics/food-nutrition-assistance/food-security-in-the-us/. Accessed 11/23/2018
22. Feeding America (2017) 2017 feeding America annual report: a hunger for a brighter tomor-row. Feeding America, Washington, DC
23. Saltelli A, Funtowicz SO (2017) What is science's crisis really about? Futures 91:5–11
24. Saltelli A, Giampietro M (2017) What is wrong with evidence based policy, and how can it be improved? Futures 91:62–71

25. Bremer S (2017) Have we given up too much? On yielding climate representation to experts. Futures 91:3
26. Ravetz JR (2005) The no-nonsense guide to science. No-nonsense guide. New International Publications, Ltd., Oxford, UK
27. Sarewitz D (2015) Science can't solve it. Nature 522:413–415
28. Moermond CTA, Janssen MPM, de Knect JA, Montforts MHMM, Peijnenburg WJGM, Zweers PGPC, Sijm DTHM (2011) PBT assessment using the revised annex XIII of REACH – a comparison with other regulatory frameworks. Integr Environ Assess Manag 8(2):359–371. https://doi.org/10.1002/ieam.1248
29. Maynard Smith J (1982) Evolution and the theory of games. Cambridge University Press, Cambridge
30. von Neumann J, Morgenstern O (1944) Theory of games and economic behavior. Princeton University Press, Princeton
31. Bunn DW (1984) Applied decision analysis, McGraw-Hill series in quantitative methods for management. McGraw-Hill Book Company, New York
32. Keeney RL (1981) Analysis of preference dependencies among objectives. Oper Res 23(6):1105–1120
33. Binder CR, Feola G, Steinberger JK (2010) Considering the normative, systemic and procedural dimensions in indicator-based sustainability assessments in agriculture. Environ Impact Assess Rev 30:71–81
34. Matthies M, Giupponi C, Ostendorf B (2007) Environmental decision support systems: current issues, methods and tools. Environ Model Softw 22(2):123–127. https://doi.org/10.1016/j.envsoft.2005.09.005
35. Elsawah S, Guillaume JHA, Filatova T, Rook J, Jakeman AJ (2015) A methodology for eliciting, representing, and analysing stakeholder knowledge for decision making on complex socio-ecological systems: from cognitive maps to agent-based models. J Environ Manag 151:500–516. https://doi.org/10.1016/j.jenvman.2014.11.028
36. Rouse WB, Morris NM (1986) On looking into the black box: prospects and limits in the search for mental models. Psychol Bull 100(1):349–363
37. Merkhofer MW, Keeney RL (1987) A multiattribute utility analysis of alternative sites for the disposal of nuclear waste. Risk Anal 7(2):173–194
38. Keeney RL (1992) Value-focused thinking: a path to creative decision making. Harvard University Press, Cambridge, MA
39. Fischhoff B, Sternberg RJ, Smith EE (2003) Judgment and decision making. In: The psychology of human thought. Cambridge University Press, Cambridge, UK, pp 153–187
40. Kleinmuntz B (1990) Why we still use our heads instead of formulas: toward an integrative approach. Psychol Bull 107(3):296–310
41. Jasanoff S (1990) The fifth branch: science advisors as policymakers. Harvard University Press, Cambridge, MA
42. Harremoes P, Gee D, MacGarvin M, Stirling A, Keys J, Wynne B, Vaz SG (2001) Twelve late lessons. In: Late lessons from early warning: the precautionary principle 1896-2000. European Environment Agency, Copenhagen, pp 168–194
43. Funtowicz SO, Ravetz JR (1993) Science for the post-normal age. Futures 25:739–755
44. Merkhofer MW, Covello VT, Mumpower J, Spicker SF (1987) Criticisms and limitations of decision-aiding approaches. In: Covello VT, Mumpower J, Spicker SF (eds) Decision science and social risk management, Technology, risk and society: an international series in risk analysis. D. Reidel Publishing Company, Dordrecht, pp 142–186
45. Jeong B, Oguz E, Wang H, Zhou P (2018) Multi-criteria decision-making for marine propulsion: hybrid, diesel electric and diesel mechanical systems from cost-environment-risk perspectives. Applied Energy 230:1065–1081. https://doi.org/10.1016/j.apenergy.2018.09.074
46. Carli R, Dotoli M, Pellegrino R (2018) Multi-criteria decision-making for sustainable metropolitan cities assessment. Journal of Environmental Management 226:46–61. https://doi.org/10.1016/j.jenvman.2018.07.075

47. Green AW, Correll MD, George TL, Davidson I, Gallagher S, West C, Lopata A, Casey D, Ellison K, Pavlacky DC, Quattrini L, Shaw AE, Strasser EH, VerCauteren T, Panjabi AO (2018) Using structured decision making to prioritize species assemblages for conservation. Journal for Nature Conservation 45:48–57. https://doi.org/10.1016/j.jnc.2018.08.003
48. Manupati VK, Ramkumar M, Samanta D (2018) A multi-criteria decision making approach for the urban renewal in Southern India. Sustainable Cities and Society 42:471–481. https://doi.org/10.1016/j.scs.2018.08.011
49. U.S. Environmental Protection Agency (2018) Recovery potential screening: the RPS methodology: comparing watersheds, evaluating options. https://www.epa.gov/rps/rps-methodology-comparing-watersheds-evaluating-options. Accessed 11/23/18
50. Ison R, Collins K (2008) Public policy that does the right thing rather than the wrong thing righter. Australia National University, Canberra
51. Bond A, Morrison-Saunders A, Gunn JAE, Pope J, Retief F (2015) Managing uncertainty, ambiguity and ignorance in impact assessment by embedding evolutionary resilience, participatory modelling and adaptive management. J Environ Manag 151:97–104. https://doi.org/10.1016/j.jenvman.2014.12.030
52. deLeon P, Steelman TA (2001) Making public policy programs effective and relevant: the role of the policy sciences. J Policy Anal Manage 20(1):163–171
53. Glover SM, Prawitt DF, Spilker BC (1995) Decision aids and user behavior: implications for inappropriate reliance and knowledge acquisition. http://www.usc.edu/dept/accounting/midyraud/glover.html
54. Saaty TL (1980) The analytic hierarchy process: planning, priority setting, resource allocation. McGraw Hill International Book Co., New York
55. U.S. Environmental Protection Agency (2019) Guidelines for Human Exposure Assessment (EPA/100/B-19/001), Chapter 8 Uncertainty and Variability. Washington, DC. Risk Assessment Forum, US EPA.
56. Sarewitz D (2004) How science makes environmental controversies worse. Environ Sci Pol 7:385–403
57. Lackey R (2007) Science, scientists, and policy advocacy. Conserv Biol 21(1):12–17

Chapter 2
Decision Uncertainty in a New Public Policy Paradigm

Abstract All policy decision making follows the Requisite Steps. However, simply following these steps does not ensure effective policy decision making. Conducting iteration through the Requisite Steps counters the otherwise reductionist practice of dividing the decision making process into those steps. Furthermore, an iterative process allows for the evaluation of Decision Uncertainty, which is the analytical component that is missing from other decision approaches. Decision Uncertainty includes not only the traditional, quantitative uncertainties associated with data and modeling-related information (scientific uncertainties) but also those uncertainties related to the clear articulation of the problem, or disagreements among experts, and lack of stakeholder understanding about their own and other stakeholders' values (non-scientific uncertainties). Decision Uncertainty is the total uncertainty that influences the confidence of stakeholders in their decisions. Typically, the primary emphasis of science-focused analysts is to minimize scientific uncertainties. However, understanding Decision Uncertainty is necessary if stakeholders want to forge better policy agreements.

Keywords Post-normal science · Wicked problems · Science integrity · Clumsy solutions · Decision traps · Sustainability assessment

2.1 Introduction

All policy decision making follows the Requisite Steps. However, simply following these steps does not ensure effective policy decision making. Conducting iteration through the Requisite Steps counters the otherwise reductionist practice of dividing the decision making process into those steps. Furthermore, an iterative process allows for the evaluation of Decision Uncertainty, which is the analytical component that is missing from other decision approaches. Decision Uncertainty includes not only the traditional, quantitative uncertainties associated with data and modeling related information (scientific uncertainties) but also those uncertainties related to the clear articulation of the problem, or disagreements among experts, and lack

© Springer Nature Switzerland AG 2020

C. H. Stahl, A. J. Cimorelli, *Environmental Public Policy Making Exposed*,
Risk, Systems and Decisions, https://doi.org/10.1007/978-3-030-32130-7_2

of stakeholder understanding about their own and other stakeholders' values (non-scientific uncertainties). Decision Uncertainty is the total uncertainty that influences the confidence of stakeholders in their decisions. Typically, the primary emphasis of science-focused analysts is to minimize scientific uncertainties. However, understanding Decision Uncertainty is necessary if stakeholders want to forge better policy agreements.

When the hunger problem, introduced in Chap. 1, can be described alternatively as a problem about school children, families, or the homeless, this is indicative of an uncertainty about problem formulation. None of these three problem formulations are wrong but each of these formulations would potentially consider different data, different stakeholders, and different policy solutions. Problem formulation uncertainty contributes to Decision Uncertainty. At each of the Requisite Steps, there are uncertainties that contribute to Decision Uncertainty. For example, when experts disagree about whether demographic or income data are more, or less, significant given the problem formulation, or whether to require 95% confidence or 80% confidence in their data, these contribute to Decision Uncertainty. When stakeholders disagree about whether to consider hungry children as more important than the hungry homeless, Decision Uncertainty is increased.

In this book, unless otherwise specified, uncertainties refer to both scientific and non-scientific forms. Each of these uncertainties contributes to a decision maker's insecurity about which option to choose. While decision makers want assurances that accurate data is used, this is not their main interest. Decision makers want to know how the combination of all relevant uncertainties influences the attractiveness of one policy option over the others. This requires that they understand Decision Uncertainty, which the current decision making approaches fail to consider. A new policy paradigm, the "open solution approach," focuses on decision makers' needs by including Decision Uncertainty, which is a gauge by which decision makers can assess their readiness to make a decision.

2.2 Reductionism Tradition

Reductionism pervades how science policy is approached and is one of the most difficult barriers to overcome in science-based public policy making. Therefore, being serious about improving public policy making starts with a discussion of the role of reductionism in stakeholder-involved processes. This sets the stage for Funtowicz and Ravetz's concept of science and post-normal science as contrasted with normal science [1]. Post-normal science reexamines the role of stakeholder values in public policy problems. The integration of the discussions on reductionism, the concepts of post-normal science, and Rittel and Webber's tame and wicked problems provides the basis from which good public policy practices can emerge. While this is a chapter grounded by a practitioner's perspective, it is also important that public policy practices are conceptually sound. Aligning concepts with practice provides more solid foundation with which to examine and evaluate appropriate approaches for more inclusive and effective public policy making.

Reductionism can be avoided by moving deliberately through the Requisite Steps and returning to discuss and reevaluate any Requisite Step when stakeholders need to better understand their problem and the data used to evaluate it. In the new paradigm, stakeholders have multiple opportunities to satisfy their curiosity about the effects of relevant facts, data, expert judgment, and values on potential solutions.

2.3   Science Is Not Science Policy

Funtowicz and Ravetz observed that the problem regarding the use of science in public policy making has been evolving over time. In their view, the traditional or normal science, with its "reductionist analytical worldview, divides systems into ever smaller elements that are studied by ever more esoteric specialism." This is particularly problematic when carried into contemporary policy making, even as such policy making requires the utilization of science [1]. Funtowicz and Ravetz describe a new kind of science, post-normal science, as complementary to normal science. They assert that post-normal science is necessary if science is to be used in democracies where there are many legitimate perspectives and criticisms that should be aired and discussed. In contrast to normal science, post-normal science involves an interpretation about how the science is viewed and used.

Normal and post-normal science are distinguished primarily by their organizing principles [2]. For normal science, the organizing principle is truth, while for post-normal science, it is quality. Funtowicz and Ravetz explain what this means with this example. When there is a clear market for the sale of a songbird, the songbird has clear value as an economic commodity (a truth in normal science). However, when the value of the songbird is considered also for its value as a species that is part of an ecosystem, which is assessed as part of the quality of life, this is post-normal science. In this example, the songbird valuation under normal science's organizing principle is over-simplified and reduced to a single view, an economic commodity. One popular area of study in environmental public policy is ecosystem services, where the value of ecosystems is typically monetized [3]. The public policy question is not how to monetize ecosystems but rather whether valuing ecosystems in this way shortchanges how stakeholders want to characterize them.

Another difference between normal and post-normal science is scale. Funtowicz and Ravetz assert that post-normal science is more important in today's public policy making than in earlier times. This is because policy decisions today have farther-reaching global impacts, both socially and economically, than those made long ago. These far-reaching effects are not typically considered in most public policy making, but the stakes today are higher and the consequences more dire. These relative scales in normal versus post-normal science influence how natural resources are viewed and the kinds of infrastructure (built, social, and economic) that are developed. Using normal science thinking, science and technological solutions appear to fix many of the most troublesome policy problems in the short term. However, as time passes many of these same technological solutions are shown to be detrimental.

For example, when US Midwestern wheat farmers in the 1930s had difficulty plowing the tough semi-arid Midwestern soils, many of them employed new technology. In 1837, John Deere invented a new self-cleaning plow blade that could cut furrows through tough Midwestern soils that helped farmers tame those lands for wheat farming [4]. However, the efficiency gained by using Deere's innovative plow blade ultimately contributed to the social, economic and ecological misery of the Dust Bowl, considered one of the worst US environmental disasters. During the 1930s in the USA, the expansion of wheat farming to the semi-arid lands of the Midwest was supported by US homesteading and agricultural policies [5]. As a country that was first founded on its eastern coast with relatively plentiful rainfall, settlers of the Midwestern plains did not consider that the weather and soils in the Midwestern plains could be different. Therefore, they applied the same knowledge and technologies that contributed to successful farming on the east coast to grow wheat in the Midwest. Between 1879 and 1929, the acreage of harvested crops increased by 91 million acres in the eight states defined as the Midwestern Great Plains. With the help of technological invention, even as wheat crop yields decreased year after year, the Midwestern farmers doubled-down on plowing more land. This strategy produced a wheat glut in 1931 but did not stave off the Dust Bowl disaster to follow [6]. For example, in 1830, 58 hours of labor were needed to send an acre of wheat to the granaries. In 1920, only 3–6 hours of labor per acre were needed. When the Midwestern soils blew away, exacerbated by the existing social and economic infrastructure, farms and businesses went out of business and banks failed, forcing President Franklin D. Roosevelt to create new federal programs such as the Civilian Conservation Corps and the Agricultural Adjustment Act to mitigate the social and economic miseries [5].

After World War II, hunger in developing countries was tackled with a biotechnological focus. The use of biotechnology in agriculture has permitted the movement from small farms to corporate farms. This increased the scale of food production, as well as reducing the number of people needed for its production [7]. Hunger was defined as a problem of growing adequate quantities of food. Breakthroughs in agricultural research created new hybrid seed varieties of wheat, corn, and rice to produce unprecedented yields. However, this came at the cost of requiring near-perfect growing conditions, consisting of irrigation, fertilizer, and pesticides [8]. Generic manipulation of seeds, farm-raising fish, and the widespread application of fertilizers and pesticides used to increase yields are examples of applying normal science to post-normal policy problems. More food was grown but the hunger problem was not solved.

Resolving these kinds of public policy problems requires another approach that views and uses science differently. In this new approach, rather than using science to define the public policy problem, stakeholders first define the problem and then consider what science is helpful for finding possible solutions. For example, the hunger public policy problem is not a problem of the amount of food grown but rather one of affordability and access. However, affordability and access are not science problems. The USA has more food than it needs to feed its own population, but there are still more than 40 million US citizens who are food insecure. In 1943, in the Indian state of Bengal, 3 million people starved for reasons other than insufficient food. Food was plentiful but it was hoarded by those who owned it in order to increase prices, which the poor could not afford to pay [8]. In many ways, increased yields have masked the need to address the original problem of food redistribution by making it appear as if something was being done.

To address the food distribution and similar policy problems, good science policy making rather than just good science is needed. For example, once stakeholders agree on the problem formulation, good science policy makers facilitate stakeholder discussion about what factors are important to consider and the trade-offs that must be made among those factors. Trade-off questions, such as whether the social, economic, and environmental costs of the adoption of the new science or technology are worth the potential benefits, can be explicitly discussed. Farmers in many countries learned that in exchange for increased crop yields, their land eventually became unusable from the build-up of salt deposits and the depletion of groundwater due to irrigation. Furthermore, the adoption of monoculture, required for their spectacular crop yields, resulted in the expungement of indigenous biodiversity. These changes contributed to social and economic disruptions that shifted the agricultural economies away from small farmers to large agribusinesses, which further exacerbated the hunger problem by reducing access to nearby-grown food. A new policy making paradigm is needed that describes science as thinking about how science is used. In this new paradigm, the emphasis is on the social context rather than on the type of science. This is an important distinction that preserves science[1] as knowledge, usable in many different policy contexts, but recognizes that how science is used depends on the context.

Post-normal thinking reveals what the normal science reductionist view had ignored as externalities: increased pollution, imbalance, inequality, and increased stakeholder dissent. Post-normal thinking requires that we have a better understanding of how science gets used as distinguished from what the science tells us. Science is blind to the societal consequences of science-based policy decisions that are now more far-reaching than ever before. The concept of the post-normal science as a non-reductionist, externality-embracing way of thinking is also supported in the social science literature. These social scientists also argue, as do Funtowicz and Ravetz, that today's natural resource management problems require that we break away from relying on the technological fixes as the sole means to solve our policy problems [9–12]. In the view of these social scientists, the side effects from not properly addressing yesterday's problems have created many of today's problems, which include global warming, social dislocation, and ecological collapse. Therefore, there is conceptual agreement among these physical and social scientists regarding the problematic practice of applying technological fixes to social

---

[1] As this is a book about science-based policy making, the definition and utility of science purely for knowledge's sake (research science) is not tackled here.

and public policy problems [13]. Although these public policy thinkers do not use the same terminology or come from the same disciplines, they agree that science alone fails to solve public policy problems. In many cases, the use of science alone exacerbates or creates additional problems. The consequences of applying the science to a social problem need to be understood and considered if there is to be effective public policy making.

2.4   Post-normal Science and Decision Uncertainty

Funtowicz and Ravetz use uncertainty to distinguish normal from post-normal science. Post-normal science is characterized by different types of uncertainties that are larger than what is found in normal science. In the post-normal science world, the facts are uncertain, values are in dispute (a form of qualitative uncertainty), stakes are high, and decisions are urgent [1]. Normal science exhibits uncertainties about the data, including compounded uncertainties when multiple pieces of data are used. Under normal science, problems such as growing crops or providing food to the food insecure typically limit the consideration of uncertainties to a reduced and isolated set of facts. In contrast, effective policy making conditions under post-normal science involve acknowledging externalities and embracing uncertainties that go beyond the traditional question of how to make decisions when there are data (or scientific) uncertainties [14]. Under post-normal science, an examination of the uncertainty about what to do and how to approach the policy problem is important. In other words, what is important for post-normal science is the uncertainty that stems from differing stakeholder perspectives and values (i.e., non-scientific uncertainties), which is a component of Decision Uncertainty.

Resolving data uncertainties, even in those cases where it is possible, does not resolve other important uncertainties. These include uncertainties about stakeholder values and perspectives that relate to how the problem is defined, whether the criteria chosen adequately characterizes that problem, and stakeholder differences on the relative importance among those criteria. Addressing these uncertainties requires embracing the diversity in perspectives and stakeholder values. While subjective, these are important to understand. Without such understanding, policy makers and other stakeholders lose a critical component of public policy making, which is the ability to articulate the rationale for their decision [15].

2.5   Choice: Same Facts, Different Decisions

To distinguish the practice of science from how science is used, it is helpful to borrow the concept of tame and wicked problems from Rittel and Webber. This is not simply a semantics issue but rather a conceptual and practical one with consequences to the public policy practitioner. An understanding of this concept leads to a better understanding of how science-based policy decision making can be more effectively conducted, particularly with diverse stakeholders in democratic societies. Rittel and Webber define a wicked problem as having the following characteristics:(1) there is no definitive formulation of what the problem is, (2) every wicked problem is a symptom of another wicked problem, (3) there are no single right answers but only better or worse conditions, (4) there are no stopping rules, (5) there is no objective measure of success, (6) acceptable solutions need to be discovered, and (7) they often involve a moral dimension [16]. Therefore, problems are wicked not because they are evil, or policy analysts have malicious intent, but because they are values-based and are difficult to tackle for all the reasons given above. Wicked problems are incorrigible in the sense that the aim in tackling them is not to find the truth but to "improve some characteristics of the world where people live" [16]. Each of the characteristics of wicked problems described by Rittel and Webber can be synthesized to stakeholder choice. Wicked problems have always existed. However, today, the consequences are more far-reaching because of increased population, decreasing resources and more global interconnectedness.

Wicked problems are often complex because scientific understanding and data are frequently necessary components to describe them. However, complexity is not what makes a problem wicked. Highly complex science problems can be solved with enough knowledge. A solution to a highly complex science problem can be verified because it must comport with observations of the physical world. What makes a problem wicked is choice and the values that influence that choice. Where values play a role there can be no "right" answer, only answers that are consistent with those values. There are no independent observations or experiments that can adjudicate the correctness of the chosen solution. A problem is wicked because people have varying points of view that result in different choices. Therefore, although science does not make the problem wicked, how to use science is a wicked problem. Often when the focus is on environmental problems, there is confusion about whether a difference in philosophy or the lack of knowledge makes the problem wicked. If choice is involved in determining a course of action, the problem is wicked. In a wicked problem, the same facts can support a different decision.

2.6 Tame Questions vs. Wicked Problems

Distinguished from wicked problems, tame problems have a clear definition, operating rules, and a clear recognizable nature with a single, right answer. Math and science problems are tame problems; albeit many are quite complex. For tame problems, the uncertainties pertain to data and model uncertainties. Tame problems, for the sake of clarity in the discussions that follow, will be referred to as questions. Using this phraseology, questions seek to satisfy curiosity or accomplish a specific unambiguous task such as "how do I fix a leaking faucet?"or "what is needed to establish basic living conditions suitable for people to live on Mars?" These are often basic science or engineering questions where the goal is to better understand the world or construct some specific object that follows physical laws. In contrast, we chose to characterize wicked problems as problems rather than questions, to distinguish them as containing values and reflecting choices that need to be made for some action or direction. In wicked problems, values determine how science is applied toward a social outcome.

For example, when we want to know what to do about sea level rise, we need to understand the science of how and when sea levels rise. This science is the tame component to the wicked problem of what we want to do about the sea level rise. For the wicked sea level rise problem, it is important to answer the tame question regarding whether sea level rise is due to natural or anthropogenic influences. The answer determines the options available to address it. If sea level rise is primarily due to natural influences, we may focus more on policies directed at avoiding the effect of sea level rise such as moving people away from low-lying areas or prohibiting new development in vulnerable coastal zones. If sea level rise is primarily due to anthropogenic influences, the options to be evaluated are also likely to include policies directed at the sources exacerbating the problem such as reducing carbon dioxide (CO_2) emissions (the main contributor to climate change). This distinction between tame questions and wicked problems is important to understand because the approaches to addressing them must necessarily be different.

2.6.1 Formulating the Wicked Problem

Rittel and Webber cautioned that when approaches used to tackle tame questions are applied to wicked problems, they are typically ineffective or produce bad results. When the US EPA evaluated DDT, they focused on whether the scientific evidence regarding the harm to wildlife from DDT exposure was sufficiently certain. The wicked public policy problem regarding the use of DDT, to manage human disease while considering environmental and other benefits and disbenefits, was tamed to the science question of the certainty with regard to DDT's adverse effects on wildlife. The resulting DDT ban seemed an obvious conclusion, but the ban removed the option to use DDT for public health control of today's mosquito-borne diseases. Similarly, when the hunger problem was tamed to the question of how to increase crop yield, growing more food was the obvious solution. However, this did not eradicate hunger and, additionally, ruined agricultural lands. When the problems seen among the incarcerated population was tamed to overcrowding, isolation seemed an obvious solution, but prisoners were not reformed and, as a result of being held in the solitary confinement, many were left in worse condition. Taming the wicked problem occurs when decision makers shortcut the policy making process by seeking to streamline discussions about the problem scope, eliminate conflicting viewpoints, and limit the number of alternatives examined. Often, this results in exacerbating the original problem, solving the wrong problem, or creating a new problem.

Consider the public policy problem of determining how power plants can comply with particulate matter standards. This is a wicked problem because compliance can be achieved in many different and legitimate ways, even as each of them reduces pollution. The Clean Air Act's strategies center around the establishment of ambient air quality standards, which is a policy tool to measure compliance with the air quality standard [17]. The underlying science about pollutants such as particulate matter suggests that there is a continuum of increased adverse health impacts from zero or very low levels to higher levels, meaning that the science does not draw a bright line between no effect and adverse effects. This requires that science policy makers determine how to implement a science policy without a bright line established by the science. Administratively, it is more feasible to establish a standard, which offers an absolute determination of compliance and simplifies implementation of the health standard. Therefore, the US EPA established the National Ambient Air Quality Standards (NAAQS) for particulate matter and seven other pollutants [18]. The establishment of the NAAQS was in response to the wicked problem of controlling pollutants to protect public health and welfare, a choice problem.

To achieve the ambient air quality standards for particulate matter under the Clean Air Act, states administer industry-specific requirements, including those for power plants [17, 19]. Most states developed regulations that require power plants to install pollution control equipment to collect the particulate matter before it is released to the atmosphere. This is done to ensure that the air quality standards would be achieved and thereby prevent area residents from inhaling unhealthy concentrations of particulate matter, including toxic particulate matter. On its face, this seemed to be a straightforward, even obvious, solution to the environmental public policy problem of reducing human exposure to particulate matter to meet the health-based NAAQS.

In the mid-1970s, there were no regulations developed to prevent the particulate matter (PM) collected from the pollution control equipment from being hauled in large uncovered trucks and driven to unlined landfills for disposal. In establishing the power plant requirements for PM, the US EPA's analysis considered only what reductions would be needed from those power plants to ensure that ambient PM air concentrations would be below the standard. This created two additional environmental problems. The first was the transport of toxic PM in uncovered trucks, creating another source of air pollution caused directly by the implementation of the power plant regulations. The second environmental problem was leaching of the toxic material into groundwater where the landfills were not lined [20, 21]. The problem of power plants complying with air quality requirements was initially solved by applying a technological fix, (i.e., installing pollution control devices to collect excess PM at the power plants). However, decision makers did not understand that in narrowly defining the problem, they had exacerbated the original wicked problem, which resulted in greater human exposure to harmful pollutants.

Holling and Chambers have also made observations about policy analysis, noting the inappropriate use of technical methods, which were designed to solve technical problems. In their view, when applied to policy problems, these analytical tools result in us "learning more and more about less and less" [22]. Holling and Chambers' approach, which they called "The Game," was developed to nurture interdisciplinary research to address the relationship between man and his environment. It includes the development of a non-optimizing (i.e., not a single goal), value-variable (not objective) model that would allow participants to experiment with different scenarios under different assumptions. Holling and Chambers were advocating for a different and non-technological approach for wicked problems. While conceptually advancing a new paradigm to address wicked problems, "The Game" still leaves a methodological gap. This gap is the capability to address important aspects of wicked problems such as the analytical consideration of uncertainties in problem definition, data, expert judgment, and stakeholder values.

In describing the problem of food sovereignty and distribution, Patel asserts that giving people the increased capability to understand the multiple facts of the hunger problem leads to developing more policy alternatives and more control over those choices [8]. Patel is also advocating for a non-technological approach for wicked problems. Without offering a methodological solution, Patel appeals for a process that allows for the social construction of solutions to increase opportunities for successful public policy making.

2.6.2   Collateral Damage of Taming Wicked Problems: Impugning Science Integrity

Funtowicz and Ravetz and others have argued that science has become tainted with politics, bias and sloppy methodologies, particularly as society reflects greater political and ideological polarization. When stakeholders are at loggerheads, but still interested in finding common ground, these values must be considered and reconciled outside of the scientific debate, if wicked problems have a chance of being solved.

Neither the science nor the facts "speak for themselves." While greater scientific understanding of the facts informs possible solutions, human values ultimately determine what choices are made. If this reality is ignored or simply not understood by public policy decision makers, stakeholders with strong opinions about how to use the science are left with no room for discussions regarding the non-scientific social context and values. With the science discussion deemed as the only legitimate basis for dispute, these stakeholders are left with the binary choice of either accepting or denying the scientific findings. Could this be a significant part of what is occurring in controversial discussions regarding climate, genetically modified organisms, and similar hot topics? Except in those cases where stakeholders are uninterested in finding common ground, the void created by expecting science alone to address wicked problems diminishes stakeholder experience and perspectives and eliminates opportunities for stakeholder discussion. Furthermore, this allows the infusion of politics, bias, and sloppy methodologies to obfuscate the conflicts. When there is no forum to discuss the social context and values behind the policy choice, stakeholders opposed to a policy choice often resort to becoming deniers of the science on which the policy is based.

Therefore, the lack of recognition in the differences and relationship between science and values degrades science integrity. This is not simply an issue created by non-scientific stakeholders, since scientists themselves often exacerbate this problem. Lackey asserts that science integrity is impugned by scientists who knowingly or unknowingly present their work as policy neutral when, in fact, their work is colored by their personal policy preferences. The obfuscation of science and advocacy leads to the marginalization of scientific contributions overall [23]. Expectations that scientific study or model results will direct or force a certain policy decision are often dashed when scientists and other stakeholders are confronted with the realization that other factors must be considered. Using post-normal thinking, it is the collective reconciled values, perspectives, and opinions of the stakeholders that must frame the problem and its potential solution. This framing must be consistent with scientific understanding, rather than expecting or allowing science to dictate the wicked problem solution.

2.6.3   Another Wicked Problem: Science Research Guided by Public Policy Problems

Wicked problems are also involved in determining what science research is pursued. The system of science research consists of industry, academic, non-profit, and governmental institutions that fund or utilize funds to make new scientific discoveries. In this system, there is an established procedure and hierarchy regarding what science questions are legitimate to ask, which studies will get funding, and how the research is perpetuated. This procedure is not science; it is science policy (a wicked problem). As such, the same wicked problem issues related to problem definition and values are found [24]. What research should get funded and which scientific questions should be pursued? The choices made in response to these questions determine the kind of scientific knowledge that becomes available. The decision about what science research to do is made in several ways. These are as follows: (1) scientists utilize their training to determine which questions are ripe for pursuit because the theoretical constructs or the practical experimental tools make it feasible; (2) scientists determine which questions are pursued based on whether they find it relevant or interesting; (3) scientists choose to pursue what they believe has the greatest chance of getting funded; (4) scientists choose to perform research that they believe will fill a gap needed to make a current or future decision.

The social construct within which science questions get asked, which questions are funded, and how the answers to these questions are developed are important public policy problems. Problems of this nature could also benefit from the same kind of post-normal approach that is being proposed for more traditionally conducted public policy decision making. Since the question of what science should get done is a policy problem, it becomes possible to consider directing scientific research from the perspective of what public problems need to be resolved. The science needed for public policy making requires a different model than that used for science research, which is conducted to satisfy academic curiosity. Public policy makers can reduce barriers to their decision making when they can identify policy problems that will benefit from specific science information or greater scientific certainty. In these cases, problem-directed research becomes an effective means to ensure that the results of the science research will be used. Compared with the situation where the public policy community hopes that scientists conduct policy-relevant research, public policy directed research is an effective means to help ensure that funding for scientific research will produce the type of information that is needed in the development of effective public policy.

2.6.4   Wicked Problems: High Stakes

For most public policy problems, the stakes are high because of their far-reaching consequences and because decisions that are made, generally, cannot be reversed. A decision to implement a public policy action has consequences across the economic, social, and environmental landscape that cannot be undone. In the scientific vernacular, we cannot do the controlled experiment because, once the experiment is done, we have forever altered the original condition. In contrast to answering a tame question that can be reversed if wrong, a wicked problem solution alters the original condition so that even if a mistake is recognized, the original condition cannot be restored and a new problem, generally of greater complexity and impact, replaces the original problem.

Consider the situation of the US EPA's approval of methyl tert-butyl ether (MTBE) as a gasoline additive under the air quality program [25]. By defining the approval of MTBE as an air quality issue, the Agency failed to consider other environmental issues such as frequent spillage at gas refueling stations. The US EPA was caught off-guard by strong public outcry when groundwater and drinking water were contaminated by the highly odiferous and hygroscopic MTBE [26, 27]. Once the broader environmental problem was recognized, the US EPA reversed the MTBE approval and removed it as a required fuel additive in 2006 [28]. However, as a wicked problem, there is no complete reversal. Groundwater wells were contaminated and needed to be cleaned up while the original air quality problem had to be resolved in another way. Public trust in the US EPA's decision making was shaken. Money spent by gasoline refiners to add MTBE, and then to remove it, could not be reimbursed.

A wicked problem was tackled as a tame question when the US Army Corps of Engineers contemplated choices in levee building to prevent flooding in New Orleans, Louisiana [29]. On average, New Orleans lies six feet below sea level. In 2005, Hurricane Katrina made landfall in New Orleans as a Category 3 hurricane. The heavy rains and storm surge from Hurricane Katrina resulted in breached and failed levees, which forced residents to evacuate their homes and, for many, eventually lose their homes. After Hurricane Katrina, the US Army Corps of Engineers evaluated levee designs as a problem of engineering specifications. Even though how to build a levee is a well-understood tame engineering question, how to build a levee that will prevent future flooding is a wicked problem. Stakeholders' value judgments are required to determine the degree of risk that stakeholders are willing to tolerate and the degree of uncertainty that stakeholders are willing to accept [30]. Effectively addressing the wicked problem will depend on answering tame questions such as those related to the expected severity of future storms. This, in turn, could depend on the degree to which scientists understand how changing climate influences the timing and severity of those storms. Uncertainties affect those answers as well as answers to questions stakeholders ask, such as the following: How high will the levee need to be? How strong will they need to construct it? How much money, as a nation, are taxpayers willing to invest? And are those with the greatest risk of home loss and displacement part of these discussions? Clearly, this is a high-stakes wicked problem whose solution will strongly depend on the assessment of uncertainties pertaining to future forecasting (scientific) and societal willingness to accept certain degrees of risk (non-scientific).

If wicked problems are characterized by choice, it is possible to reexamine Funtowicz and Ravetz's post-normal science description ("facts uncertain, values in dispute, stakes high, and decisions urgent") in this landscape. While all post-normal science problems are choice problems, not all choice problems are post-normal science problems. Choice problems such as what car to buy are relatively low-stakes problems and are neither the kind of post-normal problems contemplated by Funtowicz and Ravetz nor the wicked problems contemplated by Rittel and Webber. While there may be some uncertainty about some car buying facts and members of your family may not equally value a car capable of high speed or one with luxury components, the consequences are not globally, or societally, far-reaching. Unlike public policy decisions, it is relatively easy to reverse this type of personal wicked problem decision. Dissatisfaction with your car purchase could be reversed by selling it and buying another, perhaps not a cheap personal decision but at least a reversible one. These are wicked problems where decision stakes are sufficiently low that using a stakeholder-focused approach methodology to reach a decision does not make sense. Likewise, when facts are highly uncertain, or values are highly disputed, but the stakes are not high, over-complicating the decision making by using a stakeholder-focused methodology is most likely not warranted. In contrast, for example, a stakeholder-focused approach is appropriate to address the high-stakes problem of hunger, where stakeholders do not all agree as to what the major root cause is nor how to address the problem. Different governmental and non-profit organizations have chosen to focus on different aspects of the problem whether hungry children in schools, homeless families, or increasing the minimum wage [8, 31, 32]. Each of these perspectives is legitimate but if we are determined, as a society, to eradicate the hunger problem, we will need to determine where to focus our limited resources to be more effective.

While Funtowicz/Ravetz, Rittel/Webber, and others have described the use of science in policy making slightly differently, none have methodologically addressed how solutions to public policy problems are to emerge. A practical, stakeholder-focused approach for tackling science-based public policy problems is needed.

2.6.5   Disentangling the Tame, the Wicked, and the Tame Within the Wicked

When there is an understanding of a public policy problem's Decision Uncertainty, it is possible to fill a conceptual gap about the relationship between and among tame questions, wicked problems, and wicked problems with tame components embedded within them. This conceptual gap pertains to navigating the interface between tame questions and wicked problems when the tame is embedded within the wicked. As discussed, the current tendency for environmental public policy makers is to treat both tame questions and wicked problems as tame and to apply tame approaches to the entire policy problem. There are exceptions such as when social scientists (typically) are put in charge of engaging stakeholders to find solutions for wicked problems. But oftentimes, even social scientists utilize approaches that substantively mimic tame approaches. At other times, non-tame approaches are used but these do not provide the necessary openness and capability to appropriately handle expert data of the tame components to allow for the development of fact-driven alternatives. Therefore, without a full understanding of the tame-wicked conceptual gap, public policy making is stymied because the ability to appropriately engage stakeholders and to discover solutions is not available.

As manifest in public policy making, normal science is characterized by the domination of tame questions (and "low-stakes" wicked problems) and post-normal science is characterized by the domination of high-stakes wicked problems with embedded complex tame questions. Rittel and Webber's description of wicked problems includes a component of a moral or ethical dilemma, implying a high-stakes characteristic in the manner of Funtowicz and Ravetz. Therefore, in this sense, Rittel and Webber limited their definition of wicked problems to those public policy choice problems where the context and impacts are large scale. It is unlikely that they would have contemplated the concept of a "low-stakes" wicked problem. Tame questions can contribute to a variety of high- or lower-stakes wicked problems. Highly complex tame questions that are embedded in wicked problems can contribute added complexity and uncertainty to wicked problems. Likewise, high uncertainties in tame questions contribute to higher Decision Uncertainties in wicked problems.

2.7   From Clumsy to Open Solution Approaches

Recognizing the Requisite Steps of decision making, performing them well and with iteration, provides the foundation for a new paradigm for improved public policy making. Wicked problems require a different approach that allows stakeholders to evaluate and understand the differences among conflicting and divergent values and perspectives. Verweij et al. and others use the term "clumsy" to describe approaches that embrace a "dynamic, plural and argumentative system of policy definition and policy framing" [33, 34]. Benessia et al. define clumsy solutions as socially constructed solutions where stakeholders "work deliberately within imperfections." In a world of many stakeholders with diverse ways of perceiving the world, clumsy solution approaches are needed. In contrast to reductionism, where imperfections are ignored, clumsy solutions are achieved by working deliberately within those imperfections. Clumsy solutions embrace conflicting views of the problem by not requiring standardization of values or a single problem definition [35].

The descriptions of a clumsy solution approach have remained primarily conceptual, and therefore, public policy makers have been unable to move from concept to practice. Consequently, a next-generation public policy making approach is needed, which is grounded conceptually in the clumsy solution paradigm, but is methodologically practical and feasible. This book introduces a new paradigm, the "open solution approach." The open solution approach is an expansion and integration of insights from Vermeij and others about clumsy solutions, Funtowicz and Ravetz about normal and post-normal science, and Rittel and Webber about tame and wicked problems, with the addition of Decision Uncertainty and the structured iteration through the Requisite Steps.

An open solution approach is designed to guide stakeholders through a discovery and learning process to find common ground, rather than finding the "right" or optimal solution. This kind of approach may seem chaotic to some because diversity is embraced rather than simplified. However, it is not a random process or without guidance. An effective open solution approach is a highly structured process that encourages the consideration of diverse perspectives, uncertain values and facts, and multiple constructions of the problem definition while providing a process to reach an agreed upon course of action. The resulting benefits include helping stakeholder discover solutions that are not apparent at the beginning of their deliberative process, building understanding and public trust, and finding common ground for sustainable policy decisions. An open solution approach embraces diverse perspectives, transparency, and iteration and offers a feasible alternative to tackling these wicked problems without having the process become usurped by traditionally applied reductionism.

2.8 Handling Uncertainties in an Open Solution Approach

Wicked problems, as contrasted with tame questions, raises an important difference in thinking about uncertainty. Tame question uncertainties represent an understanding about the amount of deviation from a truth that is independent from any subjective context. In contrast, wicked problems have no single right answer, only a better or worse condition. Unlike tame questions, all the uncertainties associated with wicked problems cannot be quantified but they can be considered qualitatively, in a relative sense. Therefore, for wicked problems, rather than establishing an intrinsic quantitative uncertainty, the ability to conduct a relative comparison of uncertainties among alternatives is needed. In a wicked problem, stakeholders are determining whether an alternative represents a better or worse condition. The open solution approach provides stakeholders with a relative comparison of the alternatives and not the single best alternative. What stakeholders want to know is if Alternative B is better or worse than Alternative A and the uncertainty inherent in that determination (the Decision Uncertainty).

A computational laboratory is needed to evaluate policy alternatives. An open solution approach is such a computational laboratory. It is designed to use iterative techniques to focus on helping stakeholders understand and use Decision Uncertainty to develop and compare alternatives. With iteration, each pass through the Requisite Steps allows stakeholders to experiment with different combinations of problem formulation, data, expert judgment, and their values and minimize Decision Uncertainty.

2.9   When Is Open Best?

Open solution approaches can be very helpful for addressing two types of classic policy questions each time a wicked problem is tackled. These questions are (1) how much scientific uncertainty is too much and how do I know? and (2) how do I reconcile all these diverse stakeholder opinions into an agreed-upon plan? The first question pertains to whether decision makers think that more science research, or what kind of science knowledge, is needed before they are comfortable in making a decision. If the decision stakes are not high or there is plenty of funding and a solution is not urgent, it may not be necessary to contemplate too long about what to do. However, when stakes are high enough to matter to stakeholders or funding is tight and competitive, stakeholders may benefit from the use of an open solution approach. The second question is the public policy decision maker's classic wicked problem dilemma. When stakeholders have widely diverse, and often conflicting, opinions, decision stakes also tend to be high and a process that allows stakeholders to better understand each other and find common ground is needed.

2.9.1   How Much Uncertainty Is Too Much?

Uncertainties in science abound whether in sampling methods, data collection, experimental design, model performance or statistical applications. However, whether these uncertainties are significant to the resolution of a wicked public policy problem depends on the nature and formulation of the problem. It is not possible to know if the uncertainty is too much until we ask "too much *for what?*" Scientific uncertainties often fuel disagreements among scientists about whether the observation is real, an artifact of the scientific method or a misinterpretation by the scientist. Scientists deal with uncertainties as part of their hypothesis building and testing. These uncertainties, when sufficiently characterized, enhance the discovery of scientific knowledge and the degree to which we confidently understand it and the constraints associated with its use.

In decision making, uncertainties are only important if they have the potential to change decision makers' selection of the alternative. Suppose a measurement is known to be accurate within 10%. In isolation, some scientists or stakeholders may consider this uncertainty high. Also suppose that there is the capability to evaluate the influence of this measurement uncertainty in a computational policy laboratory. Additionally, suppose that this evaluation shows that the chosen alternative no longer remains the most attractive option only if the measurement uncertainty reaches 50% or more. Therefore, in this case, 10% measurement uncertainty is low (acceptable) because it does not change the decision maker's choice of the decision alternative.

Using an open solution approach, the US EPA was able to make a similar determination in its 2004 regulatory action on air quality. In 2004, the US EPA was required to consider industrial air emissions as 1 of 11 CAA-mandated criteria to decide whether a county was achieving the ozone NAAQS [36]. Air emission inventories are notoriously laborious to compile, and state resources to complete them are often limited. Therefore, when the US EPA used an emission inventory for some counties that some stakeholders found implausible and, therefore, unacceptable, it was important to demonstrate whether these county emissions could change the US EPA's decision. By examining the range of data possibilities within the overall context of the problem formulation, the US EPA demonstrated that further refinement of those emission inventories would not have altered the final decision. After seeing this demonstration, the stakeholders agreed. This is an example where data uncertainty could affect the final decision. Without a means to determine if the uncertainty mattered, there were stakeholders who would not otherwise have been convinced that US EPA was sufficiently informed to make a final decision. In addition, in determining that refinement of the emission inventory was not necessary, the states avoided spending limited resources on unnecessary data collection.

Unavoidable and irreducible uncertainties are often used by stakeholders to avoid making final decisions. Particularly when policy discussions do not explicitly distinguish between the science and the values components in policy making, science deniers can point to these uncertainties as a reason why the policy choices should be rejected. Within this paradigm, extremists argue that any amount of scientific uncertainty is too much, preventing any decision from being made until these uncertainties are eliminated. The response to this paradigm is twofold. First it is important to provide a space for stakeholders to discuss their values, as distinct from a discussion of the science. Second, it is necessary to examine whether the actual scientific uncertainties associated with the data being used in the decision have a significant impact on the relative attractiveness of the options being considered. The question regarding the importance of uncertainty can only be viewed specific to the context of the problem being considered. The uncertainty of a datum may be acceptable for one decision and completely unacceptable for another. Decision makers want to understand whether Decision Uncertainty is too large to make their specific policy decision. If the answer is yes, then decision makers may choose to support another decision or wait for targeted research to mitigate or resolve policy-relevant uncertainties.

2.9.2 Convergence to Agreement with Diverse Opinions

In public policy decision making, in contrast to scientific or data uncertainties that are used when answering scientific questions, uncertainties about stakeholder values are not something to be resolved but rather something to be understood, considered, and converged upon. Convergence to agreement among diverse stakeholders requires the capability to find common ground. Guided by the learning offered in an open solution approach that iterates through the Requisite Steps, there are greater opportunities for establishing mutual understanding among stakeholders, which can then lead to agreement. In such an approach, diverse stakeholder opinions may be discovered to result from misunderstandings about the problem definition, the set of relevant facts, misconceptions about what is important to other stakeholders, or the discovery that the alternatives being analyzed are insensitive to certain opinion differences. When stakeholders come to a common understanding about problem definition or understand whether scientific uncertainties affect the selection of the alternative, they are reducing Decision Uncertainty and converging on agreement.

2.9.3 Urgent Timeframes

After determining whether engaging an open solution approach could be helpful based on the kinds of policy problems that need to be addressed, potential users will likely also need to consider the urgency of the problem's timeframe. When highly complex and high-stakes decisions must be made in tight or urgent timeframes, ad hoc application of an open solution approach is unlikely to be timely. However, just as there are plans constructed for emergency evacuation or other urgent situations, it may be possible to have some highly complex, high-stakes wicked problems partially constructed for these kinds of uses.

Consider the example of a train derailment where a chlorine tanker car has overturned resulting in damage to the structural integrity of the chlorine container. Although there is no gas being released now, there is the possibility that at any moment it could rupture and emit a cloud of chlorine gas that if inhaled at a high enough concentration would be lethal. Suppose that if a release occurs under present meteorological conditions, chlorine concentrations to the north of the car will remain high enough to be lethal within approximately one kilometer where there is a hospital. The wicked problem is do you evacuate the hospital? To address such a problem, the decision maker will need to consider things such as (1) the risk of injuries, and death, to a certain number of the patients if there were a full and immediate evacuation of the hospital, (2) the cost to implement the evacuation, (3) the probability that the chlorine car may not rupture, and (4) probability that the meteorology could change so that the hospital is no longer in the danger zone. Knowing that such situations may occur, decision makers could construct an open solution approach analysis with some parameters already contemplated (e.g., toxicity of specific compounds based on whether the exposure is to air or water), leaving placeholders for specific exposure concentrations to accommodate the emergency scenario. However, this kind of upfront emergency planning is often not possible. Therefore, urgent or emergency situations do not generally lend themselves to open solution approaches.

Suppose, instead, that the situation is the same as described above, except that a rupture of the chlorine tank was possible only upon initiating its removal. Utilizing an open solution approach for this scenario may be an effective way to prepare for and resolve a dangerous situation by considering a variety of different local perspectives and community-derived alternatives. Uncertainties, stakes, and urgency should be used by stakeholders to determine whether their policy problem will benefit from open solution approaches.

Having now separated the normal from the post-normal, the science from the non-science, and the tame from the wicked, it becomes easier to examine and determine where the disagreements in environmental issues lie. Just as community organizers would not consult an air quality model to determine what their constituents value more, a scientist would not have a discussion of values as a substitute for running a computational model. Tame solutions do not solve wicked problems. If attention is paid to how the problems are defined, the clues for which approaches are more appropriate are often revealed, allowing for the discussion and choice of more relevant and effective strategies.

2.10 Summary

Examining the differences between science and science-based, but value-laden, policy problems, through a variety of examples and concepts, provides a means to better understand the mechanics of the public policy making process. However, practitioners require actionable ideas and not merely sound concepts. The open solution approach with Decision Uncertainty is a new public policy paradigm designed to address the wicked problems of public policy decision making. This is a computational policy laboratory for stakeholders to develop, experiment, and compare decision alternatives. Understanding Decision Uncertainty is a learning process. Stakeholders want to know whether Decision Uncertainty is too large to rationalize choosing one decision alternative over the others. With Decision Uncertainty, employed as a convenient gauge for stakeholders to assess their level of confidence to make their decision, it becomes possible for stakeholders to rationalize their final decision. For readers interested in how this is accomplished, the next chapter introduces Multi-criteria Integrated Resource Assessment (MIRA), an open solution approach.

References

1. Funtowicz SO, Ravetz JR (1993) Science for the post-normal age. Futures 25:739–755
2. Funtowicz SO, Ravetz JR (1994) The worth of a songbird: ecological economics as a post-normal science. Ecol Econ 10(3):197–207. https://doi.org/10.1016/0921-8009(94)90108-2
3. Costanza R, d'Arge R, de Groot R, Farber S, Grasso M, Hannon B, Limburg K, Naeem S, O'Neill RV, Paruelo J, Raskin RG, Sutton P, van den Belt M (1997) The value of the world's ecosystem services and natural capital. Nature 387:253–260
4. Macmillan D, Jones R (1988) John Deere tractors and equipment, volume 1: 1837–1959. American Society of Agricultural Engineers, St. Joseph
5. Cooke ML (1937) The future of the great plains: the report of the Great Plains Committee. Committee on Agriculture, Washington, D.C.
6. Worster D (1979) Dust bowl: the Southern Plains in the 1930s. Oxford University Press, Inc., New York
7. Hauter W (2012) Foodopoly: the battle over the future of food and farming in America. The New Press, New York
8. Patel R (2007) Stuffed and starved: the hidden battle for the world food system. Protobello Books Ltd, London
9. Woodhill J (2002) Sustainability, social learning, and the democratic imperative. In: Leeuwais C, Pyburn R (eds) Wheelbarrows full of frogs. Kninklijke Van Gorcum BV, Assen, pp 317–331
10. Woodhill J, Roling NG (1998) The second wing of the eagle: the human dimension in learning our way to more sustainable futures. In: Facilitating sustainable agriculture: participatory learning and adaptive management in times of environmental uncertainty. Cambridge University Press, Cambridge, UK, pp 46–71
11. Beck U (1992) Risk society: towards a new modernity. Sage Publications, London
12. Beck U (1994) The reinvention of politics: towards a theory of reflexive modernization. In: Beck U, Giddens A, Lash S (eds) Reflexive modernization: politics, tradition and aesthetics in the modern social order. Stanford University Press, Stanford, pp 1–55
13. Curtin CG (2015) The science of open spaces: theory and practice for conserving large complex systems. Island Press, Washington, D.C.
14. Funtowicz SO, Ravetz JR, von Schomberg R (1993) The emergence of post-normal science. In: Leinfellner W, Eberlein G (eds) Science, politics and morality: scientific uncertainty and decision making. Theory and decision library, series A: philosophy and methodology of the social sciences. Kluwer Academic Publishers, Dordrecht, pp 85–123
15. Winner L (1986) The whale and the reactor: a search for limits in an age of high technology. University of Chicago Press, Chicago
16. Rittel HWJ, Webber MM (1973) Dilemmas in a general theory of planning. Policy Sci 4:155–169
17. Clean Air Act (1990) 42 U.S.C. 7401 et seq
18. U.S. Environmental Protection Agency (2019) National ambient air quality standards. Vol 40 C.F.R. Part 50
19. U.S. Environmental Protection Agency (2016) Fine particulate matter national ambient air quality standards: state implementation plan requirements; final rule. Fed Regist 81:58009
20. U.S. Environmental Protection Agency (2018) Hazardous and solid waste management system; disposal of coal combustion residuals from electric utilities; final rule. Fed Regist 83(146):36435
21. U.S. Environmental Protection Agency (2015) Technical amendments to the hazardous and solid waste management system; disposal of coal combustion residuals from electric utilities—correction of the effective date. Fed Regist 80(127):1
22. Holling CS, Chambers AD (1973) Resource science: the nurture of an infant. Bioscience 23(1):13–20
23. Lackey R (2007) Science, scientists, and policy advocacy. Conserv Biol 21(1):12–17

24. Jasanoff S (ed) (2004) States of knowledge: the co-production of science and social order. Routledge, London
25. U. S. Environmental Protection Agency (1994) 40 CFR part 80: regulation of fuels and fuel additives; standards for reformulated and conventional gasoline: final rule. Fed Regist 59(32):7716–7878
26. U. S. Geological Survey National Water Quality Assessment (1995) Occurrence of the Gasoline Additive MTBE in Shallow Ground Water in Urban and Agricultural Areas. FS-114-95. Rapid City, SD.
27. Blue Ribbon Panel on Oxygenates in Gasoline (1999) Achieving clean air and clean water: the report of the Blue Ribbon Panel on Oxygenates in Gasoline. Washington, D.C.: United States Environmental Protection Agency
28. U.S. Environmental Protection Agency (2006) Regulation of fuels and fuel additives: removal of reformulated gasoline oxygen content requirement. Fed Regist 71(88):26692–26702
29. Rogers JD, Kemp GP, Bosworth HW Jr, Seed RB (2015) Interaction between the US Army Corps of Engineers and the Orleans Levee Board preceding the drainage canal wall failures and catastrophic flooding of New Orleans in 2005. Water Policy 17:16. https://doi.org/10.2166/wp.2015.077
30. U.S. Army Corps of Engineers (2000) Design and construction of levees. US Army Corps of Engineers, Washington, D.C.
31. Schwartz-Nobel L (2002) Growing up empty: the hunger epidemic in America. Harper Collins, New York
32. Eisinger PK (1998) Toward an end to hunger in America. Brookings Institution Press, Washington, D.C.
33. Verweij M, Douglas M, Ellis R, Engel C, Hendriks F, Lohmann S, Ney S, Rayner S, Thompson M (2006) The case for clumsiness. In: Vermeij M, Thompson M (eds) Clumsy solutions for a complex world: governance, politics and plural perceptions. Palgrave Macmillan, New York City, pp 1–27
34. Benessia A, Funtowicz SO, Giampietro M, Guimaraes Pereira A, Ravetz JR, Saltelli A, Strand R, van der Sluijs JP (2016) The rightful place of science: science on the verge. Consortium for Science, Policy and Outcomes, Arizona State University, Charleston
35. Vermeij M, Thompson M, Engel C (2006) Clumsy conclusions: how to do policy and research in a complex world. In: Vermeij M, Thompson M (eds) Clumsy solutions for a complex world: governance. Politics and Plural Perceptions. Palgrave Macmillan, New York City, pp 241–249
36. Stahl CH, Fernandez C, Cimorelli AJ (2004) Technical support document for the Region III 8-hour ozone designations 11-factor analysis. U.S. Environmental Protection Agency, Region III, Philadelphia

Chapter 3
Introduction to MIRA, an Open Solution Approach

Abstract The Multi-criteria Integrated Resource Assessment (MIRA) open solution approach is a framework and a process. MIRA is a transparent policy framework that embraces diversity, adversity, and discovery, with a process that allows for the policy-specific evaluation of Decision Uncertainty and the emergence of stakeholder agreement. The three major unique aspects of an open solution approach are the inclusion of Decision Uncertainty; stakeholder-directed, but structured, iterations of the Requisite Steps; and trans-disciplinary learning. Trans-disciplinary learning becomes key to ensuring that each iteration is informed by the previous. These components distinguish the open solution approach from conceptual social science approaches that lack methodologies to compare specific policy alternatives and the reductionist decision-analytic approach that is currently the dominant public policy making paradigm. In this chapter, the Requisite Steps as applied in a MIRA approach are described, including terminology and guiding principles specific to MIRA.

Keywords Trans-disciplinary learning · Adaptive management · Transparency in policy making · Multi-criteria assessment

3.1 Guiding Principles

The Multi-criteria Integrated Resource Assessment (MIRA) open solution approach is a framework and a process. MIRA is a transparent policy framework that embraces diversity, adversity, and discovery, with a process that allows for the policy-specific evaluation of Decision Uncertainty and the emergence of stakeholder agreement. The three major unique aspects of an open solution approach are the inclusion of Decision Uncertainty; stakeholder-directed, but structured, iterations of the Requisite Steps; and trans-disciplinary learning. Trans-disciplinary learning becomes key to ensuring that each iteration is informed by the previous. These components distinguish the open solution approach from conceptual social science approaches that lack methodologies to compare specific policy alternatives and the reductionist decision-analytic approach that is currently the dominant public policy

© Springer Nature Switzerland AG 2020   47
C. H. Stahl, A. J. Cimorelli, *Environmental Public Policy Making Exposed*,
Risk, Systems and Decisions, https://doi.org/10.1007/978-3-030-32130-7_3

making paradigm. In this chapter, the Requisite Steps as applied in a MIRA approach are described, including terminology and guiding principles specific to MIRA.

3.1.1   Inclusiveness and Transparency

In an open solution approach, stakeholder inclusiveness is not simply an invitation to the table but includes practices that result in meaningful participation and honest discussion. Ensuring that stakeholders can fully participate in the open solution process can be addressed, in part, by using a facilitator. Good facilitation throughout the process can be very helpful, and is often necessary, to build trust, ensure that all stakeholders have an opportunity to contribute, and ensure that the discussions are productive and promote social learning [1, 2]. Guided by such facilitation, stakeholders can explicitly select, examine, and test each factor that is being considered for inclusion in the policy deliberation. Each stakeholder may examine and inquire about other perspectives and positions. Questioning and critical thinking are encouraged. Allowing stakeholders to engage in this manner presents unique opportunities for targeted discussions. This approach facilitates learning about paradigms, perspectives, perceptions relative to how alternative options are selected or discovered and how they impact the overall attractiveness of the options being considered.

This is a process that is designed not only to better understand others' thinking but also to challenge one's own thinking. Particularly when issues are polarizing, stakeholders believe that it is the other stakeholders that are the barrier in the public policy process. However, an open solution approach's trans-disciplinary learning can help to bring clarity to issues and reduce barriers to understanding, even among those who originally believed that they understood all aspects of the policy problem.

Communication within the stakeholder group, and with outside parties, is facilitated when the built-in transparency of both information and rationale serves as the basis for thorough documentation of the process. Uniquely, such documentation is developed as part of the process rather than an afterthought. An open solution approach provides an organized framework and a guided process that is designed to increase and improve the opportunities to find common ground through the interactive stakeholder experience.

3.1.2   Trans-disciplinary Learning

Most decision making processes utilize, but do not specifically feature, learning. In an open solution approach, three kinds of learning are embraced but trans-disciplinary learning characterizes the open solution process. These three types of learning, multi-disciplinary, inter-disciplinary, and trans-disciplinary, are compared in Fig. 3.1.

Expert knowledge and Types of Learning

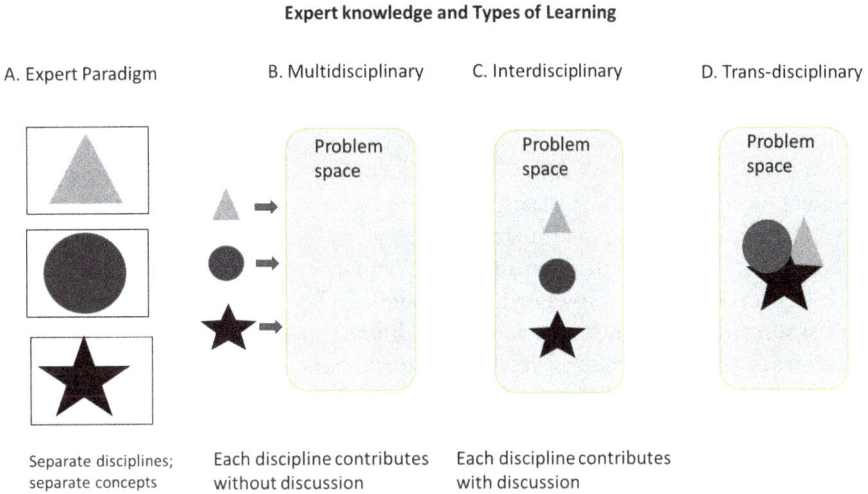

Fig. 3.1   Expert knowledge and different types of learning

   Most commonly, experts (Fig. 3.1a) build and exchange knowledge within their own disciplines, i.e., separated from other disciplines. In multi-disciplinary learning (b), experts from different disciplines are separately consulted as references. However, within the problem space, the focus for experts is to present this information to stakeholders. An example of multi-disciplinary learning is the research and presentation of study results found in the literature to the stakeholder group discussing a related issue. With inter-disciplinary learning (c), there is increased focus to include experts from different disciplines in the common discussion for direct information exchange. The single discipline experts are together, but their role is focused on teaching others about what they view as relevant knowledge from their own disciplines. In contrast, trans-disciplinary learning (d) not only involves sharing information with experts in the same room but also actively seeks to build knowledge not previously known to any of the single disciplines. This is new expertise and, now, shared knowledge. Trans-disciplinary learning is a term used by many different people, but for this book, Lawrence's definition is most useful [3]. Lawrence defines it this way: "…transdisciplinary contributions enable the cross-fertilization of ideas and knowledge from different contributors that promotes an enlarged vision of a subject, as well as new explanatory theories. Transdisciplinarity is a way of achieving innovative goals, enriched understanding and a synergy of new methods."

   The possibility of trans-disciplinary learning and new knowledge creation occurs when experts interact with other experts outside their disciplines, but within a specific policy analysis, and work toward mutual understanding through open discussions and cross-pollination of ideas. Additionally, an understanding of organizational authorities and constraints (formal and informal governance) and relationships among various stakeholder interests is often important to the success of the public policy process and policy implementation. Therefore, the use of the term trans-disciplinary learning here is not meant to constrain learning within traditionally recognized disciplines, but refers to any knowledge development that helps to improve stakeholder understanding in the policy-specific context. In this process, all stakeholders have relevant knowledge to contribute.

While we have chosen to describe the kind of learning necessary in public policy making as trans-disciplinary learning, others have used different terms. For example, Röling and others advance the concept of social learning in support of Funtowicz and Ravetz's assertions that human interaction rather than science and technology is the key toward successfully negotiating public policy making [4]. These sociologists observe that when stakeholders with different paradigms, perceptions, or disciplines (e.g., farmers to ranchers to institutions) have difficulty understanding each other, policy making is confounded. Specialized terminology and presumed (and unexplained) ways of thinking, which are used within each of these sub-populations, pose barriers to communication and understanding across disciplines.

Trans-disciplinary learning is a specific type of social learning. Simply speaking, social learning is about how to get people to interact to get things done [5, 6]. As described by these sociologists, social learning is situational learning or situational problem solving. In that sense, these sociologists are placing social learning within the realm of problem formulation and context. Similarly, trans-disciplinary learning occurs among stakeholders who are engaged in tackling a problem together. However, trans-disciplinary learning is more than contextual learning for the problem at hand. It is transformative learning that creates new knowledge that can become the basis for engaging in long-term working relationships. Therefore, social learning and trans-disciplinary learning are not equivalent. Consequently, in this book, the term trans-disciplinary learning is used, as it more specifically refers to the transformative type of learning that characterizes open solution approaches when stakeholders build knowledge, trust and relationships.

3.1.3 Adaptive Management

One critique of adaptive management has been the observation of a trial and error implementation approach. When there are many different investigators and stakeholders who have distinctive ways to define the problem, the adaptive management approach can become problematic. This is because there is no structured adaptive management methodology designed to capture relationships among individual components of a problem. Therefore, it can be difficult to learn why the previous trial might have failed or produced unexpected results. Without a clear record of the problem definition, what alternatives were evaluated, what criteria were used, and how those criteria were used, even the best efforts to conduct adaptive management are thwarted. The result is trial and error implementation where each new trial is developed independent of the previous trials, without fully understanding why the previous trial failed to produce the desired result [5, 7, 8]. An open solution approach offers adaptive management practitioners the capability to track linkages among perspectives and information. This allows them to develop subsequent trials based on an increasing understanding of where the previous one may have gone wrong. In this manner, trans-disciplinary learning and adaptive management are focused toward addressing common goals.

3.2 Trans-disciplinary Learning to Decision Uncertainty

In public policy making, stakeholders that use the trans-disciplinary learning process are teaching themselves about Decision Uncertainty and how to minimize it in their wicked problem analysis. Decision Uncertainty, when understood, allows stakeholders to compare the results of the various iterations to (1) learn which aspects of the problem (values, expert judgments, or criteria and their organization) are most influential to the attractiveness of the options and (2) assess how Decision Uncertainty can be minimized and subsequently arrive at an action plan.

Suppose decision makers want to determine what policy action to pursue to build a new coastal residential development. In the process of defining the problem, they determine that the available information about coastal erosion is inadequate for distinguishing between the two alternatives being considered. These alternatives are (1) to build residences close to the water if elevated by 12 feet or (2) to build residences 500 yards farther inland but at ground level. Depending on which alternative is chosen, the market value of the homes as well as the community's costs may be substantially different. Deciding on which option to choose is further complicated because current information about coastal erosion does not indicate, with adequate certainty, whether the problem will be flooding (potentially handled by raised homes) or land erosion (potentially handled by moving homes inland). Decision makers also realize that they do not know how coastal residents view flooding risk or how willing non-coastal taxpayers, who would share the financial burdens of emergency resources spent, view financially helping flooded coastal communities. These decision makers may choose to proceed with raising homes or moving homes if other criteria such as expiring funding or other time-constrained criteria compel them to proceed. In this situation, in the stakeholders' view, Decision Uncertainty has been sufficiently minimized. Alternatively, if they decide that they are not comfortable proceeding even if funding is expiring, Decision Uncertainty is still too high to make a decision. This example illustrates the qualitative component of Decision Uncertainty.

3.3 Requisite Steps in MIRA

As stakeholders work toward a final decision, the MIRA process allows for rethinking and re-defining the problem, changing the initially selected relevant criteria (and, hence, which experts are involved), and examining the implications of different stakeholder values and uncertainties and, ultimately, their combined effect on Decision Uncertainty. The Requisite Steps within the MIRA open solution approach provide needed structure for the stakeholder-directed process. Iteration of the Requisite Steps ensures that stakeholders retain flexibility and control of the process.

3.3.1 Requisite Step 1: Engage Stakeholders

Considering whether the "right" stakeholders are engaged and how to engage them is a non-trivial effort. Since the operative feature of the open solution approach is iteration, the composition of the stakeholder group (as well as each subsequent step in the process) is intended to be reexamined based on what has been learned in the previous iteration. In most decision processes, stakeholder participation is not part of a trans-disciplinary learning process as stakeholder groups, once formed, are fixed. Typically, stakeholders may self-select around a common interest when tackling a specific wicked problem if they have determined that the complexity and decision stakes are high enough to warrant forming a group. Engaging stakeholders in the open solution approach includes both stakeholders who self-select and others who are invited to join at any step when issues warrant re-evaluating the composition of the stakeholder group. The open solution approach is a framework that facilitates the identification of the wicked problem, which helps to identify the engagement of the appropriate stakeholders. Detailed guidance on stakeholder engagement, facilitation, and group dynamics techniques is not the focus of this book but can be found in the social science and leadership literature [9–13]. Therefore, policy practitioners are encouraged to learn about and consult experts of these stakeholder approaches.

3.3.2 Requisite Step 2: Define the Problem

Although stakeholders, especially technical experts, generally prefer to move quickly and directly to data gathering or identifying possible solutions (Steps 3 and 4), the importance of clearly defining the problem first cannot be minimized. Without clearly understanding how the problem is defined, stakeholders could be wasting time and other resources gathering data and contemplating possible solutions that are not pertinent or helpful for tackling their problem.

When stakeholders begin to define the policy problem in Step 2 of an open solution process, it is common to discover that other stakeholders may have very different perspectives about how to describe the problem. At this early stage in the process, stakeholders have many ways to describe the *Problem Narrative* (PN). At the start of the open solution process, there can be multiple problem definitions (multiple PNs), which represent different stakeholder understandings of the problem. These definitions are shared with the group by words, phrases, paragraphs, or visual aids. These are used as reference points in subsequent steps to determine how closely the data and indicators represent those narratives. By returning to this step after gaining a better understanding of the problem by experiencing some or all of Steps 3 through 8, stakeholders have multiple opportunities to refine their PN to ensure that their work will meet their goals.

Problem Narrative
The Problem Narrative (PN) is a statement by the stakeholders that defines their policy problem. The PN provides context for the analysis, including guiding the selection of data.

Eventually, multiple PNs become refined to one PN in Step 2. How the problem is defined determines the kinds of alternatives or solutions that will be developed and considered. Refinement of the PN is not necessarily a narrowing of the scope. When stakeholders refine their PN, they are determining which facets of the problem they want to address and how these are connected to their PN. Consider a group that is interested in assessing the effects of mining in an area. After discussing and exploring ideas among themselves, they decide that they want to determine the effects of mining to evaluate and recommend mitigation strategies to the permitting authority. If the permitting authority is constrained by sole reliance on the Clean Water Act, the stakeholders may decide to focus their PN on only those problems that are regulated under the Clean Water Act. This will constrain their recommendations to statutorily based alternatives for mitigation or restoration of water quality and quantity. However, if the stakeholders determine that other statutes, regulations, or guidelines are applicable, they may choose to define their problem more broadly. For example, they could consider teratogenic impacts on ecological species, the impact of air pollution on workers or impacts related to the physical and mental health of the community. Therefore, although there are many possible facets to the mining problem, in defining their PN, stakeholders explicitly determine which of these facets they want to include and how to include them.

Problem definition is an iterative process that requires stakeholder interaction. Stakeholders are expected to return to refining the original PN after deliberating criteria, data, indicators, values and uncertainties. For example, stakeholders may discover that the kind of data needed by their original PN is unavailable or too uncertain. By examining the available data, they may discover that another PN is equally or better suited to their needs. Through the learning process, stakeholders are encouraged to evaluate multiple versions of the PN against discrepancies with their own understanding of the problem and available data. The process facilitates the convergence to a single PN that more accurately reflects their corporate understanding. For the stakeholders, this minimizes PN uncertainty and, hence, Decision Uncertainty.

3.3.3 Requisite Step 3: Identify Alternatives

The first pass through the Requisite Steps, in an open solution approach, is initialized with an upfront identification of known alternatives. These alternatives are generally those that the stakeholders know they want to compare. However, through the trans-disciplinary learning that occurs in an open solution approach, the opportunity to discover additional solutions is left open. These alternatives are referred to as the *Problem Set Elements* (PSEs) in MIRA. The stakeholders' initial brainstorming of policy alternatives could include one or more favored solutions, which are those initially preferred by the stakeholders before the problem is clearly defined or data completely assessed. To avoid stakeholder disenfranchisement during these early stages, inclusion of stakeholder-suggested alternatives is encouraged, even if proffered without a convincing argument. Rather than relying on the loudest, most forceful, or articulate voice to determine which alternatives are analyzed, the efficacy of all the alternatives is evaluated in an open solution approach. Devising a solution to a wicked problem requires the capability to perform relative comparisons among alternatives to assess which results in a better or worse condition, not which is the "right" solution.

Problem Set Elements

Problem Set Elements (PSEs) are the decision alternatives, geographic areas, or other items that the stakeholders will analyze to prioritize or rank.

Because policy problems can be described as symptoms, causes, or results (such as the Hunger in America example from Chap. 1), there is a high likelihood that additional PSEs will be discovered as stakeholders think more carefully and clearly about their PN. The process of iterating through the remaining steps allows stakeholders to (1) view the problem in different ways, (2) learn about relevant new information, (3) change their minds about certain ideas, and (4) gain insights that can lead to additional possible solutions. To add, remove, or refine alternatives, stakeholders proceed through Steps 4 through 8 and return to Step 3 as comparisons among iterations reveal which alternatives look more attractive. This opportunity to create additional possible solutions helps to avoid the reductionism trap of the current public policy making paradigm.

3.3.4 Requisite Step 4: Describe and Organize Criteria

In Step 4, stakeholders refine the PN by describing the criteria or attributes of the PN that will be used to distinguish one policy alternative from another. Each of these criteria descriptions represents a different *Criteria Narrative* (CN). For example, depending on whether the problem is defined as evaluating policy alternatives such as elevating homes, building sand dunes, or requiring building setbacks in order to protect coastal homes from flooding and erosion, there are many different CNs that could be chosen. Some of the criteria that could help distinguish these different policy choices (i.e., PSEs) include (1) the economic cost differences among elevating homes, building dunes, or requiring setbacks, (2) the knowledge of the geographic extent of the floodplain, (3) the probability of severe storms, and (4) an assessment of the burden of any of these policies on existing and future homeowners. During this process, stakeholders find that their PN can be further refined to more directly tackle issues they want to address.

Criteria Narratives
A Criteria Narrative (CN) is a specific description of an effect, condition, or impact that stakeholders want to use to analyze the PSEs.

Articulating a clear and specific CN is important for a number of reasons. First, it allows stakeholders to precisely describe each CN without the constraints that are presented by the data. This, in turn, avoids making the discussion technical or expert too quickly, which often happens when data is introduced too early in the discussions. Second, it provides a platform for interaction that supports trans-disciplinary learning across CNs and data that is needed from the different disciplines to evaluate the policy problem. And finally, well-constructed CNs provide a blueprint for the next step (Step 5), the collection of data and metric construction. By having developed appropriate CNs, stakeholders and experts can engage in meaningful discussions regarding the type of data that will best represent these narratives and allow for the collection and use of the most appropriate data.

Selection of the CNs is not the only way to shape the PN. How the CNs are organized into an analytical Hierarchy also influences the degree to which the PN is reflected in the current policy analysis. An analytical Hierarchy can be simple or complex. The simplest Hierarchy is one that has a linear construction while more complex Hierarchies will have one or more branches on which the criteria reside. Fig. 3.2 compares a linear Hierarchy (a) with one example of a branched Hierarchy (b), using the same number of criteria.

The linear Hierarchy in Fig. 3.2a reflects a PN with seven independent considerations (Criteria 1 through 7) that are used to define the problem in specific terms. In many existing decision analytic or other index building methodologies involving multiple criteria, there is an implicit linear Hierarchy that is used as the default. In an open solution approach, stakeholders choose the structure of the Hierarchy, with a linear Hierarchy as but one explicit option. If the stakeholders' PN is intended to consider seven criteria that are independent and directly comparable, the linear Hierarchy is a good representation of their PN. If, however, the stakeholders' PN is intended to consider three independent criteria, with a more comprehensive treatment (e.g., more science known) for some of those criteria, the branched Hierarchy in Fig. 3.2b is a better representation of their PN. If stakeholders intend the latter

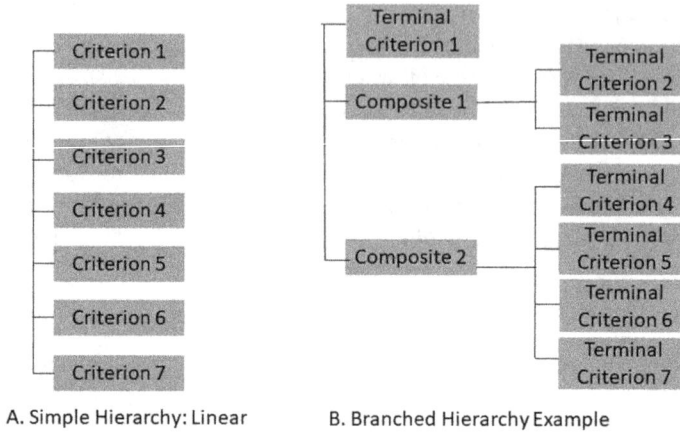

A. Simple Hierarchy: Linear   B. Branched Hierarchy Example

Fig. 3.2 Comparison of linear Hierarchy and one example of a branched Hierarchy

but unknowingly organize their criteria into a linear Hierarchy, they introduce an unjustified bias into the analysis. This bias arises from the inadvertent assignment of greater importance to the independent considerations that are described by the greater number of criteria, simply due to the Hierarchy's construction. By constructing a branched Hierarchy, this problem is avoided. As is often the case in multi-criteria problems, some considerations of the analysis have more detail than others.[1] However, just because more information exists about one consideration compared with another does not necessarily mean that it is more important than the other; it simply means that there are more data available to describe it.

The branched Hierarchy in Fig. 3.2b reflects a PN with three independent primary considerations. One of the primary considerations (Criterion 1) utilizes a single data metric[2] to define it, while the other two utilizes two and four data metrics, respectively. Those criteria at the far right, or terminal ends of the branch, are "terminal" criteria while those criteria that are defined by other criteria are "composite" criteria. In this manner, all criteria in the linear Hierarchy are terminal while there are both terminal and composite criteria in the branched Hierarchy. Only the rightmost criteria in the Hierarchy (the terminal criteria) are directly represented by data.

The construction of the analytical Hierarchy is an important first step for stakeholders to consider whether their PN is accurately represented. This requires stakeholders to think about criteria independence as well as whether the named independent criteria can be meaningfully traded off in a series of questions about which is more important and by how much (performed in Step 7). Consider a Hierarchy that describes the state of the environment by using seven water quality

---

[1] Often, the differing levels of detail reflect the availability of data due to the limitations in knowledge, time, or expertise at the time of the analysis.

[2] Data metrics are those data that are either represented by a single datum or are constructed from more than one datum and are used to represent the stakeholders' CN.

criteria and three land use criteria. Now suppose that stakeholders decide that water quality and land are the two primary considerations in their decision making. In other words, there are only two independent considerations even though one consideration can be described with more than twice as many data metrics.

If the Hierarchy were constructed linearly with the ten criteria, then the water quality consideration would improperly be weighted 7/3 times more than the land use criteria. In contrast, the stakeholder intended problem definition would now be properly represented if the ten criteria were organized in a branched Hierarchy. In this branched Hierarchy, the two composite criteria (water quality and land use) are placed at the first level. Each of the seven water quality terminal criteria is placed at the second level under the composite water quality criterion. Finally, the three land use terminal criteria are placed under the composite land use criteria. Using the branched Hierarchy, stakeholders explicitly decide which of the two primary considerations is more important to them and by how much. Furthermore, the seven water quality criteria are directly compared to determine their relative importance without including the land use criteria. Similarly, the land use criteria are compared without interference from the water quality criteria. With the branched construction, stakeholders can avoid trading-off each of the land use criterion against each of the water quality criteria. The current reductionist public policy making paradigm typically uses a linear Hierarchy, which can appear either implicitly or explicitly. This denies stakeholders an opportunity to better understand and shape their PN but also imposes an implicit PN onto the stakeholders without their explicit consent.

To foster trans-disciplinary learning and examine how the degree of Decision Uncertainty varies from one PN to another, stakeholders are encouraged to construct multiple Hierarchies and to experiment with different criteria descriptions within each alternative Hierarchy. By doing so, stakeholders use the Hierarchies as an explicit and visual discussion tool to refine the PN. Knowing that multiple Hierarchies can be created, and the results of their analysis compared (via iteration through Step 8), helps to reduce the pressure on stakeholders to get the initial Hierarchy "right." The initial Hierarchy is the start of the open solution process that is designed to allow stakeholders to experiment, consider, and then decide on how they want to describe their problem.

3.3.5 Requisite Step 5: Gather Data/Build Metrics

In Step 5, guided by the articulated PN (Step 2), possible PSEs under consideration (Step 3), and articulated CNs (Step 4), stakeholders now must determine which data best represents each CN.

An open solution approach distinguishes the determination and organization of CNs (Step 4) from identifying data that may be used to construct the metrics in Step 5. The *Criteria Metrics* (CMs) are the data metrics that quantitatively represent the CNs. The direct connection between the PN, the CNs, and the CMs establishes the context of the policy problem. In Step 5, stakeholders and experts work together to customize the creation and selection of CMs that best reflect the CNs and PN. This is an important distinction between an open solution approach and existing decision analytic tools. Many current tools provide data catalogues that are populated with pre-existing data from which the data for the present problem is to be selected, even though the catalogued data were usually collected for other purposes. Customization in Step 5 is a means to avoid falling into the reductionism public policy making trap. For example, while a CN can be described as the frequency of severe weather events, the data that may be chosen by stakeholders to represent it can be the probability of a Category 4 or higher hurricane. This probability is an example of a CM. However, the severe weather event CN may be better or more inclusively described by the probability of flooding, or high wind events or lightning strikes. All these conditions can occur with heavy thunderstorms, even when there is no Category 4 hurricane. Therefore, for stakeholders to choose data that best represent their criterion, it is important to fully explore what data is available to achieve that purpose and to ensure that the CN is specific enough to properly communicate the concern. When stakeholders jump to agreement on using certain data before explicitly defining their CN, they are relinquishing control over their PN and reducing their changes of successfully tackling their wicked problem.

Criteria Metric

The Criteria Metric (CM) is a quantitative representation of the CN. A CM can be a single datum such as a measured pollutant concentration, or a metric that is constructed from multiple pieces of data.

To better understand the open solution process of linking the PN with CNs and CNs with relevant data, it is helpful to define the differences between data and metrics. In the discussion of Step 4, the term data metric was used and, in the examples provided, referred to the use of raw data (e.g., probability of a Category 4 hurricane). However, more complex policy analyses often require metrics to contain a variety of data. Some of the data can be used in raw form to construct the needed metric while others need to be constructed from multiple data types. Data, more generally, are numeric values that result from specific measurements, mathematical operations, modeling or the quantification of some qualitative operation (e.g., the results of a survey questionnaire). Metrics, on the other hand, are a conscious use or manipulation of data by stakeholders to construct a numeric that better represents or reflects a CN. For example, ozone data are hourly monitored readings from an air quality monitor (a piece of equipment that measures ozone to a required level of certainty). In contrast, the National Ambient Air Quality Standard (NAAQS) for ozone is a metric constructed by averaging the fourth highest monitored hourly ozone measurement in each of three consecutive years [14]. The ozone NAAQS is also defined spatially by metropolitan statistical areas. These areas can be larger than counties but smaller than states. If stakeholders choose to use data that identifies air quality using this ozone metric, they have decided to choose a statutory metric. They have moved from the broad concept of considering direct human air quality exposure to a metric that is designed to determine standards compliance, which is an administrative metric used to implement the Clean Air Act. The choice of this metric is reasonable only if the metric is constructed in a manner that satisfies the intent of the CN.

Step 5 is a familiar activity for many scientists and decision analysts because this is when expert data is gathered and indicators are created. However, in the MIRA process, the transparency of the process and the encouragement to iterate provides stakeholders with new capabilities to use the best available data that aligns with their PN. Through iteration, stakeholders can develop an understanding of what data are being used, how they are being used, and why they are being used. With this information stakeholders can have a full and learned discussion about the appropriateness of the data relative to the PN. In this discussion, stakeholders contemplate whether other types or forms of data could be used, by also considering their associated uncertainties in the context of their PN. Conversely, if the stakeholders determine that there is no adequate data available that will allow for a proper analysis of the PN, they can either change the PN or table the problem until appropriate data becomes available.

Suppose that the PN provides a broad charge to characterize the overall condition of the environment. In this case the CNs could consist of air quality, water quality and forests. Note that while the PN is defined broadly as overall environmental condition, the CNs refine and narrow the scope of the PN. In Steps 4 and 5, stakeholders, working with experts, discuss CNs and choose the data metrics to populate them. For example, the forest CN could be populated with several different forest CMs such as forest area, forest edge, or the degree of forest connectivity or a composite criterion consisting of a suite of all three. Each of these CMs relates to a different issue or aspect of a forest. Deciding which ones to use and how to use them needs to be resolved with additional discussion to more clearly establish the PN. As part of this discussion, issues of spatial and temporal scale may also need to be discussed. For example, typical spatial resolution may be too large or too small or of limited coverage. As a result, the stakeholders may need to reassess their original PN (Step 2), revise the PSEs (Step 3), or change the initial set of CNs (Step 4) which in turn may require that they reassess the make-up of the stakeholder group (Step 1). This back and forth progress through the Requisite Steps, using iteration followed by refinement, is characteristic of an open solution approach and is in direct contrast with the linear, one-pass reductionist paradigm.

The wider the gap between what the stakeholders consider as representative of the CNs and the actual CMs that are being used, the greater the Decision Uncertainty. Testing the extent of Decision Uncertainty between the CNs and the PN also occurs when stakeholders work through Steps 6 through 8. By applying these steps, stakeholders can directly examine the effect of the CMs on the results of the analysis. If stakeholders determine that there is too much Decision Uncertainty, it is possible to return to Step 5 to discuss and select other CMs to evaluate or to return to Step 1 and revisit the PN.

3.3.6   Requisite Step 6: Judge the Significance of the Data

In Step 6, all CMs used in the analysis are placed on a common decision scale. This process requires expert judgment in which experts determine the *significance* of the data based on the context that is established by the PN. In Step 6, experts become engaged to help other stakeholders understand how the CMs constructed in Step 5 specifically relate to the stakeholder-defined PN. In Step 6, experts determine:

1. The significance of the data magnitudes within the context of what is being decided.
2. The consistent assessment of significance across all data. This determination especially benefits from trans-disciplinary learning.

> **Significance**
> In MIRA, significance refers to the CM's significance as it relates to the context provided by the PN. Significance is typically determined by experts and the result is to convert all the CMs to the contextual decision scale. Significance is assessed through the Indexing process.

3.3.6.1   Significance and Data Range

In Step 5, stakeholders (with experts) link CNs with CMs. In Step 6, experts first assess the significance of the individual expert CMs relative to the decision scale and then each CM to the others. In this step, experts convert the CMs from the natural units of their discipline-specific metrics to a common decision unit. This *Indexing* process requires that they determine which values of each of the CMs are equivalent to the same integer values on the decision scale, of which there are eight (1–8), by MIRA convention. Indexing is accomplished both between and among all metrics and is understood and agreed upon across all disciplines. Successful Indexing reflects trans-disciplinary learning. In MIRA, to distinguish between data (i.e., CMs) that have been gathered in Step 5 and CMs that have been expertly judged for significance relative to the PN, the term indicator is exclusively used. An *Expert Judgment Set* (EJS) is produced when all the CMs have been indexed and cross-indexed for consistency in significance. The EJS quantitatively describes how all the CMs associated with the Hierarchy are converted to decision units.

> **Indexing**
> Indexing is the process in MIRA by which experts judge the significance of the CMs relative to the context of the problem as described in the PN. In MIRA, the CM data are converted to indexed data that are placed on a non-dimensional decision scale, ranging between 1 and 8.

Expert Judgment Set
An Expert Judgment Set (EJS) is the result of completing the Indexing process for all CMs in the analysis. The EJS is a unique set of translational schemes that convert CMs to indicators for a specific Hierarchy.

Typically, when data are presented, practitioners presume that their maximum and minimum values not only represent the extremes in the data but also translate to the extremes in the degree of significance that the data have to the decision. When stakeholders spread the data uniformly across the span of significance, they are making a number of erroneous assumptions. They are presuming that (1) the range of the data for each CM, based on CM values from the set of PSEs, defines the actual range of significant values that is consistent with the PN and (2) the significance ascribed to all data are both consistent within and across disciplines and in line with the PN. More specifically, spreading data uniformly across a decision scale, whether implicitly or explicitly, potentially creates a problem when the data are a subset of a larger data set. For example, in the USA, ozone air quality varies across the country with the Los Angeles metropolitan area on the west coast and metropolitan areas on the east coast having the largest ozone values. Suppose an analyst in the US Midwest limits the use of ozone data to those values occurring in their geographic area. If their data are spread uniformly from minimum to maximum values across the decision scale, the Midwestern analysts would have a substantially different assessment of ozone significance than if east or west coast ozone data had been included. This is problematic since an assessment of significance, performed this way, is not based on ozone's environmental impacts but rather on the arbitrary range of the data that happens to be present in the US Midwest or, more generally, within the range of the PSEs' data that are being considered. Finally, and most importantly, spreading the data uniformly disregards the role of experts in the process, who are needed to judge the contextual significance of the data.

Data do not need to be expertly judged in a completely linear fashion. For example, consider a PN that includes a CN whose associate CM is ground-level ozone concentrations that pertain to the Clean Air Act's (CAA) regulatory requirements of determining whether areas are complying with the ozone NAAQS. Now suppose that an ozone expert will be asked to judge the relative significance of ozone values, ranging from 65 parts per billion (ppb) to 90 ppb. The expert will be able to explain why the relative significance between ozone values that are considerably above 70 ppb (the ozone NAAQS) versus other ozone values that are just above the NAAQS should be judged differently. For a PSE whose CM is 90 ppb, which is substantially above the standard, the expert is likely to judge its significance as only slightly more than a value of 85 ppb, which is also substantially above the standard. However, the same expert is also likely to find a CM of 72 ppb to be considerably more significant than 69 ppb, even though there is a 5 ppb difference between the first pair and only 3 ppb difference between the second pair. This is because the second pair crosses the ozone standard threshold of 70 ppb [15].

3.3.6.2   Significance Across Data

Step 6 also involves the interaction and trans-disciplinary learning among experts of different disciplines. This requires experts across disciplines to articulate their rationale for mapping the CM range to the 1 to 8 decision scale. Suppose stakeholders choose to include ground-level ozone concentrations and the distance one lives from an airport (a CM related to stress from noise pollution) in the definition of their PN. Experts indexing the ozone air quality CM and those indexing the airport distance CM are asked to explain how they were able to ensure consistency, relative to the PN, between their approaches in assessing significance. The ozone expert may inform the group that the US ozone NAAQS is 70 ppb. A psychologist could explain why the distance between neighborhoods and airports affect not just mental and physical health but also the nature of those communities. Based on their discussions about the effects on physical and mental health, the two experts could come to the judgment that 72 ppb of ozone is equivalent in significance to living 10 miles from an airport. The cross-Indexing between ozone and airport distance informs others relying on this analysis that, for the purposes stated in the PN, living 10 miles from the airport has been judged by the experts to be equivalent to an area exhibiting ozone slightly over the NAAQS.

In MIRA, the decision scale ranges from 1 to 8. Also, by MIRA convention, a value of 8 has been established as the most significant indexed data value. The import of this is (with all other variables being equal) the PSEs who have more of their data indexed at or close to 8 will be relatively more attractive than other PSEs whose data lies at the lower end.

3.3.7   Requisite Step 7: Apply Values

When more than one criterion is needed to help answer a question, it becomes necessary to compare the relative importance of the indicators against each other. Determining relative importance is not a scientific but a values question. Applying stakeholder *Preferences (i.e., Preferencing)* to the set of stakeholder-selected indicators in Step 7 is an exercise in determining the relative importance among the indicators, which is distinct from judging significance (Step 6).

> **Preferencing**
> Preferencing is the act of weighting the relative importance among indicators used to evaluate a PN. Stakeholder opinions about relative importance are contextual and based on the PN.

Returning to the ozone and airport distance example, when values are applied to these indicators (ozone levels and distance to airports) in Step 7, stakeholders are being asked to determine which of these indicators is more important to them. It is not necessary for them to understand the expert data or how the experts judged the relative significance of the data since the MIRA open solution approach intentionally separates science from values. This does not mean that there is no interaction between experts and other stakeholders during Step 7. On the contrary, because of MIRA's transparency, stakeholders may examine what data was used and how they were indexed. However, once the Indexing is complete, stakeholders should be confident enough with how the science has been incorporated into the process that they can now concentrate primarily on how they will apply their opinions about the relative importance of the indicators. Therefore, in Step 7, stakeholders focus on determining their relative Preferences among the indicators, such as whether the impacts to human health from ozone or the stress from noise pollution is more important.

In Step 7, stakeholders may decide that the negative impacts to health from ozone are twice as important as psychological impacts caused by plane noise. Also suppose that the experts for these two CMs have indexed the ozone CM at 4 on the decision scale while the stress-related CM is indexed at 8. Because the ozone value is expertly judged to be half as significant as the actual distance to the airport, the two pieces of data contribute equally to the overall attractiveness of the decision option (Ozone = 2 × 4, which is equal to an Airport Distance value of 1 × 8). In this way, Preferences and data are treated separately but are combined in a way that appropriately impacts the decision. This is a contrast to traditional decision making processes that lack transparency, which often contributes to misunderstandings and confusion when Preferences and data are comingled. The open solution approach embraces the diversity of perspectives and grants stakeholders the flexibility to design their analysis in any way that best reflects their needs. However, flexibility in the open solution approach requires accountability from stakeholders to explain their perspectives and positions in a process that, although subjective, is structured.

Consistency and subjectivity are not alike. In Step 7, even as subjective values are applied, internal consistency among a stakeholder's Preferences is desired and sought. For example, stakeholders may choose that A is more important than B and B is more important than C. However, once these choices are established, they would be inconsistent to now determine that C is more important than A. In Step 7, MIRA provides stakeholders the additional capability to examine if the Preferences they assign to the indicators are both aligned with their values and are internally consistent among themselves. This capability is provided by MIRA through the use of the Analytic Hierarchy Process (AHP). AHP is a mathematical tool designed to solicit Preference weights through pair-wise comparisons and to inform stakeholders of consistency in conducting pair-wise comparisons [16]. Rather than using AHP as a purely reductionist tool to rank the best or top alternative, in MIRA AHP is used as an analytical tool for stakeholders to better understand their own Preferencing. If assigned weights seem to be misaligned or when choosing weights through pair-wise comparisons inconsistencies are found, stakeholders can reexamine their choices. This provides additional possibilities for self-critique, revelation and trans-disciplinary learning that could lead to finding common ground for agreement.

Even when analytical tools are used to aid decision making, either by tool design or due to a lack of understanding regarding Preferencing, indicators are often simply valued equally. When this occurs, stakeholders lose the power to choose based on their own Preferences. Many analytical tools default Preferencing to simply assigning equal Preference weights (either explicitly or implicitly). The implication is that this equality represents a balanced view. However, balance is not solely reflected by equality because balance in decision making requires trade-offs. Often, decision makers want to consider some indicators as more important than others. For example, when the US EPA makes a county-based determination of compliance with the ground-level ozone NAAQS, it is required by the Clean Air Act to consider the level of air quality in the county and surrounding area together with other factors. These other factors include population density, commercial and industrial development, traffic congestion and pollutant transport. Although it is expected that these criteria are considered prior to making such a determination, they are not necessarily weighted equally [17]. In the 2004 US EPA decision that designated US counties based on their ozone NAAQS compliance status,[3] air quality data were weighted more heavily than any of the other indicator, even though these other factors were also required to (and did) influence the decision outcome. Although the Clean Air Act requires consideration of many factors, it did not contain any mandates on how those factors are to be considered or relatively weighted. In 2004, the US EPA decided that an air quality decision such as ozone attainment could be better supported if ozone air quality was weighted more heavily than other indicators [18, 19]. Weighting all indicators as equally important does not reflect the balance needed to justify the kind of decision being made. An unequal, but not random, values-informed weighting represents the necessary trade-offs that often better reflect the policy decision maker's needs.

An open solution approach provides the flexibility for stakeholders to explicitly discuss and determine this weighting structure. Based on the PN, stakeholders decide how best to weight the relative importance among a selected set of criteria; equal weighting is but one possibility. In MIRA, each set of relative stakeholder Preferences is a *Value Set* (VS). A VS consists of the Preference weights applied to all indicators in a Hierarchy to relatively rank the decision alternatives or the PSEs.

---

[3]While assessment of the compliance or attainment status of a county for ozone appears to be purely factual (i.e., whether an air quality monitor is measuring complying ozone readings), this assessment is complicated by many factors, including the lack of monitors in every county, the regional transport of ozone precursor emissions, and the atmospheric chemistry of ozone formation.

Value Set
A Value Set (VS) is the complete set of Preference weights that are assigned to each indicator in a specific Hierarchy.

The initial discussions among stakeholders related to values (Step 7) are expected to uncover and highlight differences in opinion. Creating a variety of VSs may initially appear to be unhelpful in finding common ground. However, the creation and discussion about multiple, even conflicting, VSs has the effect of focusing the disagreements and helping stakeholders determine which of their differences are meaningful within the context of the decision. In the open solution approach, stakeholders are not required to share identical Preferences to reach agreement. On the contrary, an open solution approach embraces the diversity of opinion among stakeholders. The process is designed to find overlapping Preferences that can help them reach agreement on an actionable alternative. The relative attractiveness of alternatives is determined by a combination of data, expert judgment, and Preferences. Overlap among these combinations that produce the same or equivalent results is where stakeholders can find common ground. Stakeholders are likely to find ranges of Preferences that produce the same result regarding the attractiveness of the policy options. Often stakeholders may discover that their initial VS does not accurately represent their perspective. As they better understand the PN and the relationship between the PN and the data, stakeholders have the flexibility to change their VS and explore the influence of different Preferences on the relative attractiveness of the alternatives. When stakeholders begin to eliminate differences that do not change how the PSEs are ranked, they are reducing Decision Uncertainty and creating opportunities for a consensus solution to emerge.

3.3.8 Requisite Step 8: Combine All of the Above

In Step 8, stakeholders begin the process of learning how each of the choices made in the previous seven steps influences the relative attractiveness of the alternatives. This is accomplished by calculating a single valued metric, the *Criteria Sum* (CS) for each PSE. The CS is a linear weighted sum which is the product of expertly judged data (Step 6) and the stakeholder Preference weights (Step 7). For a simple branched Hierarchy, the MIRA CS is calculated, for a given PSE, as displayed in Fig. 3.3.

Criteria Sum
The Criteria Sum (CS) is a linear weighted sum that is calculated for each PSE, which is used to rank the PSEs.

Criteria Sum $= (I_1 \times W_1) + (I_2 \times W_2 + I_3 \times W_3) \times W_7$
$+ (I_4 \times W_4 + I_5 \times W_5 + I_6 \times W_6) \times W_8$

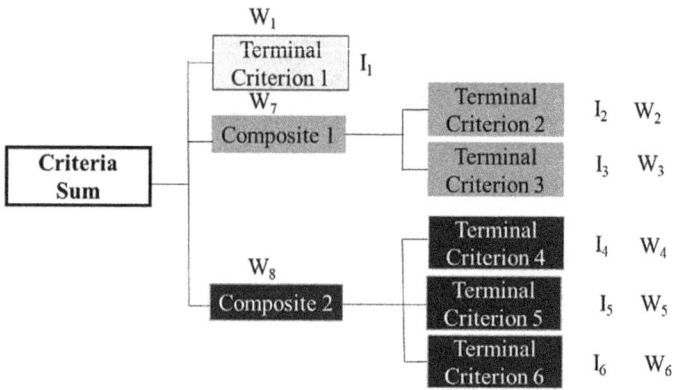

I = Indexed value of indicator
W = Stakeholder preference weights

Fig. 3.3 Calculation of the MIRA Criteria Sum

Calculation of the CS numerically collapses the Hierarchy from right to left. A CS is calculated for each PSE, allowing that PSE to be relatively ranked among the full set of PSEs. By convention in MIRA, the higher the CS, the more attractive is the PSE. A "run" in MIRA produces a ranking of the PSEs. Each run represents an iteration of the Requisite Steps, where some aspect or parameter of the process has been changed. Stakeholders conduct these experiments to examine how changes to one or more components influence the results of a run.

For any single Hierarchy, the first time stakeholders calculate the CS in Step 8, they are calculating a baseline against which they will compare additional runs. This *Baseline Run* is constructed by choosing a Uniform Indexing EJS and an Equal Preference VS.[4] After each run, stakeholders evaluate the Decision Uncertainty and decide on what further actions need to be taken. Decision makers and other stakeholders now have a computational laboratory with which to conduct experiments. By strategically varying important components of the open solution process, stakeholders can compare and evaluate the policy alternatives initially identified, as well as newly discovered ones.

Baseline Run
For each constructed Hierarchy, the Baseline Run establishes a benchmark against which subsequence iterations are compared. All Baseline Runs use a Uniform Indexing EJS and an Equal Preference VS.

---

[4]Simplifying the EJS and VS for the Baseline Run allows stakeholders to focus on refining their PN, CNs, and CMs. Customization of their analysis with a non-uniform EJS and an unequal VS follows with iteration.

3.3.9 Iteration of the Requisite Steps: Learning and Minimizing Decision Uncertainty

Decision Uncertainty is the compounded effect of all quantitative and qualitative uncertainties that impact the relative attractiveness of the PSEs. These uncertainties include how well the PN reflects the stakeholders' intended policy problem, the degree to which the CNs reflect the PN, how well the CMs reflect the CNs, the quantitative uncertainties related to the data used from either measurements or models, uncertainties about the extent to which the experts agree on the EJS, and the degree to which stakeholders are accepting of the VS used to make the final policy decision.

When all the components are combined into a CS for each PSE and the PSEs have been ranked, decision makers and other stakeholders can evaluate the Decision Uncertainty of the ranking. Furthermore, since these are wicked problems, there is no pre-determined "stopping point" for what constitutes an acceptable degree of Decision Uncertainty. However, with sufficient iteration, stakeholders develop their own sense of Decision Uncertainty, and when it is low enough, they choose a decision alternative.

To illustrate what is meant by developing a sense of Decision Uncertainty, consider, for example, the Dow Jones Industrial Average (DJIA). Imagine the first time the Dow Jones Industrial Average, a price-weighted average of select corporate stocks, was used in 1896 [20]. The first time it was used, it was likely that most people did not understand the significance of the DJIA. In 1920, the DJIA was below 100. In 1987, prior to Black Monday which occurred in October of that year, the DJIA was about 2700. On Black Monday, the DJIA plunged to approximately 1750 in one day. In 2019, the DJIA hovers around 27,000. Although these values have some meaning in the absolute, more importantly, its utility is in understanding the significance of its relative change over time. After many years of utilizing the DJIA as a measure of the US economy, consumers and economic analysts have come to realize that a one-day drop of 3% is problematic but a one-day drop of 22%, as exhibited on Black Monday in 1987, is catastrophic. This is a developed sense of the relative meaning of the DJIA that could not have been known or understood until people were exposed to it over time. This is akin to the trans-disciplinary learning that takes place as stakeholders iterate multiple times through the Requisite Steps, developing a sense for what constitutes a meaningful relative change in the Criteria Sums of the PSEs. In that process, stakeholders discover how changes made to the Requisite Steps affect the rankings of the PSEs. The DJIA is understood and useful as a measure of the US economy when considered in context and with learned experience. In similar fashion, estimating Decision Uncertainty has meaning in terms of the combination of uncertainties that are related to construction and implementation of the steps as conducted by the stakeholders for their policy problem.

Decision Uncertainty is only partially quantitative. It contains data and model uncertainty, but it also includes the uncertainty associated with many of the other steps such as defining the problem, which is qualitative by its nature. However, as a measure used in a relative comparison among alternative decision options, Decision Uncertainty (including its qualitative components) can be effectively used quantitatively. For example, in Step 7, when judging the relative importance among the indicators, it is initially difficult to determine whether differences among stakeholder Preferences will result in agreement or not (high Decision Uncertainty). However, through the open solution process, the impact of each stakeholder's VS on the attractiveness of the PSEs can be consistently and unambiguously compared. Similarly, differences in data, EJSs, and alternative PNs that result in revised Hierarchies can be examined as to their single or combined effect on the relative attractiveness of the PSEs. With each iteration, the stakeholder group explores and compares combinations of different Hierarchies, EJSs, data ranges and VSs. With this process, common ground is revealed and convergence to a solution is possible.

3.3.9.1   Determine When the Process Ends

Often in highly technical analyses, stakeholders think that more information is needed before any choice can be made. In some cases, the lack of complete information may be used as an excuse to avoid making a decision. However, guided by Decision Uncertainty, an open solution approach can be used to help stakeholders determine which types of information are critical for decision making. In particular, stakeholders can assess whether the barrier to decision making lies in the lack of data, in data uncertainty, or in areas not typically examined such as uncertainties in CNs or stakeholder Preferences. In an open solution approach, it is possible to test the sensitivity of the results by altering one or more components at a time within the steps. Each such iteration is an additional round of learning for the stakeholders. For example, it becomes possible to ask the following: If this datum were 5% more certain, how would it change the relative attractiveness of the alternatives? Or what is the combined effect on the relative attractiveness of the alternatives if the value for one indicator is 10% higher while another is 4% lower? It is also possible to examine alternative CMs or CNs to determine whether and how the PSE ranks change. As stakeholders imagine and test the possibilities, the open solution approach serves as their computational laboratory to build and examine the impact of each one of these possibilities on the relative attractiveness of the PSEs.

Because wicked problems have, among other characteristics, no single right answer, no objective measures of success, and no stopping rules, stakeholders together determine when the open solution process ends. When the stakeholder group is satisfied that Decision Uncertainty, as a combined quantitative and qualitative measure, is sufficiently minimized, the process ends.

3.4   Practitioners' Reality

Policy practitioners often feel left out when academicians seek to apply theories and insights to observations about public policy practice that seem impractical for implementation. Even for highly motivated practitioners, traveling the path from concept to implementation is daunting and often, seemingly impossible. Time and resource constraints, in addition to political and social pressures, frequently discourage the use of a methodical process, or a methodical examination of the current process, to determine whether goals are being met. However, where statutory or regulatory requirements or other circumstances require transparency and stakeholder engagement, an open solution approach can be welcome and effective.

Many in public policy organizations experience impediments to public policy making such as the lack of stakeholder inclusiveness, transparency in decision making, and efforts to be more goal-oriented. Exacerbating these impediments are power relationships among stakeholders, the nature of which may not be initially evident but need to be negotiated. Pre-determined choices made by the more powerful stakeholders are one way that power is exhibited. These choices may include pre-selection of stakeholders, indicator sets or the selection of the decision support tool. The open solution approach is designed to work in imperfect and contentious policy making situations. For dedicated public policy decision makers, identifying barriers to more inclusive, transparent, and learning-based policy making is the first step toward better public policy making. Nothing here regarding open solution approaches is intended to dismiss or minimize the import of these significant, and sometimes insurmountable, barriers. Effective public policy making is not magic but requires attention and nurture in areas that are potentially unfamiliar and strikingly different from purely scientific or technical pursuits. While not specifically discussed here, as many sociologists have addressed these issues, how policy making is done influences stakeholder behavior [5, 10, 21]. There is no substitute for good policy leadership, honest discussion and a genuine desire to find mutually agreeable solutions. Currently, however, even stakeholders willing to engage honestly to seek mutually agreeable solutions are thwarted by conceptual approaches that lack methodological rigor and reductionist methodologies designed to optimize and objectify the process. The result of this is often disenfranchisement and frustration. In contrast, using an open solution approach could help to explicitly identify areas where stakeholders can manage these relationships in a more inclusive and transparent way than is presently available within the current policy making paradigm.

Building trust among stakeholders is an important component in any public policy making process. However, this is too often neither mentioned nor addressed, especially if scientists or technical experts, whose focus and strengths tend to lie elsewhere, are leading the process. While there are no absolute assurances of success with any public policy process, an additional benefit of engaging in an open solution approach is the opportunity to build more trustworthy relationships among stakeholders. Each step of the process allows stakeholders to focus on understanding different perspectives, experience, and expertise more effectively as well as how these components contribute toward tackling the wicked problem. If open solution approaches are used by public policy making organizations or other long-standing policy stakeholder groups, there is the potential for long-term benefits. The trusting relationship started by working collaboratively on one wicked problem could become the foundation for collaboration on other complex policy problems. In this way, the use of an open solution approach could build greater capacity within these public policy organizations to practice more effective policy making beyond a single wicked problem.

3.5 Summary

The open solution approach is a necessary, feasible, and practical means to avoid the reductionist trap in the current public policy making paradigm. The MIRA open solution approach fills a conceptual and methodological gap in public policy decision making. This holistic approach to policy analysis invites stakeholders to use trans-disciplinary learning to articulate and clarify their objectives and perspectives as they proceed through the Requisite Steps. The unique use of iteration, with Decision Uncertainty as a gauge, provides policy practitioners with previously unavailable capabilities to engage diverse stakeholders and find common ground. While not every policy situation is suited to considering an open solution approach, when the decision stakes and complexity are sufficiently high, it is time to consider a more reasoned, inclusive, and methodical approach to public policy decision making. MIRA makes this possible.

References

1. Groot A, van Dijk N, Jiggins J, Maarleveld M (2002) Three challenges in the facilitation of system-wide change. In: Leeuwais C, Pyburn R (eds) Wheelbarrows full of frogs. Koninklijke Van Gorcum, Assen, pp 199–213
2. Benessia A, Funtowicz SO, Giampietro M, Guimaraes Pereira A, Ravetz JR, Saltelli A, Strand R, van der Sluijs JP (2016) The rightful place of science: science on the verge. Consortium for Science, Policy and Outcomes, Arizona State University, Charleston
3. Lawrence RJ (2010) Beyond disciplinary confinement to imaginative transdisciplinarity. In: Brown VA, Harris JA, Russell JY (eds) Tackling wicked problems: through the transdisciplinary imagination. Earthscan, London, pp 16–30
4. Röling N (2002) Beyond the aggregation of individual preferences. In: Leeuwais C, Pyburn R (eds) Wheelbarrows full of frogs: social learning in rural resource management. Koninklijke Van Gorcum, Assen, pp 25–47
5. Leeuwais C, Pyburn R (eds) (2002) Wheelbarrows full of frogs: social learning in rural resource management. Koninklijke Van Gorcum, Assen

6. Guijt I, Proost J (2002) Monitoring for social learning: insights from Brazilian NGOs and Dutch farmer study groups. In: Leeuwais C, Pyburn R (eds) Wheelbarrows full of frogs. Koninklijke Van Gorcum, Assen, pp 215–232
7. Curtin CG (2015) The science of open spaces: theory and practice for conserving large complex systems. Island Press, Washington, D.C.
8. Bond A, Morrison-Saunders A, Gunn JAE, Pope J, Retief F (2015) Managing uncertainty, ambiguity and ignorance in impact assessment by embedding evolutionary resilience, participatory modelling and adaptive management. J Environ Manag 151:97–104. https://doi.org/10.1016/j.jenvman.2014.12.030
9. Endter-Wada J, Blahna D, Krannich R, Brunson M (1998) A framework for understanding social science contributions to ecosystem management. Ecol Appl 8(3):891–904
10. Isaacs W (1999) Dialogue and the art of thinking together: a pioneering approach to communicating in business and in life. Doubleday, New York
11. Bell S, Morse S, Shah RA (2012) Understanding stakeholder participation in research as part of sustainable development. J Environ Manage 101:13–22. https://doi.org/10.1016/j.jenvman.2012.02.004
12. Andrews CJ (2002) Humble analysis: the practice of joint fact-finding. Praeger, Westport
13. Webler T, Renn O (1995) A brief primer on participation: philosophy and practice. In: Renn O, Webler T, Wiedemann P (eds) Fairness and competence in citizen participation: evaluating models for environmental discourse, Technology, risk and society, vol 10. Kluwer, Dordrecht, pp 17–34
14. U.S. Environmental Protection Agency (1998) Guideline for data handling conventions for the 8-hour ozone NAAQS. U.S. Environmental Protection Agency, Office of Air Quality Planning and Standards, Research Triangle Park
15. U.S. Environmental Protection Agency (2015) National ambient air quality standards for ozone. Fed Regist 80(208):177
16. Saaty TL (1990) Multicriteria decision making: the analytic hierarchy process. RWS Publications, Pittsburgh
17. Clean Air Act (1990) 42 U.S.C. 7407 et seq
18. Stahl CH, Fernandez C, Cimorelli AJ (2004) Technical support document for the Region III 8-hour ozone designations 11-factor analysis. U.S. Environmental Protection Agency, Region III, Philadelphia
19. U.S. Environmental Protection Agency (2004) Air quality designations and classifications for the 8-hour ozone National Ambient Air Quality Standard; early action compact areas with deferred dates. Fed Regist 84(69):23858–23951
20. Corporate Finance Institute (2018) Dow Jones Industrial Average (DJIA). https://corporate-financeinstitute.com/resources/knowledge/trading-investing/dow-jones-industrial-average-djia/. Accessed 12/13/18
21. Susskind LE, Levy PF, Thomas-Larmer J (2000) Negotiating environmental agreements: how to avoid escalating confrontation, needless costs, and unnecessary litigation. Island Press, Washington D.C.

Chapter 4
The MIRA Approach: Initialization

Abstract In the MIRA open solution approach, seeking common ground and, ultimately, consensus is facilitated by iterating through each of the Requisite Steps while using Decision Uncertainty as a gauge to help stakeholders assess when they are ready to decide. The exact progression and number of times each step is revisited will be different for different public policy analyses. These decisions are controlled by the stakeholder group as they refine their Problem Narrative and minimize Decision Uncertainty. This chapter 4 describes the MIRA initialization process, designed to produce the first run. This Baseline Run provides a benchmark against which subsequent iterations are compared. With policy practitioners in mind, Chaps. 4 and 5 describe the process, analytical strategy, and methodological aspects of the Multi-Criteria Integrated Resource Assessment (MIRA), an open solution approach.

Keywords Case example · Indicators · Stakeholder inclusiveness · Uncertainty analysis · Multi-criteria assessment

4.1 Introduction

In the MIRA open solution approach, seeking common ground and, ultimately, consensus is facilitated by iterating through each of the Requisite Steps while using Decision Uncertainty as a gauge to help stakeholders assess when they are ready to decide. The exact progression and number of times each step is revisited will be different for different public policy analyses. These decisions are controlled by the stakeholder group as they refine their Problem Narrative and minimize Decision Uncertainty. This chapter 4 describes the MIRA initialization process, designed to produce the first run. This Baseline Run provides a benchmark against which subsequent iterations are compared. With policy practitioners in mind, Chaps. 4 and 5 describe the process, analytical strategy, and methodological aspects of the Multi-Criteria Integrated Resource Assessment (MIRA), an open solution approach.

Effective public policy making requires the meaningful engagement of stakeholders. While the description of the social science approaches regarding these

© Springer Nature Switzerland AG 2020   73
C. H. Stahl, A. J. Cimorelli, *Environmental Public Policy Making Exposed*,
Risk, Systems and Decisions, https://doi.org/10.1007/978-3-030-32130-7_4

practices is beyond the scope of this book, the chances for success in any public policy making process, including the MIRA process, are improved with the use of such approaches. Therefore, practitioners are encouraged to consult those social science approaches and practices in accomplishing the first Requisite Step in decision making and at appropriate places throughout the MIRA process. This book focuses on the other aspects of the Requisite Steps that are not discussed in decision analytic approaches, in the social sciences, or elsewhere in the public policy making literature. Hence, this chapter begins with Step 2, Build the Problem Narrative.

A realistic but hypothetical policy problem example is introduced and utilized in Chaps. 4 and 5 to help describe the MIRA open solution approach. The use of a hypothetical example provides the flexibility to ensure that all important aspects of MIRA are adequately discussed and illustrated within a single example.

The Assignment The Gaia federal government has allocated a limited amount of financial and staffing resources for a project targeted to improve the environmental condition of Gaia. The Gaia elected officials direct the formation of a Committee consisting of key stakeholders, representing a range of competing interests. Given the resource constraints, the Gaian Committee is directed to choose a single district as the beneficiary of this project. In total, there are fifteen (15) districts in Gaia. The Gaia officials also mandated that the Committee provide a detailed rationale describing the environmental issues considered and how the district was chosen. This detailed rationale is essential since the Gaia government must address public comments on the selected district that is to receive extra resources.

4.2 Initializing the MIRA Analysis

Initialization requires entering information for each of the Requisite Steps. However, for the Baseline Run, not all the steps require equally detailed stakeholder attention. The Baseline Run is the first run of MIRA for each defined Hierarchy and any subsequent revision. For all Baseline Runs, a Uniform Indexing Expert Judgment Set (EJS) and an Equal Preference Value Set (VS) are used to produce the initial ranking of the Problem Set Elements (PSEs). This provides a benchmark against which subsequent iterations are compared. At the beginning of the MIRA process, use of these simplified Expert Judgment and Value Sets allows stakeholders to focus their attention on the Problem Narrative (PN). Stakeholders use the process of constructing the Baseline Run to identify the next set of trials that include changing the EJS and VS to evaluate how these changes impact the ranking of the PSEs. This kind of learning from the previous run informs how each of the following iterations will be constructed.

4.2.1   Requisite Step 2: Build the Problem Narrative

Stakeholders often find that, because of competing interests and an initial inability to prioritize problem-relevant issues, they have a difficult time articulating a concise PN. Early in the process, the PN tends to be broad and somewhat fluid and, for some stakeholders, seemingly chaotic. However, the structure of the Requisite Steps within the open solution approach guides this process. Through this structured process, stakeholders methodically examine their PN using any newly discovered insights that arise as they progress through each step. This structure, accompanied by the learning gained in each step, helps stakeholders develop a better understanding of the various perspectives that exist within the group and refine their PN. This allows them to refine and eventually settle on a final PN that is a concise, clear description of the problem. Starting with their charge from the Gaia government, the Gaian Committee initiates a discussion for the development of a PN.

The Charge   The Gaia government's charge directs the Committee to abide by three major instructions. These are as follows: (1) it must avoid the exclusive consideration of a single environmental media (air, water, or land), (2) it must consider future environmental impacts due to population growth, and (3) given the Gaia government's new treaty with its global neighbors, government officials requested that the Committee also consider at least one globally relevant issue that will demonstrate its good faith with its new partners.

Based on their discussions with the federal officials, the Gaian Committee agrees on its initial PN.

Problem Narrative   Identify the District within Gaia that can be justified as the single-most vulnerable District, with vulnerabilities that can be addressed within the Gaia federal government's authorities. The stakeholders are to consider district-level air quality, water quality, and land use vulnerabilities, including those that pertain to future conditions such as population growth and flooding due to climate change.

4.2.2   Requisite Step 3: Identify Problem Set Elements

Normally, after initializing the PN, stakeholders make a list of the decision alternatives that they want to analyze. Depending on the problem, these could be public policy alternatives, control options, geographical areas or industrial sectors. These are the PSEs that are ranked based on stakeholder-defined criteria. For the Gaian Committee, the federal officials have already determined that all 15 districts are to be analyzed. Although the PSEs are pre-determined to be all 15 districts in this simple example, in many other applications, defining the PSEs is more fluid. For example, in applications where stakeholders are seeking to develop an approach to solving a policy problem that has no clear solution, defining the PSEs is a discovery process. Based on stakeholder discussions, an initial set of alternatives is determined, which represents the collective opinions of the stakeholders. As the iterative process proceeds and trans-disciplinary learning occurs, the composition of the PSEs may change and, commonly, alternatives emerge that were not initially conceived.

4.2.3 Requisite Step 4: Build MIRA Hierarchy

In Step 4, stakeholders determine and organize the initial criteria (also referred to as decision criteria by decision analysts), which are the effects, conditions, or impacts that stakeholders want to consider in the analysis of the policy problem. A *Hierarchy* is the organizing structure that is used for those criteria that stakeholders determine to best represent the PN.

> **Hierarchy**
> The Hierarchy reflects the organization of criteria into a specific configuration. It represents the PN and influences how it is analyzed. Stakeholders may organize criteria into either a linear or branched Hierarchy.

As stakeholders build the Hierarchy, they discuss and articulate a Criteria Narrative (CN) for each of the criteria[1] in the analysis. These CNs represent the stakeholder-chosen, independent components of the analysis used to evaluate the wicked problem as described by the PN. The CNs have a twofold purpose: (1) to provide a description of the nature of the criteria, what they are intended to represent, and why they were chosen and (2) to provide enough specificity such that experts, with stakeholder consultation, can identify the appropriate data to consider in the analysis.

---

[1] Analysts typically use the term "criteria" regardless of how they may be described. However, in an open solution approach, any criterion in the analysis must be described as a CN, which often starts as a simple description of the issue or concern as related to the PN. The open solution approach requires stakeholders to consider how criteria are organized in a Hierarchy, which requires specifically articulating what these criteria represent to the PN. Therefore, in an open solution approach, all criteria are CNs. For example, water quality is a CN because a stakeholder who suggests its inclusion in the analysis has some specific concerns in mind, even if she cannot immediately articulate them. The water quality CN can be refined to represent morbidity due to lead levels in drinking water or the survival risk of an ecological species due to higher concentrations of contaminants resulting from reduced stream flows. Each of these water quality CN refinements is itself a CN. Often, moving to Step 5 (Build CMs) helps stakeholders to refine their original CN. For this reason, throughout this book, the term CN is used in place of the term criteria when pertaining to an open solution approach.

For example, at the beginning of the stakeholder discussions, a general CN about air pollution may take many forms. These could include considerations pertaining to human health impacts from ground-level ozone and haze (such as smoke from wildfires or the light-scattering effects from fine particle atmospheric concentrations) or ecosystem impacts from poor air quality such as ozone impacts on vegetation. Stakeholders that include air quality as part of their PN will work through the steps and produce one or more specific descriptions of air quality. If stakeholders have multiple air quality concerns, they are asked to identify each of them, which may require constructing a more complex Criteria Metric or creating multiple CNs.

> As the Gaian Committee proceeds with developing the CNs and constructing the initial Hierarchy, they have agreed to start by focusing on five environmental issues. These environmental issues fit into the following categories: air quality, local impacts from global issues, water quality, children's health, and the potential for future environmental impacts.

Table 4.1 lists the CNs for each of the stakeholder-identified criteria that will populate the initial Hierarchy. The Gaian Committee determines that these CNs are responsive to concerns expressed in the PN and represent a reasonable first set for initializing the MIRA process. As they proceed through the MIRA process, they will revisit the PN and the CNs to ensure that there continues to be consistency between them.

Once the CNs are defined, the Gaian Committee organizes them into a Hierarchy, which will be used for their Baseline Run. This initial linear Hierarchy in presented in Fig. 4.1.

Table 4.1 Initial CN for each of the five criteria

Stakeholder criteria	Criteria Narrative[a]
Air quality	The air quality impacts on human health in each district
Global environmental issue	Environmental issue of a global nature such as climate change
Water quality	The water quality impacts on human health in each district
Children's health	A pollutant with selective adverse health impacts on children
Potential for future environmental impacts	The resulting impact from a future condition that adversely affects human health or ecological integrity

[a]In general, the first drafting of a CN may not adequately satisfy the two purposes that it should be designed to fulfill. This is remedied as the stakeholders proceed and iterate through the remaining steps.

Fig. 4.1 Gaian initial Hierarchy

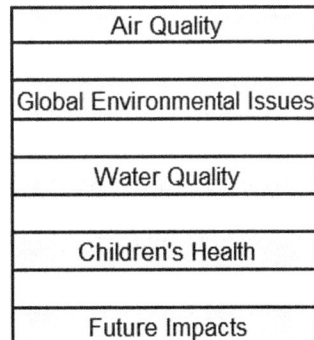

Air Quality

Global Environmental Issues

Water Quality

Children's Health

Future Impacts

4.2.4   Requisite Step 5: Gather Data and Constructing Criteria Metrics

In Step 5, guided by the CNs, stakeholders engage with experts to select CMs. A CM is the metric constructed from one or more datum that represents the CN. For example, the Gaian Committee considered the percentage of land flooded during Category 5 hurricanes as a possible CM to represent their initial CN regarding a global environmental issue. However, in contemplating their CN, the Gaia stakeholders also consider other CMs, such as vulnerable coastal property or vulnerable tourist-focused coastal businesses, which are also consistent with their CN. Therefore, to choose one CM, the Gaia stakeholders need to think more specifically about whether their CN is about coastal personal property, coastal businesses, increased flooding, or possibly a cause of, rather than an effect due to, climate change. Alternatively, the Gaian stakeholders could decide to choose more than one CM to describe a CN.

CMs can be either continuous or discrete. Continuous metrics are those that are represented by a set of real or integer numbers that have no restriction on the size of their range. Ozone concentrations and the number of housing units built before 1979 are both examples of continuous data whose values are ranges of real and integer numbers. Discrete data that are used to construct a CM generally take the form of a small set of integer values that each represent a specific category or condition. For example, a binary choice such as whether a PSE is a coastal area or not is a discrete datum.

The use of data for a single CM is not limited to a single datum. Because CNs often reflect complex issues and selectively target certain vulnerable populations, it can be useful to consider combining two or more pieces of data to reflect a complex CN. There are multiple ways to combine data. One common example that uses two pieces of data is to weight one piece of data with another to produce a different CM. For example, stakeholders seeking to quantitatively represent ozone's impact on vegetation, in an area, could construct a CM that is the product of an area's ozone concentration and its acreage of ozone-sensitive vegetation. In this case, the stakeholders may determine that the newly created weighted CM more accurately reflects the CN about vegetative exposure to ozone concentrations than a CM of ozone concentration alone.

Generally, a CM can be constructed from many pieces of data. Consider an analysis designed to examine the impact from surface coal mining. For example, when environmental analysts want to assess the efficiency of surface coal mining, they consider not just the amount of coal extracted but also how much area is disturbed due to its extraction. Therefore, from these two pieces of data, a CM describing mining efficiency can be created by taking the ratio of the tons of coal per year extracted to the hectares of land disturbed. However, if stakeholders also want to consider the number of people displaced due to the mining activity, they could construct a CM that combines both considerations. This CM could take on a variety of forms; the following is but one example. Since the higher values of mining efficiency produce less environmental impact, while a larger number of displaced people indicate greater environmental impact, dividing the number of people displaced by the mining efficiency creates a reasonable CM. Such a CM considers both concerns and numerically reflects greater environmental impact as its magnitude increased. This is an example of a CM constructed with three pieces of data. Alternatively, the stakeholders may decide to have two separate CMs, mining efficiency and the number of people displaced. Whether the stakeholders choose to use one CM or two is dependent on the PN, the CNs, and the ability of an expert to index the CM.

The Gaian Committee's experts considered several different options for CMs based on the initial five CNs. Each of these discussions is summarized below:

Air Quality Air quality can be described in many ways. Under the US Clean Air Act, air quality is defined by the National Ambient Air Quality Standards (NAAQS) for six (6) pollutants: ground-level ozone, particulate matter ($PM_{2.5}$), carbon monoxide (CO), nitrogen dioxide (NO_2), lead (Pb), and sulfur dioxide (SO_2). However, in Gaia, although $PM_{2.5}$ concentrations are frequently the highest, relative to its standard, ozone presents the most pervasive issue facing the Gaia districts. The US EPA uses the 8-hour design value for ground-level ozone as its metric to determine whether an area is complying with the NAAQS. Unable to conceive of a better metric for its air quality CN, the Gaian Committee chooses the ozone design value as its air quality CM. The statistical form of this ozone standard is the fourth highest monitored 1-hour value of ozone over a 3-year period [1].

Global Environmental Issues After the Gaian Committee's initial discussions, two CMs relative to the local impacts of global climate change remained for consideration. The first was emissions of CO_2 from the burning of fossil fuels in each of the Gaia districts. However, the large reduction in CO_2 emissions required to affect global climate change impacts is more than Gaia's total CO_2 emission inventory. Furthermore, among the international community, Gaia's total CO_2 emissions are exceedingly small and CO_2 reductions in Gaia would not provide a meaningful indicator for effective Gaia government action. Therefore, the Gaian Committee turns from mitigation to adaptation measures. They considered alternative CMs such as the percentage of land flooded due to Category 5 hurricanes. However, since a large fraction of Gaia contains low-lying areas that are prone to flooding, even during less severe storms, the Gaian Committee chooses a district's total land area with a 100-year flooding risk as the CM that is more representative of this CN. The CN to evaluate areas in the 100-year floodplain reflects a government strategy to adapt to the effects of global climate change. These CMs represent consistent but different CN refinements of the initial CN that represented a very broad statement regarding global environmental issues such as climate change.

Water Quality Nitrogen and phosphorus are the two major pollutants from agricultural fertilizer runoff that contribute to nutrient loading in Gaian water bodies. Originally, the Gaian Committee considered using both nitrogen and phosphorus as indicators. However, the results of a previous water quality study at the State University of Gaia concluded that there is a high correlation between the two pollutants. Therefore, only one of these indicators is necessary to provide the relative difference in nutrient loading among districts. Given the greater public understanding relative to nitrogen as a contributor from agricultural fertilizer runoff, the Gaian Committee selects nitrogen from agricultural runoff as the CM to represent the water quality CN. However, if there were some controversy about the study results regarding phosphorus's correlation with nitrogen, the open solution approach would encourage its inclusion. Stakeholders will be able to determine whether including both phosphorus and nitrogen will meaningfully impact the PSE rankings. If phosphorus is later found to be a redundant indicator, it can easily be removed. The ability to examine minority stakeholder opinions is an essential feature of an open solution approach.

Children's Health In the process of choosing a CM for children's health, the Gaian Committee considered a child's increased vulnerability due to active growth, body size, and behavior. Children's bodies are actively developing, and their smaller size also makes them more sensitive to lower doses of toxins. Many different pollutants, including ozone, affect children more adversely than the general adult population. However, given that ozone is already included in the analysis, the Gaian Committee consider ingested pollutants that affect developing neurological systems as important for examining environmental impacts to children's health. Both mercury and lead are toxic to children and would be good choices as they more adversely impact children compared with adults. After discussing the merits of both pollutants, the Gaian Committee determines that the potential impact from lead presents a much greater risk in Gaia. A large percentage of Gaia's housing was built prior to 1979 when lead-based paint was commonly used. Deterioration of the paint and subsequent ingestion by children is a high risk in Gaia because of the age of their homes. Although Gaia experiences mercury deposition, the experts advise the Committee that the risk to children from ingesting soil laden with mercury is far less than the risk from ingesting lead-based paint. Additionally, the data needed to construct an appropriate metric for lead ingestion is readily available within the decision timeframe and is expected to provide a rational means to differentiate the districts. Therefore, the Gaian Committee chooses the number of houses built prior to 1979 as the CM to represent the Children's Health CN.

Potential Future Environmental Impacts Based on a previous study performed by the Gaian Planning Commission, Gaia is expected to experience a substantial amount of commercial growth in the next 25 years. With such growth comes the potential for exacerbating a variety of environmental problems (such as drinking water supplies and waste management), reduced green spaces, increased traffic emissions, the expansion of urban centers, and increased industrial and transportation pollutant emissions. Also, since commercial growth is highly correlated with increasing population, and population data is more readily available, the Gaian

Table 4.2 Description of the CMs for the Gaia analysis

Criteria Narrative	Criteria Name	Criteria Metric
Air quality	OZONE	The ozone concentration within each district, expressed as the US EPA's calculated 8-hour design value in parts per billion (ppb)
Global environmental issue	CLIMATE_ CHANGE	The percentage of the district's land area that is located within the 100-year floodplain
Water quality	NUTRIENT_ LOAD	The total mass per unit area of nitrogen applied via agricultural fertilizers in each district per year, in kilograms per hectare per year (kg/ha/yr)
Children's health	LEAD	Total number of housing units built before 1979 in each district (# houses)
Potential for future environmental impacts	POPGROWTH	Projected increase in population in each district over the next 25 years (% people)

Committee chooses forecasted population growth as the CM to represent the future impacts CN.

The CMs chosen by the Gaian Committee for the initialization of the MIRA analysis are summarized in Table 4.2.

The data used for the CMs that are described in Table 4.2 are presented in the Appendix as Table A.1. For the Baseline Run, the Committee decides to use these CMs without modification as they have no current information to suggest doing otherwise.

4.2.5 Requisite Step 6: Indexing – Establish the Significance of the Data

In Step 6, experts judge the significance of the data through an Indexing process that places a CM's value on a non-dimensional decision scale. Each of the CMs that appear in the Hierarchy has its units converted in this way.

In the Indexing process, experts assign a small number of non-dimensional integers to certain values associated with each CM. These integers have a contextual non-physical unit, the decision unit. These integers comprise a measure that represents how significant the magnitude of each CM is to the decision being made. When thinking about the meaning of the decision units, it is helpful to consider the goal of the PN. For example, if the PN is to evaluate and determine which PSEs have the poorest environmental condition, then the decision units are the degree of environmental degradation or vulnerability. Alternatively, if PSEs are ranked from most attractive to least attractive based on the PN, the decision units can be defined as attractiveness.

In the Indexing step, experts assess the relative significance of a CM within and across all CMs. The interaction between and among experts across the disciplines is part of the trans-disciplinary learning that is key to the success of an open solution approach. Non-indexed CMs have no meaning relative to the PN. However, once indexed, CMs have been contextually translated by the experts and become *Terminal Indicators* (TIs), which are now meaningful within the specific context of the problem. Distinguishing between CMs and TIs helps to avoid unnecessary confusion, and sometimes discord, when stakeholders otherwise ascribe their own meanings to the data. This clarity will be important in Step 7 when stakeholders are asked to trade-off the relative importance among the indicators.

> **Terminal Indicator**
> A Terminal Indicator results from the conversion of a CM from its native units to decision units. This is accomplished when experts evaluate the significance of the CM's value relative to the PN. Terminal Indicators are the connection between the Hierarchy and the data.

An Expert Judgment Set (EJS) is created when all the CMs in a Hierarchy have been converted to TIs. Although decision units are non-dimensional, they reflect the relative magnitudes of the TIs for every PSE. In this manner, decision units directly relate to the degree of vulnerability (or attractiveness) of every PSE.

To properly distinguish one PSE from another, it is important to consider the length of the decision scale. When considering discrete CMs, a scale that is too short does not offer enough separation or distinction among the PSEs, making it difficult for decision makers to choose an alternative. Conversely, a scale that is too long offers unnecessary separation (or resolution) among the PSEs. Social scientists generally agree that a range between five and nine in length provides the necessary distinction without the unnecessary resolution [2]. An example of a decision scale that is five categories in length can commonly be found in surveys. Surveys generally ask respondents whether they Strongly Agree, Agree, are Neutral, Disagree, or Strongly Disagree. Five categories appear to be the minimum length of a decision scale for respondents to feel satisfied that their views are sufficiently represented. In MIRA, to accommodate complex environmental analyses that may likely need more decision scale resolution, the decision scale ranges from 1 to 8. As a matter of convention in MIRA, 8 on the decision scale represents either the poorest condition or the most attractive decision alternative, depending of the nature of the decision being made.

Each district has data that populate each of the five CMs. For example, the ozone design value (a continuous data metric) has a metric associated with each of the districts, which is shown in Table 4.3.

To perform the Indexing task, the experts construct a functional relationship for each CM that maps the values of that CM in its native units (one value for each PSE) to the 1–8 decision scale. If the relationship is determined to be linear throughout the decision scale, then only two values of the CM will need to be mapped to two of the eight integers on the scale. In the most general case, up to eight metric values need to be mapped to each of the eight integers that make up the decision scale. For any metric value that falls between any two of the values being mapped, a linear

Table 4.3 Ozone data for the
Gaia districts

Criteria Metric =>	OZONE
Units =>	**ppb**
Min =>	**63.3**
Max =>	**90.6**
District 1	77.7
District 2	73.4
District 3	67.9
District 4	73.1
District 5	77.6
District 6	75.8
District 7	76.6
District 8	81.5
District 9	79.9
District 10	90.6
District 11	77.4
District 12	63.3
District 13	83.8
District 14	80.2
District 15	80.6

interpolation is used to determine its indexed value. Generally, this functional form provides an adequate degree of non-linearity (accomplished in a step-wise manner) for the range of problems that are expected to be addressed. However, if found to be warranted, any manner of translational functions (e.g., logarithmic or exponential) could be used in the Indexing process. For CMs that are based on discrete, categorical data, each discrete metric value is mapped to one of the eight integers on the decision scale.

The five Gaia CMs are linearly transformed via Indexing. The equation used to index them (Eq. A.1) is presented in the Appendix.

To perform a Baseline Run, the EJS and VS are intentionally simplified. This allows stakeholders to focus on developing their PN, CNs, and CMs. The tasks of creating the PN-specific EJS and VS are postponed until stakeholders have completed the Baseline Run, which includes any iterations needed to better define the PN, CNs, CMs, and the Hierarchy. For the Baseline Run, the range in values for each CM is spread uniformly across the full decision scale. The minimum and maximum data values of each CM are assigned to the lowest and highest value on the decision scale, respectively, while the intermediate CMs are spread uniformly (i.e., linearly interpolated) across that range. By starting with this simple approach to Indexing, a baseline is developed against which the results from the next iteration of Indexing can be evaluated. There is no expectation that this uniform approach to Indexing will be utilized in the final MIRA analysis. Therefore, all MIRA Baseline Runs use an EJS that is based on a process that initially excludes expert judgments.

Table 4.4 presents the Uniform Indexing EJS for the five CMs used for the Baseline Run for the 15 Gaia districts

Table 4.4 Uniform Indexing for the Gaia analysis

Index →									
TERMINAL INDICATORS	UNITS	1	2	3	4	5	6	7	8
OZONE	ppb	63.28	67.19	71.10	75.01	78.92	82.83	86.74	90.65
CLIMATE_CHANGE	%	0.02	10.49	20.95	31.42	41.88	52.35	62.81	73.28
NUTRIENT_LOAD	kg/ha/year	1.00	1.49	1.97	2.46	2.94	3.43	3.91	4.40
LEAD	# houses	2	21,430	42,859	64,287	85,715	107,143	128,572	150,000
POPGROWTH	%	0.00	7.14	14.29	21.43	28.57	35.71	42.86	50.00

District 12's ozone value of 63.28 ppb is the lowest in the range and is therefore indexed to a value of 1, while the highest ozone value of 90.65 ppb, belonging to District 10, is indexed to 8. The ozone data for Districts 12 and 10 represent the minimum and maximum indexed values on the decision scale for ozone. District 4's ozone metric value of 73.11 ppb, located between the minimum and maximum ozone values, is linearly interpolated on the decision scale resulting in an index value of 3.5. The other four CMs are indexed in a similar fashion.

In contrast to MIRA, where a Uniform Indexing EJS is a starting point in the analysis, many current decision analytic approaches use Uniform Indexing implicitly and offer neither a means to examine the EJS nor to make changes to it. The result of this default practice of Uniform Indexing, without the opportunity to consider EJSs designed by subject matter experts, is to effectively remove contextual expert judgment from the process. Without understanding and articulating context, the significance of the data cannot be meaningfully ascribed or determined.

4.2.6 Requisite Step 7: Determining the Relative Importance of the Indicators

In Step 7, stakeholders are asked to determine the relative importance of the indicators through the explicit solicitation of stakeholder Preferences within the context of a specific PN. Importance is based on the problem context and refers to the stakeholders' relative Preferences among the indicators. In MIRA, assessing CM significance and eliciting stakeholder Preferences are two completely different activities. When stakeholder Preferences for each indicator are elicited, they are being asked to indicate their vote across the indicators within each indicator cluster[2] in the Hierarchy. Just as with Hierarchy construction and Indexing, the stakeholders will create multiple sets of Preferences (the Value Sets) as they explore the impact and their degree of resolve regarding the trade-offs that they have made.

In the Preferencing process, stakeholders are asked to determine the relative importance among a set of indicators. This is accomplished by assigning a value between 0.0 and 1.0 to each of the indicators such that the sum of these weights equals 1.0 (or 100%). For example, with three indicators, stakeholders could choose to distribute their Preferences equally, assigning each indicator one-third (or 33.3%) of the total weight. Provided that 100% of the weight is distributed among the indicators, stakeholders have many different Preferencing options.

Step 7 is complete when all the indicators contained in the MIRA Hierarchy have been weighted in this manner. A MIRA Value Set (VS) is a unique set of relative weights that represent a specific perspective regarding the Hierarchy. The Baseline

---

[2] In this chapter, only linear Hierarchies are discussed. Linear Hierarchies contain only a single indicator cluster (i.e., the entire set of indicators being utilized). More complex indicator clusters will be discussed in Chap. 5.

Table 4.5 Equal Preference VS for the Gaian Baseline Run

Indicator	Equal Preferences
CLIMATE_CHANGE	0.20
OZONE	0.20
NUTRIENT_LOAD	0.20
LEAD	0.20
POPGROWTH	0.20

Run uses an Equal Preference VS, which assigns the same degree of importance to all indicators that are being traded-off. An Equal Preference VS is used to allow stakeholders to develop a better understanding of the problem without having to initially reconcile likely differences of opinion.

Stakeholders can use the PSE ranking to more closely examine individual PSEs, indicators, and CMs, to learn about the characteristics of the PSEs that cause them to rank the way they do. This type of learning is possible because the Baseline Run is a wholly constructed analysis, even if simplified and not yet finalized. Stakeholders can vary one component, such as Preference weight, while keeping all others constant and examine the effect it has on the results. In this way, the sensitivity of the results to a single variable, which is under the control of the stakeholders, can be observed. Through this process, stakeholders may find that the data for the CMs are correct but their understanding about that data is wrong.

The Equal Preference VS for the five-indicator Gaian Hierarchy is shown in Table 4.5.

As stakeholders iterate through the MIRA process, they will experiment with many different VSs as they examine how variations in Preferencing influence the resulting changes in PSE rankings. It is true that stakeholders often have a good idea of their general Preferences but, when confronted with making choices within a specific PN, these Preferences often change. Therefore, VSs should be considered less as established stakeholder opinions and more as possibilities for seeking common ground among the stakeholder group. The final VS represents the common ground that has been reached by the stakeholders.

4.2.7 Requisite Step 8: Criteria Sum – Collapsing of the Hierarchy

A single-valued metric, the Criteria Sum (CS), is calculated for each PSE when the indicators' Preference weights are multiplied by their indexed values and summed across all indicators.[3] The CS for each PSE, in a linear Hierarchy, is calculated as follows:

---

[3]The process described in Chap. 4 applies only to linear Hierarchies. In Chap. 5, this process is generalized for more complex Hierarchies.

$$CS = \sum_{i=1}^{N} p_i I_i \qquad (4.1)$$

Where:

$CS \equiv$ the Criteria Sum
$I_i \equiv$ the value of the ith indicator in the Hierarchy
$N \equiv$ the total number of indicators in the Hierarchy
$p_i \equiv$ the Preference weight assigned to I_i

The Baseline Run is completed when the PSEs have been ranked using their CSs. The CS is the measure of the relative vulnerability or attractiveness of each PSE. The ranking of PSEs in the Baseline Run combines choices that the stakeholders have made regarding the PN, CNs, Hierarchy construction, and CMs, coupled with the simplified Uniform Indexing EJS and Equal Preference VS.

Because of the convention established in Indexing, the higher the value of a PSE's CS, the more vulnerable it is. Therefore, for the Gaia example, those districts with poorer environmental conditions are ranked closer to the top.

The numerical value of the CS for any PSE ranges between 1.0 and 8.0 because it is limited to the range of the decision scale. Suppose the indexed value for every indicator for a PSE is equal to 8.0. Therefore, regardless of what VS is applied, that PSE's CS will also equal 8.0, since the sum of the weights must equal 1.0 and all TIs are equal to 8.0. However, in an actual application, the TIs (i.e., the indexed values of the CMs) will not be the same. When this is the case, as different VSs are applied, the CSs will vary in response to changes in the VS. This is because the CSs are the sum of the products of the TIs and VS weights.

4.3   Analyzing the Baseline Run

4.3.1   Additional Analysis Methods Using the Criteria Sum

Ordinal rankings alone cannot adequately discern the differences in vulnerability among the PSEs. Often, stakeholders will want to know how far apart the top ranked PSE is from the next highest ranked PSE, or any other PSE. Therefore, MIRA contains analysis methods that utilize a PSE's CS and the numerical distance between each calculated PSE's CS and the CS of another. These are (1) Decision Distance (DD), (2) Indicator Contribution (IC), (3) Indicator Drivers (ID), and (4) Binning. When calculated for the Baseline Run (or any subsequent run), these analysis methods help stakeholders to decide what additional iterations are needed to clarify or refine the PN, CNs, EJSs or VSs. The CSs for the PSEs, ranked from most to least vulnerable, from the Gaian Baseline Run are presented in Table 4.6.

Table 4.6 PSE ordinal
rankings for the five-indicator
linear Hierarchy Baseline Run

Baseline Run	Criteria
Ranked PSEs, most to least vulnerable	Sum
District 2	4.74
District 6	4.42
District 14	4.07
District 8	3.99
District 13	3.96
District 10	3.76
District 15	3.75
District 11	3.66
District 12	3.12
District 9	3.02
District 4	2.60
District 7	2.51
District 5	2.49
District 1	1.87
District 3	1.35

4.3.2   Decision Distance

Calculated using the PSEs' CSs, *Decision Distance* (DD) is the percent difference
in CS that any PSE is separated from the top ranked PSE. For the DD, the greater
the separation, the greater is the difference in vulnerability or attractiveness between
those two PSEs.

> **Decision Distance**
> The Decision Distance (DD) is the percent difference in CS between the top
> ranked PSE and any of the other PSEs.

The Decision Distance between the top ranked PSE and another PSE is calculated as follows:

$$\mathrm{DD}_j = 100\left(\frac{CS_{\max} - CS_j}{CS_{\max} - CS_{\min}} \right) \tag{4.2}$$

Where:

$\mathrm{DD}_j \equiv$ the Decision Distance (in units of %) between the first rank and the jth ranked
PSE, where $j \geq 2$

$CS_j \equiv$ the value of the CS, in decision units, of the jth ranked PSE

$CS_{\max} \equiv$ the maximum CS from $\{CS_j\}$

$CS_{min} \equiv$ the minimum CS from $\{CS_j\}$
$\{CS_j\} \equiv$ the set of Criteria Sums for all PSEs

The results of using DD and Binning (discussed in Sect. 4.3.3) for the Gaian five-indicator Baseline Run are presented in Table 4.7.

The DDs for the second and third ranked districts (Districts 6 and 14) are 9.3% and 19.7%, respectively. In other words, Districts 2, 6, and 14 are about 10% apart from each other. Mindful that this is the Baseline Run and no final decision is being made, if the Gaian Committee were to decide based on these results, they may lean toward selecting District 2 as the sole recipient of the additional funding if they are convinced that a 9–10% separation between districts is significant. The DD provides additional information to help stakeholders assess Decision Uncertainty.

4.3.3 Binning

When there are many PSEs, Binning is a useful analysis method to help stakeholders sort PSEs into equally spaced bins. This kind of sorting imposes a uniform structure across the PSEs, using the CSs, which can help stakeholders assess the similarity of PSEs by how they fall into these groupings. Stakeholders choose the number of bins to apply. Each bin is then delineated by equally spaced CS ranges that cover the full range of CSs. Each PSE is placed into the bin whose CS range brackets its CS. Therefore, while choosing the number of bins is arbitrary, Binning allows stakeholders to apply a consistent rule across the CSs. Once stakeholders determine the number of bins to apply, the bin number for each PSE is determined as follows:

Table 4.7 Decision Distance and Binning for the Gaia Baseline Run

Baseline Run		10 bins	11 bins
Ranked PSEs, most to least vulnerable	Decision Distance (DD)(%)	(Bin #)	(Bin #)
District 2	0.0	1	1
District 6	9.3	1	2
District 14	19.7	2	3
District 8	22.0	3	3
District 13	22.9	3	3
District 10	28.8	3	4
District 15	29.0	3	4
District 11	31.8	4	4
District 12	47.7	5	6
District 9	50.9	6	6
District 4	63.2	7	7
District 7	65.7	7	8
District 5	66.4	7	8
District 1	84.7	9	10
District 3	100.0	10	11

$$BN_j = T_b - \text{Trunc}\left[T_b \left(\frac{CS_j - CS_{min}}{CS_{max} - CS_{min}} \right) \right] \quad\quad (4.3)$$

Where:

$BN_j \equiv$ the bin number of the jth ranked PSE
$T_b \equiv$ the total number of user-defined bins
$CS_j \equiv$ the value of the CS, in decision units, of the jth ranked PSE
$CS_{max} \equiv$ the maximum CS from $\{CS_j\}$
$CS_{min} \equiv$ the minimum CS from $\{CS_j\}$
$\{CS_j\} \equiv$ the set of Criteria Sums for all PSEs

When stakeholders are seeking to minimize Decision Uncertainty, fewer PSEs in a single bin, and larger bin separations between PSEs, coupled with a smaller number of overall bins signal lower Decision Uncertainty. Because users can quickly change the number of bins to examine the PSE groupings, this can provide another useful technique to analyze a PSE rank.

In Table 4.7, when the Gaian Committee chooses to group the districts into 10 bins, Districts 2 and 6 share Bin 1, District 14 is alone in Bin 2, and Bin 3 is shared by Districts 8, 13, 10, and 15. Note that there will not always be PSEs in every bin. In Table 4.7 there are no PSEs in Bin 8 when ten bins are used. By increasing the number of bins to 11, Districts 2 and 6 are now in separate bins. Whether PSEs share bins, or not, does not pre-determine how the Gaian Committee should select districts. Instead, Binning provides them with additional information about the distinctiveness between Districts 2 and 6.

Although the use of Binning in an application with only 15 PSEs is of minimal use, in applications where there are many more PSEs, it can be very useful. For example, stakeholders could continue to increase the number of bins to determine what it would take to have the top three PSEs reside in separate bins. Used with the other MIRA analysis measures, the Binning method can provide stakeholders with more certainty about making their final choice.

4.3.4   Indicator Contribution

Often, stakeholders want to understand which indicators are most influential in the production of the PSEs' ranks. In MIRA, the analysis measure *Indicator Contribution* (IC) is used for this purpose. An indicator's IC does not correspond to any single PSE; it relates to the importance of the indicator considering all PSEs. When there are questions or challenges to the data behind the CMs, the EJS chosen, or the VS used, it is important for stakeholders to know the degree of influence that each indicator plays in producing the PSEs' ranks. By using IC, stakeholders can compare the overall relative contribution from each of the indicators to the sum of the CSs from all PSEs (Total CS).

Indicator Contribution
Indicator Contribution (IC) is the contribution of any single indicator to the total Criteria Sums from all the PSEs (Total CS).

The IC of an indicator[4] for a specific MIRA run is calculated as follows:

$$\mathrm{IC}_i = 100 \left(\sum_{j=1}^{m} I_{ij} p_i \prod_{k=1}^{n} p_k \bigg/ \sum_{i=1}^{N} \sum_{j=1}^{m} I_{ij} p_i \prod_{k=1}^{n} p_k \right) \tag{4.4}$$

Where:

$\mathrm{IC}_i \equiv$ the Indicator Contribution (in units of %) of the ith indicator in the Hierarchy
$I_{ij} \equiv$ the value for the ith indicator for the jth PSE
$p_i \equiv$ the Preference weight of the ith indicator
$p_k \equiv$ the Preference of the kth Composite Indicator (CI) – where p_1 = the Preference weight for the Composite Indicator that contains I_{ij} and p_{2-n} = the Preference weights of the remaining CIs that form the direct connection from I_{ij} to and including its primary level CI in the Hierarchy (for a linear Hierarchy $p_k = 1$)
$m \equiv$ the number of PSEs
$n \equiv$ the number of levels in the Hierarchy between I_i's level and the primary level (for a linear Hierarchy, $n = 1$)
$N \equiv$ the total number of Terminal Indicators (TIs) in the Hierarchy

The IC for any single indicator is the ratio of the sum of its collective contribution from each PSE to the Total CS, in units of percent. This percentage is a function of the indicator's indexed values and the Preference weights ascribed to it from all PSEs. The IC values for each of the five indicators used in the Baseline Run are presented in Fig. 4.2.

The OZONE and NUTRIENT_LOAD indicators, shown in Fig. 4.2, provide the largest contributions of the five indicators in the Gaian Baseline Run, each contributing approximately 28% to the Total CS (for a total of 56%).

The IC can also be used to prioritize resources. Stakeholders can use information about IC to assess whether more resources need to be employed to reduce data uncertainty. For example, in the Gaian Baseline Run, the IC of the OZONE and NUTRIENT_LOAD indicators may signal to the ozone and nutrient experts to consider the uncertainty of the ozone and nutrient data used for those indicators. Understanding that these two indicators contribute 56% toward the Total CS may support the rationale to further examine and reduce data uncertainties for the ozone and nutrient data. Suppose the Gaia stakeholders determine that data uncertainty for ozone and nutrients is very high and cannot be satisfactorily reduced within their

---

[4]Equation 4.4 calculates ICs not only for simple linear Hierarchies but also for more complex Hierarchies that include Composite Indicators, which will be discussed in Chap. 5.

Indicator Contributions (IC)

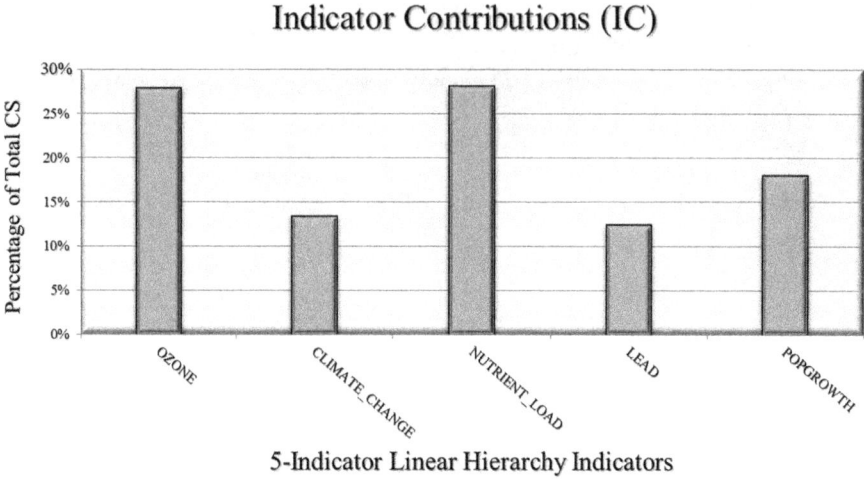

5-Indicator Linear Hierarchy Indicators

Fig. 4.2 The IC for each indicator in the five-indicator linear Hierarchy

decision making timeframe. In this case, they may want to consider the use of other data as a substitute or reformulate the PN.

With the capability to understand which data are more influential to the results, stakeholders can avoid the delays often created when some stakeholders, without an understanding of the relative influence among indicators, insist on further research or for more data to be collected before allowing a decision to be made.

4.3.5 Indicator Drivers

When stakeholders want to understand which indicator contributes the most to the rank of any individual PSE, they want to know the *Indicator Driver* (ID). This analysis measure is used to examine the relative contribution of each indicator (either Composite or Terminal) to a PSE's CS. For any PSE, an indicator's ID is calculated as follows:

$$\mathrm{ID}_{ij} = 100 \left(I_{ij} p_i \prod_{k=1}^{n} p_k \Big/ \left(\sum_{i=1}^{N} I_{ij} p_i \prod_{k=1}^{n} p_k \right) \right) \tag{4.5}$$

Where:

$\mathrm{ID}_{ij} \equiv$ the Indicator Driver (in units of %) for the ith indicator of the jth PSE in a Hierarchy

$I_{ij} \equiv$ the value for the ith indicator of the jth PSE

$p_i \equiv$ the Preference weight of the ith indicator

$p_k \equiv$ the Preference of the kth Composite Indicator – where $p_1 =$ the Preference weight for the CI that contains I_{ij} and $p_{2-n} =$ the Preference weights of the remaining CIs that form the direct connection from I_{ij} to and including its primary level CI in the Hierarchy (for a linear Hierarchy, $p_k = 1$)

$n \equiv$ the number of levels in the Hierarchy between I_{ij}'s level and the primary level (for a linear Hierarchy, $n = 1$)

$N \equiv$ the total number of TIs in the Hierarchy

Indicator Driver
The Indicator Driver is the contribution of an indicator to a PSE's Criteria Sum (in units of %).

While it is informative to examine the ID for an individual PSE, stakeholders may find the targeted comparison of IDs among a select subset of PSEs also helpful in increasing their understanding of a PSE's ranked position. For example, by comparing the contribution of each indicator to the CSs for Districts 2 and 8, the first and fourth ranked PSEs in the Baseline Run, help the Gaian Committee to better understand how District 2 is the top ranked District. The comparison of IDs for Districts 2 and 8 is shown in Fig. 4.3.

Figure 4.3 shows that LEAD and POPGROWTH drive District 2's ranking, even though both indicators were shown to only moderately contribute to the Total CS (i.e., a relatively moderate IC, shown in Fig. 4.2). Therefore, the same indicator that is shown to have a moderate contribution to the Total CS can still have a large contribution to an individual PSE's Criteria Sum.

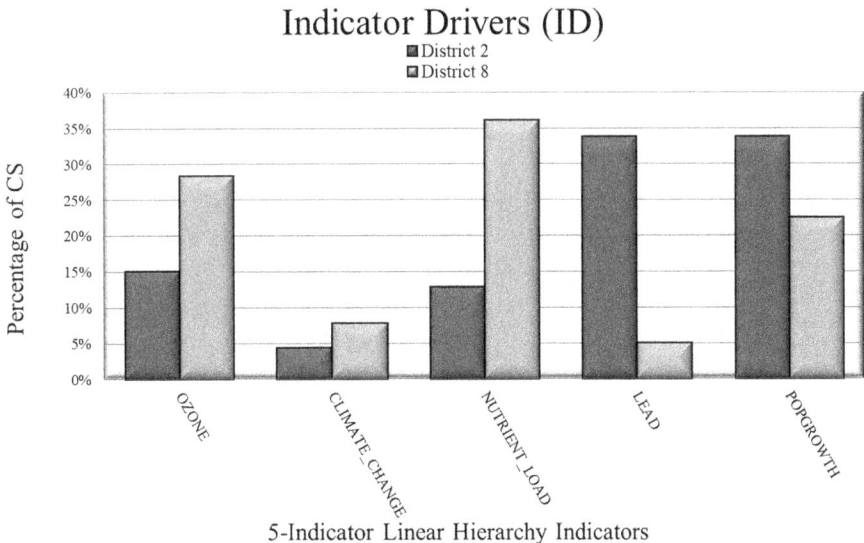

Fig. 4.3 Comparison of the District 2 and 8's IDs

4.4   Summary

Initialization in a MIRA analysis begins when stakeholders perform the Baseline Run of the first constructed Hierarchy. As stakeholders refine their PN and experiment with alternative Hierarchies, the first run of each newly constructed Hierarchy is a Baseline Run. The use of a Uniform Indexing EJS and an Equal Preference VS in the Baseline Run permits stakeholders to focus initially on articulating and understanding their Problem Narrative (PN) and ensuring that their Criteria Narratives (CNs) and Criteria Metrics (CMs) are consistent with that PN. Customization of the EJS and VSs is postponed until the stakeholders are satisfied with the Hierarchy, CNs, and CMs in their Baseline Run. The Baseline Run becomes the benchmark against which comparisons of additional runs (the iterations) are made. The four analysis methods (Decision Distance (DD), Indicator Contribution (IC), Indicator Drivers (ID), and Binning) are designed to facilitate stakeholder discussion and experimentation. The next chapter describes MIRA's iterative process, where stakeholders customize their analysis and apply additional analysis methods and strategies to assess Decision Uncertainty.

References

1. U.S. Environmental Protection Agency (2015) Appendix U to part 50 – interpretation of the primary and secondary national ambient air quality standards for ozone. 40 C.F.R. Part 50. Washington, DC
2. Likert R (1932) A technique for the measurement of attitudes. Arch Psychol 140:1–55

Chapter 5
The MIRA Approach: Iterate to Stakeholder Agreement by Minimizing Decision Uncertainty

Abstract This chapter provides a detailed description of how iteration is used in the MIRA open solution approach to foster trans-disciplinary learning. This approach culminates in a consensus decision that is supported by a comprehensive evaluation of various types of uncertainty and how they relate to Decision Uncertainty. After performing the Baseline Run in Chap. 4, stakeholders are ready to proceed with further experimentation via iteration runs to better understand alternative rankings. This chapter describes those aspects of MIRA that makes it a practical open solution process. The process maximizes the use of these iterations until an analysis is produced with acceptable Decision Uncertainty, as determined by the stakeholders. The Gaia example used in Chap. 4 is expanded in Chap. 5 to illustrate the strategies and analysis methods used in a realistic MIRA application.

Keywords Uncertainty analysis · Risk management process · Quality assurance methodology · Consensus building · Multi-criteria assessment

5.1 Introduction

This chapter provides a detailed description of how iteration is used in the MIRA open solution approach to foster trans-disciplinary learning. This approach culminates in a consensus decision that is supported by a comprehensive evaluation of various types of uncertainty and how they relate to Decision Uncertainty. After performing the Baseline Run in Chap. 4, stakeholders are ready to proceed with further experimentation via iteration runs to better understand alternative rankings. This chapter describes those aspects of MIRA that makes it a practical open solution process. The process maximizes the use of these iterations until an analysis is produced with acceptable Decision Uncertainty, as determined by the stakeholders. The Gaia example used in Chap. 4 is expanded in Chap. 5 to illustrate the strategies and analysis methods used in a realistic MIRA application.

© Springer Nature Switzerland AG 2020

C. H. Stahl, A. J. Cimorelli, *Environmental Public Policy Making Exposed*,
Risk, Systems and Decisions, https://doi.org/10.1007/978-3-030-32130-7_5

Using the Gaia example, Chap. 5 illustrates and discusses the strategy for:

1. Revising the initial Hierarchy by adding new Criteria Narratives (CNs) and examining Hierarchy constructions relative to the Problem Narrative (PN)
2. Creating multiple indicators from limited data
3. Examining the degree to which data and other uncertainties contribute to Decision Uncertainty
4. Understanding whether it is possible to rationalize the results you want
5. Conducting an analysis when some Terminal Indicators do not apply to all Problem Set Elements (PSEs)

In Chap. 5, as the analysis becomes more complex, much of the data, mathematical formulations, and intermediate calculations are placed in the Appendix for readers interested in examining these details.

5.2 Revising the Hierarchy: Adding and Reorganizing Criteria

5.2.1 Expand Linear Hierarchy

Having the capability to add new CNs when constructing a policy analysis is an important feature for any open solution approach. How these CNs are added affects the PN. Stakeholders who understand Hierarchy construction have the advantage of ensuring that, as new CNs are added, their PN continues to reflect their intended policy problem.

In Chap. 4, the Gaia federal government, tasked with evaluating the environmental condition of their 15 districts, initialized their analysis with five criteria, organized in a linear Hierarchy. Analysis of the Baseline Run showed that OZONE was one of two indicators contributing the most to the Criteria Sums of the districts. The Gaian Committee is aware that there is much more air quality data available and believes that including these data will make their analysis more robust. To start the iterative process of refining their PN, the Gaian Committee decides to include several more detailed components of air quality. This results in a refinement of their original linear Hierarchy, which includes these more detailed components of air quality.

> **Refined PN:** The Gaian Committee's refined PN includes a more detailed description of air quality. The four additional CNs are 1) cardiovascular and respiratory disease from inhaling fine particle matter (FINE_PART) and three concerns from the inhalation of various air toxic pollutants, 2) carcinogenic risk (AIR_TOX_CANCER), 3) respiratory risk (AIR_TOX_RESP), and 4) neurological risk (AIR_TOX_NEURO).

After consultation with experts, including the consideration of available data, the Gaian Committee also settles on the Criteria Metric (CM) for each of these four new CNs, which are described in Table 5.1. The data for the four new CMs are found in Table A.2 in the Appendix.

Table 5.1 Criteria Narratives and Criteria Metrics for the five additional criteria

Criteria Name	Criteria Narrative	Criteria Metric
FINE_ PART	The potential for non-lethal respiratory impacts to the general population and excess deaths in populations having cardiovascular and respiratory disease due to the inhalation of fine particulates	Fine particles are 2.5 microns in size or smaller, $PM_{2.5}$ concentrations in the district as predicted using an air quality model. The spatially averaged concentration across the district in units of parts per billion (ppb) is used
AIR_ TOX_ CANCER	An individual's cancer risk due to the inhalation of air toxic pollutants	The average individual cancer risk predicted to exist in the district. This risk is based on the sum of the products of modeled ambient air concentrations of each carcinogen impacting the district and the unit risk of each of the carcinogens, in units of 10^{-6} risk.
AIR_ TOX_ NEURO	An individual's risk of a neurological problem due to the inhalation of air toxic pollutants	The total of the modeled Hazard Quotients (HQ)[a] predicted and summed over each air toxic pollutant that impacts the neurologic system. A calculated HQ less than 1 is determined to cause no adverse neurologic effects. The units are non-dimensional.
AIR_ TOX_ RESP	An individual's risk of a respiratory problem due to the inhalation of air toxic pollutants	The total of the modeled Hazard Quotients (HQ)[a] predicted and summed over each air toxic pollutant that impacts the respiratory system. A calculated HQ, having non-dimensional units, that is less than 1 is determined to cause no adverse respiratory effects.

[a] $HQ = C_a / RfC$ where $HQ \equiv$ Hazard Quotient, $C_a \equiv$ total average daily air concentration of a given pollutant ($\mu g/m^3$), and $RfC \equiv$ reference concentration of the pollutant [1]

The Committee decides to add these four CNs to the initial five-indicator linear Hierarchy. The addition of the four CNs and their associated CMs changes the district rankings when compared with the Baseline Run of the five-indicator linear Hierarchy. Table 5.2 presents the results of the Baseline Run when the linear Hierarchy is increased to nine indicators and compares these results to those from the Baseline Run of the five-indicator linear Hierarchy.

Whenever stakeholders add or change the organization of a Hierarchy, Uniform Indexing coupled with an Equal Preference VS is used to produce a new Baseline Run. Therefore, although stakeholders may have performed a previous run with a different Hierarchy, each new Hierarchy requires initialization with a Baseline Run because the new Hierarchy represents a different PN.

Compared with the five-indicator linear Hierarchy, the impact of four additional indicators in this nine-indicator linear Hierarchy Baseline Run results in District 8 moving to the top rank (i.e., appearing more vulnerable than the other districts) while reducing the relative vulnerability of District 2, which moves from the top rank to the seventh rank. When the PN is refined, how the districts are evaluated is changed, and different results are obtained.

Table 5.2 Comparison of the five-indicator versus nine-indicator linear Hierarchy Baseline Runs

Five-indicator linear Hierarchy			Nine-indicator linear Hierarchy		
Ranked PSEs, most to least vulnerable	DD (%)	Bin #	Ranked PSEs, most to least vulnerable	DD (%)	Bin #
District 2	0.0	1	District 8	0.0	1
District 6	9.3	1	District 6	19.4	2
District 14	19.7	2	District 14	41.8	5
District 8	22.0	3	District 13	41.9	5
District 13	22.9	3	District 5	46.5	5
District 10	28.8	3	District 12	50.3	6
District 15	29.0	3	District 2	55.4	6
District 11	31.8	4	District 4	56.2	6
District 12	47.7	5	District 9	59.4	6
District 9	50.9	6	District 10	64.9	7
District 4	63.2	7	District 11	66.8	7
District 7	65.7	7	District 15	74.5	8
District 5	66.4	7	District 3	75.9	8
District 1	84.7	9	District 7	87.9	9
District 3	100.0	10	District 1	100.0	10

5.2.1.1 Analysis Using MIRA Analysis Methods

The MIRA analysis methods make it possible to do more than simply view the ordinal ranks of PSEs. Table 5.2 presents a comparison of the two Baseline Runs. Between the five-indicator and nine-indicator Baseline Runs, District 2 dropped from the top rank to the seventh rank. District 6 is the second ranked district in both runs, but its Decision Distance (DD) increased significantly in the nine-indicator run (9.3–19.4%). This indicates that District 6, without changing rank, is less vulnerable under the nine-indicator Baseline Run. Had this been a Final Run, the degree of separation between the top two ranked Problem Set Elements (PSEs) would be part of a broader discussion among decision makers. With a DD of that magnitude, the stakeholders would now have to consider all other aspects of Decision Uncertainty and decide whether the Decision Uncertainty is now at an acceptable level to develop a rationale for choosing the top ranked PSE.

The Binning analysis method is another means that validates the larger separation between the top two ranked PSEs. When ten bins are used in each run, the top ranked District 2 shares Bin 1 with District 6 in the five-indicator Baseline Run, while in the nine-indicator Baseline Run, District 8 resides there alone. The second ranked DDs can also be compared between the two runs, showing that in the nine-indicator Baseline Run, District 8 stands out more distinctively as the top ranked PSE than does District 2 in the five-indicator Baseline Run. When using the Binning method, the smaller the number of bins required to separate the PSEs, the smaller the Decision Uncertainty.

Recall that in the five-indicator Baseline Run (see Fig. 4.2), two indicators, OZONE and NUTRIENT_LOAD, have the largest Indicator Contributions (ICs) of any of the five indicators, contributing the most to the Total Criteria Sum (i.e., Total CS = the sum of the CSs from all the PSEs). Figure 5.1 presents the ICs for each of the nine indicators in the nine-indicator Baseline Run. If all nine indicators contributed equally to the Total CS, each would contribute 11.1% (100%/9). With the addition of four more air quality indicators, OZONE and NUTRIENT_LOAD remain relatively high contributors (nearly 14% each), but other air quality indicators such as AIR_TOX_RESP (14%) are also large contributors.

To assess the contribution of any single indicator on a PSE, Indicator Drivers (IDs) are used. The Gaian Committee decides to compare District 2, the top ranked PSE in the five-indicator Baseline Run, with District 8, the top ranked PSE in the nine-indicator Baseline Run. This comparison is shown in Fig. 5.2.

As in the five-indicator Baseline Run (see Fig. 4.3), when the Indicator Drivers (IDs) are used to examine District 2, in the nine-indicator Baseline Run, LEAD and POPGROWTH contribute the most to District 2's CS, even as District 2 loses its top ranked position. In the nine-indicator Baseline Run, District 8 has six of its nine indicators having higher IDs than those for District 2. Recall that in the five-indicator Baseline Run, District 8's largest IDs came from OZONE and NUTRIENT_ LOAD. These also contribute highly to its CS in the new Baseline Run along with relatively high contributions from the four new air quality indicators. In the new PN that includes a more detailed and refined picture of air quality, District 2 is no longer the most vulnerable (highest ranked) district among the 15 districts.

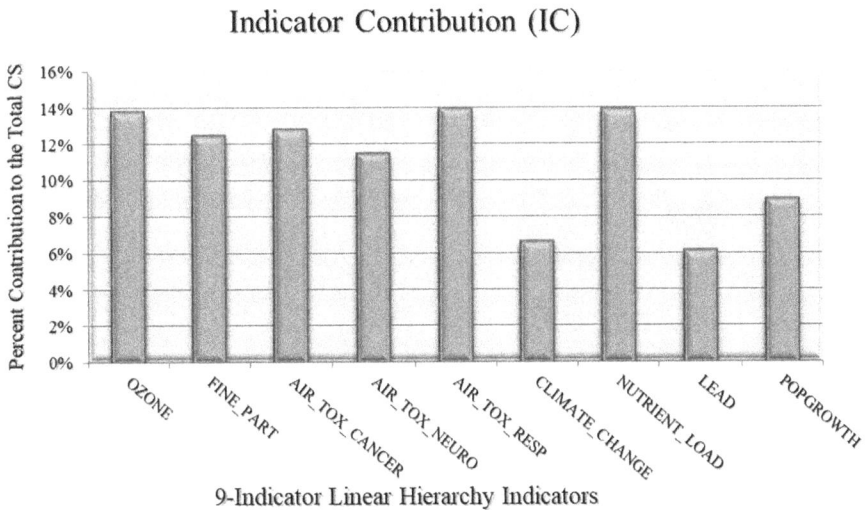

Fig. 5.1 The Indicator Contribution (IC) for the nine-indicator linear Hierarchy Baseline Run

Indicator Drivers (ID)

■ District 8
▨ District 2

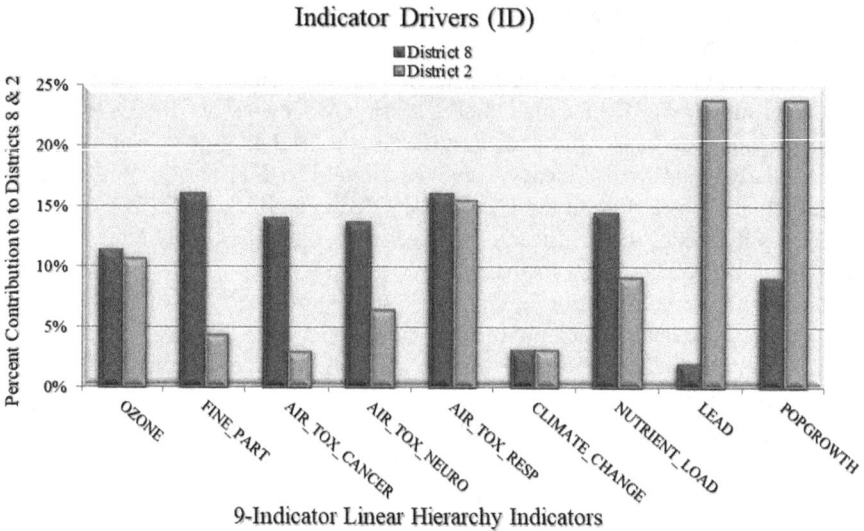

9-Indicator Linear Hierarchy Indicators

Fig. 5.2 Comparison of District 8 and 2's IDs from the nine-indicator linear Hierarchy Baseline Run

5.2.2   Branched Hierarchy

Increasing the size of the linear Hierarchy by adding four indicators is one means to expand the analysis. This method is not improper if the stakeholders' intent is to consider each of the nine indicators as independent concerns. However, if the four additional air quality indicators represent a more in-depth consideration of air quality rather than four additional independent concerns, then the nine-indicator linear Hierarchy construction is flawed. By adding four indicators, which are not meant to be independent, an unintended analytical bias in favor of air quality impacts is now introduced. When each of the five air quality indicators is analyzed as independent considerations, they are placed on an equal footing with the other four areas of interest. This biases the analysis toward air quality concerns by a ratio of 5 to 1 if all indicators contributed the same amount to the total CS. Removing this bias requires constructing a branched Hierarchy.

A branched Hierarchy allows the Gaian Committee to include the four air quality indicators as a single consideration, equal in status to the other four non-air quality indicators. Placing the air quality indicators into a separate branch in the Hierarchy means that stakeholders are choosing to more comprehensively define the impacts from air quality, while maintaining the integrity of an analysis designed to consider five independent environmental concerns. This concept is particularly useful when there are some areas of science more developed than others and stakeholders want to use as much of the available science as possible without introducing analytical bias.

In a branched Hierarchy, there are two types of indicators, *Composite Indicators* (CIs) and Terminal Indicators (TIs). CIs are used to organize those sets of indicators (indicator clusters), which form the branches of the Hierarchy. Each branch defines a specific area of stakeholder concern that requires more than one indicator to describe it. Each CI cluster can include both TIs and other CIs, depending on the complexity of the Hierarchy. While TIs are directly associated with CM values, hierarchically, CIs are one level removed from TIs. The value of the CI is the sum of the products of the TIs, in the cluster, and their Preference weights (a linear weighted sum). The far-left column, or primary level, of a branched Hierarchy is itself a CI from which a Criteria Sum is calculated for each Problem Set Element.

Composite Indicators

Composite Indicators (CIs) are used in a branched Hierarchy and consist of one or more Terminal Indicators and/or other CIs.

The Gaian Committee's decision to maintain five major areas of concern, but also to expand the definition of air quality, results in the creation of a CI, "air quality." The air quality CI replaces the OZONE TI at the primary level of the nine-indicator linear Hierarchy and creates a second level in the Hierarchy. In the second level of the Hierarchy, the five TIs for air quality are ozone, fine particles, carcinogenic impacts from air toxics, respiratory impacts from air toxics and neurological impacts from air toxics (see Fig. 5.3).

5.2.2.1 Preferencing in a Branched Hierarchy

In a branched Hierarchy, stakeholders perform Preferencing within each CI cluster. Organizing these related indicators using CI clusters allows the indicators, within each cluster, to be meaningfully traded-off. Starting with the far-right cluster of each branch (there are as many branches as there are CIs at the primary level), a fractional weight between 0.0 and 1.0 is assigned to the indicators within each cluster with the constraint that the sum of all indicator weights within the cluster must equal 1.0 or 100%. This process is applied to all clusters in the Hierarchy until every indicator in the Hierarchy has been assigned a Preference weight.

The Equal Preference Value Set used in the Baseline Run for the nine-indicator branched Hierarchy is presented in Fig. 5.3.

For the Gaia example, the second level of the Hierarchy consists of only the TIs for air quality. Therefore, the TIs[1] are multiplied by 0.2 (100%/5), the Preference weights shown in the air quality portion of Fig. 5.3, and summed to equal the value of the air quality CI (i.e., linear weighted sum). Moving to the primary level of the Hierarchy, for each PSE, the Gaia stakeholders perform a similar linear weighted sum with the Preference weights shown in Fig. 5.3. At this level, stakeholders multiply the value of the air quality CI by the Preference weight of 0.2 and the values of each of the TIs by the 0.2 Preference weights, which, once summed, produces the CS for each PSE.

---

[1] TIs are indexed CMs. Therefore, the values of the TI lie on the 1 to 8 decision scale.

9-Indicator Branched Hierarchy		
Primary Level	**Weights**	
Air Quality	0.20	
CLIMATE_CHANGE	0.20	
NUTRIENT_LOAD	0.20	
LEAD	0.20	
POPGROWTH	0.20	
Total =	1.00	

	Air Quality	**Weights**
	OZONE	0.20
	FINE_PART	0.20
	AIR_TOX_CANCER	0.20
	AIR_TOX_NEURO	0.20
	AIR_TOX_RESP	0.20
	Total =	1.00

For each PSE, the value, in decision units, of each CI in a branched Hierarchy is calculated as follows:

$$CI_{kl} = \sum_{i=1}^{N} p_{i,l+1,k} I_{i,l+1,k} \tag{5.1}$$

Where:

$CI_{kl} \equiv$ the value, in decision units, of the kth CI at the lth level in the Hierarchy

$I_{i,l+1,k} \equiv$ the value, in decision units, of the ith indicator (either Composite or Terminal) located at the $l + 1$ level of the Hierarchy and associated with the kth CI which contains N indicators

$p_{i,l+1,k} \equiv$ the Preference weight assigned to $I_{i,l+1,k}$

The process represented by Eq. 5.1 reduces all the CIs located in branches at the second and higher levels of the Hierarchy to a single CI, located at the primary level of the Hierarchy. This process is repeated until all primary-level CIs have been resolved. The last linear summation is to calculate the CS (itself a CI), which represents all the indicators in the Hierarchy, and is calculated for each PSE. Once completed, the PSEs can be ranked by their CSs.

The CS, for each PSE, is calculated as follows:

$$CS_j = \sum_{i=1}^{N} p_{i,j} I_{i,j,l=1} \tag{5.2}$$

Where:

$CS_j \equiv$ the value of the CS, in decision units, of the jth PSE

$I_{i,j,l=1} \equiv$ the value of the ith indicator, in decision units (either Composite or Terminal), of the jth PSE located at the primary level of the Hierarchy

$p_{i,j} \equiv$ the Preference weight assigned to i^{th} indicator of the jth PSE at the primary level of the Hierarchy

5.2.2.2 Importance of Indicator Location in the Hierarchy

Hierarchy location and the number of indicators within that indicator's cluster influence the IC of any single indicator. The IC's value considers all components of an indicator: data, Expert Judgment Set (EJS), Value Set (VS), and Hierarchy location. Although the use of a branched Hierarchy is necessary to compensate for inadvertent biasing, creation of a branched Hierarchy introduces another influence on the IC of the indicators in those branches that is important for stakeholders to understand.

If the IC is a single indicator's contribution to the Total CS, the position-based Indicator Contribution (IC_{pb}) is a single indicator's contribution to the Total CS due solely to its location in the Hierarchy. The IC_{pb} is independent of any data, EJS, or VS and, therefore, its magnitude is determined exclusively by where stakeholders place that indicator in the branched Hierarchy. Conceptually, it is useful to think of IC_{pb} as one of three terms that influences an indicator's IC. An indicator's IC starts with a value of IC_{pb}, which then either increases or decreases once the value of its CM, Indexing, and Preferencing are considered. The following is a conceptual representation of the relationship between IC and IC_{pb}:

$$IC\left(\% \text{ of total CS}\right) = IC_{pb} + \left(EJS_{PSE} - EJS_{same\ data}\right) + \left(VS_{EJS} - VS_{Equal}\right)$$

Where:

$(EJS_{PSE} - EJS_{same\ data})$ = percent contribution to the PSEs' CS from indicators as indexed using the expert determined, or uniform, EJS minus the percent contribution that the indicators would have if they all had the same value

$(VS_{EJS} - VS_{Equal})$ = the influence that the indicator's Preference weight would have on the second term minus what the influence that the Preference weight from an Equal Preference VS would have on the second term, where the EJS is not necessarily uniform, the VS is not necessarily Equal Preferencing, and the sum of the second and third terms can be either positive or negative.

The IC_{pb} is calculated, using Eq. 4.3, by assigning that each indicator has the same value (rendering the EJS irrelevant) in the Hierarchy and applying an Equal Preference VS. Doing this removes any influence from the differences among indicators that are due to their data (as indexed by the EJS) and the VS used. This causes the last two terms in our conceptual equation above to both equal zero. By calculating IC_{pb} for each indicator in a Hierarchy, an analyst can determine the contribution to the Total CS, for each of the indicators, that results entirely from its location in the Hierarchy. This means that an indicator that is located at the primary level of the Hierarchy (holding all else constant) will have an advantage over an indicator at a higher level in the Hierarchy, prior to any consideration for that indicator's data (Indexing) or Preferencing. There are times when an indicator's position in the Hierarchy plays a dominant role in determining its IC.

Figure 5.4 presents a comparison of both ICs and IC_{pb}s from the Baseline Runs of the Gaian Committee's nine-indicator linear and branched Hierarchies.

For both Baseline Runs represented in Fig. 5.4, the percentage to the right of each indicator is its IC_{pb} value, while the percentage above the indicator represents its IC. In the linear Hierarchy, with nine indicators, the IC_{pb} value for each indicator is 11.1%. In contrast, in the branched Hierarchy (right panel of Fig. 5.4), the IC_{pb} for an indicator at the first level of the Hierarchy is 20%, while the IC_{pb} for an indicator at the second level is 4.0%.

Comparing primary-level indicators between the linear and branched Hierarchies, the difference in IC_{pb} of 8.9% (20% − 11.1%) is due solely to the difference in the number of indicators at the primary level between the two Hierarchies (nine vs. five). However, for the indicators that moved from the primary level of the linear Hierarchy to the secondary level of the branched Hierarchy, the change in their IC_{pb} of 7.1% (11.1% − 4.0%) is due to both their change in Hierarchy level and the number of indicators in their cluster. For example, the IC for the OZONE indicator resulting from the Baseline Run of the nine-indicator linear Hierarchy is 13.7%,

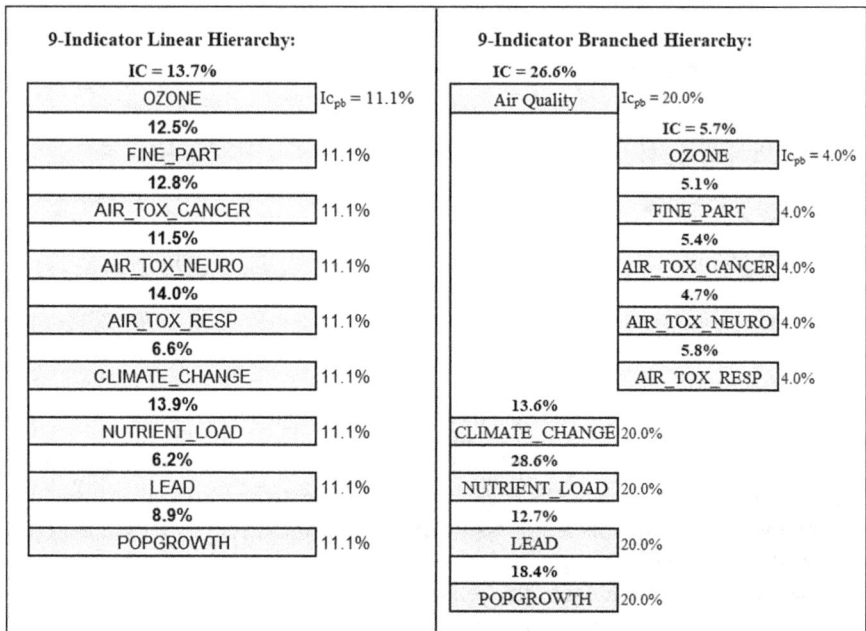

Fig. 5.4 Comparison of linear and branched Hierarchies' IC and IC_{pb}

while it is 5.7% in the equivalent run of the nine-indicator branched Hierarchy. The difference in the OZONE IC of 8.1% is due to the combination of the change in its location (from an IC_{pb} of 11.1% to 4.0%) and the difference in the relative values among the indicators. However, since the relative values among the indicators represented in Fig. 5.4 are the same in both Hierarchies, the change in the ICs is due to the change in IC_{pb}s.

When an indicator's IC is greater than its IC_{pb} value, that indicator has an increased contribution to the Total CS compared with other indicators whose ICs are less than their IC_{pb}. For example, for the branched Hierarchy, the OZONE IC is 5.7% while the OZONE IC_{pb} is 4.0%. The difference of 1.7% informs the stakeholders that, based on a Uniform Indexing EJS and Equal Preference VS, OZONE contributes more to the Total CS than it would have due solely to its location. This is in contrast to the CLIMATE_CHANGE indicator whose difference in IC and IC_{pb} (13.6% − 20% = −6.4%) shows that it contributes less to the Total CS than would have been expected if only examining the contribution due to its location.

The sum of IC percentages and IC_{pb} percentages at the primary level must each independently add to 100%. Therefore, if there is an indicator with an IC greater than its IC_{pb}, there will be at least one other indicator with an IC less than its IC_{pb}. The indicator's IC_{pb} tells the stakeholders the magnitude of the relative importance that they have assigned to that indicator based solely on how they decided to construct their Hierarchy. Using this information, the stakeholders can decide whether they may have inadvertently given any of the indicators in the Hierarchy an importance that is inconsistent with their PN. With such information, stakeholders learn the role of Hierarchy construction on the PSE rankings. If this construction proves to be problematic relative to their intended PN, they can change the Hierarchy.

5.2.3 Refining the PN

Based on more in-depth discussions to ensure that important factors are included in the analysis, the Gaian Committee also decides to revise the Hierarchy to add ten more CNs. The Gaian Committee knows that the impacts from fine particulate matter and acid rain (one of the newly added CNs) include a variety of human health and ecological concerns. As a result, they decide to revise the CNs for fine particulate matter and acid rain by creating another CN for each pollutant. By doing so, stakeholders accomplish two things: (1) more specifically defining different aspects of fine particulate matter or acid rain and (2) controlling the relative importance of each of those aspects. This flexibility and control is not possible if all the effects from the fine particulate matter or acid rain CN are represented by a single CM.

The Gaia stakeholders construct four distinct indicators, two for fine particulate matter and two for acid rain. The basis for constructing two indicators for fine particulate matter is that exposure to this pollutant causes two distinctly different effects: (1) premature deaths primarily among those that suffer from pre-existing cardiovascular and respiratory disease and (2) decreased lung function, aggravated asthma, and other non-lethal respiratory problems in the general population. The basis for constructing two indicators for acid rain comes from its differential impacts on trees versus aquatic species. For acid rain, the first CN pertains to the depletion of soil nutrients that in turn has an adverse impact on forest viability, while the second CN pertains to the impact that acid rain has on the toxicity to fish and other aquatic organisms. The stakeholders decide that two CNs are needed because (1) acid rain facilitates the depletion of nutrients and minerals in the soil necessary for trees and other vegetation and (2) acid rain also causes leaching of aluminum from soils into lakes, streams, and wetlands, which adversely impacts aquatic organisms.

All of the newly added CNs are described in Table 5.3. There are now 21 CNs in the Gaian analysis, which the Gaian Committee decides to group into six independent considerations.

This major revision to the Hierarchy is based on information gained from the previous iterations plus a stakeholder desire to more comprehensively evaluate the environmental condition of the Gaia districts. Having learned the importance of the linear versus branched Hierarchy construction, the Gaian Committee decides to group 19 of the 21 CNs into 4 general categories, 3 based on their environmental media (air, land, water) and 1 that represents future concerns. The two remaining independent CNs represent two health impacts whose route of exposure does not fit into any of the other four categories being considered. The resultant Categorical Hierarchy, with 21 indicators, is shown in Fig. 5.5. The "R" in Fig. 5.5 indicates that IBI is a Reverse Indicator. This is an indicator where the largest CM value in the range indicates the best or least vulnerable condition and, therefore, is equated to an indexed value of 1 and the smallest value in the range is equated to 8. This is the reverse of other indicators where large values signal poorer condition or less attractiveness.

By refining the particulate matter and acid rain CNs (now four separately defined CNs), the Gaia stakeholders are more satisfied that this change allows them to assess the environmental issues consistent with their current thinking regarding the PN. The descriptions for the new particulate matter and acid rain CMs and the CMs corresponding to the other ten new CNs are provided in Table A.3 of the Appendix. Once these CMs are indexed, their indicators are organized into the 21-indicator Categorical Hierarchy shown in Fig. 5.5.

Although four CNs typically call for four CMs, there is not necessarily a one-to-one relationship between a CN and a CM.[2] It is possible to use the same CM to reflect multiple CNs by Indexing the CM values differently for each associated CN. Differential Indexing allows for the creation of multiple TIs from a single CM. The Gaian Committee decides that even though the data for particulate matter and acid rain are limited, they are suitable for the four new indicators. As they are still constructing the Hierarchy and working toward a Baseline Run, they postpone establishing a non-uniform EJS for these CMs.

---

[2] In a US EPA study using MIRA, multiple indicators were also constructed from the same CM [2].

Table 5.3 Description of the 21 CNs contained in the 21-indicator Categorical Hierarchy

Criteria Name	Criteria Narrative
OZONE	The respiratory impacts from ozone on human health in each district
FINE_PART-PD	The potential for excess deaths in populations having cardiovascular and respiratory disease in the districts that result from the inhalation of fine particulates
FINE_PART-R	The potential for non-lethal human respiratory effects from the inhalation of fine particles
AIR_TOXIC_CANCER	The cancer risk to individuals based on the inhalation of air toxic pollutants that are emitted in the districts
AIR_TOX_RESP	The risk that an individual will have a respiratory health problem based on the inhalation of air toxic pollutants
AIR_TOX_NEURO	The risk that an individual will contract a neurologically based health problem based on the inhalation of air toxic pollutants that are emitted in the districts
RADON	The potential for lung cancer from the inhalation of radon gas inside an individual's domicile
CLIMATE_CHANGE	The potential for increased flooding within a district due to the increase in rainfall that will result from climate change. Since accurate estimates of increased rainfall are not available, the vulnerability of a district can be used as a measure of flooding potential.
ACID_RAIN-F	Deleterious impact of acid rain on trees/forests due to the soil leaching of nutrients and minerals
ACID_RAIN-AS	Deleterious impact of acid rain on aquatic species, particularly from the leaching of aluminum from soils into lakes and streams
MERCURY	Deleterious impact of mercury on soil microbes, which are important to the terrestrial food chain
INSECTICIDES	Deleterious impact to waterways due to insecticides in rain runoff
OZONE_VEGETATION	The impact that ozone has on trees within each of the districts
NUTRIENT_LOAD	Deleterious impact of fertilizers on contamination of surface waters within the district
IMPERVCOV	Deleterious impact to the management of flooding and water percolation to ground water due to concrete surfaces and other impervious surfaces that result from development
COAL_ASH	Deleterious impacts from landfills containing coal ash, which leak toxic contaminants (including arsenic, lead, mercury, and selenium, as well as aluminum, barium, boron and chlorine) into the ground water of each of the districts
IBI	The Index of Biotic Integrity, a measure of the biological health of macroinvertebrates in the streams within each district. Macroinvertebrates are the bottom of the food chain for fish in streams. Biologists use the IBI as an indicator of a stream's ecosystem integrity.
LEAD	Impact of lead exposure to children's neurological health
POPGROWTH	The 25-year forecasted population growth for a district
URBGROWTH	The 25-year forecasted urban population growth for a district
MENTAL_HEALTH	Deleterious impact from environmental stresses or pollutant exposure on mental health such as depression, aggression and negative cognition

Fig. 5.5 The Gaian
21-indicator Categorical
Hierarchy

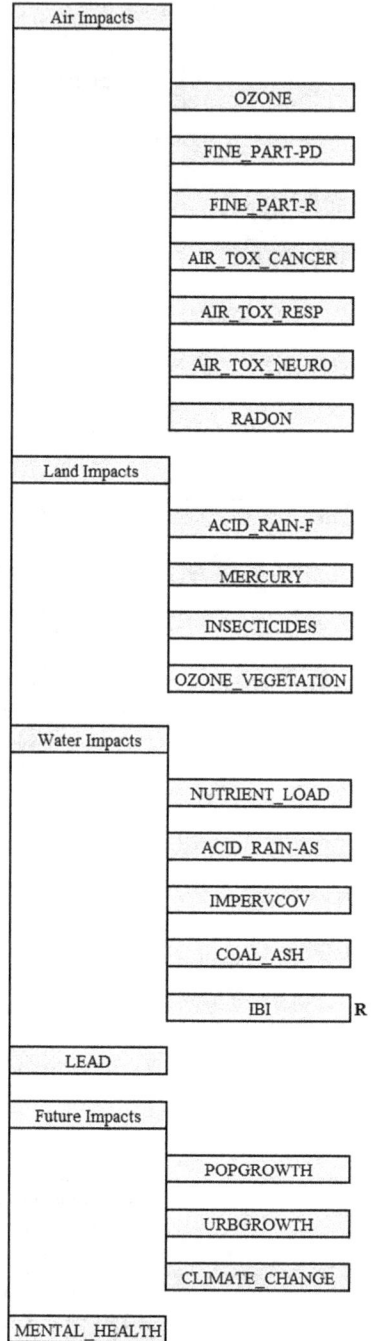

The data for the ten additional CMs[3] in the 21-indicator Categorical Hierarchy are found in Table A.4 in the Appendix. The Uniform Indexing EJS for this Baseline Run is presented in Table A.5 in the Appendix.

5.2.4 Complex Branched Hierarchies

5.2.4.1 Category-Based Hierarchy

The change from the linear to the branched Hierarchy is not the only consideration for stakeholders constructing a MIRA Hierarchy. Throughout the process, stakeholders not only discuss the potential for new CNs but also whether the new CNs are organized in a manner that allows for the kinds of trade-offs that must be made when determining the relative importance for the indicators in the Final Hierarchy (Requisite Step 7).

Both the previous nine-indicator Hierarchy and the expanded 21-indicator Hierarchy are Categorical Hierarchies. When the Baseline Runs are made for these two Hierarchies, the 12 new CNs (grouped primarily by environmental media) result in a change of ranks among the PSEs. The top four ranked PSEs from these two Baseline Runs are presented in Table 5.4.

Within the top four ranked PSEs, Districts 6 and 2 switch places between the two Baseline Runs. Additionally, there is a large corresponding change in DD. The nine-indicator Baseline Run indicates that Districts 2 and 6 are similar in vulnerability with a second rank DD of 1.9%. In the 21-indicator Baseline Run, this DD is 20.2%, indicating that District 2 is considerably more vulnerable than District 6.

When nine indicators are used, the small DD of the second rank may make it more difficult for the Gaian Committee to clearly select only the top ranked District 6 for additional funding. This is because some stakeholders could compellingly argue that the small separation of 1.9% makes Districts 6 and 2 indistinguishable.

While DD is a specific analysis measure used to assess the separation between the top ranked PSE and any other PSE, a related metric, ΔDD_{adj}, can be used to assess the separation between any two adjacent PSEs. The separation between Districts 2 and 8 in the nine-indicator Hierarchy, ΔDD_{adj}, is 9.7% (11.6% − 1.9%). Whether 9.7% is a small or large difference in DD is a subjective assessment. Stakeholders will conduct a more detailed evaluation through iteration to develop their understanding sufficiently to make the assessment. Based on the nine-indicator Baseline Run, some stakeholders might think it reasonable to advocate for provid-

---

[3] The discussion to identify CMs for the corresponding CNs is an important one. For this Gaia example, this discussion is simplified. In decision analytic approaches, this is the selection of decision criteria and indicators, which commonly settles on criteria and indicators used by previous analyses and, hence, are readily available. In an open solution approach such as MIRA, ensuring a good match between the CN and the CM includes stakeholders (and experts) asking each other whether the CM is the best available measure or gauge of the CN. When necessary, CMs are custom-constructed to ensure good representation of the CNs.

Table 5.4 Comparison of the Baseline Run results for the top four ranked PSEs

Nine-indicator Categorical Hierarchy		21-indicator Categorical Hierarchy	
First four ranked PSEs, most to least vulnerable	DD (%)	First four ranked PSEs, most to least vulnerable	DD (%)
District 6	0.0	District 2	0.0
District 2	1.9	District 6	20.2
District 8	11.6	District 8	25.0
District 14	23.7	District 12	38.7

ing Districts 8, 2, and 6 with additional funding, since the ΔDD_{adj} between Districts 2 and 8 could be considered small enough to pressure the Gaian Committee to also include District 8. However, in the 21-indicator Baseline Run, with a DD of the second rank equal to 20.2%, the Gaian Committee can more confidently choose only District 2, as it is more clearly distinct from the other districts. When more indicators are included, DD does not always increase. However, for the Gaia example, DD has increased and, hence, the inclusion of 12 more indicators has reduced Decision Uncertainty.

Although there is nothing improper with categorizing indicators, certain forms of categorization present problems for stakeholders in Step 7 (Apply values). In this step, stakeholders are asked to make trade-offs to assign relative importance across the indicators. The Gaian Committee members discover that, whether the nine-indicator or 21-indicator version of the Categorical Hierarchy is used, they cannot meaningfully trade-off the indicators. In making trade-offs, stakeholders answer the question "Which criteria do I think is more important and by how much?" For example, when this question is posed to the Gaian Committee between the air impact and land impact indicators, the members cannot make this trade-off, since it would require indicating whether, and by how much, air is more important than land. To address this problem, they need a different type of Hierarchy. Therefore, although building the Hierarchy is Step 4 and assessing Preferences is Step 7, there is typically a feedback or iteration required to re-examine the Hierarchy as stakeholders proceed through the MIRA process.

5.2.4.2   Functional-Based Branched Hierarchy

While categorizing Criteria Narratives is often an easy, and seemingly natural, way to construct a Hierarchy, this typically presents a dilemma for stakeholders when they begin to discuss and evaluate the relative Preferences for each of these indicators that result from the CNs. The Gaian Committee finds itself in a similar dilemma and chooses to return to Step 4 to re-think the construction of the Hierarchy.

A functional branched Hierarchy addresses the problem of ensuring that the PN is appropriately defined with meaningful trade-offs between and among the CNs.

Re-visiting their PN and Hierarchy, the Gaian Committee re-examines their real concerns and re-articulates them as human health, ecological integrity[4] and future impacts. They find that trading-off human health and ecological integrity at the primary level is possible as their responsibilities and authorities require that when resources are limited, they must focus on directly mitigating adverse impacts to human health. For the Gaian Committee, trading-off human health or ecological integrity against future impacts represents a meaningful trade-off between present and future effects. Depending on the PN, there are many ways to organize a functional Hierarchy. For example, the Gaian Committee chooses to consider only MERCURY's ecological impacts but another group of stakeholders could have decided to also consider MERCURY's human health effects or its human health effects exclusively.

The modified Gaian functional Hierarchy using the current suite of 21 indicators is shown in Fig. 5.6. The 21 indicators of the functional Hierarchy are organized by how each of these indicators differentially affects the Gaia environment. In a functional Hierarchy, rather than trading-off different facets of the ecological system, stakeholders trade-off among the potential adverse impacts.

For the Gaian Committee, the 21-indicator functional branched Hierarchy, represented in Fig. 5.6, presents a rational reflection of their refined PN. This is a Hierarchy that defines the problem as having three independent primary-level general considerations (CIs): human health, ecological integrity, and future impacts. Human health includes pollutants that affect human health regardless of whether their exposure comes via the air, water, or land. Ecological impacts are those that result from air pollution, water pollution, and land use, which directly and indirectly affect human habitation, human health, and ecosystems.

The Gaian Committee's refined PN is:

Identify those districts whose local environments cause, as well as exhibit, adverse impacts to human health and ecological integrity. Human activities (including land use) that contribute to increased pollution to air, water and land are to be considered. The Gaian analysis will include the human health and ecological impacts of pollution from major anthropogenic sources (including the forecasted growth of these sources) in Gaia. The local impacts of climate change on human well-being and ecological integrity are also to be considered.

The MIRA process facilitates the creation and experimentation with multiple Hierarchies, until the final Hierarchy is settled upon, which reduces Decision Uncertainty. The construction of multiple Hierarchies is an important transdisciplinary learning component of MIRA's iterative process, which is necessary for good policy decision making.

---

[4]The term ecological integrity is used as a general term in this example to refer to those impacts that pertain to ecological systems and ecological species and to distinguish these impacts from direct human health impacts.

Fig. 5.6 The Gaian
21-indicator functional
Hierarchy

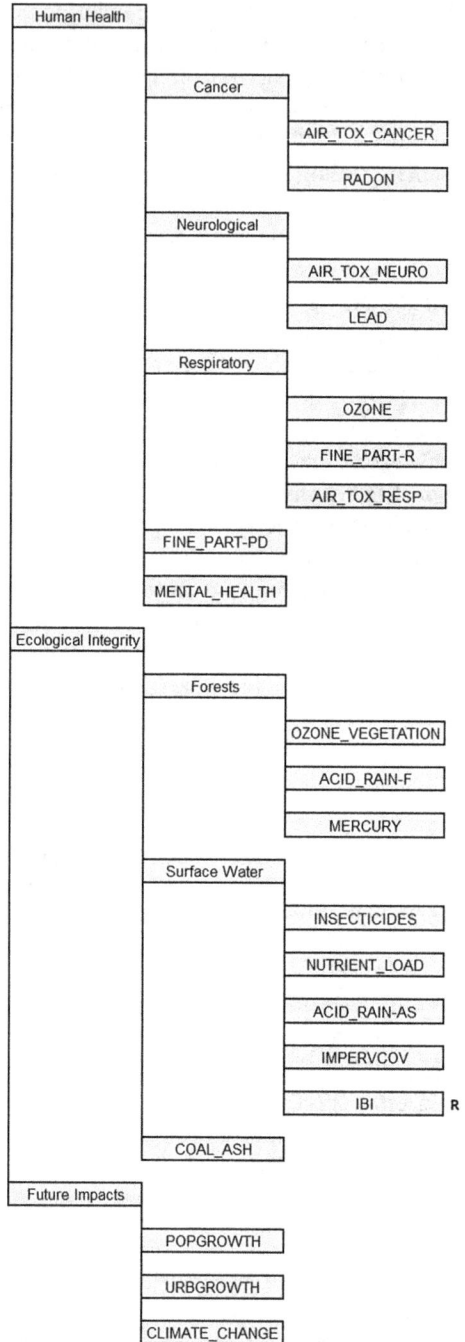

5.3   Weighted Indicators

For any multi-criteria analysis, the discussion, selection, and construction of indicators is important because the suite of selected indicators should represent the closest agreement possible with the stakeholders' PN. A close agreement minimizes Decision Uncertainty. For this reason, stakeholders may also want to consider the use of weighted indicators. When an indicator is selected that places some stress on a resource (i.e., the recipient of the adverse impact whether human or ecological), stakeholders may want to consider the magnitude of the insult (e.g., does this level of ozone impact 50,000 or 100,000 people). This can be accomplished by weighting (i.e., multiplying) the indicator's value by the magnitude of the resource it is affecting.

Particularly when a pollutant differentially impacts various segments of a resource, weighting can ensure that the indicators accurately represent the problem's effect on the resource. For example, while high ozone concentrations are not good for anyone, they selectively and more adversely affect the young, the elderly and those with compromised respiratory systems. Stakeholders concerned about the extra burden these pollutants exert on these sensitive populations may decide to construct indicators designed specifically to account for such differences. Public health officials know that any given ozone concentration has different health impacts on young and otherwise healthy adults as compared with children and the elderly. The elderly and children with limited or no access to health care have even more difficulty coping with air pollution. Stakeholders can define the CN as the potential impact that a district's ozone concentrations have on a particularly sensitive population, such as poor children or the elderly. An indicator of this type can be constructed by using both the ozone concentrations and the number of poor children (and/or elderly) in that district. If study results were available that predicted exposure to that cohort, then utilizing those results would produce a far better CM. However, without such information, experts may decide that the differential impacts to poor children cannot be considered, or they might decide that an indicator as simple as the product of a single-valued ozone indicator and the number of children will suffice. In this latter situation, greater uncertainty is introduced because the CM is not as aligned with the CN as the stakeholders would have preferred.

When weighted indicators are used, a transformation is necessary to keep them within the decision scale, which allows them to remain comparable to other indicators. This transformation is described in more detail in the Appendix.

The Gaian Committee decides to weight eight indicators (OZONE, AIR_TOX_CANCER, AIR_TOX_NEURO, AIR_TOX_RESP, FINE_PART, LEAD, OZONE_VEGETATION, and NUTRIENT_LOAD) by the magnitudes of their adversely affected resources. For each of the human health air pollutant-related indicators, the Committee decides to weight them by total population and low-income children. The LEAD indicator is weighted by low-income children only. These children are disproportionately affected by lead ingestion since they are more likely to occupy unremediated older homes where lead paint was commonly used. The Gaian

Committee weights ozone for its ecological impact because of its disproportionate effect on certain tree species. When considering the ozone vegetation indicator, the Gaian Committee decides to weight it by the percentage of a district's area occupied by ozone-sensitive trees. Finally, the Committee decides to weight the indicator for nitrogen by a district's stream density. They reasoned that the magnitude of the nutrient runoff problem, from the application of nitrogen fertilizers, depends on both the amount of nitrogen applied and the number of streams into which nitrogen can migrate.

Weighting these five indicators increases the number of indicators in the Hierarchy to a total of 26. The 26-indicator weighted functional Hierarchy can be found in the Appendix as Fig. A.1. Having changed the Hierarchy again, the Committee performs a new Baseline Run. The effect of using weighted indicators compared with unweighted indicators on the ranking of the PSEs is shown in Table 5.5.

In the unweighted Baseline Run, District 6 is ranked #2, behind District 8 and separated by a DD of 15.8%. When weighted indicators are added, District 6 moves from the second rank to the top rank but only nominally, as District 2 is a close second with a DD of 0.9%. Weighting the indicators has made the sole selection of the top ranked district significantly less certain, as is reflected in the DD. Therefore, if the Gaian Committee is convinced that inclusion of weighted indicators is more reflective of their PN, then their decision has been made considerably more difficult. This is not a reason to choose not to weight the indicators, but these kinds of decisions will affect how distinctive each of the PSEs appears in the analysis and change the stakeholders' estimate of Decision Uncertainty.

Table 5.5 Comparison of ranked PSEs in the unweighted versus the weighted functional Hierarchy

Unweighted Hierarchy		Weighted Hierarchy	
Ranked PSEs, most to least vulnerable	DD (%)	Ranked PSEs, most to least vulnerable	DD (%)
District 8	0.0	District 6	0.0
District 6	15.8	District 2	0.9
District 2	19.3	District 8	22.2
District 11	31.4	District 11	29.8
District 4	33.4	District 12	35.1
District 12	36.7	District 4	44.0
District 13	37.9	District 13	46.3
District 3	44.1	District 14	48.8
District 10	47.4	District 10	56.2
District 14	53.5	District 9	57.0
District 9	61.6	District 3	59.3
District 5	70.0	District 5	64.3
District 7	80.6	District 7	75.6
District 15	86.5	District 15	89.6
District 1	100.0	District 1	100.0

When stakeholders consider the possibility of using weighted indicators in a Hierarchy, they also need to consider whether experts will index the weighted or the unweighted CMs. For example, whether to index an unweighted CM prior to weighting it or after depends on what makes more sense to the experts who are assessing its significance. For example, for oncology professionals, assessing the significance of the air toxic cancer risk CM before it is weighted by population makes more sense because, for them, it is a more commonly used metric than the population-weighted cancer risk. Therefore, stakeholders will need to consider whether to perform the weighting before or after Indexing.

5.4   Beyond Baseline Runs: Iterate to Understand Decision Uncertainty

The Gaian Committee has progressed from its original five-indicator linear Hierarchy to a 26-indicator weighted functional branched Hierarchy, which, for simplicity, will be referred to as the weighted functional Hierarchy. Until now, the Gaian Committee has been occupied with ensuring that the proper Criteria Narratives are included and appropriately reflected in their Hierarchy. To accomplish this, their analysis was simplified by using a Uniform Indexing Expert Judgment Set (EJS) and an Equal Preference Value Set (VS). With the understanding that the Hierarchy can be changed again, once the stakeholders are satisfied with a Hierarchy and its Baseline Run has been completed, they proceed to customize their analysis. The customization begins with the subject matter experts generating an initial non-uniform EJS.

5.4.1   Expert Judgment Sets

The Baseline Run for the weighted functional Hierarchy is now the benchmark from which to compare the impacts of different Expert Judgment Sets. For a PN-specific assessment of significance, progressing to a contextually informed EJS requires that the subject matter experts develop a corporate understanding across all CMs. Experts in their single disciplines generally have little or no sense of the relative significance of their data when compared with data from the many other disciplines represented in the analysis. Therefore, a trans-disciplinary discussion among experts ensures that the approach to determining significance is consistently applied in the Indexing process.

Suppose that the Gaian Committee initially hosts a small group of subject matter experts to discuss the effects to the ecosystem that result from mercury in soil (MERCURY), the Index of Biological Integrity (IBI), and urban growth (URBGROWTH). Before these experts can appropriately interact, there is a problem pertaining to the disparity between the available data for mercury and the stakeholders' CN that needs to be addressed. This is a fairly common problem when conducting environmental or other analyses. At the time when the Gaian Committee decided to include a mercury CN pertaining to impacts on soil microbes, they did not understand that the kind of soil data needed is not readily available. The only mercury (Hg) data available to construct the MERCURY CM is modeled atmospheric Hg deposition (in units of milligrams per hectare per year, mg/ha/yr) at locations throughout each of the districts. Since the CN for mercury requires the use of Hg concentrations in soil, the experts need to estimate what soil concentrations of Hg will result from Hg deposition. Fortunately, the soil experts acquire the physico-chemical properties of the soil across all districts at the locations where the Hg deposition estimates were made. This allows the experts to establish a functional equivalence between Hg deposition and Hg soil concentrations. This equivalence is reflected in the EJS for the MERCURY deposition data. Had it not been feasible to establish this functional equivalence, the Gaia stakeholders would have had to consider whether to change their mercury CN or remove it from consideration.

The mercury subject matter expert knows that a concentration of 0.07–0.3 mg of mercury per kilogram of soil (mg/kg) harms the soil microbes, which corresponds to a range of approximately 0.1–2.0 mg/ha/yr of Hg deposition. This reflects the functional equivalence established by the experts. The Hg deposition data in Gaia ranges from 0.198 to 2.3 mg/ha/yr. The initial response of the subject matter expert is to move the Gaia districts' lowest CM to an index value of 7 or 8, reasoning that a value of 0.198 mg/ha/yr is twice as high as the concentration of Hg that begins to adversely impact soil microbes. To the mercury experts, 0.198 mg/ha/yr is a highly significant value. Next the IBI subject matter expert considers how she will index her data. As an expert she knows that an IBI non-dimensional score of 100 is a pristine water body and that IBI scores in the 70s are also indications of good water quality. However, once the IBI score drops below 30, a water body is considered in very poor condition. Furthermore, it is her experience that as industry and development encroach on previously pristine water bodies, IBI experts start to become concerned when IBI scores are in the 40s.

Before either expert establishes their Indexing for each CM, they confer to ensure that they are being consistent in how they assign significance. As these two subject matter experts explain their rationale for assigning significance for mercury deposition and IBI, the two experts conclude that a mercury level of 0.2 mg/ha/yr should be considered ecologically similar in significance to an IBI score of 42. In this interaction between these experts, the focus on the adverse ecological impacts is checked against the PN to ensure that these CMs, and the way in which the experts are assessing significance, are consistent with that PN. The relative Indexing of MERCURY and IBI provides a good example of why any approach that defaults to Uniform Indexing can produce misleading results because it ignores the contribution of experts.

Recall that in the Hierarchy discussion, the Gaian Committee could have placed the mercury CM in the human health cluster, had they wanted to consider mercury in that way. However, the Gaian Committee constructed their Hierarchy with MERCURY as an ecological integrity indicator. For this reason, in the Indexing step, experts assess mercury deposition as an ecological rather than human health indicator. The urban growth subject matter expert establishes the relative significance of her data in a process like the one used by the mercury and IBI experts. She knows that less than 5% growth over 25 years is not considered problematic, but growth rates over 40% are. Therefore, for this PN, the CMs of 0.198 mg/ha/yr, 42, and 44%, for mercury deposition, IBI, and urban growth, respectively, are set at the index value of approximately 5 out of 8. Based on the trans-disciplinary learning achieved by these expert interactions, Table 5.6 shows the resulting Indexing schemes for these three CMs.

As other subject matter experts explain the significance of their CM values and the discussion begins to include all the subject matter experts, it becomes possible to explicitly index all the data used in the analysis. Since Indexing is transparent in MIRA, experts and other stakeholders are free to experiment with different EJSs. Different EJSs can also be constructed when there are disagreements among experts regarding either uncertainty about their judgments or uncertainty about the science. Sensitivity analyses using Indexing alternatives also allow experts to evaluate whether and how sensitive the PSE ranking is to those judgments. When experts do not agree, constructing and presenting alternative EJSs, with the resulting PSE ranks, also helps to inform decision makers whether and when expert disagreements impact the results. The differences in EJSs are part of a comprehensive evaluation of Decision Uncertainty.

Earlier, when the Gaian Committee added two new CNs for fine particulate matter and acid rain, they understood that the particulate matter and the acid rain experts would be Indexing the two CMs differently to create two TIs from one CM. For the fine particulate matter CM, the subject matter expert informs the Gaian Committee that a lower ambient air concentration of fine particulate matter (in microgram per cubic meter, $\mu g/m^3$) triggers respiratory distress across a wider population than the concentration that elicits concerns regarding premature death. The same concentration of fine particles (i.e., an equivalent degree of significance) will produce adverse respiratory impacts before it produces premature death. Therefore, the experts indexed the same particulate matter data with higher index values (i.e., more significant) for respiratory impacts than for premature death. Additionally, if the fine particle CM had not been split into two indicators, it is likely that the subject matter expert would have centered the NAAQS for $PM_{2.5}$ (12 $\mu g/m^3$) on the index value of 5. However, since the impacts from premature death are not considered in the fine particle respiratory indicator, the index value of 5 is equated to a value of 15 $\mu g/m^3$.

Suppose the subject matter experts for acid rain know that the Gaia forest soils have less buffering capacity for sulfur deposition than do the Gaia surface waters. Then, when the same amount of sulfur compounds is deposited (in kilograms per

Table 5.6 First three subject matter experts Indexing data non-uniformly

Index →		1	2	3	4	5	6	7	8
TERMINAL INDICATORS	UNITS								
MERCURY	**mg/ha/ yr**					0.20	0.70	1.20	1.69
IBI	**non-dim**	71.09	65.31	59.54	47.99	42.21	36.43	30.66	24.88
URBGROWTH	**%**		4.36	17.51	30.66	43.81	56.96	83.26	109.56

Table 5.7 Expert Indexing same CM for Two TIs

Index →									
TERMINAL INDICATORS	UNITS	1	2	3	4	5	6	7	8
FINE_PART-R	μg/m³		5.00	8.50	12.00	15.00	23.33	31.67	40.00
FINE_PART-PD	μg/m³	0.00	6.00	12.00	24.60	37.20	49.80	62.40	75.00
ACID_RAIN-F	kg/ha/yr	1.53	5.60	9.66	13.73	17.80	21.87	25.93	30.00
ACID_RAIN-AS	kg/ha/yr	5.00	9.85	14.71	19.56	24.41	29.26	34.12	38.97

hectare per year, kg/ha/yr) across terrestrial and aquatic surfaces, the impacts on aquatic species (ACID_RAIN-AS) are less significant compared to forests impacts (ACID_RAIN-F).[5] The differentially indexed fine particulate matter data and acid rain data are shown in Table 5.7 and reflect the significance of these data as guided by each of their four CNs.

The first non-uniform EJS for the weighted functional Hierarchy, settled on by the Gaian Committee, is presented as Table A.6 in the Appendix. This EJS represents the judgments of the subject matter experts and may well become the EJS used in making the Committee's final decision. The Gaian Committee may reconsider and revise this EJS as other facts or issues arise, including challenges to the experts' judgments from other stakeholders.

One common way that an EJS may change is when new information becomes available. For example, in 2015, the US EPA updated its Ozone National Ambient Air Quality Standard (NAAQS) to 70 parts per billion (ppb), making it more stringent than the 75 ppb Ozone 2008 NAAQS that was used in the first non-uniform EJS. Therefore, an additional EJS that stakeholders may want to examine is one in which the ozone Indexing is based on the 2015 standard. One possible way to index the 2015 versus the 2008 ozone NAAQS is shown in Table 5.8. Centering the value of the ozone NAAQS on a single index value (5 in this case) sets a benchmark by which to compare either of these ozone indicators with other indicators in the analysis.

With all else remaining constant and applying the Equal Preference Value Set, the single change made to the ozone Indexing scheme has little effect on the ranked PSEs. This is shown in Table 5.9. A different EJS and VS may produce more significant changes in the results; this is an example where disputes among experts or stakeholders can easily be examined and resolved using MIRA.

For the Gaian analysis, the change in ozone Indexing did not change the results in a significant way because of (1) the relatively small impact that the change has on the indexed ozone values for all the 15 PSEs and (2) the location of the ozone indicators in the Hierarchy (the third level).

If the Gaian Committee had been troubled by the lack of sensitivity to the changed EJS, they could have revised the Hierarchy, which would reflect a different PN. Instead, the Gaian Committee decides that the insignificant change in both

---

[5] For this illustration, presume that the difference in adverse impacts to forests and aquatic species is due solely to the difference in exposure to sulfur compounds because of soil vs. aquatic buffering capacity.

Table 5.8 Re-Indexing the Ozone NAAQS based on new science

Index → Terminal Indicators	Units	1	2	3	4	5	6	7	8
OZONE 2008	ppb	62.00	65.25	68.50	71.75	75.00	82.50	90.00	120.00
OZONE 2015	ppb	60.00	65.25	68.50	67.50	70.00	76.67	83.33	90.00

Table 5.9 Comparison of the PSE ranks and DDs with updated ozone data

Using Indexing for 2008 Ozone Standard		Using Indexing for 2015 Ozone Standard	
Ranked PSEs, most to least vulnerable	DD (%)	Ranked PSEs, most to least vulnerable	DD (%)
District 6	0.0	District 6	0.0
District 2	3.3	District 2	3.0
District 11	20.3	District 11	20.2
District 8	23.6	District 8	23.4
District 12	24.7	District 12	24.2
District 14	27.8	District 14	27.8
District 4	31.0	District 4	30.8
District 10	41.5	District 10	41.5
District 13	42.6	District 13	42.5
District 9	57.1	District 9	57.0
District 5	69.5	District 5	69.5
District 7	70.8	District 7	70.8
District 3	72.7	District 3	72.5
District 15	74.8	District 15	74.8
District 1	100.0	District 1	100.0

ordinal rankings and DDs means that, for this specific problem, the change in the NAAQS for ozone is of little consequence. Although in this circumstance, the updated significance for OZONE did not appreciably change the results, the Gaian Committee chooses to use the most recent ozone information on effects. The availability of new data or information on pollutant effects in environmental analyses is frequently a possibility. Therefore, having the capability to demonstrate whether and how this new information affects the results can be important to ensuring that the results are based on the best available information.

With the Gaian Committee decision to use the 2015 Ozone NAAQS, the EJS is updated to the final non-uniform EJS, which is presented in Table A.7 in the Appendix.

5.4.2   Value Sets

In the MIRA approach, creating Value Sets is both an analysis method and a means of articulating stakeholder Preferences. To facilitate greater stakeholder inclusiveness, MIRA users can comfortably include additional CNs without the common analytical concern that there will not be a means to evaluate their usefulness. Agreeing to more indicators than may ultimately be useful is part of the MIRA learning process. In this process, stakeholders see for themselves the import of including each indicator. Recall that in each branch of the Hierarchy, stakeholders assign relative Preferences by distributing 100% of the weight. As the number of indicators increase, these Preference weights, if every indicator is to receive some weight, can become quite small. As such, a Hierarchy that includes unnecessary indicators dilutes the contribution of those that are deemed more pertinent to the problem context. Therefore, although stakeholders are seeking inclusivity, the discussion about the number and type of indicators should include determining whether some of these indicators are redundant and unnecessary or whether several indicators could be combined into fewer indicators. Through iteration and discussion, stakeholders refine the number and nature of the indicators.

When there is a complex set of indicators, experimenting with Value Sets can help stakeholders evaluate the contribution and usefulness of these indicators in analyzing their policy problem. Stakeholders construct multiple experimental MIRA runs designed to explore the influence of each of the indicators on determining the PSE rankings when the VS changes. By placing large, small and no Preference weight on an indicator or groups of indicators, stakeholders can evaluate whether or how the PSE rankings change. The use of Value Sets in this way can be a powerful analysis method that helps to instruct stakeholders as to which indicators are most influential. Knowing this, stakeholders can deliberate whether these indicators' influence is appropriate based on their PN and CNs. Even if suggested indicators are ultimately determined to be unnecessary or redundant, the open solution approach allows for the evaluation of the potential indicators and, if appropriate, provides a justification for their removal. In other decision analytic approaches, stakeholders are typically required to convince the decision makers to include the indicators without the benefit of evaluating their utility.

Furthermore, experimentation with different Value Sets is also useful in exploring overlapping perspectives among diverse stakeholders. Commonly, stakeholders are led to believe that there is a single right answer, and consensus means that some stakeholders must acquiesce to the more popular perspectives. In the MIRA process, stakeholders often find that a variety of perspectives support the same decision. In this way the use of Value Sets provides a practical method for finding common ground.

5.4.2.1   Value Sets as an Analysis Method and for Quality Assurance of Data

After having settled on an EJS, the Gaian Committee proceeds to apply some simple Value Sets. This starts the process that will lead them to an understanding of the relationship between the data and the district rankings. Some of the questions that the Gaian Committee can ask include (1) which districts look more vulnerable when human health is the primary focus and how much more vulnerable are they than the other districts, (2) what change in Preferences must occur for these districts to appear less vulnerable, and (3) does my understanding of the data comport with the results I see?

The weighted functional Hierarchy has three CIs at the primary level. This allows the stakeholders to use some simple Value Sets to compare the PSEs ranking and DDs when all the weight is alternatively placed on human health, ecological integrity or future impacts. In this exploratory analysis, equal weights for the indicators in the second and third levels are maintained (i.e., maintaining the Equal Preference Value Set elsewhere in the Hierarchy). Using the final EJS, a comparison of the Equal Preference VS and a VS that places all the weight on the human health cluster of indicators is presented in Table 5.10.

Although District 6 is the top rank in both runs, its separation from the second ranked district (i.e., second rank's DD) is significantly smaller in the Equal Preference run than in the 100% human health run (3.0% vs. 20.8%). Additionally, District 2 has fallen from second to fifth place while its DD has changed from 3.0% to 35.2%. This shows that District 2 is considerably less vulnerable from a human health perspective than when the concerns of the full Hierarchy, under the Equal Preference VS, are considered.

Other simple Value Sets can also be examined to help the Gaia stakeholders understand what drives the vulnerability of the other districts. Table 5.11 presents the resulting district rankings based on using five additional simple Value Sets. These new VSs involve only the three CIs at the primary level of the Hierarchy. They are (A) 100% ecological integrity (Eco), (B) 100% future impacts (Future), (C) 50/50 split between human health (HH) and Eco, (D) 50/50 split between Eco

Table 5.10 Comparison of ranked districts using the Equal Preference VS and 100% human health VS

Equal Preference Value Set			100% human health Value Set		
Ranked PSEs, most to least vulnerable	DD (%)	Δ DD$_{adj}$ (%)	Ranked PSEs, most to least vulnerable	DD (%)	Δ DD$_{adj}$ (%)
District 6	0.0	0.0	District 6	0.0	0.0
District 2	3.0	3.0	District 8	20.8	20.8
District 11	20.2	17.2	District 14	24.0	3.2
District 8	23.4	3.3	District 9	34.9	11.0
District 12	24.2	0.8	District 2	35.2	0.3
District 14	27.8	3.5	District 12	48.7	13.4
District 4	30.8	3.1	District 7	49.9	1.3
District 10	41.5	10.7	District 5	53.7	3.8
District 13	42.5	1.1	District 13	54.6	0.9
District 9	57.0	14.4	District 3	60.5	5.8
District 5	69.5	12.5	District 11	79.7	19.2
District 7	70.8	1.3	District 1	87.7	8.0
District 3	72.5	1.8	District 15	92.4	4.7
District 15	74.8	2.3	District 4	94.7	2.3
District 1	100.0	25.2	District 10	100.0	5.3

Table 5.11 Change in district ranks with simple Value Sets

100% human health (HH)		A 100% eco integrity (Eco)		B 100% future impacts (Future)		C 50/50 HH/Eco		D 50/50 Eco/Future		E 50/50 HH/Future	
Ranked PSEs	DD (%)	Ranked PSEs	DD (%)	Ranked PSEs	DD (%)	Ranked PSEs	DD (%)	Ranked PSEs	DD (%)	Ranked PSEs	DD (%)
District 6	0.0	District 4	0.0	District 2	0.0	**District 6**	0.0	District 11	0.0	**District 6**	0.0
District 8	20.8	**District 10**	23.2	**District 6**	15.1	District 9	4.8	District 4	1.7	District 2	10.4
District 14	24.0	District 11	38.5	District 11	18.2	District 5	9.2	**District 10**	11.6	District 14	29.5
District 9	34.9	District 5	44.8	District 12	20.2	District 4	11.0	District 2	13.9	District 8	29.5
District 2	35.2	District 15	49.2	District 13	34.5	District 8	11.9	District 12	31.2	District 12	33.9
District 12	48.7	District 9	61.8	District 14	34.6	District 14	23.2	**District 6**	39.2	District 13	48.3
District 7	49.9	District 7	72.2	District 8	37.3	District 2	25.8	District 13	50.7	District 11	51.6
District 5	53.7	District 12	78.6	**District 10**	40.7	District 7	31.6	District 8	53.3	District 3	70.3
District 13	54.6	District 2	83.0	District 4	47.6	District 11	31.8	District 14	56.4	District 9	75.4
District 3	60.5	District 8	85.2	District 3	58.9	District 12	36.4	District 15	62.4	**District 10**	81.0
District 11	79.7	District 13	85.5	District 15	72.2	**District 10**	39.5	District 3	85.9	District 4	82.9
District 1	87.7	District 14	93.2	District 1	73.2	District 13	49.8	District 9	86.3	District 7	83.6
District 15	92.4	**District 6**	96.7	District 7	85.2	District 15	56.3	District 5	87.4	District 5	97.0
District 4	94.7	District 1	99.4	District 9	86.7	District 3	70.3	District 7	92.3	District 1	97.8
District 10	100.0	District 3	100.0	District 5	100.0	District 1	100.0	District 1	100.0	District 15	100.0

and Future, and (E) 50/50 split between HH and Future. The rankings that result from applying these VSs are shown in direct comparison to the rankings that resulted from use of the 100% HH VS. These simple VSs are designed to experiment with just the primary-level indicators to learn the nature of a district's vulnerability from the general perspectives of human health, ecological integrity and future impacts.

Experimentation with simple VSs is the beginning of PN-specific transdisciplinary learning. With this kind of learning, stakeholders discover how the PSE ranks change due to the compounded effects from changing VSs, when the data and EJS are held constant for a specific Hierarchy. Districts 6 and 10 are bolded in Table 5.11 to emphasize their rank movements. District 6 is the top ranked PSE when all the weight is placed on human health (100% HH). However, when all the weight is placed on Eco (A), District 6 drops to the 13th rank. This informs the Gaian Committee that District 6's greatest vulnerability is related to human health while its ecosystem appears to be less vulnerable than most of the Gaia districts. When the 100% Future Value Set (B) is applied, District 6 again rises to be among the top ranked PSEs.

District 10 appears to be District 6's "opposite." It ranks the lowest when all the weight is placed on human health but near the top with regard to ecological integrity. District 10's vulnerability from the perspective of future impacts is relatively moderate, placing it in the middle of the rankings. It is important to note that the Hierarchy, CMs, and Indexing have remained unchanged across all these runs while the VS is varied but consistently applied.

Given these results, the Gaian Committee expects that District 6 will rank somewhere in the middle when the 50/50 HH/Eco (C) Value Set is applied. However, District 6 is the top ranked PSE in this run. This suggests that District 6's indicators for human health overwhelm its very low Eco vulnerability. Although the results using the 100% future impact VS suggest that District 6's vulnerability to future impacts is very high, when the 50/50 Eco/Future (D) VS is applied, District 6 lies in the middle of the PSE ranks. This somewhat unexpected result suggests that the stakeholders examine District 6's human health versus ecological integrity Terminal Indicators (TIs)[6] and CMs. From this result, the Gaian Committee is beginning to learn that District 6 is very vulnerable when using human health indicators, somewhat vulnerable when using future impact indicators, and considerably less vulnerable when using ecological integrity indicators. Similarly, when the 50/50 HH/Future (E) VS is applied, District 6 is, again, the top ranked PSE. This implies that the value of the indicators that define the future impacts CI for District 6 is also high but not as high as the indicators that define human health. Due to the transparency of the analysis, this can be confirmed with an examination of the CMs and EJS.

Using the Indicator Driver (ID), stakeholders can identify the indicator(s) that contribute the most to District 6's rank, relative to the other districts for any given VS. Using the 50/50 Eco/Future (D) VS, Fig. 5.7 compares IDs for the Eco and

---

[6] TIs are CMs that have been expertly indexed.

Comparison of Indicator Drivers (ID)
50/50 Eco/Future VS (D)

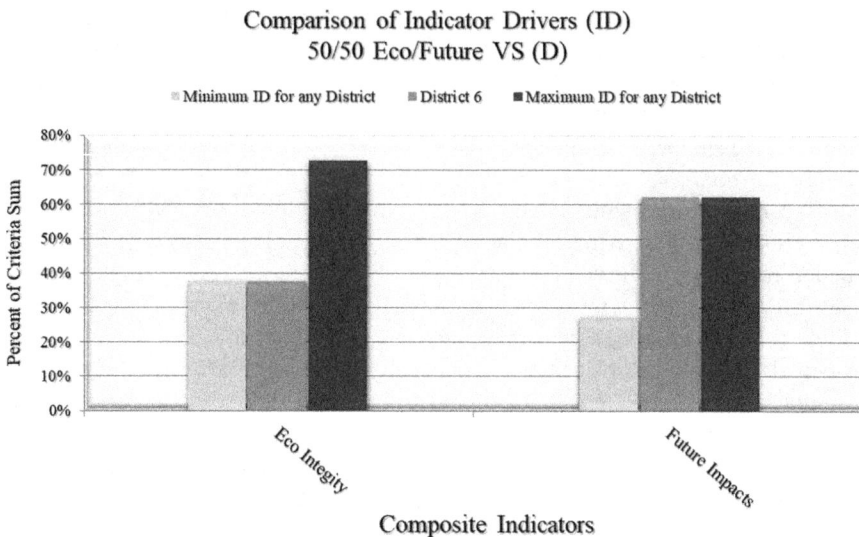

Fig. 5.7 Eco and Future CIs for District 6, compared to maximum and minimum IDs

Future Composite Indicators for District 6 with the minimum and maximum IDs found in the set of all districts. This comparison allows the stakeholders to quickly place District 6 among the other districts.

In the 50/50 Eco/Future VS, the 14 HH IDs are zero because the HH indicators are effectively removed from the analysis when no Preference weight is applied to the HH CI. In Fig. 5.7, District 6 represents the minimum ID for the Eco Composite Indicator cluster and the maximum ID for the Future Composite Indicator cluster. This explains District 6's sixth ranked position when 50% of the Preference weighting is split between Future and Eco (D). This is a simple illustration of the kinds of experiments that stakeholders can conduct to learn about what is driving the PSE ranks. As the Gaia stakeholders gain greater understanding about the ranking and characterization of the districts using the simple VSs, they may choose to experiment with increasingly more complex and nuanced VSs involving indicators at the second or third levels of the Hierarchy. The extent to which any stakeholder group chooses to explore these VSs is driven by the kinds of questions they wish to answer. Examining individual districts and conducting these kinds of experiments allows the Gaian Committee to develop the context-specific information that will form the basis of the rationale for their final decision.

Another benefit of the MIRA approach is the added capability to conduct quality assurance. Quality assurance of data is best done in context. Assuming the data has been properly collected, context determines whether the data is "good enough." Currently, when decision making is conducted, many stakeholders often apply their own standards of data quality (and potentially also in ad hoc ways) to their analysis (such as whether the uncertainty is below some threshold). Applied in these ways, without understanding how the data are to be used, creates problems. There is a good likelihood that, if standards are applied in this way, they will be applied inappropriately, incompletely or inconsistently. Using exploratory VSs, different datum can be isolated. This permits stakeholders to examine the resulting PSE ranks and assess whether these results comport with their knowledge and understanding of the datum. For example, if a PSE seems to rank unexpectedly high or low, stakeholders can examine the raw data, consider how it is indexed, and determine whether the data were in error. The stakeholders have gained a better understanding of that PSE relative to the other PSEs because of the MIRA process. Using exploratory VSs in MIRA to quality-assure the data allows stakeholders to quickly assess whether the data have errors. In this process, as stakeholders delve more deeply into examining the relationship between any single datum and the PSEs, they may also conclude that other data are more suitable for analyzing the PN and make a change to the CM. Therefore, in addition to learning how Preferences affect the PSE rankings, the use of exploratory VSs is an effective means to perform some initial quality assurance of the PSE data in the context of the PN.

The Gaia stakeholders may question the data used for District 6 as a result of evaluating PSE ranks that were produced using the previously described simple VSs. Suppose the appearance of District 6 as the top rank in the 100% HH VS versus the 13th rank in the 100% Eco VS alerts some of the stakeholders who believe that they know District 6 well. These stakeholders, based on their general knowledge of District 6, expected it to rank lower than it did (i.e., less vulnerable) under the 100% HH VS. Having flagged District 6 through these simple VSs, stakeholders can specifically examine its human health CMs and Indexing relative to that of the other districts. Suppose the stakeholders find that there was a typographical or other error in District 6's LEAD, CLIMATE_CHANGE or other data. They can make the correction and re-run the analysis. Alternatively, these stakeholders may discover that their understanding of District 6 was wrong. In addition, stakeholders can re-examine how the HH CMs (and other CMs) were indexed. Discussions with human health experts may indicate that adjustments need to be made to the EJS. Therefore, the Baseline Run can be repeated with all the agreed changes.

5.4.2.2 Value Sets to Find Common Ground

Frequently, the single most contentious area in the policy deliberation process is finding a way to reconcile the diversity of stated Preferences that exist among the stakeholders. Contrary to personal principles such as honesty, punctuality or morality, values that are used in policy decision making are reflective of the problem context. When stakeholders ask themselves "What is more important to you (and by how much?)," their answers depend on the context (i.e., the PN). Therefore, stakeholders are really asking themselves, "*Given the choices before me*, what is more important and by how much?" In decision making, the available choices depend on how the problem is described. Recall the hunger policy example from Chap. 1. When hunger is defined as a school-aged children's problem, the available choices include free school lunches and weekend food backpacks. Alternatively, when hunger is defined as a homelessness problem, the choices are likely to include soup kitchens and volunteers passing out food on the streets. The PN provides the necessary context, which makes it possible to develop and discover wicked problem solutions.

Most people believe that their perspectives are a fixed set of values that are generally held. Many current decision analytic approaches also presume this to be so. It is from this perspective that the standard practice of using economic utility curves across many different applications emerged. However, by proceeding through the MIRA process, people demonstrate considerably more flexibility when asked to make choices within a specific context. Conducting a MIRA open solution analysis engages stakeholders by eliciting their Preferences in context. This separates a stakeholder's opinion relative to a specific policy issue from how she might otherwise feel about the same or a similar issue that is presented generally or without context. When a questioner asks a stakeholder to opine on a policy issue without providing context, she will implicitly insert her own context to answer the question.

Stakeholder flexibility in Preferencing is not a compromise but rather a discovery process pertaining to the specific PN and a recognition that opinions are generally held with implicit context. Once a specific context is presented, the Preferences that are expressed are those specific to that context, which remain consistent with their basic, more generally held values. When stakeholders discover that the impact on the ranked set of PSEs changes very little as the weight of the indicator is changed, they are learning about the relationship between the PN, data, how the PSEs rank and the import of the strength of their own opinion regarding the indicator when used in context. The creation of VSs through iteration in the MIRA approach reveals overlaps in stakeholder perspectives. In these overlaps, stakeholders that do not have identical opinions (i.e., VSs) about indicator trade-offs find that they can still agree on a PSE ranking or a final decision (selection of one or more PSEs).

The Gaian Committee notices that as they make simple changes to the VS, certain districts do not change their relative ranks or those ranks change very little. For example, District 6 remains among the top ranked districts when any of the following simple Value Sets are used: (1) 100% HH, (2) 100% Future, (3) 50/50 HH/Eco, and (4) 5050 HH/Future. Starting with an initial understanding of PSE rank movements, based on these simple VSs, the Gaian Committee proceeds to find common ground among its stakeholders by continuing to experiment with more complex Value Sets.

As the Gaian Committee creates different Value Sets, they begin to discover patterns in district ranks when certain sets of Preference weight ranges are applied to certain indicators. For example, District 6 ranks at or near the top when human health indicators are Preferentially weighted. However, District 6 also maintains its top rank whether the VS weights for HH/Eco/Future CI are 100% HH, 100% Future, 50/50 HH/Eco, or one of the following combinations: 30/40/30, 33/22/45, 40/15/45, or 35/10/55. In fact, the Gaian Committee finds that the HH CI may be weighted as low at 30% (in certain combinations with Eco and Future) and District 6 will continue to retain the top rank. These alternative VSs that produce District 6 as the top rank are illustrated in Fig. 5.8.

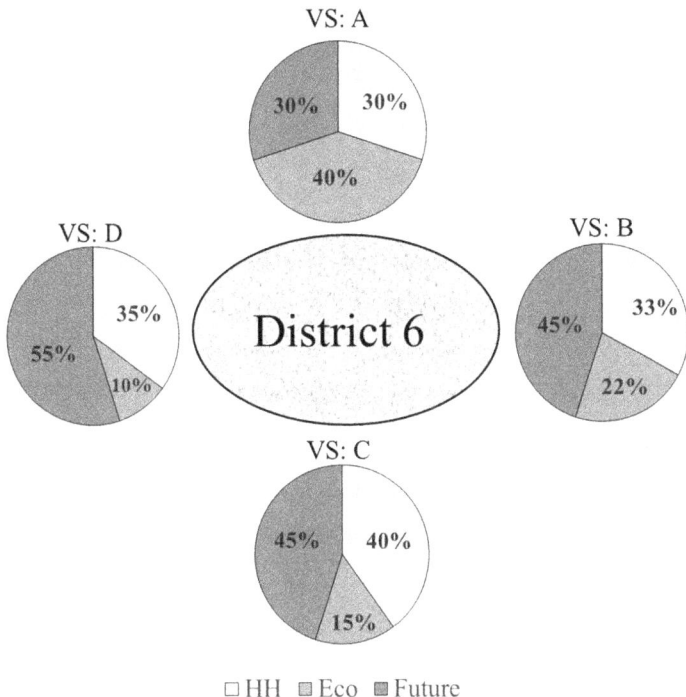

Fig. 5.8 Some of the Value Sets producing District 6 as the top ranked PSE

The Gaian Committee begins to discern a pattern that District 6 remains at the top rank until the combined weight for HH and Eco falls below 50%, unless Future is substantially more preferred than Eco. In other applications, finding common ground may be even easier as stakeholders may find many more than three or four VSs that can produce the same or similar rankings.

In experimenting with Value Sets, the Gaian Committee members learn that District 6 ranks first whenever the HH CI is given Preference over the Eco and Future CIs. However, in many of the Value Sets, District 2 is a close second to District 6 and District 2 becomes the top ranked PSE when the Eco and Future CIs are more preferred over HH. This balance is reflected in Fig. 5.9.

While Fig. 5.8 illustrates the variety of Value Sets resulting in District 6 holding the top rank, Fig. 5.9 illustrates that when a slight change is made to HH and Eco (moving from A to B), Districts 2 and 6 jointly occupy the top rank (DD = 0%) (B). Prior to making that change, District 6 is ranked first with District 2 in second place having a DD = 1.4% (A). But the tipping point occurs when enough weight is placed on the future impacts CI. At that point District 2 takes over as the top rank from District 6 as shown in Fig. 5.9 (C) (District 6's DD = 1.5%).

Suppose, in a separate line of inquiry, some members of the Gaian Committee argue that ecological integrity is basic to good human health. Therefore, while human health should remain important, these stakeholders argue that the Eco

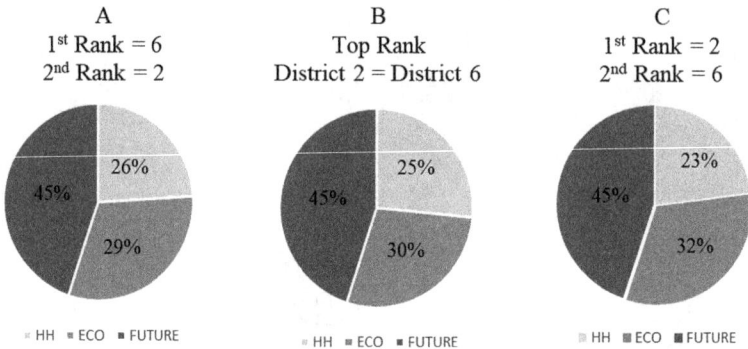

Fig. 5.9 Progression of Value Sets showing point at which District 6 loses top rank to District 2

indicators should be given higher Preference weights than the HH indicators. The Gaian Committee wants to know if shifting the VS to prefer Eco over HH affects the district rankings. Figure 5.10 illustrates a range of Eco-Preferred VSs that continue to produce District 6 as the top ranked PSE.

Still limiting the VS changes to the primary level of the Hierarchy (i.e., Equal Preference VS remains for indicators at the other levels of the Hierarchy), the Committee members notice that when Eco is weighted more Preferentially than HH, District 6 remains the top rank even when Eco is weighted as high as 50%. Furthermore, it is also possible to keep District 6 as the top ranked PSE, even as the HH Preference weight decreases from 50% to 41%, provided additional weight is placed on the future impacts CI. This is consistent with information the Gaia stakeholders discovered in one of the earlier experiments with the simple VSs (see the comparison among Value Sets C, D, and E in Table 5.11). In that experiment, the Gaia stakeholders discovered that District 6's top rank can be assured if more weight is placed on the future impacts CI as the weight on HH is decreased. This is also consistent with the results shown in Fig. 5.8.

From the experiments shown in Table 5.11, Figs. 5.8, and 5.9, in addition to other variations, Gaia stakeholders learn how and why the districts rank as they do. These and many more experiments can be conducted to find the ranges among the Preference weights that allow stakeholders to better understand where different perspectives result in different PSE ranks and where they do not. This is an example of how the MIRA process facilitates finding common ground among diverse stakeholder perspectives. The Final VS agreed to by the Gaian Committee is presented in Fig. A.2 in the Appendix.

5.4.2.3 Backcasting

Particularly when policy issues are contentious and the stakes are high, some stakeholders may worry that other, more powerful, stakeholders or decision makers will game the system. They worry that these individuals may inconsistently or selec-

Fig. 5.10 Four Eco-Preferred VSs producing District 6 as the top rank

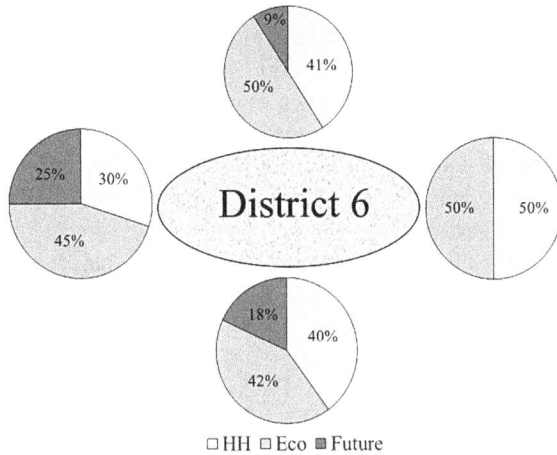

□ HH □ Eco ■ Future

tively use data and apply expert judgment or Preferences. In many of these types of situations, decision makers have a preferred alternative in mind and expect the analytical results to support their choice. Working backward to determine what VS is required to produce a chosen result is another way for stakeholders to learn about what influences the PSE rankings.

In backcasting, stakeholders use the given data and Indexing and iterate through combinations of Preference weights to find one or more Value Sets that produce, or produce as closely as possible, their desired PSE ranking. Note that neither the data nor the EJS has been altered. In other words, the science, facts, and expert judgments regarding the relationships between the data and its significance to the decision context are not changed and continue to be consistently applied across all PSEs. However, since VSs are subjective, these can be explored to determine which perspectives result in producing the decision maker's preferred result. By performing the analysis this way, the stakeholders are, in effect, approaching the problem with a presumed final decision. With a specific result in mind, they then determine the set of values needed to produce that decision. Once the VS is found, the stakeholders ask themselves if they can agree with the Preferences that are implied by their desired outcome and, if so, can they justify their decision using those values. Stakeholders may surprise themselves when they discover what VS is necessary to produce their desired outcome. For example, US EPA officials assessing an area's compliance status with the ozone NAAQS are required, by the Clean Air Act, to consider many factors [3]. These factors include air quality data, industrial source and vehicle emissions, population, and contributions to downwind air quality. While these officials have flexibility in what VS is used in their decision, they may find it easier to justify their air quality decision if their VS reflects a larger Preference for air quality rather than another indicator, such as population.

Suppose a Gaia stakeholder felt strongly that District 12 should be chosen while others in the group did not agree. Rather than having a protracted general discussion about the validity of this advocate's views, backcasting can be used to determine what set of values is needed to justify choosing District 12. Once this VS is found, the stakeholder group is informed and more prepared to discuss the acceptability of the implied values rather than simply arguing about whether District 12 should be chosen.

After having performed many Value Set iterations, the Gaia stakeholders find that there is no VS that produces District 12 at the top rank. However, they do find a single Value Set that results in District 12 rising to the second rank with a relatively small DD. The comparison of the District 12 Backcast VS and the Final VS is presented in Table 5.12.

Forcing District 12 toward the top of the rankings requires drastically altering the Preferences from the Final VS. To move District 12 near the top, very little weight is placed on future impacts, while substantially greater weight is placed on ecological integrity. Furthermore, among the human health cluster of indicators, mental health must be more heavily weighted than in the final stakeholder VS to produce the desired result. Table 5.13 presents a comparison of the district rankings when the Final VS is used versus the District 12 Backcast VS. The greatest differences are bolded in Table 5.13.

Although the stakeholders' run using the Final EJS and VS results in District 5 ranking below District 12, when stakeholders change the VS to move District 12 as high up in the rankings as possible, this also affects District 5's relative vulnerability. District 5 moves to the top rank even as it is only slightly more vulnerable than District 12 (DD = 2.31%). With the current data and EJS, even as stakeholders may try to move District 12 to the top rank, it is not possible. It is also interesting to note that when the Binning method is used with the Backcast VS, Districts 5, 12, and 9 all appear in the top bin, with the largest separation (District 9's DD = 4.05%) indicating that these districts are quite similar.

Table 5.12 Comparison between the Final VS and the District 12 Backcast VS

Primary level	Preference weight applied (fraction out of 1.0)	
	Final VS	District 12 Backcast VS
Human health	0.3	0.25
Eco integrity	0.45	0.74
Future impacts	0.25	0.01
Human health		
Cancer	0.15	0.05
Neurological	0.3	0.1
Respiratory	0.2	0.15
FINE_PART-PD	0.1	0.15
MENTAL_HEALTH	0.25	0.55
Eco integrity		
Forests	0.5	0.9
Surface water	0.4	0.1
COAL_ASH	0.1	0
Future impacts		
POPGROWTH	0.2	0.2
URBGROWTH	0.4	0.4
CLIMATE_CHANGE	0.4	0.4

Table 5.13 Comparison of PSE ranks using Final VS versus District 12 Backcast VS

Final VS			District 12 Backcast VS		
Ranked PSEs, most to least vulnerable	Bin (#)	DD (%)	Ranked PSEs, most to least vulnerable	Bin (#)	DD (%)
District 2	1	0.0	District 5	1	0.0
District 6	1	6.55	*District 12*	1	2.31
District 8	1	9.05	District 9	1	4.05
District 12	2	15.16	District 7	2	16.85
District 13	2	15.76	District 4	3	24.09
District 11	2	15.87	District 3	4	30.05
District 14	4	30.58	**District 2**	4	34.74
District 10	4	39.75	District 10	4	34.92
District 9	6	50.09	District 8	4	36.77
District 4	6	52.03	District 1	4	36.93
District 3	6	54.64	District 11	4	38.65
District 7	6	54.84	District 13	4	38.70
District 5	7	64.74	District 14	6	52.28
District 1	10	99.29	District 6	7	63.54
District 15	10	100.00	District 15	10	100.00

Because the data and the EJS remain the same and the VS is consistently applied, any PSE's relative vulnerability also affects the positions of the other PSEs. Therefore, the use of backcasting in this manner does not game the system. In this process, stakeholders may find some results impossible to obtain given the data and the EJS. In other situations, stakeholders may find themselves unwilling to accept the VS required to justify the desired outcome. The transparency of the MIRA process makes backcasting an analytical asset in policy decision making rather than a means to obfuscate the decision. The Gaian Committee need not choose the top ranked district. They could still select District 12, but if they do, they are indicating that they are comfortable with the alternative VS. Additionally, they may be compelled by other stakeholders to explain why District 5 is bypassed, as it ranks higher than District 12 under the alternative VS.

Stakeholders can construct the Hierarchy, select the criteria, index the data, and apply Preferences in any way that is consistent with how they choose to define the PN. However, once done, the analytical components are applied consistently across all PSEs. This is in direct contrast to many decision making processes where decision makers, often implicitly, apply one set of criteria to some PSEs and a different set of criteria to others, or apply the same criteria differently across the PSEs.

5.4.2.4   Using AHP to Explore Preference Uncertainty

As discussed earlier, assessing Preferences within a specific problem context requires stakeholders to consider trade-offs among the indicators within each cluster of the Hierarchy. In most situations, stakeholders assign Preference weights by examining the cluster of indicators and simply deciding how to distribute the weights among the indicators. When they do this, they believe that their applied weights represent their contextual Preferences or opinions about these indicators. However, as the number of indicators in a cluster increases, it becomes progressively more difficult to confidently assign weights that properly represent the stakeholders' contextual Preferences. In such situations, the Analytic Hierarchy Process (AHP) can be used to provide stakeholders with a more structured alternative. AHP is a tool that helps stakeholders examine whether their assigned Preference weights accurately reflect their opinions about the relative importance of the indicators, within a cluster. Furthermore, AHP significantly increases the prospects of developing indicator weights that are more closely aligned with stakeholder contextual Preferences by using the revealed contradictions (through its consistency ratio metric) as a tool in their analysis [4]. Consequently, AHP is a useful method within the MIRA process that can help stakeholders reduce Decision Uncertainty regarding the Preferencing process.

In AHP, stakeholders conduct pair-wise comparisons of indicators within a cluster. Particularly when there are many indicators, stakeholders may find it daunting or arbitrary to simply distribute the weights. Using a method whereby a stakeholder has only to consider two indicators at a time is simpler and considerably more understandable. When the pair-wise comparison approach is used, stakeholders have a greater chance to examine whether these distributed weights are being made consistently. Within AHP, the consistency ratio is a tool used to flag inconsistent pair-wise responses. MIRA provides the stakeholders the option of using the pair-wise comparison approach to Preference weight one or more clusters in a Hierarchy. Although there is no "right" set of Preferences, using the AHP option within MIRA allows stakeholders to rethink their Preferences if the consistency ratio suggests that their pair-wise value judgments are inconsistent.

Even if stakeholders initially choose not to utilize AHP's pair-wise comparison approach, they may use AHP subsequently to evaluate the degree to which their chosen set of weights represents their opinion about the relative importance of the indicators. The Appendix provides an example of how MIRA applies AHP to discern the set of pair-wise comparisons, which represent perfect consistency, when the original Preferences were established without AHP.

Having a means to evaluate the consistency of stakeholder Preferences when AHP is used to perform the Preferencing, or when AHP is used to expose the pair-wise comparisons that are implied by the assigned Preferences, helps address uncertainties in the VS, which facilitates a better understanding of Decision Uncertainty.

5.4.3   *Iterating Combinations of Uncertainties*

Iteration to examine combinations of uncertainties is a powerful analysis methodology for any policy practitioner. When uncertainty is evaluated within an open solution approach, such as MIRA, practitioners have the capability to understand the impact that multiple uncertainties have on the PSE rankings. MIRA's analysis methods that generate quantitative information about Indicator Contributions and Indicator Drivers can also be used to evaluate uncertainties related to data, expert judgments or VS, either singly or in combination. Another analysis method, the Similarity Index, is introduced to facilitate this evaluation.

Up until this point, stakeholders have entered quantitative information (data, expert judgments, Preferences) without explicitly evaluating the influence of uncertainty in their analysis. Conducting uncertainty analyses is an essential part of the process, providing stakeholders and decision makers with information that is necessary to support their decision. Iterating to reduce Decision Uncertainty in MIRA is the experimental process that permits the discovery of how PSE rankings are impacted when data, EJS, and VS uncertainties are considered.

5.4.3.1 Similarity Index – Evaluating Combinations of Uncertainty

When many iterations are performed, stakeholders are uncertain about many things, including how similar (or dissimilar) one iteration is from another. In MIRA, the *Similarity Index* (SI) is a convenient analysis measure that allows stakeholders to quickly and quantitatively compare one run against another run. With the SI, stakeholders are comparing two entire ranked sets of PSEs, rather than examining individual PSE ranks or DDs within a run.

Similarity Index
MIRA's Similarity Index is a single-valued metric that provides a quantitative measure of how similar one ranked set is to another, considering both differences in ordinal rankings and DDs.

The SI is a non-dimensional number that is designed to range from 0 to 1000. A value of 0 represents the greatest possible dissimilarity between two PSE rankings, while a value of 1000 represents perfect agreement in both ordinal ranking and Decision Distances (i.e., identical results). The mathematical construction of the SI contains two terms: the first represents the sum of the changes in ordinal ranks over all PSEs, and the second represents the sum of the changes in Decision Distance among all PSEs. The SI also provides flexibility relative to how similarity is defined. In practice, one run, represented by a specific Hierarchy, Expert Judgment Set (EJS), and Value Set (VS), is compared against other runs that could have a different Hierarchy and/or EJS and/or VS. The mathematical formulation of the SI is described in the Appendix.

In Sect. 5.4.1, the Gaian Committee evaluated the PSE ranks when the OZONE CM was updated from the 2008 Ozone NAAQS to the 2015 Ozone NAAQS. This change did not change the ordinal ranks of the PSEs (see Table 5.9). The SI quantifies this comparison. When the weight on the ordinal ranking term is set to 1.0 (i.e., the SI is calculated by considering only ordinal ranking), the SI = 1000 since there is no change in ordinal ranking. However, when the weight for the DD term is set to 1.0, SI equals 998, which confirms the very small change in the DDs between the two runs, which was reflected in Table 5.9.

It is possible to quantitatively analyze some of the previous iterations on the simple VSs conducted by the Gaian Committee. Six pairs of these runs from Table 5.11 are compared in Table 5.14 using the SI.

With SIs ranging from 149 to 518, the six simple VSs are shown to be substantially different from each other which quantitatively confirms the results of the previous analyses. Although still quite dissimilar, the last pair of simple VSs, 50/50 Eco/Future versus the 50/50 HH/Future, is the most similar to each other, with an SI = 518.

Generating the range of VSs that produces District 6 as the top rank, as shown in Figs. 5.8 or 5.10, is a good start for stakeholders seeking to reduce Decision Uncertainty. Table 5.15 presents another set of experiments using SI. In this table, the four VSs from Fig. 5.10 that produce District 6 as the top rank are compared with the Equal Preference VS. Conducting these experiments allows stakeholders to use District 6 as the focal point to evaluate the similarity of these four runs. Table 5.15 shows the four VSs that produce District 6 as the top rank and the corresponding SI that results when their ranked sets are compared with the ranked set that results from applying an Equal Preference VS.

Although District 6 is the top rank in all five VSs, the results from applying VS A are most similar to those from applying the Equal Preference VS. However, whether an SI of 903 makes the runs from the Equal Preference and VS A effectively the same is for stakeholders to decide. After performing enough iterations, stakeholders can become sufficiently knowledgeable regarding how the SI measure performs within their problem context.

The District 12 Backcast VS that results in District 12 ranking second can also be evaluated with the SI, using the Final VS as a comparison. Although these two VSs appear to be substantially different, stakeholders may want to quantify the results of the VS differences that are shown in Table 5.12. By setting the SI terms to equally weight ordinal ranks and DDs and setting p equal to 1.0 (i.e., no Preference for large vs. small differences in rank or DD), the SI for the District 12 backcast results compared with the final ranked PSEs is 332. Therefore, the District 12 Backcast VS not only appears to be substantially different from the Final VS but, by calculating SI, this is both confirmed and quantified in terms of the magnitude of the differences in the resulting PSE ranks.

Table 5.14 Similarity Index (SI) for six simple VS runs from Table 5.11

Comparison of simple VS run pairs	SI $w_r = w_{DD}$ $p = 1.0$
100% HH vs. 100% Eco	149
100% HH vs. 100% Future	425
100% Eco vs. 100% Future	192
50%/50%: HH/Eco vs. Eco/Future	393
50%/50%: HH/Eco vs. HH/Future	462
50%/50%: Eco/Future vs. HH/Future	518

Table 5.15 SI Comparisons between Equal Preference VS and four VSs that produce District 6 as top rank

Comparing Equal Preference VS with the following	SI $w_r = w_{DD}$ $p = 1.0$
VS A: 30% HH, 40% Eco & 30% Future	903
VS B: 33% HH, 22% Eco & 45% Future	873
VS C: 40% HH, 15% Eco & 45% Future	829
VS D: 35% HH, 10% Eco & 55% Future	800

5.4.3.2   Effect of Data Uncertainty

Data uncertainty is one component of Decision Uncertainty. Typically, scientists quality-assure the data independently from how the data may be used. This is a practice that is consistent with reductionism whereby independent components can be disassembled and reassembled without affecting the whole. Although it is important to have an appropriate level of basic quality assurance (e.g., data are collected using appropriate methods and are appropriately documented), when policy practitioners use these data, it is important to determine whether and how these data uncertainties affect their policy choices. Reducing data uncertainty becomes necessary only when a decision maker's policy choice may change because of it.

Quality-assuring data out of context presents two major problems. The first problem is that the data may be used for a purpose that was not originally intended or contemplated when the data were generated or collected. This creates the potential for misunderstandings about the utility and constraints of the data. For example, consider a set of water flow data in a river that was collected to estimate the potential for flooding in the near term. The flow gauge that is needed for that purpose does not need to be accurate during low flow events. However, if one needed to estimate the concentration of pollutants in the river, the use of the data collected from this flow gauge is not likely to be accurate enough to determine concentrations within an acceptable uncertainty range, since it is the low flow events that are of interest in this context (i.e., pollutant concentrations increase with less water). If this flow gauge data is used by stakeholders to create a CM for pollutant concentration, it introduces a potentially large uncertainty into their analysis. More appropriately in this situation, stakeholders should look for more suitable data or, if not available, consider whether to change the PN or suspend the decision until appropriate data can be collected.

The second problem with using the data out of context is that scientists or others may expend large amounts of time or money reducing data uncertainty when it is unnecessary for the present analysis. Data that may seem highly uncertain, when considered out of context, may not be problematic for inclusion in the analysis when considered contextually. The evaluation of data uncertainty in MIRA (based on the PN) gives stakeholders the capability to examine how data uncertainty (together with other uncertainties) affects the PSE ranks.

Table 5.16 Results of three separate uncertainty runs for the data: (1) POPGROWTH, (2) URBGROWTH, and (3) MENTAL_HEALTH

Exploring Decision Uncertainty: Single datum uncertainty								
	Uncertainty band	Top five ranked districts						PSEs in Bin 1
Simulation	(± %)	1st	2nd	3rd	4th	5th	DD (%)	(#)
Final Run	NA	2	6	8	12	13	6.5	3
POPGROWTH	+20%	2	6	8	12	11	4.5	3
	−20%	2	6	8	13	11	3.1	3
URBGROWTH	+20%	2	6	8	12	13	3.2	3
	−20%	2	6	8	11	12	9.6	2
MENTAL-HEALTH	+20%	6	2	8	13	12	1.1	2
	−20%	6	11	2	13	8	9.4	2

Evaluating data uncertainty within the MIRA approach is accomplished by performing runs in which a CM value is varied across its uncertainty band. In doing so, the stakeholders can determine the degree to which the CM's uncertainty impacts the attractiveness of the options being considered. Suppose the experts, advising the Gaian Committee, do not know the precise uncertainty for each of the CMs for the POPGROWTH, MENTAL_HEALTH, and URBGROWTH, but they know that it is unlikely that they would be greater than ± 20% (the 20% uncertainty band).[7] Based on this expert opinion the Committee decides to test the effects of a 20% uncertainty band with each of the three CMs on the PSE ranks and hence each CM's contribution to Decision Uncertainty. Table 5.16 presents the resulting top five ranked districts and two MIRA analysis methods, DD and Binning, for each of these three separate runs. Stakeholders are not limited to examining only these analysis measures and are encouraged to use other measures such as IC, ID, and SI to explore answers to their questions about the results.

The first row in Table 5.16 presents the results of the Final Run made by the Gaian Committee using their Final EJS and Final VS. Comparing these results against the results of the bracketing analysis provides stakeholders with a quantitative means to assess the impact of a datum's uncertainty on Decision Uncertainty.

When POPGROWTH and URBGROWTH are individually changed at the extremes of the 20% uncertainty band, the top three districts remain the same as in the Final Run. However, in three out of four runs, Districts 2 and 6 appear to be more similar when these data are more uncertain, as indicated by the DDs. The exception is when the URBGROWTH datum is overestimated (i.e., its actual value is 20% less than the CM value used in the Final run), the DD increases by 3.1% (9.6% vs. 6.5%). This indicates that, with this uncertainty, District 6 is less similar to District 2 than it was in the Final Run. Furthermore, in this URBGROWTH

---

[7] Probable uncertainty ranges can often be derived by interviewing experts, by starting with a small uncertainty (such as 1%), and by incrementally increasing this value while asking the expert the likelihood of the uncertainty approaching these increasingly higher values. In this manner, even for CMs where experts have not previously considered uncertainty quantitatively, it is often possible to obtain a realistic uncertainty band.

uncertainty run, the number of districts in Bin 1 is reduced from 3 to 2, indicating that Districts 2 and 6 are less similar to District 8 compared with the Final Run. Whether the URBGROWTH datum is over- or underestimated by 20%, District 2 is still the top ranked district. However, the DD is larger (9.6%) when it is overestimated and smaller (3.2%) when the datum is underestimated. In contrast, District 6 replaces District 2 as the top ranked district when the MENTAL_HEALTH CM is either over- or underestimated by 20%.

Understanding how single datum uncertainty affects the results is useful, but real-world analyses must address multiple data uncertainties within a single analysis. Furthermore, a realistic examination of data uncertainty requires more than simply bracketing the uncertainty band. What is needed is a Monte Carlo simulation that generates many runs, which each use random values within a CM's uncertainty band for all the CMs where uncertainty is evaluated. Such a simulation allows analysts to examine the full effect of the collective uncertainties of the CMs for each PSE. By performing a Monte Carlo simulation, a clearer and more realistic view of the effect of data uncertainty on Decision Uncertainty emerges.

To illustrate the simultaneous analysis of the collective uncertainties for multiple CMs, five are selected for an abbreviated Monte Carlo analysis consisting of ten simulation runs. Table 5.17 presents these CMs and their accompanying uncertainty bands. In a real-world application, the uncertainty analysis is expected to include all CMs.

In this abbreviated Monte Carlo simulation, each of the five CMs is assigned a randomly generated value, selected from within the defined error bands shown in Table 5.17. Note that to perform an adequate Monte Carlo simulation of this type, there need to be many more runs to expose the full extent of the impact that data uncertainty has on Decision Uncertainty. However, with that caveat in mind, the results of this uncertainty analysis are presented in Table 5.18.

Table 5.17 The five Criteria Metrics and their individual uncertainties

Criteria Metric	Uncertainty band
MENTAL_HEALTH	±20%
LEAD	±10%
ACID_RAIN[a]	±10%
URBGROWTH	±20%
INSECTICIDES	±10%

[a]For a CM such as acid rain, which is used to create two indicators in the Gaian analysis, it is possible to have two different uncertainty bands. This is because the data may be more certain when it is used for one indicator than when it is used for another indicator. In Table 5.17, for this illustration, the same uncertainty band is used for acid rain.

Table 5.18 Results of the simultaneous analysis of five uncertain Criteria Metrics

Exploring Decision Uncertainty: Combined Data Uncertainties							
	Top five ranked PSEs					DD (%) of the 2nd	
Simulations	1st	2nd	3rd	4th	5th	ranked District	# of PSEs in Bin 1
Final Run	**2**	**6**	**8**	**12**	**13**	**6.5%**	**3**
1	2	8	6	13	12	0.9%	4
2	6	2	8	13	11	1.3%	3
3	13	6	2	8	11	6.5%	4
4	2	6	8	11	12	6.9%	2
5	2	6	11	8	12	9.9%	2
6	2	6	12	11	13	4.0%	2
7	2	8	6	13	12	3.7%	3
8	6	2	13	11	8	6.1%	2
9	6	8	2	12	11	0.5%	3
10	2	6	13	8	11	13.0%	1

The first row in Table 5.18 presents the results of the Final Run made by the Gaian Committee. The remaining rows are the results from each of the ten simulation runs. Comparing the results of the Final Run against the results from the ten simulation runs provides stakeholders with a quantitative way to determine the impact of the combined influence of the five CMs' data uncertainties on Decision Uncertainty. Although Decision Uncertainty includes other more qualitative uncertainties, Table 5.18 permits a portion of the Decision Uncertainty to be examined quantitatively. Therefore, although quantitative measures are useful for assessing Decision Uncertainty, analysts are cautioned to use these measures as aids rather than as determinants in their decision making.

Six of the ten runs in Table 5.18 reproduce District 2 as the top rank, and four runs reproduce the top two ranks, but none can reproduce the top three ranks of the Final Run. Of these six simulations, four show either a similar or larger second rank DD, when compared with the Final Run DD (6.5%). One of these four simulations (Simulation 10) produces a second rank DD that is approximately twice that of the Final Run and places District 2 alone in the top bin. On the other hand, Simulations 1 and 7 show District 2 to be far closer to its second rank DD than that found in the Final Run. Three of the simulations have District 6 as the top rank (with District 2 as the second rank). However, in Simulation 2, when Districts 6 and 2 switch ordinal ranks, they are separated by a smaller second rank DD than is found in the Final Run (1.3% versus 6.5%). Had these been the results from a true Monte Carlo analysis, the Gaian Committee may be reassured that either District 2 or District 6 is a justified choice for extra funding.

5.4.3.3   Effects of Uncertainty Combinations

While MIRA can easily be used to evaluate combined data uncertainties, it is designed for much more. As with the examination of uncertainty for one or more pieces of data, it is also desirable to perform similar experiments relative to the uncertainties in stakeholder Preferences and expert judgments. For example, a CM's uncertainty in one EJS/VS combination may drastically change the ranks of PSEs while, in another, the uncertainty may be inconsequential. Because it is possible in MIRA to hold some variables constant and change others, it is possible for stakeholders to learn which uncertainties most influence the results. Decision makers and other stakeholders can then decide whether they have the need, willingness, and resources to reduce those influential uncertainties. Understanding the combined impact of data, EJS, and VS uncertainties permits decision makers to more effectively target limited organizational resources to those data and areas where the uncertainties have profound impact on the choices that are made. The computational laboratory constructed using MIRA allows stakeholders to explore many different combinations of uncertainty with the goal of defining and reducing Decision Uncertainty.

There is EJS uncertainty because there is no precise, scientifically correct, single function that translates a CM value to one of the values on the decision scale. Two experts from the same field, having the same level of expertise, may index a CM differently. In a process similar to eliciting data uncertainties, experts are asked to assess the bounds of uncertainties around their judgments of the CMs' significance. Similarly, stakeholders can evaluate the uncertainty in the Final VS resulting from differences in their perspectives. For both EJS and VS uncertainties, what may appear to be large numeric differences do not always translate into large Decision Uncertainty. Likewise, stakeholders will want to explore apparent small differences in EJS and VS uncertainties as these may translate into unacceptably large Decision Uncertainty.

The exploration of Decision Uncertainty when data, EJS, and VS uncertainties are simultaneously considered is illustrated with an abbreviated Monte Carlo simulation of ten runs. Each simulation uses randomized data values: (1) for the five CMs in Table 5.17, (2) for each Preference weight within ±10% of its actual value from the Final VS, and (3) within ±10% of each of the CM values that are associated with a value on the decision scale (i.e., the Indexing scheme). The results of simultaneously analyzing the combined effects of data, Preferencing, and Indexing uncertainties are presented in Table 5.19.

Table 5.19 presents the resulting top five ranked districts and two MIRA analysis methods, DD and Binning, showing the effect of combining uncertainties from the data, EJS, and VS. In this abbreviated Monte Carlo simulation, when these uncertainties are considered, four of the ten simulations result in District 2 as the top ranked district and, in another four simulations, District 6 is the top ranked district (with District 2 as the second in rank). Only Simulation 8 reproduces the top 2 ranks, and none reproduce the top 3 ranks of the Final Run. Simulations 5 and 6 produce widely different results from the previous runs, when data uncertainties were evaluated singly, as well as the other runs in this simulation. Had this been the result of a full Monte Carlo simulation, these results would need to be further investigated to determine if they were true outliers. Of the four combined uncertainties simulations where District 2 remains the top rank, in two of them the second ranked district shows a smaller DD, while for the other two simulations the DDs are approximately twice as large as that found in the Final Run. Had these been the

Table 5.19 Combined uncertainty analysis of CMs, EJS, and VS in ten simulations

Exploring Decision Uncertainty: Combined uncertainties of CMs, EJS, and VS							
	Top five ranked PSEs					DD (%) of the 2nd	
Simulations	1st	2nd	3rd	4th	5th	ranked District	PSEs in Bin 1 (#)
Final Run	**2**	**6**	**8**	**12**	**13**	**6.5**	**3**
1	2	8	6	11	13	3.1	4
2	6	2	8	12	11	4.4	2
3	2	8	11	12	6	4.9	4
4	6	2	12	11	8	12.6	1
5	8	2	11	12	14	3.9	3
6	13	6	2	8	11	5.1	2
7	6	2	12	11	13	22.2	1
8	2	6	13	12	11	14.7	2
9	2	8	6	12	11	11.9	1
10	6	2	13	8	12	16.0	2

results of a true Monte Carlo analysis, the Gaian Committee would be faced with larger Decision Uncertainty reflected in the greater movement among the top two ranked districts and larger DDs. This may suggest to them that they conduct a few more experiments to minimize Decision Uncertainty before making their final district choice. Alternatively, the Gaian Committee, after examining these results, may conclude that they can justify either Districts 2 or 6 as their final choice. Whether Decision Uncertainty is sufficiently minimized is a matter for the stakeholders to decide.

Decision Uncertainty is the combined uncertainty pertaining to all aspects of the Requisite Steps of decision making. In the Gaia example, the examination of the combined effect of CM values and VS and EJS uncertainties on PSE ranks, second rank DDs, and the number of PSEs in the top bin is only a portion of the overall Decision Uncertainty. There are many additional experiments that could, and should, be performed using the MIRA approach. As part of the evaluation of Decision Uncertainty, the Gaian Committee returns to the PN and CNs to consider how closely their vision of the policy problem is reflected in these components. This qualitative assessment, combined with the more quantitative assessments of CMs, EJSs, and VSs, is used to construct the rationale for the Gaian Committee's final decision.

As the Gaian Committee begins to conclude its analysis and prepares the presentation of its recommendation to the Gaia federal government, it documents the process of discovery regarding the relative characteristics of the 15 districts as described with 26 indicators, organized in a weighted functional Hierarchy. Although the Final EJS and VS represent the consensus of the Gaian Committee members, these could change based on input from federal government officials and members of the public. In the process of arriving at its recommendation, the Gaian Committee also used the MIRA process to quality-assure the data in their PN context. As a result, the Committee is well-prepared to address any questions or misunderstandings raised by the government officials or the public regarding the relevant facts of the

Gaian analysis. In a real-world application, many more simulations would likely have been performed with the stopping point determined by the stakeholders' confidence in their readiness to make a final decision (i.e., when the Decision Uncertainty is acceptable to them). The final ranked PSEs are presented in Table 5.20.

> **Gaian Committee Recommendation:** After considering the results of the combined uncertainty analysis, the Gaian Committee decides that Decision Uncertainty is sufficiently reduced such that a rational justification can be made for choosing either District 2 or District 6.

As demonstrated with the Gaia example, MIRA provides a means to evaluate the effects of data and other uncertainties, singly or in combination, to determine how they affect the ranked PSEs. The capability to conduct such analyses helps the Gaian Committee avoid a common public policy pitfall when data or model uncertainties are involved, and the policy issues are contentious. In these situations, there is a tendency for stakeholders, unhappy with the decision, to argue that the uncertainties must be eliminated or reduced, which often delays policy making. The Gaian Committee's recommendation is supported by the MIRA process that allows the exploration of alternative choices in problem definition, CMs, EJSs, VSs and the corresponding uncertainties within a single analysis. If the Gaia government's reaction to the Committee's recommendation is that they must recommend only a single district, the government may change the Committee's charge to expand the scope of the original charge. If so charged, the Gaian Committee can conduct another analysis to include additional environmental social, economic, and cultural CNs. This would be a new PN, which stakeholders will define and analyze with their new set of CNs, CMs, EJSs, and VSs.

Table 5.20 Results from the Final Run of the Gaian Committee

The Gaian Committee's Final Run				
Rank	Ranked PSEs, most to least vulnerable	10 Bins	Decision Distance (DD) (%) of the 2nd ranked District	ΔDD_{adj} (%)
1	District 2	1	0.0	0.0
2	District 6	1	6.5	6.5
3	District 8	1	9.0	2.5
4	District 12	2	15.2	6.1
5	District 13	2	15.8	0.6
6	District 11	2	15.9	0.1
7	District 14	4	30.6	14.7
8	District 10	4	39.8	9.2
9	District 9	6	50.1	10.3
10	District 4	6	52.0	1.9
11	District 3	6	54.6	2.6
12	District 7	6	54.8	0.2
13	District 5	7	64.7	9.9
14	District 1	10	99.3	34.5
15	District 15	10	100.0	0.7

5.5   When All Criteria Do Not Apply: Defining Classes

In some environmental analyses, the full set of criteria, identified for the evaluation of the PN, do not apply to every PSE. This issue can be illustrated with an example using submerged aquatic vegetation (SAV). SAV is an indicator that applies only to those areas with water bodies such as lakes or bays. The presence of SAV signals healthy water bodies with adequate dissolved oxygen and habitat for aquatic life. Because consistency in the application across PSEs is a feature of an open solution analysis, data for every PSE is input for every CM. Therefore, in an analysis where SAV is used, applying an SAV value to all PSEs, including solely terrestrial ones, is problematic. This means that in areas that have no lakes or bays, the data for the SAV CM equals zero. The zero SAV value for these areas is not equivalent to a zero SAV value for areas with water bodies. In the first case, there is no expectation, and therefore there should be no judgment made, that SAV at any level should contribute to the analysis of a district's environmental vulnerability. The inclusion of SAV as an indicator, which contributes to the CSs of these PSEs, inappropriately penalizes (or overestimates the vulnerability of) these terrestrial areas as exhibiting excessively poor environmental conditions. In the second case, some SAV level is expected because the area has water bodies. In these areas, a zero SAV value accurately signals an extremely poor or vulnerable condition.

To address this issue, MIRA offers a *Classes* feature, which allows stakeholders to analyze a PN where one or more indicators do not apply to every PSE. By defining multiple subsets of PSEs that can belong to different Classes, each Class can be evaluated by a different subset of indicators.

> **Classes**
> The use of Classes allows stakeholders to appropriately analyze PSEs that need to be evaluated with different subsets of indicators.

The SAV example represents one of the simplest applications of Classes in that only one criterion (SAV) and two Classes (areas with water bodies and areas without water bodies) are involved. Some applications may need stakeholders to define many Classes, each of which will be evaluated using a unique subset derived from the full set of indicators. For example, some policy problems may involve evaluating SAV, forests, and freshwater macroinvertebrates (measured as the Index of Biotic Integrity (IBI)), features that do not exist in every PSE. In this more complex example, the PSEs in the analysis could contain none of these features, or one, two, or all of them. Each of these combinations becomes a Class, with a total of seven in this example: (1) No SAV, forests, or IBI; (2) SAV only; (3) forests only; (4) IBI only; (5) SAV and forests; (6) forests and IBI; and (7) SAV and IBI.

In general, when MIRA's Class methodology is used, each defined Class is characterized as follows: (1) each subset of PSEs within a specific Class is evaluated by a unique set of indicators, (2) each PSE is a member of only one Class, and (3) the CSs calculated for all PSEs, regardless of Class, must be able to be compared and ranked in an unbiased manner.

When utilizing Classes, in those indicator clusters where one or more indicators do not apply, the Preference weights that are assigned by the stakeholders to the original set of indicators are reproportioned to ensure that 100% of the available Preference weights are still distributed. Alternatively, stakeholders may choose to solicit Preferences for the affected clusters. The reason for this is because when calculating the CS for a PSE that resides within any single Class, any indicator that does not apply in that Class is assigned a Preference weight equal to 0.0. Doing this effectively eliminates any contribution to the CS from this indicator for that Class. If the weights are not re-proportioned, an analytical bias will result that underweights the affected indicator cluster by the amount equal to the original Preference weight of the removed indicator. The reproportioned weights for the applicable indicators within such a cluster are determined as follows:

$$\text{Pref\_Class}_i = \left(\frac{\text{Pref}_i}{\sum\limits_{i=1}^{n} \text{Pref}_i} \right) \tag{5.3}$$

Where:

Pref_Class$_i$ ≡ the reproportioned Preference weight for each ith applicable indicator in the cluster
Pref$_i$ ≡ the Preference weight assigned by the stakeholders to the ith indicator
n ≡ the # of indicators in the cluster

Suppose that the Gaian Committee decides to include SAV as an additional indicator in the analysis because five districts contain water bodies and they believe that SAV is important for their analysis of environmental vulnerability. For districts with water bodies, more SAV indicates a healthier and less vulnerable district. Since larger values signal a better condition, SAV is added to the weighted functional Hierarchy as a Reverse Indicator in the surface water cluster. Ten of the 15 Gaia districts do not have any water bodies. Therefore, the Gaian Committee constructs two different Classes of PSEs, one for those districts that contain water bodies and a second for those districts that do not. The Gaian Committee's surface water CI cluster, with the new SAV TI[8] added, is shown in Fig. 5.11.

---

[8]As a new CM, SAV experts index the SAV CM after discussion with other discipline experts to ensure consistency in determining significance across all CMs used in the analysis. Once indexed, the SAV CM is a Terminal Indicator (TI).

Fig. 5.11 Surface water
cluster including SAV

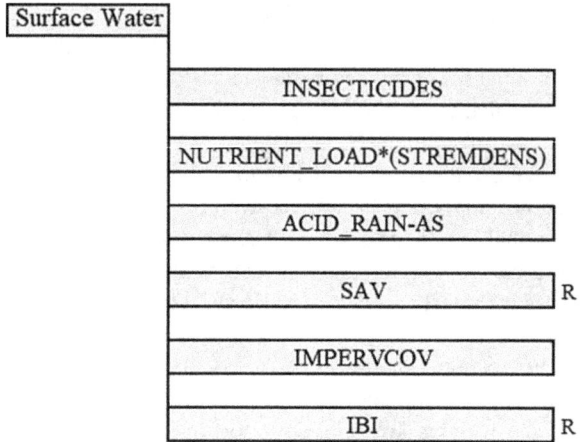

Surface Water

INSECTICIDES

NUTRIENT_LOAD*(STREMDENS)

ACID_RAIN-AS

SAV	R

IMPERVCOV

IBI	R

Since the SAV indicator is placed in the surface water CI cluster, this is the only indicator cluster in the Hierarchy that needs to have its Preference weights adjusted. Table 5.21 compares the surface water cluster VS used for those districts that contain water bodies with those that do not and need Preference weight reproportioning.

With SAV added to the surface water cluster, Column A in Table 5.21 shows the Gaian Committee's newly solicited[9] VS. When Classes are not considered, the VS in Column A is applied to all districts regardless of whether they contain water bodies. When Classes are considered, the VS in Column A is applied only to districts with water bodies, and the redistributed Preference weights (reproportioned VS) in Column B are applied to districts without water bodies.

Table 5.22 shows the five districts that have bodies of water with their accompanying SAV CM values.[10] All the other districts do not have water bodies and, therefore, the value of their CMs for SAV is zero.

With the addition of SAV to the Gaian weighted functional Hierarchy, its CM values require Indexing. The SAV experts engaged by the Gaian Committee index the SAV CM after discussion with the other experts. The SAV Indexing is presented in Table 5.23.

---

[9] When new indicators are added, stakeholders discuss and agree on a new set of Preference weights for the affected indicator cluster (Column A in Table 5.21). In Column B, as an alternative to reproportioning the Preference weights, stakeholders could newly solicit a VS for this Class (with weights adding up to 1.0). If stakeholders intend to maintain their Preferences for those PSEs where the indicator cluster effectively remains as the original cluster, reproportioning the Preference weights is acceptable. Most often, this is the case.

[10] The SAV CM is an example of a CM that requires more refinement. This SAV data in hectares is not expected to be specific enough for most CNs. A district with relatively low acreage of SAV could represent 100% coverage of its water body while a district with relatively high acreage might represent a small percentage of its water body. Indicators such as SAV are more meaningful if presented as a percentage of total water bodies or total district area. However, this SAV CM is used in this section to illustrate the Classes feature.

Table 5.21 Comparison of Preference weights for districts with and without water bodies

Indicators in surface water cluster	A. VS for surface water cluster with water bodies	B. VS for surface water cluster without water bodies
INSECTICIDES	0.15	0.25
NUTRIENT_ LOAD*(STREMDENS)	0.15	0.25
ACID_RAIN-AS	0.10	0.17
SAV	0.40	0.00
IMPERVCOV	0.10	0.17
IBI	0.10	0.17

Table 5.22 SAV data for the Gaia districts with water bodies

Districts with water bodies	SAV (ha)
District 11	67
District 10	130
District 5	226
District 4	485
District 9	688

Table 5.23 Indexing scheme for the indicator SAV

Index →									
TERMINAL INDICATORS	UNITS	1	2	3	4	5	6	7	8
SAV	**ha**	2500	2144	1789	1433	1077	721	366	10

Table 5.24 presents a comparison of the Gaian analysis with and without Classes. The stakeholder's Final EJS revised to include SAV (Table 5.23) and the stakeholders' Final VS, adjusted for the inclusion of SAV (Table 5.21) are used for the comparison runs.

The ranks of the five districts with water bodies (Districts 4, 5, 9, 10, and 11) either increased (appear to be more vulnerable) or stayed the same when Classes are considered. Of these districts, District 11 shows the largest change in rank, as it moves up three positions. Districts 5, 9, and 10 become relatively more vulnerable by two positions, while District 4's rank remains the same in both runs.

Although both Districts 11 and 14 (District 14 does not contain water bodies) changed three positions in ordinal rank, comparing their DDs presents a very different story. District 11's DD changed by 19.4% (30.2% − 10.8%) while District 14's DD changed by −0.4% (27.9% − 30.3%). This informs the stakeholders that the impact from including SAV is far more significant for District 11 than it is for District 14 even though their change in ordinal rank would imply similar impact.

Because the districts rankings are relative, districts without water bodies also changed positions. Notably, among the districts without water bodies, District 12 shows the most significant change based on its DD (11.3% − 21.0%). District 12 appears substantially more vulnerable (rank #3) when Classes are ignored. District

Table 5.24 Comparison of district rankings with and without the use of Classes

MIRA run without Classes			MIRA run with Classes		
Ranked PSEs, most to least vulnerable	DD (%)	Δ DD$_{adj}$ (%)	Ranked PSEs, most to least vulnerable	DD (%)	Δ DD$_{adj}$ (%)
District 2	0.0	0.0	District 2	0.0	0.0
District 6	**5.8**	5.80	District 6	**8.1**	8.1
District 12	**11.3**	5.50	**District 11**	**10.8**	2.7
District 8	25.1	13.80	District 8	19.8	9.0
District 14	**27.9**	2.80	District 12	**21.0**	1.2
District 11	**30.2**	2.40	**District 10**	28.5	7.5
District 13	31.9	1.70	District 13	29.0	0.5
District 10	47.1	15.20	District 14	**30.3**	1.2
District 4	50.6	3.50	**District 4**	32.2	1.9
District 3	56.2	5.60	**District 9**	40.9	8.7
District 7	58.6	2.40	**District 5**	46.1	5.2
District 9	59.0	0.30	District 3	61.1	15.0
District 5	64.0	5.00	District 7	61.2	0.1
District 1	88.8	24.90	District 1	99.6	38.4
District 15	100.0	11.20	District 15	100.0	0.4

Bolded district names indicate those containing water bodies

1, another district without a water body, retains the same ordinal rank in both runs (#14). However, its DD changes from 88.8%, when Classes are ignored, to 99.6%, when Classes are considered, a 10.8% drop, a measure of the degree to which District 1's vulnerability was inappropriately overestimated. When Classes are ignored, District 1 appears more vulnerable. These results are expected because when Classes are used, SAV's indexed value, for the districts without water bodies, changes from 8.0 to 0.0, numerically removing SAV completely from consideration in the analysis.

Additionally, from Table 5.24, the top ranked district, District 2, compared with District 6 (#2 rank) shows a relative increase in vulnerability, as indicated by a slightly larger DD that results when Classes are used (8.1% vs. 5.8%). The greater separation between Districts 2 and 6 shows that with the appropriate consideration of Classes, the Decision Uncertainty should be slightly less if the stakeholders are considering recommending District 2.

One might expect that once Classes are appropriately considered, districts without water bodies will be appropriately assessed as less vulnerable, as they are no longer inappropriately penalized for a zero SAV CM. However, District 2, a district without water bodies, is the top ranked district whether Classes are considered or ignored. This warrants a more detailed examination using Indicator Drivers to determine how this is so. Figure 5.12 compares the Indicator Drivers in the surface water cluster for Districts 2 with and without the consideration of Classes.

When Classes are not considered for District 2, the SAV CM value indexes to 8.0 on the decision scale, since its value is zero. In this case, a single VS is applied to all PSEs (Value Set A in Table 5.21) and SAV is (incorrectly) the biggest contributor to

SAV Analysis for District 2 with and without the use of Classes

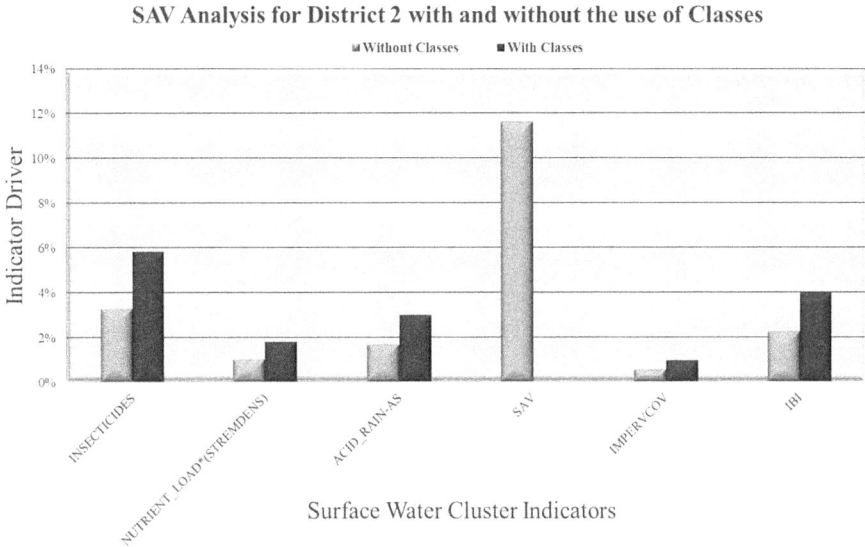

Fig. 5.12 Comparison of the ID's for District 2's surface water cluster with and without using MIRA's Class methodology

the surface water portion of District 2's CS. When Classes are considered, SAV appropriately has no contribution to District 2's CS. For District 2, the total surface water ID contribution, when Classes are ignored (the sum of the light-colored bars in Fig. 5.12), is 20% while it is 16% when Classes are considered. The District 2 IDs confirm that its ranking is the result of its proper assessment.

The results from the evaluation of District 2 with and without Classes also have a potentially important implementation implication if District 2 were selected by the Gaian Committee. Table 5.25 presents the Indicator Drivers for District 2 (with and without the consideration of Classes) for the surface water CI and the three CIs at the primary level of the Hierarchy: human health, ecological integrity and future impacts. As the top ranked PSE, the Gaian Committee is seeking to understand which indicators contribute the most to District 2's CS. The greatest contributing indicators suggest the area of focus for the Gaian Committee's mitigation efforts.

The differences in District 2's IDs with and without the use of Classes present the Gaian Committee with a different assessment. When Classes are appropriately considered, the surface water CI contributes less to District 2's CS than when Classes are ignored. This is also reflected in the change in ID at the primary level of the Hierarchy where the ecological integrity CI (of which surface water is a component) has a commensurate reduction in its contributor to District 2's CS. For the Gaia government, the import of failing to use Classes when appropriate is that the resulting analysis is misleading. Such an analysis could lead the Gaia government to spend more of its limited resources on an overemphasis of surface water issues in District 2 rather than on other issues that may be more important.

Table 5.25 District 2 comparison of Indicator Drivers without and with Classes

Select District 2 Indicator Drivers without and with Classes		
Composite Indicator	ID (%) without Classes	ID (%) with Classes
Surface water	20	16
Human health	28	30
Eco integrity	46	42
Future impact	26	28

5.6   Summary

Experimentation with different Hierarchies, Criteria Metrics, Expert Judgment Sets, Value Sets, and combinations of their associated uncertainties is an essential feature of an open solution approach. In MIRA, performing sensitivity and uncertainty analyses with each of those components and in combination is a means for stakeholders to explore and reduce Decision Uncertainty within the context of their Problem Narrative. MIRA is a new paradigm for conducting public policy analysis, with each iteration providing another opportunity for trans-disciplinary learning.

Because of the transparency of the MIRA process, the use of backcasting helps stakeholders to avoid a common pitfall of current decision making that biases the analysis by inconsistently using data, EJSs, or VSs. Stakeholders can use backcasting to determine what VS justifies their desired outcome. To justify a favored PSE ranking, stakeholders may find that they will have to agree to a VS that they cannot defend or support. Consequently, backcasting can be an effective way to convince initially unconvinced stakeholders to consider more justifiable alternatives.

Finding common ground among stakeholders, in policy analysis, is given new meaning with the use of the MIRA open solution approach. Rather than a process that permits stakeholders to more firmly hold their original positions as others express their views, a stakeholder using the MIRA process gains opportunities to test her positions and then discuss these results together with the results from the testing of other stakeholders' positions. Through the MIRA process stakeholders learn about the relationship between the PSEs and the data by exploring nuances in their own and others' perspectives while guided by the Problem Narrative and the other analytical components that they helped formulate, select, and synthesize. A process that allows the analysis of as many Value Sets as is necessary to explore the diversity among stakeholders is likely to provide the greatest chance for agreement on a final decision. Furthermore, exploring the combined effects of multiple uncertainties (data, EJSs, and VSs) gives decision makers and other stakeholders the confidence and effectiveness they need to fulfill their public policy responsibilities, transparently and rationally.

Diverse and contentious stakeholders, decision makers who want certain results, and wicked problems are difficult but common components of public policy decision making. Keeping the facts separate from stakeholder values while applying the facts and those Preferences consistently across all the decision alternatives is central to the MIRA approach. Therefore, even with broad latitude in finding many possible solutions to the wicked problem, good public policy makers want to ensure that data are not selectively manipulated to favor one decision over another and expert judgments and Preferences are not inconsistently applied across the possible alternatives. Collectively, stakeholder inclusiveness, transparency and trans-disciplinary learning, as guided by Decision Uncertainty, are the trademarks of an open solution approach to effective public policy decision making. MIRA is a practical and feasible open solution approach that represents a next-generation approach for public policy making.

References

1. U.S. Environmental Protection Agency (2018) National Air Toxics Assesssment: 2014 Technical Support Document. https://www.epa.gov/national-air-toxics-assessment/2014-nata-technical-support-document. Accessed 30 June 2019
2. Stahl CH, Cimorelli AJ, Mazzarella C, Jenkins B (2011) Toward sustainability: a case study demonstrating trans-disciplinary learning through the selection and use of indicators in a decision making process. Integr Environ Assess Manag 7(3):483–498
3. U.S. Environmental Protection Agency (2004) Air quality designations and classifications for the 8-hour Ozone National Ambient Air Quality Standard; early action compact areas with deferred dates. Fed Regist 84(69):23858–23951
4. Saaty TL (1990) The analytic hierarchy process: planning, priority setting, resource allocation. RWS Publications, Pittsburgh

Chapter 6
Open Solution Approaches: Not Just for Decision Making

Abstract Indices have policy and decision making implications as they are created to simplify the delivery of a more complex set of information. The open solution approach used in alternatives ranking and multi-criteria decision analysis also provides a solid foundation for understanding indices and their construction. Examples of commonly found indices are the Dow Jones Industrial Average (DJIA), Consumer Reports' ratings for appliances, US News and World Reports College Ranking, the Organisation for Economic and Co-operative Development (OECD) Better Life Index, and Sustainability indices. The Federal Reserve Board and investors look to the DJIA's value as one of many factors to assess US fiscal policy, determine whether and what stocks to invest, and judge the health of the US economy. A highly ranked appliance may compel consumers to choose that brand over others, resulting in increased profits for the appliance manufacturer. Therefore, stakeholders commonly use indices to help them to decide what to do but typically do so without understanding what factors comprise these indices. The current reality of indices is that they are often a convenient shortcut around a robust policy discussion to a policy conclusion. A more knowledgeable citizenry schooled in open solution approaches could change this reality, by better informing their own decision making and by making index creators more accountable.

Keywords Index construction · Multi-criteria assessment · Public policy implications · Public dialogue

6.1 Introduction

Indices have policy and decision making implications as they are created to simplify the delivery of a more complex set of information. The open solution approach used in alternatives ranking and multi-criteria decision analysis also provides a solid foundation for understanding indices and their construction. Examples of commonly found indices are the Dow Jones Industrial Average (DJIA), Consumer Reports' ratings for appliances, US News and World Reports College Ranking, the Organisation for Economic and Co-operative Development (OECD) Better Life Index, and Sustainability indices. The Federal Reserve Board and investors look to the DJIA's value as one of many factors to assess US fiscal policy, determine whether

© Springer Nature Switzerland AG 2020   151
C. H. Stahl, A. J. Cimorelli, *Environmental Public Policy Making Exposed*,
Risk, Systems and Decisions, https://doi.org/10.1007/978-3-030-32130-7_6

and what stocks to invest, and judge the health of the US economy. A highly ranked appliance may compel consumers to choose that brand over others, resulting in increased profits for the appliance manufacturer. Therefore, stakeholders commonly use indices to help them to decide what to do but typically do so without understanding what factors comprise these indices. The current reality of indices is that they are often a convenient shortcut around a robust policy discussion to a policy conclusion. A more knowledgeable citizenry schooled in open solution approaches could change this reality, by better informing their own decision making and by making index creators more accountable.

Like multi-criteria decision making, index creation follows the Requisite Steps. Indices are equivalent to the MIRA Criteria Sum (CS). Because alternatives ranking and multi-criteria decision analysis effectively create indices as an intermediate step, it makes sense to more generally consider the utility of an open solution approach in index construction. Index construction presents many of the same choices about defining the Problem Narrative (PN), selecting Problem Set Elements (PSEs), choosing Criteria Narratives (CNs), selecting data and constructing Criteria Metrics (CMs), determining the significance of the data, and applying Preferences (determining the Value Set, VS). The context and the assumptions used to create indices determine their meaning. Do we understand what they are? Using both scientific and common examples, this chapter illustrates how these complex choices can be synthesized to the Requisite Steps of decision making introduced in Chap. 1. The MIRA open solution approach, with its hallmark components of inclusiveness, transparency, trans-disciplinary learning and Decision Uncertainty, is used to dissect and examine a variety of commonly found indices.

6.2   Multi-criteria: All Around Us

Indices are important in many different facets of our lives. They can be a convenient measure to summarize a complex issue but, without transparency, they can also create problems. For example, for many years, Temple University's Fox School of Business in Philadelphia, Pennsylvania, was ranked #1 by the US News and World Reports College Ranking. However, in 2017, the Fox School of Business was widely condemned for submitting false data to the US News and World Reports College Ranking to intentionally inflate their ranking for their online MBA program [1]. Since then, Temple University's Fox School of Business was found to have submitted false data for five more MBA programs. The Dean of the Fox School of Business, Moshe Porat, was fired. Temple students sued the university. The Dean and others knew that the quality of the student body and the reputation of the Fox School of Business were determined, in part, by the school's US News and World Reports College Ranking. Considered as a measure that could boost a school's reputation among applicants, the US News and World Reports College Ranking index score became the metric to conquer for Dean Porat. The lack of transparency made it easier to deceive potential student applicants.

Temple has committed to fixing the problem and its Fox School of Business no longer participates in the US News and World Reports College Ranking or the Princeton College Survey, which uses similar data. However, the US News and World Reports College Ranking methodology continues to lack transparency for the potential students, their parents, and alumni regarding exactly what Criteria Narratives and data are used and how they are Preferenced. For the community using these rankings to select schools, the question is whether the US News and World Reports and the participating schools should be more accountable and transparent about these school rankings. What are the CNs and CMs that can distinguish between a poor, good, or excellent school? Who selects them? Whose Preferences are applied? With potentially large investments and careers at stake, there are decision making consequences for school-bound families. With so much at stake and the continued lack of transparency about what these college rankings represent, citizens should question the legitimacy of the index and the message it implies.

Indices are used when many factors need to be considered simultaneously. When conclusions are drawn about the robustness of the economic market or the well-being of a country, information about employment, new jobs created, and the number of citizens with access to affordable health care and to a high school education are among the many kinds of information used to make such assessments. In these circumstances, the open solution approach can be effectively applied, even when formal decisions are not being made. In this chapter, examples from the US EPA, the OECD, and *Newsweek* are used to illustrate how the Requisite Steps of decision making are implicitly used to construct these indices. Examining indices in this way can help interested citizens become more informed consumers, investors, and community members.

6.3   US EPA Hazard Ranking Score

The US EPA Hazard Ranking Score (HRS) is a scoring system that was established in a 1990 final rule and is used by federal program managers. Its purpose is to evaluate the potential hazards from air, soil, and water at abandoned sites that are subject to the US EPA's Comprehensive Environmental Response, Compensation, and Liability Act (CERCLA, also known as the Superfund program) [2]. Guided by the HRS, the US EPA ranks Superfund sites (the PSEs) to prioritize their cleanup. The PN for the HRS requires the evaluation of the migration of chemical contaminants at a site through air, soil, surface water and ground water.

The HRS is an algorithm designed to examine four migration pathways (groundwater, surface water, soil and air) for chemical contaminants in a defined geographic area. Each of these pathway components is a Composite Indicator (CI) where the analyst considers the following three factor categories (the CNs at the second level of the Hierarchy): (1) the likelihood of release, (2) the waste characteristics, and (3) the targets (e.g., human health or streams). The maximum numerical value (the indexed CMs) of each of these factor categories is different and is established in the

US EPA final rule. Once the numerical value for each factor category is determined for a site, the pathway CI is the product of the numerical values for those three factor categories divided by a scaling factor (82,500) to ensure that the pathway score stays between 0 and 100. This procedure is repeated for each of the four pathways.

The four pathways' indices are then combined into the HRS (also a CI), which itself ranges from 0 to 100 in value. The overall HRS is combined via a root mean square of the individual pathway scores as shown in Eq. 6.1.

$$HRS = \sqrt{\frac{S_{gw}^2 + S_{gsw}^2 + S_{sessi}^2 + S_a^2}{4}} \tag{6.1}$$

Where:

$S_{gw} \equiv$ ground water migration score
$S_{sw} \equiv$ surface water migration score
$S_{sessi} \equiv$ soil exposure and subsurface intrusion score
$S_a \equiv$ air pathway migration score

The HRS is conceptually equivalent to the MIRA Criteria Sum, with Eq. 6.1 the equivalent of Preferencing and combining primary-level Composite Indicators (the four CIs, S_a, S_{sess}, S_{gw}, S_{sw}) with those Preference weights. Having constructed the HRS algorithm in this way, the four pathways appear to be equally weighted (i.e., equally Preferenced). However, because the individual pathway terms are combined as a root mean square, the larger a pathway score (relative to another pathway score), the larger its effective weight becomes. If the HRS algorithm had been constructed by simply weighting and summing the pathway scores, instead of squaring them, the Preference weights for the higher scores would have to be larger than those for the lower scores to achieve the same result as produced by Eq. 6.1. Details for the calculation of the HRS can be found in the 1990 final rule.

Two tables from that rule (Table 2.7 from page 51532 and Table 3.2 from page 51596) are partially reproduced in Tables 6.1 and 6.2 to illustrate the transformation of two criteria to the HRS decision scale, which is the equivalent of establishing an Expert Judgment Set (EJS) in the MIRA open solution approach. These two tables pertain to the waste and ground water migration CMs.

Table 6.1 Waste characteristic EJS

Table 2.7 Waste characteristics factor category values	
Waste characteristic product	Assigned value
0	0
Greater than 0 to less than 10	1
10 to less than 1×10^2	2
...	...
1×10^{12}	1000

Table 6.2 Ground water migration EJS

Table 3.2 Containment factor values for ground water migration pathway

Source	Assigned value
Evidence of hazardous substance migration from source area: No liner	10
A. No evidence of hazardous substance migration from source area, a liner, and none of the following present: (1) maintained engineered cover or (2) functioning and maintained run-on control system and runoff management system or (3) functioning leachate collection and removal system immediately above liner	10
Any one of the three items present in A	9
Any two of the three items present in A	7
…	…
Source area inside or under maintained intact structure that provides protection from precipitation so that neither runoff nor leachate is generated	0

In these two tables,[1] the two CMs are being indexed to a decision scale that ranges from 0 to 1000. However, other Indexing ("assignment") tables in the US EPA final rule indicate that the decision scale range is much larger. For example, the maximum indexed values for the hazardous waste quantity factor and the human food chain population factor are 1 million and 3.1 million, respectively. The decision scale is something that users could examine to determine whether there is an unintentional bias in the HRS calculation. For example, for CMs such as ground water migration, the worst case scenario may not sufficiently contribute to the HRS calculation due to the constraint on their maximum value. For Superfund sites with hazardous waste quantities resulting in the maximum indexed value, no other factor may matter. If these kinds of situations are intentionally designed, the HRS is appropriate. However, these and other questions are for HRS users to ponder. Furthermore, these factors after Indexing are combined into a pathway score, which requires Preferencing. Therefore, the VS for the HRS is a combination of a portion of the methodology used to calculate each pathway score and the primary-level Preferencing reflected in Eq. 6.1.

Where regulatory requirements stipulate the use of the HRS, users typically input variables into Eq. 6.1 without considering the construct of the HRS. However, if stakeholders want to use the same criteria as the HRS but also want additional flexibility to Preference those criteria differently, a MIRA Hierarchy can be constructed. Figure 6.1 shows how such a Hierarchy could be constructed by expanding the S_{GW} branch (similar branches could be constructed for the other terms in Eq. 6.1).

Furthermore, where there is flexibility to construct the CNs and CMs differently (or to use different CNs and CMs) than those in the HRS (i.e., a different PN), the open solution approach can help stakeholders derive these alternative components. Knowledgeable users can use their PN to construct a different Hierarchy, articulate CNs and CMs, work with experts determining the significance of the CMs, and define Preference weights, rather than being constrained to what is prescribed in the HRS approach.

---

[1] There are many similar tables in the final rule that "assign values." These "assignment" tables translate the raw data for each CM onto a non-dimensional decision scale.

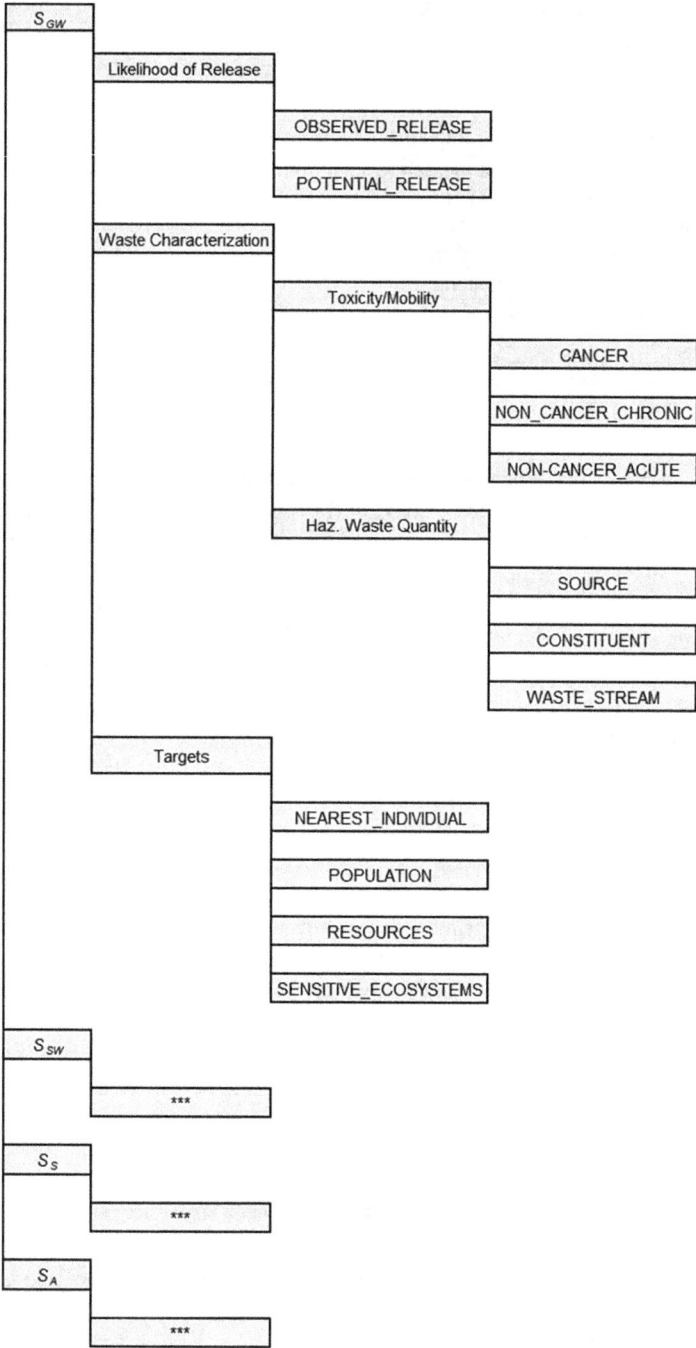

Fig. 6.1 Hazard ranking score Hierarchy with S_{GW} Composite Indicator branch detail

6.4 US EPA BenMAP

The US EPA Environmental Benefits and Analysis Mapping (BenMAP) tool is used to evaluate the economic impacts (i.e., the costs and benefits) of controlling emissions that affect the attainment of the ozone and particulate matter National Ambient Air Quality Standards (NAAQS) [3]. When the US EPA evaluates different geographic regions and compares their status based on the BenMAP results, these regions are the PSEs. In BenMAP, the CNs are not specifically or separately articulated from the CMs. BenMAP is an example of a tool with a default data catalogue as well as a default algorithm for how those data can be used.

The eight CMs in BenMAP are the valuation or economic costs of (1) mortality, (2) asthma emergency room visits, (3) asthma hospital visits, (4) myocardial infarction emergency room visits, (5) myocardial infarction hospital visits, (6) school days lost due to illness from high levels of ozone or particulate matter, (7) work days lost due to illness from high levels of ozone or particulate matter, and (8) household soiling due to high levels of particulate matter. The BenMAP index can be replicated in a MIRA Hierarchy as shown in Fig. 6.2.

In this linear BenMAP Hierarchy, all the Criteria Metrics have dollars as their units. The CN descriptions are implied by the CMs selected. Each CN is the cost of the identified effect, e.g., cost of mortality (value of a life) or cost of asthma emergency room visits. Furthermore, the significance of the CMs is set with their monetization, larger costs being proportionally more significant than lower costs. The decision scale units are in dollars with the range of the scale reflecting the range of costs across the eight CMs. Not only do costs establish the significance in BenMAP but they also determine the relative Preferences across the indicators. Monetizing criteria effectively combines Indexing and Preferencing into a single step. Therefore, in BenMAP, the effective Expert Judgment Set (EJS) and Value Set are set by the relative dollar costs.

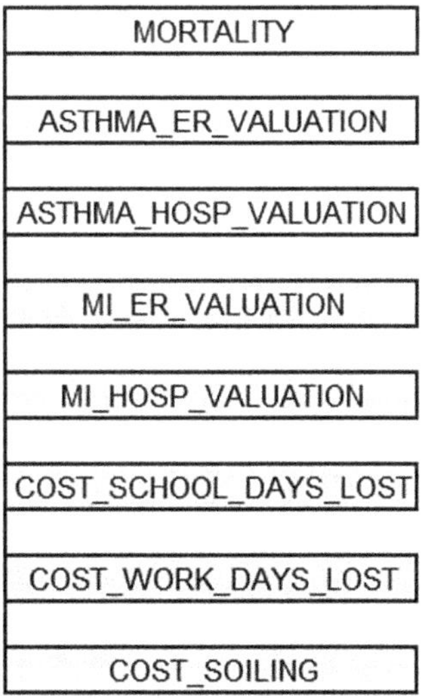

Fig. 6.2 BenMAP
Hierarchy

MORTALITY
ASTHMA_ER_VALUATION
ASTHMA_HOSP_VALUATION
MI_ER_VALUATION
MI_HOSP_VALUATION
COST_SCHOOL_DAYS_LOST
COST_WORK_DAYS_LOST
COST_SOILING

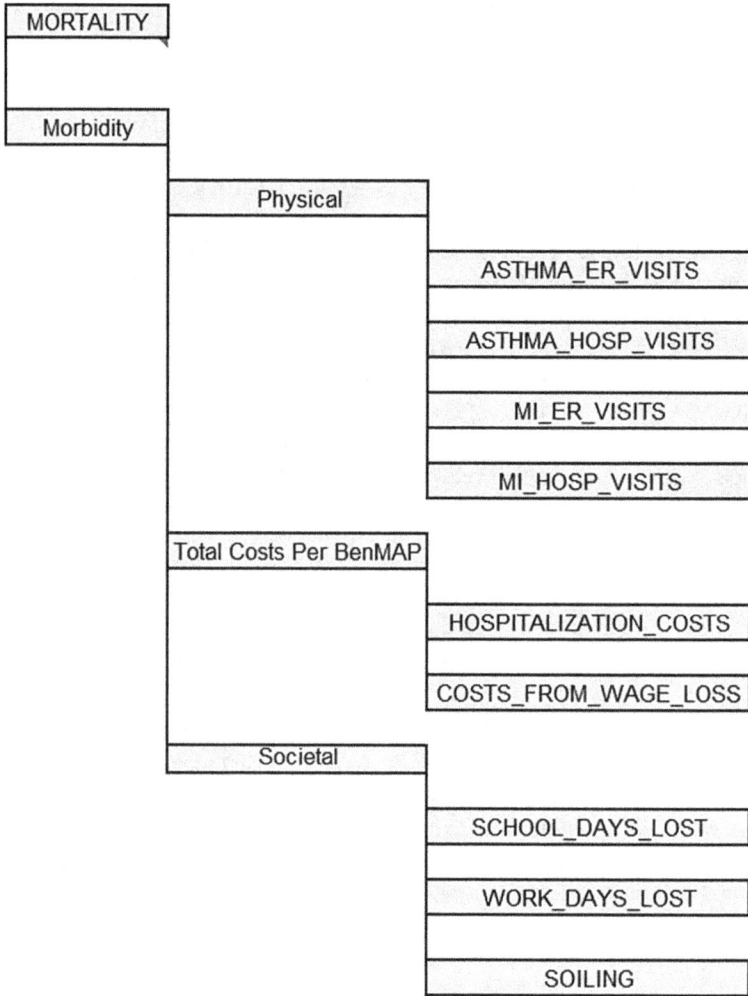

Fig. 6.3 Example alternative MIRA Hierarchy using BenMAP data

Provided users understand and agree to the implications of the choices made in BenMAP, there is nothing wrong with using the BenMAP index as programmed in the US EPA tool. For stakeholders with the flexibility to Preference the BenMAP indicators differently, it is possible to accomplish this with a MIRA Hierarchy. This alternative is shown in Fig. 6.3.

However, stakeholders desiring more flexibility to add or remove indicators or weight indicators independent of economics can consider many other possibilities for Hierarchy construction and developing non-monetized CMs. For example, cued by the BenMAP work or school days lost indicators, it is possible to consider them differently. The loss of work or school days can be viewed alternatively as a measure of inconvenience, a measure of predicted high school graduation rates, or a measure of another social or cultural concern. Considering other potentially important non-monetary factors such as the ecological damage of pollutants to national forests, the equity (social and economic) impacts on differentially sensitive populations, or the regulatory burden on state and local governments when their jurisdictions impose emission controls to meet the ozone and particulate matter NAAQS is one of the many possibilities that exist.

6.5   OECD Better Life Index

The Organisation for Economic and Co-operative Development (OECD) is an intergovernmental economic organization with 36-member countries whose purpose is to stimulate economic progress. The OECD's stated goals are to "shape policies that foster prosperity, equality, opportunity and well-being for all" [4]. The Better Life Index (BLI) was developed by the OECD to allow users to rank the quality of life in countries based on considerations other than Gross Domestic Product (GDP) or other, more traditional, economic statistics. With its BLI web dashboard, the OECD invites users to apply their own opinions regarding the relative importance of these considerations. In the 2017 version of the OECD's BLI, 38 countries are included in the rankings: 35 OECD members (for unstated reasons, Lithuania was the only member country not included) plus an addition of 3 non-member countries (Brazil, the Russian Federation, and South Africa), bringing the total to 38 countries that were evaluated and ranked.

The OECD defines 11 CNs to assess the well-being of a country's citizens in the BLI.[2] These 11 CNs are (1) housing, (2) income, (3) jobs, (4) community, (5) education, (6) environment, (7) civic engagement, (8) health, (9) life satisfaction, (10) safety, and (11) work-life balance. In total, 24 CMs specifically define the 11 CNs. Based on the report describing the BLI, the OECD does not indicate or suggest that these 11 CNs are the result of discussions outside of those within the OECD group tasked with formulating the BLI [4]. The descriptions of the CMs for each of the 24 indicators are provided by the OECD and can be viewed at https://stats.oecd.org/ Index.aspx?DataSetCode=BLI or, with the MIRA Criteria Names (for analysis in Sec. 6.5.1), in Table A.10 in the Appendix. The ranking of these countries, relative to their quality of life, or well-being of their citizens, is based on these CMs that have been collected or estimated by the OECD.

---

[2] Other organizations have different versions of a Better Life Index. For example, the United Nations' report on happiness is similar. It can be found at https://s3.amazonaws.com/happiness-report/2019/WHR19.pdf. For the UN, six major Criteria Narratives are used: (1) GDP per capita, (2) social support, (3) healthy life expectancy at birth, (4) freedom to make life choices, (5) generosity, and (6) perception of corruption. While this and other similar indices are not evaluated here, the interested reader is encouraged to explore these indices using the concepts of the open solution approach.

The data (CMs) used for the BLI are available but there are unexplained gaps and inconsistencies among the accessible data tables. There is no discussion in the BLI metadata or report regarding the data (missing, available, or estimates). Therefore, while not ideal because there is a greater likelihood of using data not intended to be used together, it was possible to find numerical values for each country's CM, even if these values do not all appear in a coherent set of OECD data tables.

Figure 6.4 is a screenshot showing the CMs organized for the first three Composite Indicators (housing, income, and jobs) and their corresponding Terminal Indicators. The screenshot shows a portion of the Hierarchy that is used to calculate the BLI.

The full Hierarchy for the 24 CMs of the BLI is shown in Fig. 6.5.

The 11 CNs used by the OECD are primary-level indicators in a MIRA Hierarchy. Of the 11 CNs, two are Terminal Indicators (TIs) and nine are Composite Indicators (CIs).

The accessibility to BLI data allows users to examine these CMs. For example, disposable income and net worth define the Income Composite Indicator. Savvy stakeholders schooled in the open solution approach can discuss to what extent income or wealth correlates or predicts well-being. While it can be argued that defining an adequate living income is location-dependent, it is also true that one's sense of well-being or fulfillment based on income is culturally, societally, and personally dependent. By comparing the direct numeric values for these indicators across the 38 countries, the OECD is equating the same level of well-being to an income level and net worth in every country.

There are also many questions about the other non-economic indicator choices used in the BLI. For example, the two environmental indicators chosen are $PM_{2.5}$

→ Inequality	Total ▼								
→ Measure	Value								
		Housing			Income		Jobs			
→ Indicator		Dwellings without basic facilities ⓘ	Housing expenditure ⓘ	Rooms per person ⓘ	Household net adjusted disposable income ⓘ	Household net financial wealth ⓘ	Labour market insecurity ⓘ	Employment rate ⓘ	Long-term unemployment rate ⓘ	Personal earnings ⓘ
Unit		Percentage	Percentage	Ratio	US Dollar	US Dollar	Percentage	Percentage	Percentage	US Dollar
		▲ ▼	▲ ▼	▲ ▼	▲ ▼	▲ ▼	▲ ▼	▲ ▼	▲ ▼	▲ ▼
→ Country										
Australia		..	20	..	32 759	427 064	5.4	73	1.31	49 126
Austria		0.9	21	1.6	33 541	308 325	3.5	72	1.84	50 349
Belgium		1.9	21	2.2	30 364	386 006	3.7	63	3.54	49 675
Canada		0.2	22	2.6	30 854	423 849	6	73	0.77	47 622
Chile		9.4	18	1.2	..	100 967	8.7	63	..	25 879
Czech Republic		0.7	24	1.4	21 453	..	3.1	74	1.04	25 372
Denmark		0.5	23	1.9	29 606	118 637	4.2	74	1.31	51 466
Estonia		7	17	1.6	19 697	159 373	3.8	74	1.92	24 336
Finland		0.5	23	1.9	29 943	200 827	3.9	70	2.13	42 964
France		0.5	21	1.8	31 304	280 653	7.6	65	4	43 755

Fig. 6.4 OECD screenshot showing some of the BLI data

Fig. 6.5 Hierarchy for the
OECD's Better Life Index

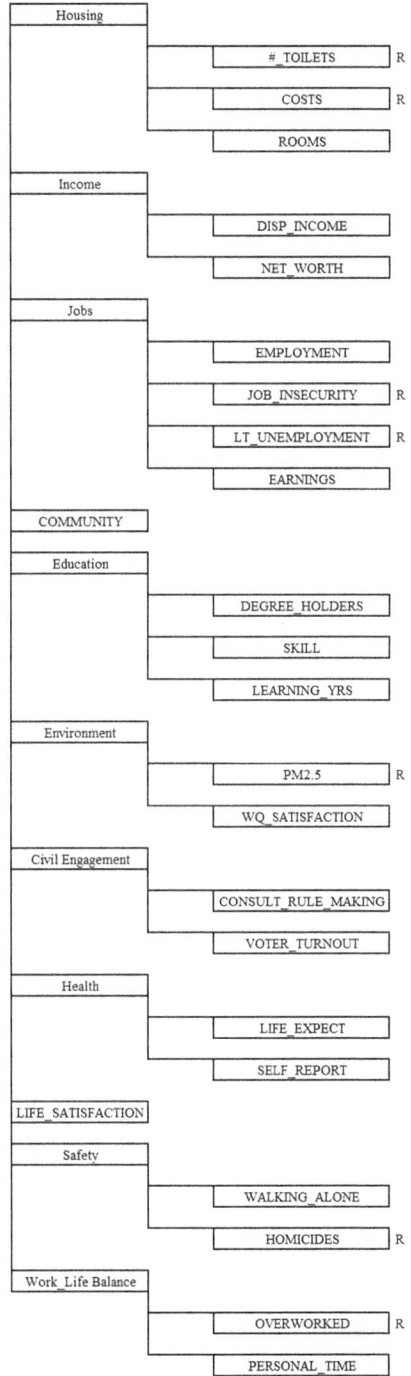

concentrations and a citizen survey response, by country, to the question of whether they are satisfied with their water quality. When constructing analyses, stakeholders should always ask whether the indicators appropriately and accurately reflect the CN. Since the BLI did not offer a specific discussion of the environmental CNs, only a description of the CMs, no conclusion can be made about whether these two environmental indicators are appropriate. However, in this case, it is reasonable to question whether the water quality survey response is appropriate as an assessment of water quality. Uninformed citizens could have responded positively, not knowing whether their water was clean. However, if the CN had been described as citizen satisfaction about water quality, independent of actual water quality, the data from the survey could represent a good Criteria Metric.

By defining Education as a component of better life, the OECD BLI suggests a better life when there are more degree holders or more years of formal education. For example, the range of values for the LEARNING_YRS data is from 15 years to 21 years. Based on this data, the fewest years of education a 5-year-old child can expect to receive, across the 38 countries evaluated, is 15 years while the greatest number of education years is 21 years. This is a relatively small data range where even the most disadvantaged country can expect to see its 5-year-old children remain in school until age 20.[3] If these data are accurate, users could ask whether another education indicator might be more useful as the import of including the LEARNING_YRS indicator is limited with respect to distinguishing among the 38 countries.

It is unclear how the BLI is specifically used by the OECD, governments, or other organizations. However, the amount of effort invested in the creation of the BLI suggests that it is important to the OECD. If this is the case, one hopes that there will be future opportunities to consider it using an open solution approach. If the concepts of the open solution approach are applied to the BLI, countries being assessed by the BLI, and other stakeholders using the BLI for policy decision making, should expect to engage in a meaningful discussion and examination of the other indicators comprising the BLI. Without such stakeholder discussion and agreement, the BLI could signal unintended policy implications and make leaps of judgment about countries without having captured the real policy issues that affect better life in those countries.

---

[3] There are many data used in the BLI that, on closer examination, appear to be either wrong or at least require greater explanation. For example, the data for LEARNING_YRS appears to be questionable. The LEARNING_YRS data indicates that, on average, the children in each of the OECD countries receive at least a high school education and perhaps some university education. Nonetheless, the availability of the data presented by the OECD for the BLI is a step toward improved transparency about how the better life status of these countries is evaluated.

6.5.1 Replicating the OECD BLI in MIRA

By using the MIRA concepts to back into the BLI, it is possible to infer the BLI Problem Narrative.

BLI Problem Narrative Assess and rank the OECD countries based on indicators for better life, which are (1) housing, (2) income, (3) jobs, (4) community, (5) education, (6) environment, (7) civic engagement, (8) health, (9) life satisfaction, (10) safety, and (11) work-life balance.

The Problem Set Elements (PSEs) are the 38 countries ranked by the BLI (Requisite Step 3). Figure 6.5 shows that the BLI Hierarchy has two levels. However, the OECD offers limited flexibility for the creation of Value Sets for the BLI by allowing website users to input their own relative Preferences at the primary level of the Hierarchy only. However, for the indicators at the second level of the Hierarchy, the OECD sets the weights to Equal Preferences.

When users come to the OECD BLI website (http://www.oecdbetterlifeindex. org), they are invited to input their Preferences to determine how they affect the country rankings. On the web interface, a user can independently set each of the nine primary-level Composite Indicators and the two primary-level Terminal Indicators to one of six values that relate to their importance. In this way, a user applies his Preferences regarding the relative importance among the 11 indicators. An equally spaced bar slide from 0 to 6 effectively lets users select increments of 16.67% (100%/6) in expressing importance. A choice of 0 for any indicator removes it from the analysis. Once the 11 primary-level indicators are Preferenced, the ranking of the 38 countries is plotted on a graph whose vertical axis is the value of the BLI (a single-valued metric that combines all considerations and is used to rank the countries, a metric akin to MIRA's Criteria Sum). With the OECD pre-determined Equal Preferences at the second level and users selecting Equal Preferences at the primary level of the Hierarchy, the results are shown in Fig. 6.6.

Users not schooled in open solution approaches may be led to think that Preferencing is the only part of the BLI index construction where choices are made. Recall that ranking Problem Set Elements (which are countries in this case) also requires establishing an Expert Judgment Set (EJS). Notable in the BLI report there is no mention of how the data are indexed or how the TIs were Preferenced at the second level of the Hierarchy.

Because indices most commonly use a Uniform Indexing EJS and an Equal Preference VS, these are applied, using MIRA, to determine whether the BLI can be replicated. Under this scenario, the MIRA ranking, as compared with the ordinal ranking of the OECD countries, is shown in Table 6.3. The ranked OECD countries in Table 6.3 are the tabular form of the results shown in Fig. 6.6.

The OECD and MIRA rankings are very similar but not identical. One possible explanation is the differences in data used since the data was scattered across a variety of tables on the OECD BLI website. It is also likely that there are slightly different assumptions and/or operations made with the BLI data as constructed by the

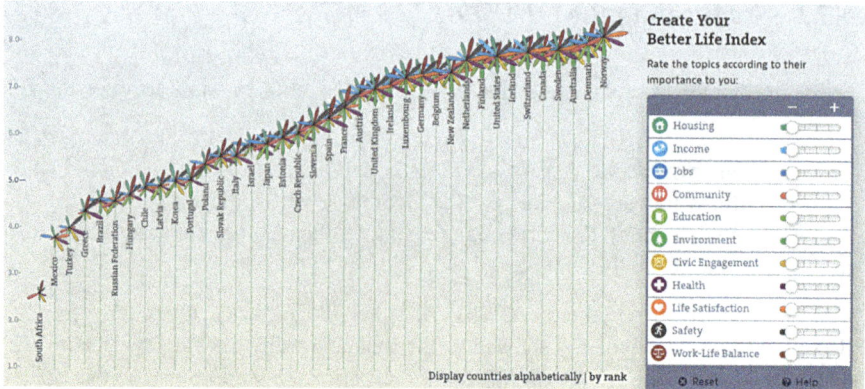

Fig. 6.6 Ranking of OECD countries using Equal Preference Value Set

OECD compared with the consistent application of the Uniform Indexing EJS and Equal Preference VS used in the MIRA Baseline Run. The lack of complete transparency in the OECD BLI makes it impossible to completely replicate or explain the results on the OECD website.

The bolded PSEs in Table 6.3 highlight those countries that occupy different ordinal ranks when the OECD and MIRA runs are compared. The average degree of separation among the PSEs that changed ranks is 1.6%, which is quite small. By using MIRA, stakeholders are offered the additional capability to evaluate the relative PSE ranks using DD and ΔDD_{adj}. The rankings obtained from the BLI website and the MIRA replication are similar enough to conclude that the OECD applied a Uniform Indexing EJS to all the CMs and an Equal Preference VS at the second level of the Hierarchy.

The Uniform Indexing EJS used to reproduce the BLI can also be examined in more detail regarding the relative significance of the indicators used. Because the OECD data is accessible, it is possible to apply the open solution approach for Indexing across the data, by, for example, comparing the LEARNING_YRS data with another, such as $PM_{2.5}$, which has a data range from 3 micrograms per cubic meter ($\mu g/m^3$) to 28 $\mu g/m^3$. Because Uniform Indexing is used to calculate the BLI, the data points that represent the worst and best conditions are deemed equally significant across the indicators. The full Uniform Indexing EJS for the BLI can be found in Table A.11 in the Appendix. Therefore, 15 years of education and 28 $\mu g/m^3$ of $PM_{2.5}$, both representing the poorest or most vulnerable condition, are determined to be equally significant in assessing the BLI. At the other end of the decision scale representing the best or least vulnerable condition, 21 years and 3 $\mu g/m^3$ are deemed equivalent in significance. To provide a benchmark for this air quality metric, in the USA, the human-health-based $PM_{2.5}$ National Ambient Air Quality Standard (NAAQS) of 12 $\mu g/m^3$ can be used. Knowing this allows users to assess the significance of the $PM_{2.5}$ data across the 38 countries as evaluated by the BLI. Using Uniform Indexing, the 12 $\mu g/m^3$ $PM_{2.5}$ US NAAQS is approximately equivalent to

Table 6.3 Comparison of OECD and MIRA Baseline Runs for the BLI

Comparison of ordinal ranking			
OECD BLI	MIRA BLI baseline run		
Ordinal ranking	Ordinal ranking	DD (%)	ΔDD_{adj} (%)
Norway	Norway	0.0	0.0
Australia	Australia	3.4%	3.4%
Iceland	Iceland	5.0%	1.6%
Canada	Canada	5.5%	0.5%
Denmark	Denmark	7.8%	2.3%
Switzerland	Switzerland	9.3%	1.5%
Netherlands	Netherlands	9.6%	0.3%
Sweden	Sweden	9.7%	0.1%
Finland	Finland	10.0%	0.3%
USA	USA	12.4%	2.4%
Luxembourg	Luxembourg	13.5%	1.1%
New Zealand	New Zealand	13.6%	0.1%
Belgium	**Ireland**	**18.5%**	**4.9%**
UK	**Belgium**	**18.7%**	**0.2%**
Germany	**UK**	**20.3%**	**1.7%**
Ireland	**Germany**	**20.9%**	**0.6%**
Austria	Austria	23.8%	2.9%
France	France	30.3%	6.5%
Spain	Spain	33.5%	3.3%
Slovenia	Slovenia	34.7%	1.1%
Estonia	**Israel**	**36.5%**	**1.8%**
Czech Republic	**Estonia**	**37.6%**	**1.1%**
Israel	**Czech Republic**	**38.5%**	**0.9%**
Italy	Italy	42.1%	3.6%
Japan	**Slovak Republic**	**45.7%**	**3.6%**
Slovak Republic	**Japan**	**47.8%**	**2.1%**
Poland	Poland	52.6%	4.8%
Portugal	Portugal	56.3%	3.7%
Korea	Korea	59.4%	3.1%
Hungary	Hungary	60.9%	1.4%
Latvia	Latvia	62.2%	1.4%
Russia	Russia	68.9%	6.7%
Chile	**Greece**	**69.1%**	**0.2%**
Brazil	**Chile**	**71.6%**	**2.5%**
Greece	**Brazil**	**72.5%**	**0.9%**
Turkey	Turkey	74.4%	1.9%
Mexico	Mexico	89.1%	14.8%
South Africa	South Africa	100.0%	10.9%

18.6 years of education. These types of benchmarks should be examined and discussed across all the indicators. It is for the BLI users to examine and question whether these and other significance levels are appropriate. Users of the BLI should assess whether a Uniform Indexing EJS is representative of what they understand to be reflective of their PN. Getting to this understanding requires trans-disciplinary learning, which is currently not available via the BLI approach.

Consequently, although the OECD provides the user with the ability to assign her values to the 11 indicators, the hidden assumptions made relative to assigning significance to the data as well as assigning importance to most of the indicators that are directly linked to the data are problematic. In fact, by allowing the users to apply their Preferences only to the primary level of the Hierarchy without explaining the other essential aspects of the analysis that have been hard-wired, the unsuspecting user may believe that the OECD has offered the capability to fully examine a country's quality of life based on his own Preferences. The question for users of the BLI is whether a BLI construction via the open solution approach would convey a different message.

6.6   Other Indices

There are a wide variety of indices, covering topics such as best corporate investment opportunities, countries with the greatest food sustainability, the "greenest" corporations, the best consumer products, the best colleges and universities, how "good" a country is, and happiness [5–8]. Regardless of the topic or the specific components that make up the index, the Requisite Steps apply, and transparency and stakeholder engagement are beneficial to the construction of the index. Therefore, armed with an understanding of the Requisite Steps and the open solution approach, curious readers are well-equipped to question and evaluate any of these indices.

For example, the Dow Jones Industrial Average (DJIA) is a price-weighted index of 30 large, publicly owned US companies. Price-weighting means that instead of stakeholders applying Preference weights, the price value of the stock (dollars) for each day is used to calculate the DJIA. More expensive per stock prices contribute more to the DJIA. The Standard & Poor's (S&P) Dow Jones Indices Board selects the stocks for the DJIA. The DJIA is one means to capture the US economy using a monetary measure. Other means include the S&P 500 Index, Gross Domestic Product (GDP), and the Consumer Price Index (CPI). When the DJIA was created in 1896, the US economy was heavily industrial. Over time, the US economy has become more technological and communications oriented. This is reflected in the 52 changes to the component stocks since the creation of the DJIA. The most recent change in 2018 was the removal and replacement of General Electric, which had the longest continuous presence in the DJIA, by Walgreens Boots Alliance. Some financial and stock analysts questioned the motivation behind the GE removal from the DJIA as GE's stock had been losing value and its removal boosted the DJIA [9]. Most Americans were likely unaware of the change in the composition of the DJIA or the import of that change. It is not expected that consumers will have any input into the selection of stocks that comprise the DJIA. However, readers now informed about open solution approaches can assess, for themselves, the utility and meaning of the daily and longer-term trends of the DJIA.

Newsweek publishes an annual "Green Ranking" Score for large corporations in the USA and across the world to measure a company's environmental commitment. In 2011, the *Newsweek* Green Ranking Scores measured a company's environmental commitment as assessed by three Composite Indicators: footprint, policies, and transparency.[4] Although *Newsweek* mentions that the footprint consists of 750 metrics, these metrics are not described. The underlying data used to assess policies or transparency are also not described. Table 6.4 shows the top four US companies with the "greenest" scores for 2011.

The Composite Indicator (CI) scores for footprint, policies, and transparency (published by *Newsweek*) are shown for the four companies in columns B–D of Table 6.4 with the corresponding *Newsweek* Green Ranking Scores in column E. Because *Newsweek* did not discuss how the three CIs were Preferenced to produce the result in column E, MIRA was used to backcast the VS that produces the closest Criteria Sum to the Green Ranking Score. The MIRA CS, for the backcast results, is shown in column G with the VS that produced those CSs shown in column H. Notable is the use of different VSs to produce the Green Ranking Scores for each of these four companies. This is potentially misleading to viewers as presenting these company scores in a single ranked list implies that they were evaluated identically. While the same VS (51% footprint, 43% policies, 6% transparency) was used to produce the Green Ranking Scores for HP and Sprint, out of these four calculated Green Ranking Scores, three different VSs were used. Without access to the data used to calculate these Composite Indicators, it is not possible to determine if, or how, the data were expertly judged for significance. Perhaps differences in how each of the company data was indexed better explain the resulting Green Rankings rather than the Value Set applied.

Recall that, in an open solution approach, the same EJS and VS are applied across all PSEs, which represents an even-handed, non-biased analysis. The use of

---

[4]The most current *Newsweek* Green Ranking is for 2017 and can be found at https://www.newsweek.com/newsweek-green-rankings-2017-methodology-739761. The 2017 version is defined explicitly as a sustainability index and penalizes corporations for deriving a part of their revenue from coal, tobacco or weapons manufacturing. The 2017 version differs from the 2011 in that there are now ten Composite Indicators (energy productivity, greenhouse gas productivity, water productivity, waste productivity, green revenue, sustainability pay, audited environmental metric, monetary fines and product/services). In 2014 (available at newsweek.com/green), there were five Composite Indicators (energy productivity, carbon productivity, water productivity, waste productivity, and reputation) comprising the Green Ranking. For 2014 and 2017, *Newsweek* did not make the data for the Green Ranking available, although it indicated only publicly available data was used. The 2011 Green Rankings use fewer Composite Indicators than 2014 or 2017, making it simpler to illustrate how such an index can be evaluated using a MIRA open solution approach. A question for users of the *Newsweek* Green Rankings is whether comparisons from year to year are valid since the composition of the index changes substantially.

Table 6.4 Top 4 US companies' *Newsweek* Green Ranking Scores and MIRA-derived Value Set

A Company	B Footprint	C Policies	D Transparency	E *Newsweek*GreenRanking Score	F MIRA Criteria Sumwith EqualPreference VS	G MIRA Criteria Sum(closest to Green Ranking)	H MIRA VS Closest to Green Ranking(footprint/policies/transparency)(%)
IBM	79	86	83	82.5	82.6	82.6	33/33/33
HP	67	88	64	75.8	72.9	75.8	51/43/6
Sprint	71	82	68	75.6	73.6	75.6	51/43/6
Baxter	60	89	78	74.9	75.6	74.2	50.8/48.2/1

different VSs for different PSEs means that there is bias in the analysis of the Green Ranking Scores. In Table 6.4 for example, transparency is considered less important in the HP Green Ranking. Column F shows the re-calculated CSs if the Equal Preference VS (used for IBM's Green Ranking Score) is consistently applied to the three CIs for each of the four PSEs. Note that, when this is done, Baxter (ranked #4 with the *Newsweek* methodology) is now ranked #2 among these four companies. In 2011, *Newsweek* published the top 25 US companies based on their Green Ranking Scores. If consumers, Baxter, or the other ranked corporations had realized that the consistent application of a single VS changes the relative rankings of the companies, it is interesting to consider what this would have meant for *Newsweek* and the *Newsweek* Green Ranking Score.

It is not known to what extent the *Newsweek* Green Ranking Scores affect a corporation's value, stewardship, or reputation, but if a high Green Score is viewed as desirable, it could encourage other companies to do more. Like many index tools, consumers are not expected to participate in the construction of the *Newsweek* Green Ranking Score. However, an informed consumer, knowledgeable about the available choices in index construction, can use this knowledge to better understand the Green Ranking Score index, assess whether the messages about the index are factually based, and determine whether the methodology used to construct it is sound.

As can be seen by these index examples, the lack of transparency can leave much room for speculation or confusion about the data and how they are used. More importantly, the underlying message in these indices are masked, which could have important consequences to consumers using them to make decisions about how to invest their money, choose a college, or reward corporate behavior. While many organizations may guard their index construction as a proprietary business asset, to the extent that these indices affect consumer and other policy decisions, it may be time to demand greater transparency. At a minimum, better-informed consumers can improve their own decision making with the increased capability to assess the presented facts and how they are used.

6.7   Summary

Each of the examples in this chapter is an illustration of the use of multiple criteria combined into a single-valued metric to help inform people about something that is complex. This can help stakeholders when the index is well-conceived and understood but can backfire and mislead stakeholders when it is not. Consumers, stock traders, and other stakeholders often use indices as policy tools, flags for policy direction, or an assessment of a corporation's stewardship or a country's reputation. As such, the construct of these indices may profoundly influence decision making. Indices are helpful and often necessary to help stakeholders understand complex phenomena. However, without understanding what criteria are included; how they are organized, expertly judged and weighted; and how each of these components supports the index's Problem Narrative, people may be misinformed or misled about the issue. In these cases, it may be more harmful to rely on an index.

The common use of indices to assess a variety of public policy issues, including stock markets, environmental stewardship, global citizenship, better life, and happiness, is an additional opportunity for stakeholders to consider the wicked problem dilemma by applying open solution approaches. Both indices and policy problems start with understanding that there are different ways to define the wicked problem, whether it is fiscal health, good citizenship, better life, happiness, hunger, incarceration, chemical safety or car safety. Open solution approaches are not just for policy decision making. Whose perspectives are captured and how they are described in the Problem Narrative and Criteria Narratives determine how the policy issue or the index is defined and then addressed.

The Requisite Steps of decision making, coupled with iteration and Decision Uncertainty, provide a useful guide for interested citizens and stakeholders to assess indices and public policy making. Used within an open solution approach, these steps help to avoid the common traps in current public policy decision making, such as reductionism and the usurpation of choices. For those interested in such engagement, MIRA is a necessary, feasible, and practical means to achieve better policy making. Even if public policy organizations fail to use more open solution approaches or are slow to adopt such approaches, stakeholders who are educated about these approaches become better personal decision makers and better citizens that can more meaningfully participate in the public policy making process.

References

1. Arvelund E, Snyder S (2018) Ousted Temple dean Moshe Porat was a rainmaker for the business school. Philadelphia Inquirer, 16 July 2018
2. U.S. Environmental Protection Agency (1990) Hazard ranking system: final rule. Federal Register 55(241):51532–51667
3. U.S. Environmental Protection Agency (2017) Environmental benefits mapping and analysis program – community edition (BenMAP-CE). U.S. Environmental Protection Agency. https://www.epa.gov/benmap. Accessed 3/29/2019
4. Organisation for Economic Co-operation and Development (2017) Better life index. OECD. http://www.oecdbetterlifeindex.org/. Accessed May 24, 2019
5. Arabesque (2019) S-Ray. Arabeque. https://arabesque.com/s-ray/. Accessed 5/24/19
6. Helliwell JF, Layard R, Sachs JD (2019) World happiness report 2019. United Nations, New York
7. RobecoSAM AG (2019) Corporate sustainability assessment. RobecoSAM. https://www.robecosam.com/csa/indices/?r. Accessed 5/24/19
8. Klemeš JJ (2015) Assessing and measuring environmental impact and sustainability. Elsevier, Oxford, UK
9. Gandel S (2018) Dow shortsighted: index curators misread economy in kicking GE to the curb. Philadelphia Inquirer, 23 June 2018

Appendix

Chapter 4

4.2.4 Gather Data and Constructing Criteria Metrics

Table A.1 Criteria Metric values used for the five-indicator Baseline Run

Criteria →	OZONE	CLIMATE_ CHANGE	NUTRIENT_ LOAD	LEAD	POPGROWTH
UNITS →	ppb	% land	kg/ha/year	# houses	% people
Min Value →	**63.3**	**0.02**	**1.00**	**2**	**−0.92**
Max Value →	**90.6**	**23.21**	**4.40**	**150,000**	**737.15**
District 1	77.7	2.59	1.00	2	39.00
District 2	73.4	0.41	2.00	150,000	737.15
District 3	67.9	0.02	1.21	195	−0.92
District 4	73.1	1.50	2.76	490	150.00
District 5	77.6	1.04	2.44	593	−0.16
District 6	75.8	73.28	1.00	94,157	−0.04
District 7	76.6	0.41	2.86	43	2.40
District 8	81.5	6.02	4.02	238	47.52
District 9	79.9	13.93	2.69	1217	20.00
District 10	90.6	7.65	2.72	6326	−0.26
District 11	77.4	17.84	3.20	14,391	−0.11
District 12	63.3	12.00	3.37	8365	−0.01
District 13	83.8	21.47	3.72	23,024	0.03
District 14	80.2	7.81	3.99	37,219	−0.05
District 15	80.6	23.21	4.40	2620	1.90

© Springer Nature Switzerland AG 2020
C. H. Stahl, A. J. Cimorelli, *Environmental Public Policy Making Exposed*,
Risk, Systems and Decisions, https://doi.org/10.1007/978-3-030-32130-7

4.2.5   Indexing: Establish the Significance of the Data

Linear Interpolation

To transform a Criteria Metric value that falls between any two adjacent expertly judged equivalencies to their indexed values, a linear interpolation, as described by Eq. A.1, is performed to calculate the CM's indexed value.

If the CM value to be indexed is within the range of CM values associated with the nth and (nth − 1) index value (where n is an integer on the decision scale and $n > 1$), then:

$$\text{Index}_i = (n-1) + \left(\frac{\text{CM}_i - \text{CM}_{n-1}}{\text{CM}_n - \text{CM}_{n-1}} \right) \tag{A.1}$$

Where:

$\text{CM}_i \equiv$ the value of the Criteria Metric being indexed
$\text{CM}_n \equiv$ the value of the Criteria Metric that is associated with the nth index value
$n \equiv$ is the first integer on the decision scale whose value is $>\text{CM}_i$
$\text{Index}_i \equiv$ the index value of the Criteria Metric value being indexed

Finally, if the CM value to be indexed is $\leq \text{CM}_1$ or $\geq \text{CM}_8$, then the index for that CM value is set equal to 1.0 or 8.0, respectively.

Chapter 5

5.2.1 Expand Linear Hierarchy

Table A.2 Criteria Metric data for the four additional air quality criteria in the Gaian analysis

	FINE_PART	AIR_TOX_CANCER	AIR_TOX_RESP	AIR_TOX_NEURO
UNITS →	$\mu g/m^3$	10^{-6} risk	non-dim	non-dim
District 1	27.00	5.00	7.00	2.00
District 2	6.44	6.43	7.93	2.99
District 3	23.67	67.44	10.33	6.70
District 4	20.37	96.07	6.43	7.03
District 5	43.95	74.41	11.88	5.17
District 6	52.36	53.65	2.78	11.80
District 7	2.61	25.00	2.28	8.78
District 8	59.88	82.30	12.47	9.99
District 9	52.85	43.23	5.30	3.48
District 10	7.72	58.21	6.14	1.26
District 11	53.21	2.03	1.11	3.80
District 12	2.52	71.71	11.76	6.73
District 13	36.63	33.72	3.80	9.51
District 14	15.82	72.62	10.36	2.09
District 15	20.29	11.29	5.98	1.23

5.2.3   Refining the Problem Narrative

Table A.3 Description of the 19 Criteria Metrics

Criteria Name	Criteria Metric
OZONE	The ozone concentration within each district, expressed as the US EPA's calculated 8-hour design value[a] within each district in parts per billion (ppb)
FINE_PART	Fine particles are 2.5 microns in size or smaller, $PM_{2.5}$ concentrations, in $\mu g/m^3$, in the district as predicted using an air quality model
AIR_TOXIC_CANCER	The average individual cancer risk predicted to exist in the district. This risk is based on the sum of the products of modeled ambient air concentrations of each carcinogen impacting the district and the unit risk of each of the carcinogens divided by the number of concentration predictions made, in units of 10^{-6} risk.
AIR_TOX_RESP	The total of the modeled Hazard Quotients (HQ)[b] predicted and summed over each air toxic pollutant that impacts the respiratory system. A calculated HQ less than 1 is determined to cause no adverse respiratory effects.
AIR_TOX_NEURO	The total of the modeled Hazard Quotients (HQ) predicted and summed over each air toxic pollutant that impacts the neurologic system. A calculated HQ less than 1 is determined to cause no adverse neurologic effects.
RADON	Average value of radon measurements taken in each district in picocuries per liter (pCi/L).The data range from 1 pCi/L to 8 pCi/L.
CLIMATE_CHANGE	The percentage of the district's land area located within the 100-year floodplain
ACID_RAIN	Wet sulfate deposition from monitoring locations within Gaia and interpolated across each of the districts in kilograms per hectare per year (kg/ha/year)
MERCURY	Mercury (Hg) atmospheric deposition as predicted by an air quality model at multiple geographic locations throughout each of the Gaia districts. The Criteria Metric is the 90th percentile of the modeled Hg deposition values that were calculated across a district in milligrams per hectare per year (mg/ha/year).
INSECTICIDES	Total pounds of insecticide applied in each district per year based on sales data
OZONE_VEGETATION	W126 index, a weighted index designed to reflect the cumulative exposures that can damage plants and trees during the consecutive 3 months in the growing season when daytime ozone concentrations are the highest and plant growth is most likely to be affected, in units of parts per million-hours (ppm-hours)
NUTRIENT_LOAD	The total mass per unit area of nitrogen applied via agricultural fertilizers in each district per year (kg/ha/year)
IMPERVCOV	The percentage land within each district that is impervious to the absorption of rain water
COAL_ASH	The amount of coal combustion products that are received by landfills within each district from power plants operating in Gaia per year (tons/year)

Criteria Name	Criteria Metric
IBI[c]	A non-dimensional index that ranges from 0 to 100 that associates anthropogenic influences on a water body with biological activity in the water body. The higher the value of the IBI, the greater the biological health of the water body.
LEAD	Total number of housing units built before 1979 in the district (# houses)
POPGROWTH	Projected increase in population in each district over the next 25 years (% people)
URBGROWTH	Projected percent increase in the urban population in each district over the next 25 years (% urban people)
MENTAL_ HEALTH	Average noise level experienced by individuals in each district in decibels (db)

[a]The US EPA uses a network of ambient ozone monitors to measure levels of ground-level ozone across large metropolitan areas. Consistent with statutory requirements and for ease of implementation, the ozone concentration obtained by measurement at these monitors is then calculated as a statistic. Using the highest of these calculated statistics, a single, homogeneous, worst-case value across an entire metropolitan area is considered representative of the area by the US EPA. This is the calculation of the ozone design value at each NAAQS monitor. The US EPA uses the 8-hour design value for ground-level ozone as its metric to determine whether an area was complying with the National Ambient Air Quality Standard. The statistical form of this standard is the fourth highest monitored 1-hour value of ozone over a 3-year period. Ozone design values vary across metropolitan areas, but without air quality modeling, the design value is most likely the best choice of a single concentration value that represents an entire area.

[b]Hazard Quotient from the National Air Toxics Assessment, available from https://www.epa.gov/national-air-toxics-assessment/2014-nata-technical-support-document [1].

[c]The IBI is a *Reverse Indicator*. As the value of IBI decreases, the more vulnerable the PSE appears. For Baseline Indexing, a Reverse Indicator is handled in MIRA by using the same Indexing approach that has been described above with the exception that the largest value in the range is equated to an indexed value of 1 and the smallest value in the range is equated to an indexed value of 8.

Table A.4 District Criteria Metric Values for the ten additional CNs

UNITS →	MERCURY	INSECTICIDES	RADON	ACID_RAIN	OZONE_VEGETATION	IMPERVCOV	COAL_ASH	IBI	URBGROWTH	MENTAL_HEALTH
	kg/ha/year	lbs/year	pCi/L	kg/ha/year	ppm-hours	%	tons/year	non-dim	% urban people	db
District 1	2.09	24,698	1	19.7	24.5	11.0	0	66.4	64.9	69.6
District 2	1.06	38,691	1	22.7	4.2	6.8	185,669	35.3	173.7	82.7
District 3	2.30	32,602	8	1.6	22.6	39.8	0	61.2	188.5	75.6
District 4	1.68	1758	1	25.1	5.2	12.4	945,789	42.8	113.5	51.9
District 5	0.95	21,400	3	36.9	14.8	16.6	0	2.4	15.1	66.5
District 6	0.32	55,167	1	30.4	20.5	1.5	0	76.7	30.3	68.4
District 7	0.59	52,468	8	13.0	21.4	28.9	228,443	94.4	42.2	98.2
District 8	1.31	53,264	1	1.5	29.6	14.7	279,934	40.4	83.4	85.5
District 9	1.12	21,925	3	21.2	37.0	27.8	0	26.6	4.4	75.2
District 10	2.04	67,537	1	28.1	14.5	23.5	610,142	94.3	95.7	50.8
District 11	0.24	60,762	1	39.0	21.7	22.6	30,696	1.7	137.7	51.8
District 12	2.08	1426	3	10.3	10.0	22.4	372,102	62.1	72.1	93.7
District 13	1.58	35,342	1	5.5	17.1	34.7	221,888	57.9	108.7	76.5
District 14	0.20	3124	1	26.5	15.5	0.7	0	9.8	114.5	90.8
District 15	0.35	47,341	3	6.8	1.3	30.2	618,360	4.6	28.6	53.0

Table A.5 Uniform Indexing EJS for the Gaia example

Terminal Indicators	Units	Index → 1	2	3	4	5	6	7	8
OZONE	ppb	63.3	67.2	71.1	75.0	78.9	82.8	86.7	90.6
POPGROWTH	% people	0.0	7.1	14.3	21.4	28.6	35.7	42.9	50.0
CLIMATE_CHANGE	%	0.0	10.5	21.0	31.4	41.9	52.3	62.8	73.3
NUTRIENT_LOAD	kg/ha/year	1.00	1.49	1.97	2.46	2.94	3.43	3.91	4.40
LEAD	#	2	21,430	42,859	64,287	85,715	107,143	128,572	150,000
AIR_TOX_CANCER	10^6 Risk	2.0	15.5	28.9	42.3	55.8	69.2	82.6	96.1
RADON	pCi/L	1	2	3	4	5	6	7	8
AIR_TOX_NEURO	non-dim	1.2	2.7	4.2	5.8	7.3	8.8	10.3	11.8
FINE_PART-R	$\mu g/m^3$	2.5	10.7	18.9	27.1	35.3	43.5	51.7	59.9
AIR_TOX_RESP	non-dim	1.1	2.7	4.4	6.0	7.6	9.2	10.8	12.5
FINE_PART-PD	$\mu g/m^3$	2.5	10.7	18.9	27.1	35.3	43.5	51.7	59.9
MENTAL_HEALTH	db	50.8	57.5	64.3	71.1	77.8	84.6	91.4	98.2
OZONE_VEGETATION	ppm-hours	1.3	6.4	11.5	16.6	21.7	26.8	31.9	37.0
ACID_RAIN-F	kg/ha/year	1.5	6.9	12.2	17.6	22.9	28.3	33.6	39.0
MERCURY	mg/ha/year	0.20	0.50	0.80	1.10	1.40	1.70	2.00	2.30
INSECTICIDES	lbs/year	1426	10,870	20,314	29,759	39,203	48,648	58,092	67,537
ACID_RAIN-AS	kg/ha/year	1.5284	6.8772	12.2261	17.5749	22.9237	28.2726	33.6214	38.9702
IMPERVCOV	%	0.70	6.29	11.87	17.45	23.04	28.62	34.21	39.79
IBI	non-dim	94.44	81.20	67.96	54.72	41.47	28.23	14.99	1.75
COAL_ASH	tons/year	0	135,113	270,226	405,338	540,451	675,564	810,677	945,789
URBGROWTH	% urban people	4.4	30.7	57.0	83.3	109.6	135.9	162.2	188.5

5.3   *Weighted Indicators*

Transformation of Weighted Indicators

In MIRA a weighted indicator is calculated by first Indexing the Criteria Metric (CM) (i.e., converting the CM values to values on the decision scale) to produce the unweighted indicator value. Then the indexed value is multiplied by its corresponding resource weighting data value. Once the indicator has been weighted, its value will no longer lie on the decision scale. For example, if an indexed value of O_3 concentration, a value between 1 and 8, were weighted by population, the resulting range of weighted values would be significantly larger than the decision scale's range of values. Therefore, to ensure that the value of the weighted indicator falls within an appropriate range on the decision scale, the weighted indexed values must be re-scaled; this is accomplished in MIRA as follows:

$$S_j = \min\{I_1 \ldots I_m\}$$
$$+ \left[\frac{W_j - \min\{W_1 \ldots W_m\}}{\max\{W_1 \ldots W_m\} - \min\{W_1 \ldots W_m\}} \right] \left(\max\{I_1 \ldots I_m\} - \min\{I_1 \ldots I_m\} \right) \qquad (A.2)$$

Where:

$W \equiv$ the weight of the indicator prior to re-scaling
$S_j \equiv$ the re-scaled value of weighted indicator W for the jth PSE
$\{I_1 \ldots I_m\} \equiv$ the set of the unweighted indicator values for all PSEs
$\{W_1 \ldots W_m\} \equiv$ the set of the weighted indicator values for all PSEs
$j \equiv$ the jth PSE
$m \equiv$ the total # of PSEs

Fig. A.1 The Gaian
26-indicator weighted
functional Hierarchy

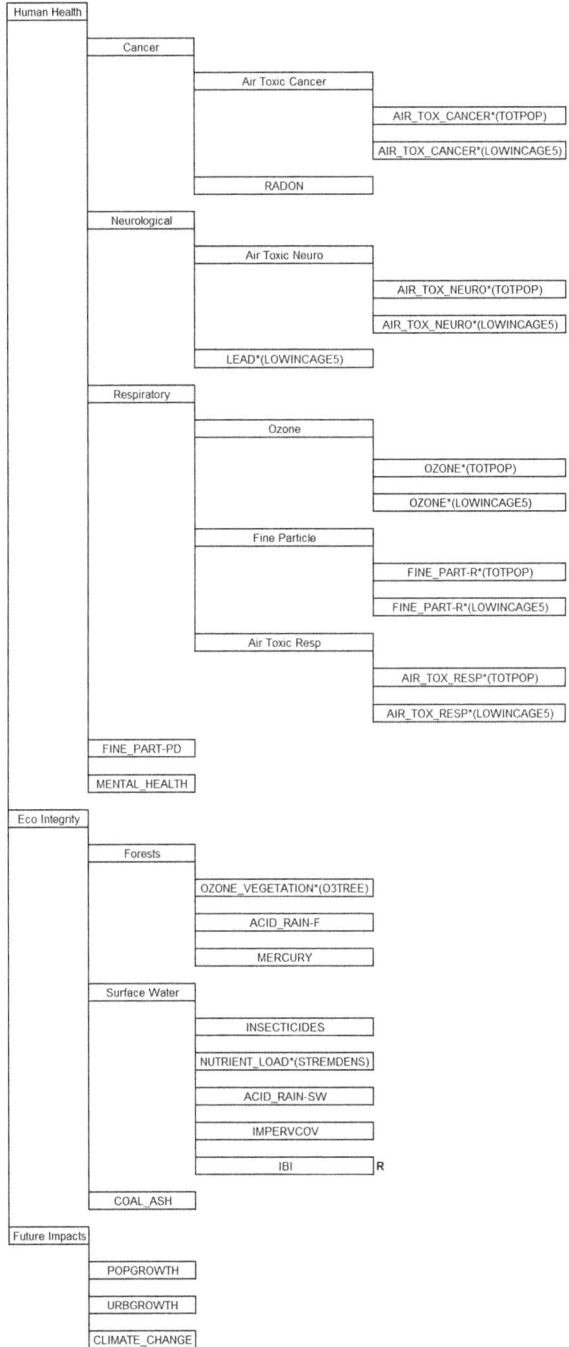

5.4.1 Expert Judgment Sets

Table A.6 Non-uniform EJS for the Gaia weighted functional Hierarchy using the 2008 Ozone NAAQS

| OZONE 2008 | | | | | | | | | |
Terminal Indicators	Units	1	2	3	4	5	6	7	8
OZONE	ppb	**62.00**	**65.25**	**68.50**	**71.75**	**75.00**	**82.50**	**90.00**	**120.00**
POPGROWTH	% people	3.00	9.71	16.43	23.14	29.86	36.57	43.29	50.00
CLIMATE_CHANGE	%	5.00	8.50	12.00	31.42	35.71	40.00	46.17	52.35
NUTRIENT_LOAD	kg/ha/year	1.50	2.50	2.92	3.33	3.75	4.17	4.58	5.00
LEAD	#	200	6400	12,600	18,800	25,000	48,333	71,667	95,000
AIR_TOX_CANCER	10^6 Risk	2.03	13.78	25.50	37.21	48.92	60.64	72.35	84.06
RADON	pCi/L	1	2	3	4	5	6	7	8
AIR_TOX_NEURO	non-dim	0.00	0.10	1.89	3.67	5.46	7.25	9.03	10.33
FINE_PART-R	µg/m³		5.00	8.50	12.00	15.00	23.33	31.67	40.00
AIR_TOX_RESP	non-dim	3.00	4.15	5.30	6.45	7.60	9.23	10.85	12.47
FINE_PART-PD	µg/m³	0.00	6.00	12.00	24.60	37.20	49.80	62.40	75.00
MENTAL_HEALTH	db	40.00	70.00	73.33	76.67	80.00	90.00	95.00	120.00
OZONE_VEGETATION	ppm-hours	1.00	7.00	11.00	15.00	20.50	26.00	31.50	37.00
ACID_RAIN-F	kg/ha/year	1.53	5.60	9.66	13.73	17.80	21.87	25.93	30.00
MERCURY	mg/ha/year					0.20	0.70	1.20	1.69
INSECTICIDES	lbs/year	2000	5000	13,333	21,667	30,000	38,333	46,667	55,000
ACID_RAIN-AS	kg/ha/year	5.00	9.85	14.71	19.56	24.41	29.26	34.12	38.97
IMPERVCOV	%	5.00	8.61	12.22	15.82	19.43	23.04	31.41	39.79
IBI	non-dim	71.09	65.31	59.54	47.99	42.21	36.43	30.66	24.88
COAL_ASH	tons/year	50,000	250,000	380,000	440,000	540,000	750,000	900,000	
URBGROWTH	% urban people		4.36	17.51	30.66	43.81	56.96	83.26	109.56

Table A.7 The final EJS for the Gaia weighted functional Hierarchy using the 2015 Ozone NAAQS

OZONE 2015

Terminal Indicators	Units	1	2	3	4	5	6	7	8
OZONE	ppb	**60.00**	**62.50**	**65.00**	**67.50**	**70.00**	**76.67**	**83.33**	**90.00**
POPGROWTH	% people	3.00	9.71	16.43	23.14	29.86	36.57	43.29	50.00
CLIMATE_CHANGE	%	5.00	8.50	12.00	31.42	35.71	40.00	46.17	52.35
NUTRIENT_LOAD	kg/ha/year	1.50	2.50	2.92	3.33	3.75	4.17	4.58	5.00
LEAD	#	200	6400	12,600	18,800	25,000	48,333	71,667	95,000
AIR_TOX_CANCER	10^6 Risk	2.03	13.78	25.50	37.21	48.92	60.64	72.35	84.06
RADON	pCi/L	1	2	3	4	5	6	7	8
AIR_TOX_NEURO	non-dim	0.00	0.10	1.89	3.67	5.46	7.25	9.03	10.33
FINE_PART-R	$\mu g/m^3$		5.00	8.50	12.00	15.00	23.33	31.67	40.00
AIR_TOX_RESP	non-dim	3.00	4.15	5.30	6.45	7.60	9.23	10.85	12.47
FINE_PART-PD	$\mu g/m^3$	0.00	6.00	12.00	24.60	37.20	49.80	62.40	75.00
MENTAL_HEALTH	db	40.00	70.00	73.33	76.67	80.00	90.00	95.00	120.00
OZONE_VEGETATION	ppm-hours	1.00	7.00	11.00	15.00	20.50	26.00	31.50	37.00
ACID_RAIN-F	kg/ha/year	1.53	5.60	9.66	13.73	17.80	21.87	25.93	30.00
MERCURY	mg/ha/year					0.20	0.70	1.20	1.69
INSECTICIDES	lbs/year	2000	5000	13,333	21,667	30,000	38,333	46,667	55,000
ACID_RAIN-AS	kg/ha/year	5.00	9.85	14.71	19.56	24.41	29.26	34.12	38.97
IMPERVCOV	%	5.00	8.61	12.22	15.82	19.43	23.04	31.41	39.79
IBI	non-dim	71.09	65.31	59.54	47.99	42.21	36.43	30.66	24.88
COAL_ASH	tons/year	50,000	250,000	380,000	440,000	540,000	750,000	900,000	
URBGROWTH	% urban people		4.36	17.51	30.66	43.81	56.96	83.26	109.56

5.4.2.2 Value Sets to Find Common Ground

Primary Level	Weights
Human Health	0.300
Eco Integrity	0.450
Future Impacts	0.250
	1.000

Human Health	Weights
Cancer	0.150
Neurological	0.300
Respiratory	0.200
FINE_PART-PD	0.100
MENTAL_HEALTH	0.250
	1.000

Cancer	Weights
Air Toxic Cancer	0.400
RADON	0.600
	1.000

Air Toxic Cancer	Weights
AIR_TOX_CANCER*(TOTPOP)	0.400
AIR_TOX_CANCER*(LOWINCAGE5)	0.600
	1.000

Neurological	Weights
Air Toxic Neuro	0.200
LEAD*(LOWINCAGE5)	0.800
	1.000

Air Toxic Neuro	Weights
AIR_TOX_NEURO*(TOTPOP)	0.400
AIR_TOX_NEURO*(LOWINCAGE5)	0.600
	1.000

Respiratory	Weights
Ozone	0.200
Fine Particle	0.250
Air Toxic Resp	0.550
	1.000

Ozone	Weights
OZONE*(TOTPOP)	0.400
OZONE*(LOWINCAGE5)	0.600
	1.000

Fine Particle	Weights
FINE_PART-R*(TOTPOP)	0.400
FINE_PART-R*(LOWINCAGE5)	0.600
	1.000

Air Toxic Resp	Weights
AIR_TOX_RESP*(TOTPOP)	0.400
AIR_TOX_RESP*(LOWINCAGE5)	0.600
	1.000

Eco Integrity	Weights
Forests	0.500
Surface Water	0.400
COAL_ASH	0.100
	1.000

Forests	Weights
OZONE_VEGETATION*(O3TREE)	0.250
ACID_RAIN-F	0.150
MERCURY	0.600
	1.000

Surface Water	Weights
INSECTICIDES	0.250
NUTRIENT_LOAD*(STREMDENS)	0.350
ACID_RAIN-AS	0.100
IMPERVCOV	0.200
IBI	0.100
	1.000

Future Impacts	Weights
POPGROWTH	0.200
URBGROWTH	0.400
CLIMATE_CHANGE	0.400
	1.000

Fig. A.2 Final Value Set for the Gaia example

5.4.2.4 Using AHP to Explore Preference Uncertainty

Using AHP in MIRA

The AHP methodology for assigning Preference weights via pair-wise comparisons, which is available in MIRA, was developed by Saaty. There are significant benefits to using this approach to Preferencing, especially for indicator clusters with many indicators. However, this process needs to be performed several times before an adequate degree of consistency is achieved. Once completed with acceptable consistency, stakeholders will either have a much better understanding of their opinions regarding the relative importance among indicators, relative to the specific decision context, or the process will have facilitated the need to develop those opinions. As a result, stakeholders will be more confident that the weights they assign represent the Preferences among the indicators that they intended to apply. To accomplish this, Saaty developed the consistency ratio (CR), which is a means to quantify the consistency of the pair-wise comparisons [2]. A perfectly consistent comparison matrix has a CR equal to 0.0. A CR of 0.1 or less is considered acceptable.

Consider an example where stakeholders are faced with a cluster of three indicators (A, B, and C), which they decide to Preference using the AHP pair-wise comparison process. Suppose a stakeholder determines B to be more important than A and A to be more important than C. To be consistent, the stakeholder will need to find that B is more important than C. As the number of indicators increase, the likelihood of maintaining consistency decreases significantly. One major reason for this is that when performing pair-wise comparisons it is easy for stakeholders to view an indicator in one way when comparing it to indicator A and in another when comparing it to indicator B.

For example, consider the three indicators in the future impacts cluster of the weighted functional Hierarchy (POPGROWTH, URBGROWTH & CLIMATE_CHANGE). When comparing the relative importance of POPGROWTH to CLIMATE_CHANGE, stakeholders may think of POPGROWTH as the future increase in the number of people impacted by CLIMATE_CHANGE (i.e., the higher the POPGROWTH, the more people will be impacted by flooding). However, when comparing POPGROWTH to URBGROWTH, the stakeholders may unknowingly change how they view POPGROWTH. In this second pair-wise comparison, they may now think of POPGROWTH as a rural area indicator and URBGROWTH as an urban area indicator related to possible future impacts to the environment. In the first case the stakeholder was trading-off the importance between the increases in flooding due to climate change versus how many people are going to be impacted by that flooding. In the second case, POPGROWTH is viewed as a surrogate for future environmental impacts that correlate with increased population. Therefore, how an indicator is viewed is often influenced by the other indicator with which it is being compared. Together with scrutinizing the definition of the indicators in relation to the Criteria Narratives, AHP becomes another tool for helping stakeholder to reduce Decision Uncertainty in their VSs.

In MIRA, stakeholders may use AHP in the following way to reduce Decision Uncertainty. Consider again the indicators within the future impacts Composite Indicator (POPGROWTH, URBGROWTH, and CLIMATE_CHANGE). The first step is to perform the three pair-wise comparisons and construct a square 3 × 3 matrix with the results. Suppose the Gaia stakeholders make the following pair-wise comparisons: (1) POPGROWTH is 2× more important than URBGROWTH, (2) POPGROWTH is 1.5× more important than CLIMATE_CHANGE, and (3) URBGROWTH is 2× more important than CLIMATE_CHANGE. The comparison matrix that would result is shown in Table A.8.

The cells to the right of the diagonal are the results of the pair-wise comparisons, while the cells to the left are their reciprocal. Therefore, if POPGROWTH is 2x more important than URBGROWTH, then URBGROWTH must be ½ as important as POPGROWTH. The diagonal entries to this matrix all have the value of 1.00 since, by definition, an indicator compared with itself must equal 1.

Saaty has shown that Preference weights that relate to pair-wise comparisons, when placed in a square matrix such as the one shown in Table A.8, are equal to the values in the eigenvector with the highest eigenvalue. For this comparison matrix the eigenvector with the highest eigenvalue was found to have the following values: 0.46, 0.32, and 0.22. These values are, by definition, the Preference weights for the three indicators in the future impacts cluster that resulted from the pair-wise comparisons that were made and appear in Table A.8. A consistency ratio of 0.093 is calculated for these pair-wise comparisons; since this is ≤0.1 the comparisons are considered to be internally consistent.

In addition to using AHP to perform the Preferencing, stakeholders can use AHP to assess what pair-wise comparisons would have had to be made to generate their assigned Preferences. Using MIRA, stakeholders can determine what perfectly consistent, pair-wise comparisons are implied by a set of Preference weights assigned to the indicators in a cluster. Consider the Preference weights used in the Gaian Committee's final VS for the surface water cluster. The Committee's selection of Preference weights for the surface water cluster is found in Fig. A.2. Since the surface water cluster contains four indicators, six pair-wise comparisons are required. Using AHP, the perfectly consistent 4 × 4 comparison matrix (i.e. having a CR = 0.0) that is implied by the assigned Preferences that appear in the surface water cluster is shown in Table A.9.

The utility in performing this analysis is to inform the stakeholders what opinions, regarding the relative importance among the indicators, are implied by the Preference weights that they applied. It is easier to understand the opinions that have been applied in Preference weighting an indicator cluster by examining the

Table A.8 Example pair-wise comparisons for the indicators in the future impacts cluster

	POPGROWTH	URBGROWTH	CLIMATE_CHANGE
POPGROWTH	1.00	2.00	1.50
URBGROWTH	0.50	1.00	2.00
CLIMATE_CHANGE	0.67	0.50	1.00

Table A.9 Implied comparison matrix for the assigned Preference weights to the surface water cluster

	INSECTICIDES	NUTRIENT_LOAD*(STREMDENS)	ACID_RAIN-AS	IMPERVCOV	IBI
INSECTICIDES	1.00	0.71	2.50	1.25	2.50
NUTRIENT_LOAD*(STREMDENS)	1.40	1.00	3.50	1.75	3.50
ACID_RAIN-AS	0.10	0.29	1.00	0.50	1.00
IMPERVCOV	0.80	0.29	0.57	1.00	2.00
IBI	0.40	0.29	1.00	0.50	1.00

implied pair-wise comparisons than it is by examining the Preference weights. Therefore, determining the implied pair-wise comparisons acts to facilitate the type of learning that is central to the MIRA process.

5.4.3.1 Similarity Index: Evaluating Combinations of Uncertainty

Mathematical Formulation of the Similarity Index

The mathematical formulation of MIRA's Similarity Index contains two terms: one that considers the differences in ordinal PSE rankings between the two runs and one that considers the differences in DDs among the PSEs between the two runs. To provide the user with the greatest degree of flexibility in defining and evaluating the similarity between the results of two runs, the user defines the relative importance between each of the two terms of the SI by weighting each term (W_R and W_{DD}). These weights range between 0.0 and 1.0 with the requirement that the sum of the weights must equal 1.0 (or 100%). This allows the user to examine similarity using each term separately or in various combinations.

Additionally, the user may increase or decrease the importance of large differences in either of the two ranked sets' ordinal rankings or DDs relative to smaller differences. This is done by allowing the user to input a value for the exponent "p," which raises each term in the mathematical expression for SI to the same power. For $p > 1$, as the differences in ordinal ranks or Decision Distance increase, the degree of dissimilarity (as expressed by the SI) increases in a manner that is greater than a simple linear proportion. Furthermore, as the value of p increases, the larger differences become more important than they would have based solely on their proportional differences. Conversely, for $p < 1$ the opposite is true. As the difference in ordinal ranks or Decision Distances increases, the degree of dissimilarity decreases non-linearly.

Finally, to maintain a consistent range in the SI, independent of the number of PSEs that are contained in the list, each term is normalized by the maximum possible change that can occur for a problem set having "n" PSEs. Once normalized, rendering it non-dimensional, the expression for SI is multiplied by 1000 to avoid a metric with decimal places. As a result, SI will range between a value of 0 and 1000. A value of 0 represents the greatest possible dissimilarity between two PSE rankings, while a value of 1000 represents perfect agreement between the two rankings.

The mathematical formulation of MIRA's Similarity Index is as follows:

$$
\mathrm{SI} = 1000 \left\{ 1 - \left[w_R \left(\frac{\sum_{i,j}^{n} \left| R_i - R_j \right|}{2\alpha \cdot \mathrm{Trunc}\left(\frac{n}{2} \right)\left(n - \mathrm{Trunc}\left(\frac{n}{2} \right) \right)} \right)^p \right. \right.
$$

$$
\left. \left. + w_{\mathrm{DD}} \left(\frac{\sum_{i,j}^{n} \left| \mathrm{DD}_i - \mathrm{DD}_j \right|}{2\alpha \cdot \mathrm{Trunc}\left(\frac{n}{2} \right)\left(n - \mathrm{Trunc}\left(\frac{n}{2} \right) \right)} \right)^p \right] \right\} \qquad \text{(A.3)}
$$

Where:

$R_i \equiv$ the rank of the ith PSE in the first ranked set

$R_j \equiv$ the rank of the ith PSE in the ranked set being compared to the first ranked set

$\mathrm{DD}_i \equiv$ the Decision Distance of the ith PSE in the first ranked set

$\mathrm{DD}_j \equiv$ the Decision Distance of the ith PSE in the ranked set being compared to the first ranked set

$n \equiv$ the # of PSEs

$\alpha = \frac{1}{n-1} \equiv$ the value of the interval of equally spaced PSE Decision Distance intervals

$w_R \equiv$ the user-specified weight for the rank difference term

$w_{\mathrm{DD}} \equiv$ the user-specified weight for the DD difference term

$w_R + w_{\mathrm{DD}} \equiv 1.0$

$p \equiv$ the user-specified exponent that changes how the magnitude of differences in rank or DD is considered

Chapter 6

6.4   OECD Better Life Index

Table A.10 Description of the OECD's BLI Criteria Metrics

OECD Name (Composite: Terminal Indicator)	MIRA Criteria Name	Criteria Metric Description (from https://stats.oecd.org/Index.aspx?DataSetCode=BLI)
Housing: Dwellings without basic facilities	# TOILETS	This indicator refers to the percentage of the population living in a dwelling without indoor flushing toilet for the sole use of their households. Flushing toilets outside the dwelling are not to be considered in this item. Flushing toilets in a room where there is also a shower unit or a bath are also counted. Percentage of the population. The value is an average of the data available between 2012 and 2016/2017, pending data availability and break in the series. The value refers to 2016 for Luxembourg and the Netherlands, 2015 for Korea and Mexico, 2013 for the USA, 2010 for Brazil, 2008 for Japan, 2002 for Chile, and 2001 for Canada.
Housing: Housing expenditure	COSTS	This indicator considers the expenditure of households in housing and maintenance of the house, as defined in the SNA (P31CP040, Housing, Water, Electricity, Gas, and Other Fuels; P31CP050, Furnishings, Households' Equipment, and Routine Maintenance of the House). It includes actual and imputed rentals for housing; the expenditure in maintenance and repair of the dwelling (including miscellaneous services) and in water supply, electricity, gas, and other fuels; as well as the expenditure in furniture and furnishings and household equipment and goods and services for routine maintenance of the house as a percentage of the household gross adjusted disposable income. Data refer to the sum of households and nonprofit institution serving households. Percentage of the household gross adjusted disposable income. The reference year is 2015 with the exception of 2016 for Canada, Denmark, and the UK; 2014 for Italy, New Zealand, and Norway; and 2013 for Switzerland.

OECD Name (Composite: Terminal Indicator)	MIRA Criteria Name	Criteria Metric Description (from https://stats.oecd.org/Index.aspx?DataSetCode=BLI)
Housing: Rooms per person	ROOMS	This indicator refers to the number of rooms (excluding kitchenette, scullery/utility room, bathroom, toilet, garage, consulting rooms, office, shop) in a dwelling divided by the number of persons living in the dwelling. Rate (number of rooms divided by the number of people living in the dwelling). The value is an average of the data available between 2012 and 2016/2017, pending data availability and break in the series. The value refers to 2016 for Canada, Luxembourg, and the Netherlands, 2015 for Korea and Mexico, 2013 for Japan and New Zealand, and 2002 for Chile.
Income: Household Net adjustable disposable income	DISP_INCOME	It is the maximum amount that a household can afford to consume without having to reduce its assets or to increase its liabilities. It is obtained adding to people's gross income (earnings, self-employment and capital income, as well as current monetary transfers received from other sectors) the social transfers in-kind that households receive from governments (such as education and health care services) and then subtracting the taxes on income and wealth, the social security contributions paid by households, as well as the depreciation of capital goods consumed by households. Available data refer to the sum of households and nonprofit institution serving households. US dollars at current PPPs per capita. The reference year is 2016 for all countries. PPPs are those for actual individual consumption.
Income: Household net financial wealth	NET_WEALTH	It considers the total wealth: financial and non-financial assets, net of liabilities, held by private household residents in the country. Non-financial assets include the principal residence, other real estate properties, vehicles, valuables, and other non-financial assets (e.g., other consumer durables). It is compiled following the OECD Guidelines for Micro Statistics on Household Wealth (OECD, 2013). The indicator excludes pension schemes related to employment, as available only for a limited number of countries. US dollars at 2016 PPPs per household. The reference year is 2014 for all countries, with the exception of 2016 for Canada and the USA; 2015 for Denmark, Korea, the Netherlands, and the UK; 2013 for Australia, Estonia, Ireland, and Portugal; and 2012 for Spain. PPPs are those for private consumption.

(continued)

Table A.10 (continued)

OECD Name (Composite: Terminal Indicator)	MIRA Criteria Name	Criteria Metric Description (from https://stats.oecd.org/Index.aspx?DataSetCode=BLI)
Jobs: Labor market insecurity	JOB_INSECURITY	This indicator is defined in terms of the expected earnings loss, measured as the percentage of the previous earnings, associated with unemployment. This loss depends on the risk of becoming unemployed, the expected duration of unemployment, and the degree of mitigation against these losses provided by government transfers to the unemployed (effective insurance). "High"/"low" refers to values for people with tertiary/below upper secondary education. Percentage of the employed people. The reference year is 2015 with the exception of 2013 for Germany.
Jobs: Employment rate	EMPLOYMENT	It is the number of employed persons aged 15 to 64 over the population of the same age. Employed people are those aged 15 or more who report that they have worked in gainful employment for at least 1 hour in the previous week, as defined by the International Labour Organization – ILO. Percentage of the working-age population (aged 15–64). The reference year is 2017 for all the countries.
Jobs: Long-term employment rate	LT_UNEMPLOYMENT	This indicator refers to the number of persons who have been unemployed for 1 year or more as a percentage of the labor force (the sum of employed and unemployed persons). Unemployed persons are defined as those who are currently not working but are willing to do so and actively searching for work. Percentage of the labor force. The reference year is 2017 for all countries with the exception of 2013 for Korea.
Jobs: Personal earnings	EARNINGS	This indicator refers to the average annual wages per full-time equivalent dependent employee, which are obtained by dividing the national-accounts-based total wage bill by the average number of employees in the total economy, which is then multiplied by the ratio of average usual weekly hours per full-time employee to average usually weekly hours for all employees. It considers the employees' gross remuneration, that is, the total before any deductions are made by the employer in respect of taxes, contributions of employees to social security and pension schemes, life insurance premiums, union dues, and other obligations of employees. US dollars at current PPPs. The reference year is 2017 for all countries.

OECD Name (Composite: Terminal Indicator)	MIRA Criteria Name	Criteria Metric Description (from https://stats.oecd. org/Index.aspx?DataSetCode=BLI)
Community: Quality of support network	COMMUNITY	This is a measure of perceived social network support. The indicator is based on the question: "If you were in trouble, do you have relatives or friends you can count on to help you whenever you need them, or not?," and it considers the respondents who respond positively. "High"/"low" refers to the percentage of people with tertiary/below upper secondary education. Percentage of people aged 15 and over. The reference period is the 3-year average 2015–2017 for all the countries.
Education: Educational attainment	DEGREE_ HOLDERS	Educational attainment considers the number of adults aged 25 to 64 holding at least an upper secondary degree over the population of the same age, as defined by the OECD-ISCED classification. Percentage of the adult population (aged 25 to 64). The reference year is 2017 for all countries with the exception of 2015 for Brazil, Chile, and the Russian Federation.
Education: Student skills	SKILLS	Students' average score in reading, mathematics, and science as assessed by the OECD's Programme for International Student Assessment (PISA). "High" and "low" are the top quartile and bottom quartile of the PISA index of economic, social, and cultural status (ESCS). Average PISA scores. The latest available year is 2015 for all countries.
Education: Years in education	LEARNING_YRS	This indicator is the average duration of education in which a 5-year-old child can expect to enroll during his/her lifetime until the age of 39. It is calculated under the current enrolment conditions by adding the net enrolment rates for each single year of age from the age of 5 onward. Number of years. The reference year is 2016 for all countries with the exception of 2014 for Japan.
Environment: Air pollution	PM2.5	The indicator is the population-weighted average[a] of annual concentrations of particulate matter less than 2.5 microns in diameter (PM2.5) in the air. Data are averaged over a 3-year period. Micrograms per cubic meter. The reference year is 2013 (3-year moving average) for all countries.

(continued)

Table A.10 (continued)

OECD Name (Composite: Terminal Indicator)	MIRA Criteria Name	Criteria Metric Description (from https://stats.oecd.org/Index.aspx?DataSetCode=BLI)
Environment: Water quality	WQ_SATISFACTION	The indicator captures people's subjective appreciation of the environment where they live, in particular the quality of the water. It is based on the question "In the city or area where you live, are you satisfied or dissatisfied with the quality of water?," and it considers people who responded they are satisfied. Percentage of people. The reference period is the 3-year average 2014–2016 for all countries.
Civic engagement: Stakeholder engagement for developing regulations	CONSULT_RULE_MAKING	This indicator describes the extent to which formal stakeholder engagement is built in the development of primary laws and subordinate regulations. The indicator is calculated as the simple average of two composite indicators (covering respectively primary laws and subordinate regulations) that measure four aspects of stakeholder engagement, namely, i) systematic adoption (of formal stakeholder engagement requirements), ii) methodology of consultation and stakeholder engagements, iii) transparency of public consultation processes and open government practices, and iv) oversight and quality control that refers to existence of oversight bodies and publicly available information on the results of stakeholder engagement. The maximum score for each of the four dimensions/categories is one and the maximum aggregate score for the composite indicator is then four. The stakeholder engagement indicator has been computed based on responses to the 2017 OECD's regulatory indicators survey for OECD countries as well as Colombia and Costa Rica and to the OECD-IDB Survey on Regulatory Policy and Governance 2015 for Brazil. Respondents to all surveys were government officials. The scores for primary laws refer exclusively to processes for developing primary laws initiated by the executive. There is no score for primary laws for the USA, where all primary laws are initiated by Congress, and Brazil. In the majority of countries, most primary laws are initiated by the executive, except for Mexico and Korea, where a higher share of primary laws are initiated by parliament/congress (respectively, 66% and 87%). The reference year is 2017 for all countries with the exception of 2015 for Brazil.

OECD Name (Composite: Terminal Indicator)	MIRA Criteria Name	Criteria Metric Description (from https://stats.oecd.org/Index.aspx?DataSetCode=BLI)
Civic engagement: Voter turnout	VOTER_TURNOUT	Voter turnout is here defined as the ratio between the number of individuals that cast a ballot during an election (whether this vote is valid or not) to the population registered to vote. As institutional features of voting systems vary a lot across countries and across types of elections, the indicator refers to the elections (parliamentary or presidential) that have attracted the largest number of voters in each country. Percentage of the population. The latest available year is 2018 for Colombia, Finland, Hungary, Italy, Mexico, the Russian Federation, Slovenia, and Turkey; 2017 for Austria, Chile, the Czech Republic, France, Korea, the Netherlands, New Zealand, Norway, and the UK; 2016 for Australia, Germany, Iceland, Ireland, Lithuania, the Slovak Republic, Spain, and the USA; 2015 for Canada, Denmark, Estonia, Greece, Israel, Poland, Portugal, and Switzerland; 2014 for Belgium, Brazil, Japan, Latvia, South Africa, and Sweden; and 2013 for Luxembourg.
Health: Life expectancy	LIFE_EXPECT	Life expectancy measures how long on average people could expect to live based on the age-specific death rates currently prevailing. This measure refers to people born today and is computed as a weighted average of life expectancy for men and women. Number of years. The reference year is 2016 for all countries with the exception of 2017 for Mexico and 2015 for Brazil, Canada, Chile, Colombia, France, and South Africa.
Health: Self-reported health	SELF_REPORT	This indicator refers to the percentage of the population aged 15 years old and over who report "good" or better health. The WHO recommends using a standard health interview survey to measure it, phrasing the question as "How is your health in general?" with response scale "It is very good/ good/ fair/ bad/ very bad." "High"/"low" refers to values for people with net disposable income in the highest/lowest quintile. Percentage of the population. The reference year is 2016 with the exception of 2017 for New Zealand, 2015 for Chile and Iceland, 2014 for Australia, and 2006 for Mexico.

(continued)

Table A.10 (continued)

OECD Name (Composite: Terminal Indicator)	MIRA Criteria Name	Criteria Metric Description (from https://stats.oecd.org/Index.aspx?DataSetCode=BLI)
Life satisfaction: Life satisfaction	LIFE_SATISFACTION	The indicator considers people's evaluation of their life as a whole. It is a weighted sum of different response categories based on people's rates of their current life relative to the best and worst possible lives for them on a scale from 0 to 10, using the Cantril Ladder (known also as the "Self-Anchoring Striving Scale"). "High"/"low" refers to values for people with tertiary/below upper secondary education. Mean value (Cantril Ladder). The reference period is the 3-year average 2015–2017 for all the countries.
Safety: Feeling safe walking alone at night	WALKING_ALONE	The indicator is based on the question "Do you feel safe walking alone at night in the city or area where you live?," and it shows people declaring they feel safe. Percentage of people aged 15 and over. The reference period is the 3-year average 2015–2017 for all the countries.
Safety: Homicide rate	HOMICIDE	Deaths due to assault. Age-standardized rate per 100,000 population. The reference year is 2015 with the exception of 2016 for Austria, the Czech Republic, Hungary, Lithuania, the Netherlands, and Sweden; 2014–2016 for Iceland; 2014 for France, Ireland, Portugal, and the Slovak Republic; 2013–2015 for Luxembourg; and 2013 for Canada, New Zealand, and the Russian Federation.
Work-life balance: Employees working very long hours	OVERWORKED	This indicator measures the proportion of dependent employed whose usual hours of work per week are 50 hours or more. Percentage of the dependent employed. The reference year is 2017 with the exception of 2015 for Brazil.

OECD Name (Composite: Terminal Indicator)	MIRA Criteria Name	Criteria Metric Description (from https://stats.oecd.org/Index.aspx?DataSetCode=BLI)
Work-life balance: Time devoted to leisure and personal care	PERSONAL_TIME	This indicator measures the number of minutes (or hours) per day that, on average, full-time employed people spend on leisure and on personal care activities. Leisure includes a wide range of indoor and outdoor activities such as walking and hiking, sports, entertainment and cultural activities, socializing with friends and family, volunteering, taking a nap, playing games, watching television, using computers, recreational gardening, etc. Personal care activities include sleeping (but not taking a nap), eating and drinking, and other household or medical or personal services (hygiene, visits to the doctor, hairdresser, etc.) consumed by the respondent. Travel time related to personal care is also included. The information is generally collected through national Time Use Surveys, which involve respondents keeping a diary of their activities over one or several representative days for a given period. Number of hours per day spent on leisure and personal care. The reference year is 2016 for Japan and the USA; 2015 for Canada; 2014–2015 for Turkey and the UK; 2013–2014 for Italy; 2013 for Belgium; 2012–2013 for Germany; 2010 for Norway and South Africa; 2009–2010 for Estonia, Finland, France, New Zealand, and Spain; 2009 for Korea; 2008–2009 for Austria; 2006 for Australia; 2003–2004 for Poland; 2003 for Latvia; 2001 for Denmark; and 2000–2001 for Slovenia and Sweden.

[a]Although for the PM2.5 data, the OECD BLI team states that this is the population-weighted PM2.5 3-year air quality average, both the units of the data and the numerical value of the data strongly suggest that the PM2.5 is not population weighted but simply the PM2.5 spatial concentration average for each country. Therefore, as discussed in Chap. 6, PM2.5 is treated as the air quality concentration of PM2.5

Table A.11 Uniform Indexing EJS producing the replication of the OECD's BLI

OECD's BLI 2017 Index →	Units	1	2	3	4	5	6	7	8
COMMUNITY	%	78.00	80.86	83.71	86.57	89.43	92.29	95.14	98.00
CONSULT_RULE_MAKING	non-dim	0.80	1.14	1.49	1.83	2.17	2.51	2.86	3.20
COSTS	% disposable	26.00	24.43	22.86	21.29	19.71	18.14	16.57	15.00
DEGREE_HOLDERS	%	38.00	46.09	54.17	62.26	70.34	78.43	86.51	94.60
DISP_INCOME	$	11,592	16,405	21,218	26,031	30,845	35,658	40,471	45,284
WALKING_ALONE	%	35.60	43.39	51.17	58.96	66.74	74.53	82.31	90.10
EARNINGS	$	12,061	19,347	26,633	33,919	41,204	48,490	55,776	63,062
EMPLOYMENT	%	43.00	49.14	55.29	61.43	67.57	73.71	79.86	86.00
HOMICIDES	rate/100,000 people	26.70	22.91	19.13	15.34	11.56	7.77	3.99	0.20
JOB_INSECURITY	%	36.10	31.04	25.99	20.93	15.87	10.81	5.76	0.70
LEARNING_YRS	years	15.10	15.94	16.79	17.63	18.47	19.31	20.16	21.00
LIFE_EXPECT	years	57.50	61.29	65.07	68.86	72.64	76.43	80.21	84.00
LIFE_SATIS-FACTION	mean value (Cantril Ladder)	4.70	5.11	5.53	5.94	6.36	6.77	7.19	7.60
LT_UNEMPLOYMENT	%	16.46	14.12	11.77	9.43	7.08	4.74	2.39	0.05
NET_WORTH	$	70,160	170,002	269,844	369,686	469,527	569,369	669,211	769,053
#_TOILETS	%	37	32	26	21	16	11	5	0
OVERWORKED	%	32.64	28.00	23.35	18.71	14.07	9.43	4.78	0.14
PERSONAL_TIME	hours/day	12.40	12.98	13.56	14.14	14.73	15.31	15.89	16.47
PM2.5	$\mu g/m^3$	28.00	24.43	20.86	17.29	13.71	10.14	6.57	3.00
SELF_REPORT	%	33	41	49	57	64	72	80	88

ROOMS	#/person	0.90	1.14	1.39	1.63	1.87	2.11	2.36	2.60
SKILL	PISA score	390.60	410.37	430.14	449.91	469.69	489.46	509.23	529.00
VOTER_TURNOUT	%	47.00	53.29	59.57	65.86	72.14	78.43	84.71	91.00
WQ_SATISFACTION	%	55.00	61.29	67.57	73.86	80.14	86.43	92.71	99.00

References

1. U.S. Environmental Protection Agency (2018) National air toxics assesssment: 2014 technical support document. https://www.epa.gov/national-air-toxics-assessment/2014-nata-technical-support-document. Accessed 6/30/19
2. Saaty TL (1990) Multicriteria decision making: the analytic hierarchy process. RWS Publications, Pittsburgh

Index

© Springer Nature Switzerland AG 2020
C. H. Stahl, A. J. Cimorelli, *Environmental Public Policy Making Exposed*,
Risk, Systems and Decisions, https://doi.org/10.1007/978-3-030-32130-7

CPSIA information can be obtained
at www.ICGtesting.com
Printed in the USA
BVHW011916201119
564393BV00001B/3/P

9 783030 321291

South Africa Contemporary Analysis

South African Review, 5

Edited and compiled by
Glenn Moss and Ingrid Obery

Published for the
Southern African Research Service

HANS ZELL PUBLISHERS
London • Munich • New York • 1990

Hans Zell Publishers
An imprint of the Bowker-Saur division of Butterworths
Borough Green, Sevenoaks, Kent, England TN15 8PH

Co-published with Ravan Press (Pty) Ltd.
P.O. Box 31134, Braamfontein 2017, South Africa

British Library Cataloguing in Publication Data
Sout Africa contemporary analysis
   1. South Africa, Politics
   I. Moss, Glen II. Obrey, Ingrid
   320.968

   ISBN 0-905450-42-6

Library of Congress Cataloging-in-Publication Data

South Africa contemporary analysis
Edited by Glenn Moss and Ingrid Obery.
   518p.23cm
   Includes bibliographical references.
   ISBN 0-905450-42-6
   1. South Africa--Politics and government--1978- I. Moss, Glenn. II. Obery, Ingrid.
DT1967.A53 1990   89-77269
968.06'3--dc20   CIP

Cover design by Robin Caira

Printed on acid-free paper.

Printed in Great Britain by
Bookcraft Ltd., Midsomer Norton, nr Bath

Contents

SECTION 2: THE FRONTLINE: REGIONAL POLICY IN SOUTHERN AFRICA

Preface

The publication of this - the fifth - *South African Review* has been tinged with sadness for many of those involved.

One figure intimately involved in previous *Reviews* - David Webster - was assassinated on 1 May 1989, shortly after he and Maggie Friedman completed their contribution to this *Review*. David compiled a number of articles for previous editions of *South African Review*, co-ordinated sections on repression and rural areas, and assisted in planning and running the contributors' seminars which are an integral part of producing the *Review*.

His death is a matter of sadness and anger for many, including the staff of the Southern African Research Service and Ravan Press, many of whom were close friends of David.

This *Review* is dedicated to the memory of David Webster, not only because he was an active participant in the project from its birth, not only because he was a friend and colleague of many of those who have contributed to this and other *Reviews*, but because his life and work represent a tradition of intellectual excellence and political engagement which guide both Ravan Press and SARS in their publishing activities.

This *Review* had its origins in a seminar of potential contributors, held in Johannesburg at the end of January 1989. Since that meeting, some fifty authors, editors, co-ordinators and others have been involved in producing a final volume.

While SARS and Ravan Press bear responsibility for the overall nature of the book, and for the selection of articles, contributors are responsible for the views expressed in their own articles.

Apart from the contributors, particular thanks are due to the section co-ordinators: Jonathan Hyslop (State and Politics), Edwin Ritchken (Rural Politics) and Jean Leger and Eddie Webster of the Sociology of Work Programme, University of the Witwatersrand (Labour). The bulk of co-ordination and production, including typesetting and design, was undertaken by Ingrid Obery, while Glenn Moss and Lauren Gower were responsible for most of the editing. Ivan Vladislavic undertook proof-reading and checking of final pages in a most efficient and competent manner.

This *Review* is the outcome of a creative interaction between its originator (SARS), publisher (Ravan Press) and contributors. Credit is due to all those involved.

List of Contributors

GLENN ADLER is a PhD candidate at Columbia University, completing a thesis on labour organisation in the Eastern Cape motor industry.

JOHN AITCHISON is deputy director of the Centre for Adult Education, University of Natal, Pietermaritzburg.

MAX BACHMAN is a medical doctor, and a research officer attached to the Industrial Health Research Group, University of Cape Town.

PAUL BENJAMIN is a senior research officer attached to the Centre for Applied Legal Studies, University of the Witwatersrand, and a practising labour lawyer.

MARK BENNETT was, at the time of writing, employed on a project investigating youth unemployment for the Department of Sociology, University of Natal, Durban.

ANDREW BORAINE is a staff member of PLANACT, Johannesburg, undertaking research into housing, local government and urban development.

DIANA CAMMACK has been working for the World Food Programme among refugees in Southern Africa and Pakistan. She is author of *The Rand at War, 1899-1902*, University of Natal Press, James Currey and University of California Press, forthcoming.

WILLIAM COBBETT is co-ordinator of PLANACT, Johannesburg, and co-editor of *Popular Struggles in South Africa*, James Currey, 1988.

JONATHAN CRUSH teaches in the Department of Geography, Queen's University, Canada.

ROBERT DAVIES is a researcher based at the Centro de Estudos Africanos, Universidade Eduardo Mondlane, Maputo, Mozambique.

BARRY DU TOIT is a committee member of the Grahamstown Rural Committee, an affiliate of the National Committee Against Removals.

ROBERT EDGAR teaches in the African Studies and Research Programme, Howard University, Washington, and has conducted extensive research in Lesotho. He is author of *Prophets With Honour. A Documentary History of Lekhotla la Bafo*, Ravan Press, 1987; and *Because They Chose the Plan of God. The Story of the Bulhoek Massacre*, Ravan Press, 1988.

ALAN FINE is labour editor of *Business Day*.

MAGGIE FRIEDMAN is a member of the Human Rights Commission.

JEREMY GREST lectures in Comparative African Government, University of Natal, Durban. He has a major research interest in Portuguese-speaking Southern Africa.

JONATHAN HYSLOP is a lecturer in the Department of Sociology, University of the Witwatersrand.

PETER KALLAWAY is associate professor in the School of Education, University of Cape Town. He edited *Apartheid and Education. The Education of Black South Africans*, Ravan Press, 1984, and is co-author of *Johannesburg: Images and Continuities. A History of Working Class Life Through Pictures*, Ravan Press, 1986.

MIKE KENYON is a fieldworker for the Grahamstown Rural Committee, an affiliate of the National Committee Against Removals.

JEAN LEGER is research officer at the Sociology of Work Programme, University of the Witwatersrand.

TOM LODGE was a senior lecturer in the Department of Political Studies at the University of the Witwatersrand until 1988. He is currently Director of the Africa Programme at the Social Science Research Council, New York, and is the author of *Black Politics in South Africa since 1945*, Ravan Press and Longman, 1983.

IAN MACUN was, at the time of writing, a researcher with the Industrial Health Research Group, University of Cape Town. He currently teaches in the Department of Sociology, University of Cape Town.

JUDY MALLER lectures in the Department of Sociology, University of the Witwatersrand. She previously worked as a researcher for the Labour and Economics Research Centre, Johannesburg.

GERHARD MARE lectures in the Department of Sociology, University of Natal, Durban, and is co-author of *An Appetite for Power. Buthelezi's Inkatha and the Politics of 'Loyal Resistance'*, Ravan Press and Indiana University Press, 1987.

JOHANN MAREE is associate professor and head of the Department of Sociology at the University of Cape Town, and an editor of the *South African Labour Bulletin* and *South African Sociological Review*. He edited *The Independent Trade Unions, 1974-1984*, Ravan Press, 1987.

GLENN MOSS is manager of Ravan Press, and a past editor of *Work In Progress*.

BILL NASSON is a senior lecturer in the Department of Economic History, University of Cape Town, and author of *Abraham Esau's War: The Black South African War, 1899-1902*, Cambridge University Press, forthcoming.

MUNTU NCUBE is a researcher at the University of Natal, Durban.

DEBORAH NEWTON is a research worker for the National Committee Against Removals (NCAR), which co-ordinates a number of regional organisations concentrating on struggles over land.

INGRID OBERY is co-ordinator of the Southern African Research Service and an editor of *Work In Progress*.

THOMAS OHLSON is a researcher attached to the Centro de Estudos Africanos, Universidade Eduardo Mondlane, Maputo, Mozambique.

MARK PHILLIPS is a research officer at the Centre for Policy Studies, University of the Witwatersrand.

MAX PRICE is a medical doctor, and senior research officer at the Centre for the Study of Health Policy, Department of Community Health, University of the Witwatersrand.

EDWIN RITCHKEN is a post-graduate student in the Department of Political Studies, University of the Witwatersrand. His current research covers rural history and politics in South Africa.

CHRIS ROGERSON is associate professor and reader in human geography in the Department of Geography and Environmental Studies, University of the Witwatersrand.

JEREMY SEEKINGS is completing a PhD thesis on black township politics in the PWV area, through Oxford University. He is currently based in the Western Cape.

BRENDAN SEERY is a Namibia-based journalist reporting for *The Star* newspaper.

LALA STEYN is a rural fieldworker for the Surplus People's Project (Western Cape), an affiliate of the National Committee Against Removals.

MARK SWILLING is a research officer at the Centre for Policy Studies, University of the Witwatersrand. He co-edited *State, Resistance and Change in South Africa*, Southern Books, 1988.

RUPERT TAYLOR is a lecturer in the Department of Political Studies, University of the Witwatersrand.

EDDIE WEBSTER is professor of Sociology at the University of the Witwatersrand. He is the author of *Cast in a Racial Mould. Labour Process and Trade Unionism in the Foundries*, Ravan Press, 1985; an editor of the *South African Labour Bulletin*; and edited *Essays in Southern African Labour History*, Ravan Press, 1978.

DAVID WEBSTER was, until his assassination on 1 May 1989, senior lecturer in the Department of Social Anthropology, University of the Witwatersrand; a member of Detainees' Education and Welfare; and a noted human rights activist. This volume is dedicated to his memory.

Abbreviations

In this South African Review, the term 'black' refers collectively to all racially oppressed groups in South Africa, ie Africans, Indians and coloureds.

AAC - Alexandra Action Committee
AAC Anglo American Corporation
Achib - African Council of Hawkers and Informal Businesses
AFL-CIO - American Federation of Labor-Congress of Industrial Organisations
AHI - Afrikaanse Handelsinstitut
ANC - African National Congress
Assocom - Associated Chambers of Commerce
Ayco - Alexandra Youth Congress
Ayo - Ashdown Youth Organisation
Azapo - Azanian People's Organisation
BBC - British Broadcasting Corporation
BBM - Bakgatla Ba Mocha Tribal Authority
BC - Black Consciousness
BCG - Beira Corridor Group
BCP - Basotholand Congress Party
BDF - Botswana Defence Force
BLA - Black Local Authority
BOSS - Bureau of State Security
CAB - Complaints Adjudication Board
Cahac - Cape Areas Housing Action Committee
CAL - Cape Action League
CBD - Central Business District
CBM - Consultative Business Movement
Ccawusa - Commercial, Catering and Allied Workers Union of SA
CED - Corporation for Economic Development
CFA - Commission for Administration
CNE - Christian National Education
COIN - Counter-insurgency
Cosas - Congress of South African Students
Cosatu - Congress of South African Trade Unions
CP - Conservative Party
CWIU - Chemical Workers Industrial Union
DCDP - Department of Constitutional Development and Planning
DDA - Department of Development Aid

Demalcom - Defence Manpower Liaison Committee
DLAHA - Department of Local Administration, Housing and
Agriculture
DMI - Department of Military Intelligence
DP - Democratic Party
DTA - Democratic Turnhalle Alliance
EAC - Economic Advisory Council
ECC - End Conscription Campaign
Edeyo - Edendale Youth Organisation
EEC - European Economic Community
EPG - Eminent Persons' Group
ESOP - Employee Share Ownership
Fapla - People's Armed Forces for the Liberation of Angola
Fawu - Food and Allied Workers Union
FBWU - Food and Beverage Workers Union
FCI - Federated Chambers of Industry
FCWU - Food and Canning Workers Union
FHA - Family Housing Association
Fosatu - Federation of South African Trade Unions
Frelimo - Front for the Liberation of Mozambique
GDP - Gross Domestic Product
GM - General Motors
GMC - General Motors Corporation
GMSA - General Motors South Africa
HNP - Herstigte Nasionale Party
Hods - Hands Off District Six Committee
ICU - Industrial and Commercial Workers Union of Africa
ILO - International Labour Organisation
IMF - International Monetary Fund
IMSSA - Independent Mediation Service of South Africa
ISA - Internal Security Act
Iyo - Imbali Youth Organisation
JIC - Joint Intelligence Committee
JMB - Joint Matriculation Board
JMC - Joint Management Centre
JSE - Johannesburg Stock Exchange
KFC - KwaZulu Finance and Investment Corporation
KLA - KwaZulu Legislative Assembly
KRO - Krugersdorp Residents' Association
KwaNacoci - KwaZulu/Natal Chamber of Commerce and Industry

Kwasa - KwaZulu Staff Association
LHDA - Lesotho Highlands Development Authority
LIFO - Last-in First-out
LLA - Lesotho Liberation Army
LRAA - Labour Relations Amendment Act
Macwusa - Motor Assemblers and Components Workers Union of SA
Mawu - Metal and Allied Workers Union
MDM - Mass Democratic Movement
Micwu - Motor Industry Combined Workers Union
MNR - Mozambique National Resistance
Mosa - Machinery and Occupational Safety Act
MP - Member of Parliament
MPC - Multi-Party Conference
MPLA - Popular Movement for the Liberation of Angola
MRA - Mgwali Residents' Association
MSMS - Mine Safety Management System
MWU - Mine Workers Union
Naawu - National Automobile and Allied Workers Union
Nactu - National Council of Trade Unions
Nafcoc - National African Chambers of Commerce
NBRI - National Building Research Institute
NCC - National Co-ordinating Committee
NDM - National Democratic Movement
Nehawu - National Education, Health and Allied Workers Union
Nest - New Era School Trust
NF - National Forum
NGO - Non-Governmental Organisation
NHS - National Health Service
NIS - National Intelligence Service
NJMC - National Joint Management Centre
NMS - National Management System
NP - National Party
NP-SWA - National Party of South West Africa
NRP - New Republican Party
NSMS - National Security Management System
NUM - National Union of Mineworkers
NUM - New Unity Movement
Numsa - National Union of Metalworkers of South Africa
OAU - Organisation of African Unity

OPM - Office of the Prime Minister
OSP - Office of the State President
PAC - Pan Africanist Congress
PC - President's Council
PFP - Progressive Federal party
Potwa - Post Office and Telecommunications Workers Association
PRE - Economic Rehabilitation Programme
PTA - Pungutsha Tribal Authority
PWV - Pretoria-Witwatersrand-Vaal
Renamo - Mozambique National Resistance (MNR)
RFDP - Rehoboth Free Democratic Party
RLDF - Royal Lesotho Defence Force
RSA - Republic of South Africa
RSC - Regional Services Council
SAAF - South African Air Force
SAAU - South African Agricultural Union
SAB - South African Breweries
SABC - South African Broadcasting Corporation
Saccola - South African Co-ordinating Committee on Labour Affairs
SACP - South African Communist Party
Sactu - South African Congress of Trade Unions
Sacwu - South African Chemical Workers Union
SADCC - Southern African Development Co-ordination Conference
SADF - South African Defence Force
Samcor - South African Motor Corporation
Samwu - South African Municipal Workers Union
SAP - South African Police
SAPDC - South African Pharmaceutical and Drug Company
Sarhwu - South African Railway and Harbour Workers Union
SAS - Special Air Services
SATS - South African Transport Services
Seifsa - Steel and Engineering Industries' Federation of SA
Sosco - Soweto Students Congress
Soyo - Sobantu Youth Organisation
SSC - State Security Council
Swanu - South West Africa National Union
Swapo - South West Africa People's Organisation

Swapo-D - South West Africa People's Organisation (Democrats)
Swapol - South West Africa Police
SWATF - South West Africa Territory Force
TEBA - The Employment Bureau of Africa
TGNU - Transitional Government of National Unity
TGWU - Transport and General Workers Union
UDF - United Democratic Front
UK - United Kingdom
UN - United Nations
UNHCR - United Nations High Commission for Refugees
UNICEF - United Nations Children's Fund
Unita - National Union for the Total Independence of Angola
Untag - United Nations Transition Assistance Group
UP - United Party
US - United States
USSR - Union of Soviet Socialist Republics
UWCO - United Women's Congress
Uwusa - United Workers Union of South Africa
Vgawu - Vukani Guards and Allied Workers Union
VVPP - Vukani Vulimehlo People's Party
WHAM - 'Winning-hearts-and-minds'
Zanu (PF) - Zimbabwe African National Union (Patriotic Front)
Zapu - Zimbabwe African People's Union
ZAR - Zuid Afrikaansche Republiek

Most abbreviations used in the text of the *Review* are spelled out in full on first usage. Newspapers referred to in the text are abbreviated in the following way:

AIM - Mozambique Information Agency
BD - Business Day
Cit - Citizen
CP - City Press
DN - Daily News
EPH - Eastern Province Herald
FM - Financial Mail
NM - Natal Mercury
NN - New Nation

NW - Natal Witness
NW Echo - Natal Witness Echo
Sow - Sowetan
SStar - Sunday Star
ST - Sunday Times
Star - The Star
STrib - Sunday Tribune
VW - Vrye Weekblad
WM - Weekly Mail

General Introduction

Glenn Moss and Ingrid Obery

South Africa moves into the 1990s in a state of uncertainty and flux not experienced for decades. The old order is changing - some argue dying - but the past endures and continues to influence the present and the future.

Many previous ways of analysing and interpreting society's dynamics, structures and interests no longer hold in the same way as before, and long-held dogmas and 'common sense' truths appear less self-evident.

In a remarkably short space of time, the style and form of government seem to have altered: prominent political prisoners have been released after decades of incarceration, and massive rallies and protest marches - to all intents and purposes in support of the still-banned African National Congress and South African Communist Party - have taken place without any state reaction. Activities which previously guaranteed arrest or detention, conviction and sentence, take place with cabinet ministers justifying them in the name of 'democracy' or 'legality'. Some of the most obvious manifestations of apartheid, including the enforcement of separate amenities for different races, are breaking down - sometimes by government decree, but more often as a result of defiance and resistance.

But appearances can be deceptive, and the task of serious commentators and social analysts is to move beyond the realm of the obvious, revealing the contradictions and structures, processes and pressures, interests and conflicts, which are the motor force of any society.

The contributions to this - the fifth - *South African Review* provide a background and context within which contemporary developments can be analysed and assessed. The present and the future do not spring, fully grown, from nowhere. They are formed and structured in the crucible of the past, and carry with them all the marks and burdens of history.

In this *Review*, attention is drawn to many of the major factors which have to be investigated as part of the ongoing attempt to interpret events: the realm of the state and resistance politics; the crucial events in Southern Africa which have given rise to a new regional conjuncture and balance of force; developments in the arena of labour, where organised working-class interests confront those of the state and capital most clearly; and the often neglected rural and bantustan areas, where politics is subject to rhythms different in many respects to urban conflicts and processes.

The state, in particular, is a major focus of analysis in the contributions which make up this volume. For the past decade, attention has been drawn to the restructuring of the South African state and the form of government. Since the era of 'total strategy' introduced by PW Botha and his generals in the late 1970s, observers have pointed to the development of a 'state within a state', a largely secret structure based on a web of interlocking institutions: the State Security Council, National Security Management System, Joint Management Centres, Defence Manpower Liaison Committees and Regional Services Councils. Government became increasingly secret as these institutions developed, and parliament, the cabinet, and the National Party itself were marginalised from the real seats of power and decision-making.

This trend was dramatically increased with the declaration of a partial state of emergency in 1985, extended to cover the whole country in 1986, and still in force as 1989 drew to a close. Emergency rule centralised power to an even greater extent in the hands of the 'securocrats' and those institutions associated with repression, notably the military.

The last years of PW Botha's rule were marked by a massive growth in the power of the Office of the State President, as the structures of real government were moved further and further away from public scrutiny and accountability. The racially-exclusive white Houses of Parliament - previously one of the National Party's instruments of government - was marginalised in this restructuring, and the party itself lost its importance as an instrument of policy and participation.

The dramatic events surrounding PW Botha's fall from power - his stroke, resignation as leader of the National Party, and subsequent reluctant resignation as state president - paved the way for FW de Klerk's appointment as party leader and state president. De Klerk won sufficient support from the NP's caucus - made up of the party's sitting members of parliament - to become new leader of the party. But his majority was slim, and the caucus required three rounds of voting before the challenge from Finance Minister Barend du Plessis - rumoured to be PW Botha's choice as successor - was defeated.

To some extent, De Klerk's victory as party leader, and subsequent appointment as state president, reflected a rebellion within the ranks of the parliamentary caucus of the National Party. MPs had been increasingly marginalised from real decision-making power under Botha's rule, as parliament as an institution was undercut by the structures constituting the 'state within a state'. Although a cabinet minister under Botha, and leader of the NP in the Transvaal, De Klerk was not directly associated with the 'inner core' comprising the State Security Council and its attendant apparatuses. His election reflected a desire on the part of parliamentary party structures to reassert centrality in government.

Within a very short time, De Klerk had undertaken a number of actions which indicated a style of leadership very different to that of the Botha era. Government granted permission for a number of large protest marches throughout the country, despite the fact that they were clearly held in contravention of emergency regulations. A number of long-term political prisoners were released without restriction, including senior ANC members who had been convicted in the Rivonia and mini-Rivonia trials. A subsequent rally held to welcome these prisoners went off without state interference, despite the massive public show of support for the banned ANC and SACP, the reading of a message from ANC president Oliver Tambo - a 'listed' person and hence quoted in contravention of the Internal Security Act - and the support which many of those present showed for the ANC and its policies.

And as the year drew to a close, De Klerk undertook to repeal the Separate Amenities Act, one of the pillars of the racial segregation which forms part of the system of apartheid.

De Klerk's administration was, of course, under the same intense international and national pressure which had confronted PW Botha. It was under Botha that South Africa had acquiesced to independence for Namibia, Africa's last remaining colony. This reflected the changing balance of forces both in the Southern

African region, and internationally. The South African military setback suffered at Cuito Cuanavale was indicative of a changing military balance within the region, while initiatives from the Soviet Union and United States to 'restabilise' Southern Africa and end the Angolan and Namibian wars demonstrated a new international context limiting options for South African militarists.

In addition, growing pressure on the government in terms of sanctions and disinvestment, foreign loan repayments and its need to reschedule payments, and internal defiance, all created the context for De Klerk's new, more 'open', style of governing.

But as South Africa enters the last decade of the 20th century, De Klerk's power base remains unclear. Unlike his predecessors, he did not come to power with the support of powerful apparatuses - the military in the case of PW Botha; the police for John Vorster. As a cabinet minister, De Klerk did not head structures providing the power and loyalty necessary to create a base within the state. Indeed, the Education and Home Affairs portfolios he held previously were inauspicious institutions from which to construct such a base.

A small majority of the National Party's parliamentary caucus supported De Klerk as party leader. This created the context for his election as state president. But the 'state within a state' created over the past decade - the web of government associated with the State Security Council and National Security Management Systems (recently renamed the National Management System) - has not withered away. Neither have those interests embodied in the military, police and other institutions which formed the basis for PW Botha's power and restructuring of the state. They continue within the institutions of government as a reminder of 'total strategy', militarism, destabilisation and internal repression. As such, their influence remains, and cannot be discounted solely because the new state president has a different style of leadership to his predecessors.

Much of the recent past will come back to haunt De Klerk's administration, creating contradictions within the state, and limiting the parameters which constrain government initiatives even further. The role of repression and the security forces in this process is central. Even if rule by state of emergency is lifted, South African law contains a battery of repressive and controlling legislation, which constitutes the legal basis for the power of the police, military and intelligence agencies. Many of the most repressive components of the emergency regulations exist in permanent statute, notably the Internal Security Act; Police, Prisons and Defence Acts; and some one hundred statutes which limit freedom of information, expression and assembly.

There are already indications of discontent within the ranks of the police. Activities clearly in contravention of the security laws as they exist take place without the state acting. The possibility of rebellion within the police over curtailment of powers - even though those powers still exist in law - cannot be ignored.

In addition, De Klerk's attempt to create a more acceptable international face for National Party rule is seriously threatened as the activities of secret police squads are revealed. Previous allegations concerning the activities of formally-sanctioned police death squads are gaining credence nationally and internationally, as cracks appear in the once seamless wall of secrecy and silence. Attacks against opponents of apartheid - including assassinations both within South Africa and elsewhere, bomb blasts, and secret police squads involved in sabotage and terrorism - have increased in the past few years.

If allegations that these activities were sanctioned at very senior levels are proved to be true - and evidence is mounting that this is the case - De Klerk's government may find this legacy of 'total strategy' and destabilisation difficult to suppress. Senior officials and cabinet ministers implicated either directly or through collective responsibility continue to serve under De Klerk. Any thorough investigation of this facet of the South African conflict could provoke open rebellion within the police force and military, and end up implicating some of those who currently occupy senior positions within the institutions of government.

One key process identified by a number of contributors to this volume involves initiatives to privatise a number of sectors of society - not only in the sphere of the economy, but also in areas of reproduction such as housing, education and health care. This involves a number of initiatives, which demonstrate a convergence of interests between elements in government and monopoly capital. They include

- state attempts to sell off assets as a means of generating income for a beleaguered budget;
- abrogation of state responsibility for health, social welfare and education, and placing of the financial burden for these services

on their users. This relates particularly to those areas of working-class reproduction which have historically been neglected, and which the state is under pressure to provide or upgrade;

- deracialisation of certain social services without direct state involvement in that process. This has allowed for racial reform in some areas, without the NP government having to confront its constituency on the issue;
- relocation of power and decision-making within the private sector, particularly big capital, as a means of ensuring that any change in state power will not necessarily lessen power held by the ruling class as a whole.

The fluidity of contemporary South African politics renders many developments difficult to interpret and analyse without reference to previous trends. The contributions to this *South African Review* aim to provide the background and depth necessary to analyse current events in South Africa. Most articles cover 1988 and the first half of 1989, although some provide earlier background or more recent information. But whatever time period covered, the issues and dynamics analysed and described in this volume point to the questions of the future, while assessing the past. It is this interaction between the dead hand of the recent past and the struggles to remould South African society - be it by the De Klerk government, capital, labour, privatisation lobbies or the popular forces broadly associated with the ANC - which constitutes the thematic unity in this volume.

Section 1: The State and Politics

Introduction

Jonathan Hyslop

This section on the state and politics reflects the complexity of a period of stalemate in the confrontation between state and mass oppositional movements. This introduction begins by characterising the period as one of political deadlock, in which a new negotiation politics has begun emerging. It contends that while our contributors provide an extremely fascinating descriptive account of the state and political struggle in this period, their work points to important inadequacies in our existing theoretical understanding of the state.

It is argued that theoretical viewpoints which break with the notion of the state as a simple instrument of the dominant classes, and which instead see it as a contested field of social relations, giving due attention to the autonomy of its internal structures, may add depth to our interpretation of current political processes. This theme is pursued through a discussion of aspects of current state policy, and the meaning of the recent military dominance of the state. The significance of white politics for the regime and opposition, and the impact on the state of the economic crisis, is discussed. The implications of a non-reductionist view of the state for oppositional strategy are considered. Finally, it is argued that all analysis of the state needs to be placed within an understanding of its context in a global state system.

The period 1986-1989

The conflicts of the 1980s have tested the limits of political possibility for both the dominant groups and the popular oppositional movements in South African society. The Botha administration's attempt to implement its reform package was brought crashing down in ruins by the revolt of 1984-86. Although the vaunted National Security Management System succeeded, through the counter-revolutionary onslaught of 1986-8, in re-establishing the state's physical control of those urban areas which had been swept by insurgency, the state has made little progress in creating a new hegemonic political order. Limited economic resources prevent it from instituting sufficiently widespread material improvements in social conditions to buy off discontent, while refusal by the Nationalists during the PW Botha era to treat with the ANC have prevented the government from drawing any significant sector of black opinion into its attempts at constitutional restructuring. Yet popular movements have also faced frustration in their endeavours.

While the 1980s saw historically unequalled popular mobilisations, the widespread belief of the middle of the decade that popular insurrection or revolutionary warfare could bring down the state has been shown to be a fantasy. Wielding an intact modern military apparatus and an administrative machine increasingly structured around security considerations, the regime had little difficulty in demolishing popular organisation, even if it could not uproot its social basis, or prevent new sprouts of resistance from constantly working their way through the cracks.

The interest in 'negotiated settlement', which has been characteristic of both dominant groups and popular movements in the late 1980s, is the result of this deadlock. It has become clear that, at least for a long historical period, the state has the repressive capacity to deny victory to the mass movement, while the mass movement has the capacity to deny legitimacy to the existing political order and to block much of its social policy and political project. In this situation the reality that both sides face is the prospect of a conflict that cannot be won in any medium-term time period, or at any bearable military and political cost.

Lodge's article shows that the ANC, while involved in various internal differences over this issue, has moved determinedly to confront the possibility of negotiation.

The National Party leadership has been much more reluctant to face the prospect of negotiating with the ANC, an organisation which it has spent years demonising; yet it seems that under FW de Klerk, that 'unthinkable' prospect has become increasingly thinkable.

The conjuncture of the period between the second state of emergency in June 1986 and the disintegration of the PW Botha leadership of the NP in 1989 is addressed by this section of the *Review*.

Our contributors survey the organisation of the state in that period, its aims and parameters; and the way in which popular movements responded. Just as the major political actors found their political assumptions contradicted by the experience of the 1980s, the events of this decade have also presented a challenge to those producing critical social science and political analysis.

That a flourishing school of such literature exists is evidenced by the articles published here. Considerable work has been done in generating detailed and precise accounts of the structures and policies of the state, and of the activities of movements of resistance.

Yet an examination of our contributions might prompt the thought that in some ways critical social science has been behind popular movements in developing a new understanding of its own assumptions. For the theory underlying social science writing on the state and politics has remained largely unquestioned; social analysts have not always fulfilled their obligation to theorise anew in the light of their observations of reality. Specifically, it seems clear that the processes described in these pieces cannot be explained within the bounds of existing South African theoretical approaches to the state.

Insofar as a dominant discourse concerning the state exists within radical social science in South Africa, it draws on the neo-marxism of the late 1960s and early 1970s, especially the work of Althusser[1] and Poulantzas.[2]

This approach to the state is characterised by instrumentalism; it is assumed that the state is the instrument of capital, that particular state policies embody particular capitalist interests, that differences within the dominant social groups or within the state reflect the interests of different capitalist fractions. Yet much of the work represented here is interesting and creative precisely because it explores processes that cannot be explained purely within the confines of such a framework. Discontinuities exist between the policies of capital and those of the state; state initiatives are often strongly autonomous from the interests of external reference groups influenced by bureaucratic interests; political conflict within government is frequently based on the rivalries of bureaucratic and security apparatuses.

But this discrepancy between social theory and practical social analysis has generally not been addressed by contemporary South African social science practitioners. Our contributors have generally not posed theoretical frameworks which account for their rich empirical findings.

To do this it is imperative for critical social science in South Africa to take cognizance of more recent international developments in state theory. Attention needs to be given to the later work of Poulantzas,[3] which sees the state not as a monolith, but a field of social relations across which various social forces and bureaucratic apparatuses battle for power. This view of the state as a battlefield also involves the view that the struggles of subordinate classes impose specific policies on the state, the structures of which may thus embody popular gains. The struggle of popular movements is not only something taking place 'outside' the state but also something that passes through it.

Such an approach enables us to understand how popular struggle also moulds state policy, and how certain changes in state policy have marked victories for popular resistance. Critical social theory in South Africa could also profitably give attention to the remarkable work of Theda Skocpol.[4] She argues convincingly that classical marxism 'failed to foresee or adequately explain the autonomous power...of states as administrative and coercive machines embedded in a militarised international states system'.[5]

Skocpol's work emphasises that the relationship between the state and the dominant social classes is not a given but varies historically. Bureaucrats may be subordinated to, or largely autonomous of, capitalist interests. We cannot tell 'theoretically' what their relationship is before specifically investigating a particular state. If we want to be able to argue that the state serves particular interests, we must be able to show concretely what the causal mechanisms are that move the state to intervene actively for those interests.

The state must be examined as a set of organisations, extracting resources from society and deploying them in coercive and administrative organisations; it is only by understanding the features and capacities of this organisation in historically specific terms that we can grasp the way it acts. The state organisation must respond to internal class conflicts in society on the one hand, and the military pressure of an international state system on the other. Its fate depends largely on how well it is able to muster resources and maintain its internal cohesion in coping with these strains.

This type of approach is clearly relevant in analysing the internal dynamics of the South African state in this period of crisis. Its beleaguered survival has everything to do with its ability to maintain its cohesion in the face of internal and external threats. Its predominant feature - the great growth in the influence of the security apparatus - has little to do with the relations between capitalist fractions. Thus, understanding the state means an end to class-reductionist state analysis. The analyses of the state in this volume provide an empirical basis for such non-reductive analysis.

Militarisation

The role of the security forces in the state has been a notable feature of the current period. The contribution by Phillips and Swilling maps out the history of how, in the crisis of the mid-1980s, the security establishment became the dominant force in policy-making, subordinating other branches of state to security goals. This article examines the way in which military might and changes in urban policy were combined in a reform-repression package in a way which attempted to provide the state with a road out of its dilemmas.

Webster and Friedman's article details the repressive element of that policy. David Webster's tragic killing provides brutal confirmation of the authenticity of the analysis provided in this piece. The contribution by Boraine examines more closely the urban policy component of state strategy. What our authors share is a serious consideration of the autonomous dynamics of military and bureaucratic structures. Their studies thus tend to underpin the argument that the actions of the state are not reducible to 'reflected' class interests.

An attempt to analyse the military role in politics raises the issue of the 'militarisation' thesis, which has been influential in recent years. This approach tends to see the military as having become overwhelmingly predominant within the state, and as having effectively marginalised other state departments and power groups in the process.

The merit of this thesis is that it correctly identifies a real accretion of power to the military in state decision-making. The problem with the thesis is that it tends to see this as a linear, irreversible process; it tends to ignore the possibility that other sectors of the state and other dominant social forces may reassert themselves against the military; and it tends to overstate the extent to which militarist values are hegemonic within the dominant social groups.

The failure of the SADF at Cuito Cuanavale seems to have led to a diminution in the degree of security force dominance of other sectors of the state; and the rise of FW de Klerk appears to involve in part a revival of civilian and constitutionalist forces in the National Party against the securocrats. There is certainly a continuing, and as yet unresolved struggle for power at the highest levels of the state. Again, only a careful analysis of state structures will yield an understanding of its likely results.

Taking white politics seriously

A central inadequacy in the militarisation thesis is that it tends to see white politics as a side-show, hiding the real location of power in the hands of the military. Previous to 1989, this view tended to go hand in hand with a deep skepticism about developments in white politics on the part of the mass oppositional movements. They often tended to feel that it did not matter much what the balance of power between different white parties was. Yet these perspectives ignore the real power which still lies in the hands of the white electorate.

Although enormously strong and centralised at the administrative and coercive level, the regime relies for its legitimation on the mechanisms of white elections, an arena in which it is threatened and under considerable pressure. It is only if we are prepared to understand concrete effects of this specific institutional mechanism of legitimation that we can appreciate how white politics affects the national political situation.

The need of government to gain white electoral legitimacy places serious constraints upon it, while the particular alignment in parliament can have a real effect on the course that the state pursues. The NP's fear of being outflanked on the right has been a major limitation on its ability to develop new and innovative policy. Taylor's article tends to suggest that the NP has retained its domination of white politics at the expense of a coherent political programme. Moreover, developments in white politics are central to future political outcomes.

The question as to whether government will be prepared to enter into negotiation with the ANC depends directly on the balance of forces in white politics, even if this is conditioned by the actions of mass movements. If extreme right-wing reaction becomes dominant in white politics, the possibility of social transformation will be tremendously delayed and won at a much greater price in suffering. An NP government which opposed negotiation or a CP government which tried to return to Verwoerdian policy could persist and maintain military power for a very long time, even though this would mean international isolation, economic decline and continual warfare.

The more flexible tactics adopted by oppositional movements in the 1989 white elections reflect a recognition of some of these factors. In the medium term the extent to which white political opinion can be shifted away from the security psychosis created by the NP in the 1980s, and toward a pro-negotiation position, is a decisive factor in South Africa's future. In the interim, the extent to which parliamentary and extra-parliamentary forces can link up may be important in demolishing legal apartheid institutions.

Economic debacle

The contributions of Phillips and Swilling, and Boraine deal to some extent with the way in which the state's limited fiscal resources are a blockage to its ability to deliver a 'securocrat' solution which circumvents the need to deal with democratic political participation simply by raising living standards. Even were this politically possible, the material resources to do it do not exist.

Just as in the bantustans, the tension between the state's political need to purchase support through public spending and the economic imperatives toward the reduction of expenditure cannot be reconciled. Nor has capital been particularly supportive to the state on this score. As a surplus extractor the state may well, as Skocpol[6] reminds us, come into conflict with the dominant classes if its tax exactions become too resented. Capital has become increasingly hostile to the bureaucratic growth of the state.

A serious consideration of the economic dimensions of the state should also remind us that ideologies can have a peculiar effectiveness of their own. Often they get in the way of, as much as they advance, their proponent's interests. A case in point is the role of the 'free market' ideology - on the one hand serving to mask capitalist greed and state parsimony, but on the other also interfering increasingly with its adherents' grasp of political reality.

The popularity of 'free market' thought was generated out of the dilemma of public policy created by the economic crises. The apparent inability of the state to provide social services and of industry to absorb unemployed labour poses major potential sources of social discontent. The free marketeers appear to believe that the issue of social services can be depoliticised by encouraging private sector initiative in this area, and emphasising individual responsibility for welfare provision, while employment difficulties can be overcome by the expansion of the informal sector.

Here we present two studies of aspects of free market policy; Kallaway's article describes the development of privatisation policy in education; while Rogerson's study on the position of hawkers reflects the complex process of deregulation of informal sector activity. The ultimate success of these state initiatives must be very much in doubt. The state is under powerful pressures from both conservative white constituencies and popular mass movements to maintain public spending in various social welfare sectors. Too dramatic a decline in the level of provision is likely to repoliticise rather than depoliticise such fields as education and health.

The history of the 'redistribution with growth' policies pursued in developing countries in the 1970s suggests that while the informal sector may provide some new jobs, the absence of expansion in formal sector industry places definite limits on overall economic growth, and thus on the informal sector's ability to absorb the unemployed.[7]

The proponents of free market magic cures have not confronted the limitations imposed by popular unwillingness to accept their prescriptions on their policies' practicability. In fact faith in such policies can blunt their followers' grasp of the inescapability of questions of *political* power; the reality is that business has to compromise with popular and labour movements which have decidedly collectivist views. The simple 'truths' of the *Financial Mail* and *Business Day* editorials are not a basis for South Africa's economic future, because they do not appear to be negotiable. Real life will force the free marketeers to settle for a mixed economy if they want a negotiated solution.

Equally, however, there are cases where the effects of ideology are no more than a crudely manipulative hiding of social reality. A particularly interesting aspect of Rogerson's article is the way in which he draws attention to the use of the dichotomy 'first world - third world' within current dominant group politico-economic discourse. By asserting that whites live in a 'first world' environment while blacks live in a 'third world' environment, and that the country needs to protect the economic resources embodied in its first world infrastructure, insidious ideological aims are achieved. The extent to which the affluence of the 'first world' sector is based on the poverty of the 'third world' sector is obscured; class and racial inequality are justified in a bland and technocratic terminology. The language of development has become the last refuge of the apartheid bureaucrat.

Dual power: the mythology of the left

If the state faces obstacles to attaining its goals at every turn, the same can be said of the oppositional movements. As Lodge's article suggests, within the ANC debates continue to rage as to how apartheid can be defeated. The belief was prevalent in the 1984-86 period that popular organisation linked to armed struggle could demolish the existing state, and substitute for it a new state founded on new structures created by a popular mobilisation. This was the local manifestation of an international strand in the thinking of the left which holds up the Paris Commune and October 1917 as models of how power may be 'seized'.

However, the failure of insurrectionary strategies in the mid-1980s should make us question the fundamental assumptions which underpin this viewpoint. In South Africa, critical social science has tended not to interrogate the basic premises of dual power theories. But it is contended here that the dual power perspective is based on historical, political and theoretical misconceptions.

Firstly, there are serious questions as to whether popular mass mobilisation can overturn a regime in the absence of internal breakdown of the state. Skocpol[8] argues convincingly that the 'classical' revolutionary events took place in the power vacuums created by disintegration of the existing state structures.

The Russian revolution of October occurred in a situation where the pressures of war had caused the collapse of the existing regime and in which the forces which tried to take the helm of the state could not reconstruct it. The revolutionary movement in China grew in the space provided by the inability of the Kuo Min Tang to create a state which controlled the national territory.

The implication of Skocpol's argument appears to be that in neither case were dual power organs responsible for overthrowing the established pre-revolutionary state; in both they grew because the state had disintegrated already. Thus for her the authorities displaced by Lenin and Mao were in fact ones which had failed to create an effective state structure.

One might add that more recent 'seizures of power' have occurred either in states where the regime had lost any real social base, even amongst the dominant groups (Cuba, Nicaragua); or in situations where the state was a colonial power or reliant on an external power and thus where the foreign power could withdraw without putting its own survival in question (Vietnam in the 1950s and 1970s, the Portuguese colonies); or in situations in which the military forces of the state could be won to support popular revolt (Iran).

None of this is to underplay the role of mass revolt in bringing about these historic transformations. But it is to say that in the absence of particular conditions which create an internal dynamic to dissolution within the state, dual power organisation is almost bound to fail.

None of the scenarios described above apply in South Africa at present; the state is not likely to disintegrate as a result of internal or external factors; it has a significant minority social base; it is not a foreign entity and therefore, being unable to withdraw, is not likely to commit suicide; the popular movement, because of the racial basis of the SADF, is unlikely to be able to cause sufficient disaffection in the ranks of the military to cause state coercive power to fall apart. In addition it might be remarked that a state founded on an internal socio-economic structure of far greater size and institutional capacity than were Batista's Cuba, Somoza's Nicaragua or Thieu's Vietnam is likely to be a correspondingly more durable one.

The prospect of the state collapsing in the face of dual power organisation seems fanciful. This is by no means to negate the centrality of popular movements in transforming South African society. Trade unions, youth movements and township community organisations bear most of the credit for having moved the regime, in 15 years, from Verwoerdian apartheid to contemplation of the legalisation of the ANC. But the end of apartheid will come at the negotiating table, even though this will be the outcome of a long history of resistance.

Anyone who wants to invoke the Paris Commune as an example to South Africans ought to remember that it only developed in the space provided by the collapse of Louis Napoleon's state in the Franco-Prussian war, and that the ramshackle armed forces of the new administration wiped it out completely.

Secondly, dual power organisations in an industrial society are likely to face considerable obstacles to their attaining the administrative capacity and legitimacy which could make them an effective basis for a new state. Some of these obstacles were apparent even in the Russian case, and Russia - as Perry Anderson reminds us - possessed a feudal state in the pre-revolutionary period,[9] and is therefore a dubious model for social change in industrial capitalist states.

The difficulties intrinsic to dual power strategies are identified by Sirianni[10] in a comparative study of dual power organs in advanced industrial societies. Sirianni points out that the improvised character of dual power organs makes it difficult to regularise their procedures. This can lead to arbitrary practices which may delegitimate these bodies in the eyes of crucial groups that a popular movement needs to ally with. Dual power organisations are likely to have limited administrative capacities, and thus may be inferior to existing state structures in terms of their material ability to deliver specific services; this incapacity is likely to have further debilitating effects on their legitimacy.

Sirianni also suggests that the role of dual power organs as alternative state structures may well be an obstacle to winning crucial layers of middle-class state employees - structurally located within state institutions - to popular movements. Such arguments are not made to cast doubt on the central importance of popular organisation, but rather to suggest that the conception that dual power organs, on their own, form the basis for an alternative state is simplistic. The real question is rather how popular organisation can be linked to a struggle for the transformation of power relations which inevitably occurs within, as well as outside, state structures.

Seekings's study of township people's courts shows that while they did respond to a popular disaffection with the state legal system, and provided a real service and focus of unity, there was significant disaffection from them in areas where courts became vehicles for the arbitrary activity of 'youth' elements. A similar point could be made with reference to education boycotts; while expressing profound popular discontents, and being a focus for mass organisation, the lack of capacity of the popular movement to provide an alternative education service threatened the social movement of youth with division between those returning to school and those remaining on boycott.

Sirianni's point on the role of state employees is confirmed by the experience of school boycotts, in that the alienation of teachers became a major problem within them. The People's Education initiative partly overcame these divisions precisely because it redirected students to struggling within state structures.

Thirdly, the dual power conception is theoretically based on an instrumental notion of the state; each state form has, simply and unproblematically, a basis in the interests of a particular social class. Yet as we have seen this viewpoint fails to understand how the internal dynamics of state organisation have real effects, or how the actions of popular movements can impose popular gains on the structures of the state. These issues have emerged particularly in current debates on the question of whether the mass democratic movement should participate in elections and local government structures.

Although it is often asserted that the existing boycott of these institutions is a strategy and not a principle, there is clearly a strong reluctance to take the risk of participation. Yet the possibility of exploring this option cannot be ignored. There are often instrumentalist assumptions about the state's underlying opposition to participation; it is assumed that structures created by the dominant classes can only serve their interests, and that participation in state structures will disorganise the working class.

The former assumption negates the possibility that creative intervention can change the functions of a political institution; it is an ideology of powerlessness. The latter assumption ignores the reality that internationally, struggles to gain democratic rights and social benefits have often been successfully pursued by labour and popular movements in part through participation in (often less-than-democratic) elected institutions.

A focus on attaining goals through participation in the state is not of itself disorganising of popular forces. It may just as well be used as a positive factor in organising them.[11]

There are political forces which reject, *a priori*, any form of participation in state structures, and moreover, the negotiation politics that are now beginning to emerge. In this approach such policies are argued to lead inevitably to a sell-out of national and working-class interests.

One tradition which takes such a view is explored with great understanding by Nasson's portrayal of the Unity Movement tradition in his beautifully written discussion of politics in the Western Cape. However valid some of this current's critiques of certain dominant political practices may be, it does seem essential to challenge the fundamentalist opposition to either participation or negotiation which exists on both the nationalist and leftist flanks of the mass movement.

Such a position assumes that there is an effective alternative to engaging with the state. It assumes that some time, some variant of dual power politics can be made to work.

As has already been extensively argued though, this is an exceedingly remote possibility. The advocacy of a position which is impossible to achieve is at best myopic and at worst a sectarian gimmick for capitalising on the difficulties facing organisations which do confront the question of negotiation. Such a stance has no real alternative to offer.

The argument that there is a danger of a negotiated settlement selling out working-class interests does not hold water. The ability of the working-class organisations to pursue their goals would be aided by the establishment of democratic political rights. Democratic political mechanisms - elections, freedom of speech and assembly, the right to form political groupings - cannot be reduced to reflections of bourgeois ideology.

As the current struggles of the Chinese democracy movement, Polish Solidarity and the forces of perestroika in the Soviet Union show, they are vital to any acceptable form of socialism. What the labour movement is able to attain for itself in the long run depends on the level of organisation of the working class and its consequent ability to impose these goals on a post-apartheid state. Workers' movements are thus more, rather than less, likely to be able to achieve their aims if their rights to independent expression are made secure; they have a material interest in political democracy.

The international context

Finally, it may be observed that South African politics is increasingly demonstrating the invalidity of analyses which take the nation-state as their only unit of analysis. The Soviet Union's new focus upon its internal problems has meant that it is more skeptical of long-term entanglements in Third World politics. Its new thinking about South African politics has made it willing to contemplate the role of broker in a South African settlement.

The blunting of the Cold War politics of High Reaganism by Gorbachev's brilliant diplomacy, and the West's desire to extricate itself from its South African debacle, have brought about a situation where the US, the UK and the USSR seem to be engaged in close co-operation with each other in imposing pressure for a settlement in South Africa. There is no doubt that these pressures played an important role in setting up the Namibian decolonisation, and that they are pushing the South African state and the ANC toward the conference table.

It is no longer possible to understand South African politics in national terms. But there is also an important need for internationalisation of social analysis of the South African state. An important way out of the theoretical emptiness of current critical South African social science would be to generate a better understanding of South African state and society through making comparisons between our own society and those which have a similar location in the world economy and similar social structure. In particular I am thinking of those societies which are also placed in a semi-peripheral position within the world economy, such as the Philippines and Brazil. Such a move would be in line with current trends in development sociology away from grandiose dependency or neo-Marxist theory, and toward the drawing of limited conclusions from substantial comparative studies.[12]

South Africa is about to undergo a new and unique historical experience. In order to contribute to our understanding of it, those engaged in analysis of this process will have to take new theoretical and methodological roads.

Notes

1 For example, L Althusser,'Ideology and Ideological State Apparatuses', in L Althusser, *Essays on Ideology*, London, 1984.
2 N Poulantzas, *Political Power and Social Classes*, London, 1974.
3 N Poulantzas, *State, Power, Socialism*, London, 1974; see also the useful survey of Poulantzas's work in M Carnoy, *The State and Political Theory*, Princeton, 1984, 98-124.
4 T Skocpol, *States and Social Revolutions, A Comparative Analysis of France, Russia and China*, Cambridge, 1985; also relevant to the issues raised here is T Skocpol, 'Political Responses to Capitalist Crisis: Neo-Marxist Theories of the New Deal', *Politics and Society*, 10(2), 1980.
5 Skocpol, *States and Social Revolutions*, 292.
6 Skocpol, *States and Social Revolutions*.
7 BT Mkandawire, 'Employment Strategies in the Third World: A Critique', *Journal of Contemporary Asia*, 7(1), 1977; J Weeks, 'Imbalance between the Periphery and the Centre and the "employment crisis" in Kenya', in D Oxaal et al, *Beyond The Sociology of Development*, London, 1975.
8 Skocpol, *States and Social Revolutions*.
9 P Anderson, *Lineages of the Absolutist State*, London, 1974, 358.
10 C Sirianni, 'Councils and Parliaments: The Problems of Dual Power and Democracy in Comparative Perspective', *Politics and Society*, 12(1), 1983.
11 Sirianni, 'Councils and Parliaments'.
12 P Evans and J Stephens, 'Studying Development since the Sixties: the Emergence of a New Comparative Political Economy', *Theory and Society*, 17(5), 1989.

Repression and the State of Emergency: June 1987 - March 1989

David Webster and Maggie Friedman

The year 1988 marked the 40th anniversary of the Universal Declaration of Human Rights, adopted by the UN General Assembly in 1948. The Declaration contains 30 Articles, and South Africa infringes every single one of them to varying degrees.

Article 9 is brief and to the point: 'No one shall be subjected to arbitrary arrest, detention or exile'. The South African authorities' systematic breach of Article 9 of the Declaration is the focus of this work.

But one should not lose sight of the other major breaches of human rights which permeate the entire social fabric. In a recently published survey of human rights around the world,[1] South Africa features sixth from bottom of a list of 89 countries, scoring only 22% on a scale of human rights. Countries with a tradition of repression do better, such as Chile with a rate of 35%, and those *betes noires* of the government media, Zimbabwe and Zambia, rating 45% and 51% respectively. South Africa is left rubbing shoulders with Rumania and Libya.

Given the long history of suppression of opposition in South Africa, it is surprising that the hydra-headed apparatus of repression assembled by the security system has been put in place on a trial-and-error basis. This has sometimes given rise to ramshackle structures, clumsily executed, and embarrassing to their constructors.

The most recent example of this has involved the inept (but brutally repressive) attempt to foist 'independence' on the people of KwaNdebele, which involved an alliance of police and collaborating politicians in mass detentions, vigilante terror groups, kidnappings and assassinations.

Over time, however, the authorities have generally built up an impressive repertoire of repressive institutions and weapons which are having their intended effects. A mood of political quiescence has existed for the last six to nine months, although there are spontaneous outbursts of protest and resistance, and the continuing low-intensity war in Natal between Inkatha members and their identified enemies in the United Democratic Front (UDF) and Congress of South African Trade Unions (Cosatu).

In the Transvaal there are explosive issues in the schools, which are currently bubbling just under the surface. But for the most part, the ponderous weight of four successive states of emergency has borne down on progressive organisations and, in some areas, completely destroyed democratic structures.

This article involves an overview of trends in various areas of repression, ranging from formal and legal methods (such as detentions, bannings, and the use of the courts to criminalise political opponents), to extra-legal and informal repression (such as the use of 'dirty tricks' squads, vigilantes, death squads and surrogate forces).

Over the last year-and-a-half, new trends have emerged in the methods of suppressing opposition to apartheid, and some previous police methods have been dropped. The pattern of police action is one of probing and blocking: as groups associated with the broad democratic movement seek new avenues for mobilisation and action, so the security forces search for ways of stopping or controlling them. Repressive methods are sometimes reactive, at other times, proactive.

The number of activists detained fell steadily in the latter half of 1988 and early 1989, and then dropped dramatically as a result of a mass hunger strike in February and March 1989. By April 1989 as few as 150 emergency detainees remained in detention, mainly in the Natal area. The puzzle to be solved is why the state allowed this form of formal repression to sink to such a low level, since it is manifest that the level of general state oppression has not declined.

In the last 18 months there has been a discernible drop in militancy of the extra-parliamentary opposition, with less open activity on display. Detentions are generally an accurate reflection of the state of struggle, as security police usually respond to intensified resistance by detaining participants.

In addition, political activists may have evolved more sophisticated and secretive methods of organisation, involving better evasive techniques, and located closer to the grassroots membership than earlier methods of mass mobilisation.

The decline in the number of detentions is further explained by the state's greater use of other methods to achieve the same results, thus limiting the almost-universal condemnation of detention-without-trial. These include:

- extensive use of the courts to criminalise opposition to apartheid, as in the marathon three-year treason trial of UDF, South African Council of Churches and Azanian People's Organisation members (the Delmas trial);
- placing punitive restrictions on key individuals and organisations, severely hampering their work;
- the emergence of surrogate organisations and vigilantes, which play a disorganising role;
- police use of pre-emptive methods, including the banning or controlling of meetings, events and funerals which could mobilise militants.

Repression as a war of attrition: towards a permanent state of emergency?

An article in the previous *South African Review*[2] analysed the emergencies of 1985/6 and 1986/7, and remarked upon the spectacular scale of the repression, most visible in the detention of about 8 000 activists in the former and over 25 000 in the latter period. But between June 1987 and December 1988, the previous policy of mass and indiscriminate detentions has given way to a more measured and targeted selection of victims, accompanied by a major decline in the numbers being detained.

Figure 1 below[3] shows the change in the tempo of detentions over a 30-month period. The first 12 months, which began in the June 1986 emergency, depict all three graphs on a very steep incline, reflecting the mass detentions of that time. From July 1987, however, the three graphs tend to level off, and the daily detainee prison population declines, to the point where fewer than 1 000 were being held in December 1988. This involved a remarkable turnabout from the highpoint of July 1986, when just under 10 000 detainees were being held at one time.

The major new development of the state's detention strategies was what can be called the Gulag syndrome: important oppositional leaders are subjected to prolonged detention, with no prospect of release. This raised the spectre of South Africa under a permanent state of emergency, with major community leaders facing indefinite

Figure 1: Emergency detentions since 12 June 1986

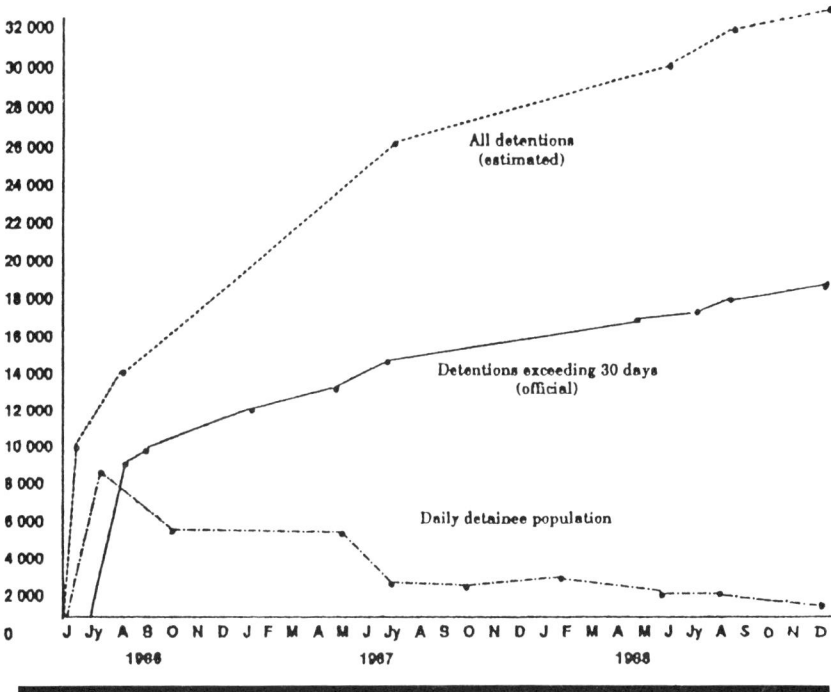

incarceration, perhaps stretching to ten years, as under Smith's Rhodesian regime, or in Stalin's labour camps.

A core of important leaders and activists, numbering about 200 and spread around the country had, by the end of 1988, spent 30 months (917 days) in prison. Many from this group had also spent eight months in Vlok's cells during the 1985/6 emergency.

The successful hunger strikes which precipitated a spate of releases in the first three months of 1989, including some of the most important community leaders, have cast doubt on the effectiveness of the state of emergency - although the police may still need emergency regulations to neutralise released detainees by means of restrictions. The full and complex array of weapons available to the state and its pro-apartheid allies is graphically portrayed in Figure 2, below.[4]

Figure 2: Anatomy of Repression

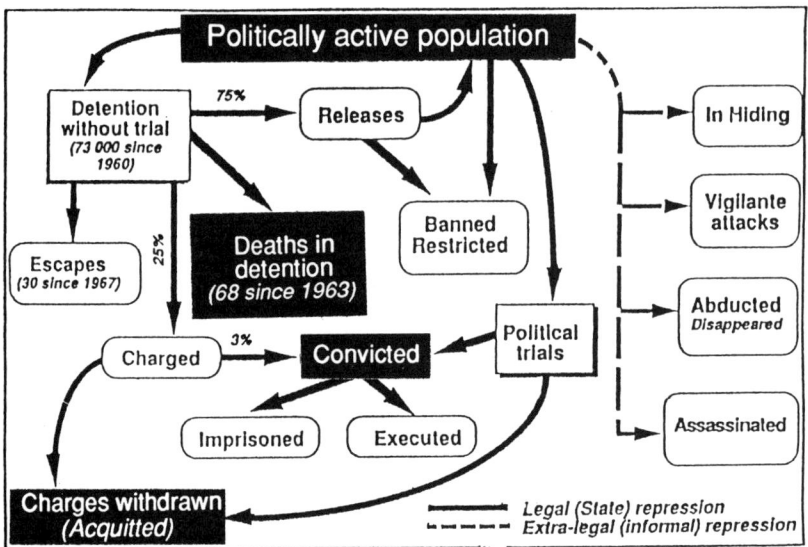

Detention trends

The collection of accurate statistics on detentions is limited by the veil of secrecy erected by security police, and Law and Order Minister Vlok's idiosyncratic interpretation of the Public Safety Act of 1953.

While the Act compels the minister to name people held under the legislation, Vlok tables the names of only those held for more than 30 days.

A large number of detainees are held for periods shorter than this and become, therefore, statistically invisible. For example, the state listed 2 986 detainees for the period June 1987 to June 1988, while the University of the Witwatersrand's Centre for Applied Legal Studies had records of 4 591 detentions for the same period, and believed the figure was closer to 5 000.[5]

The security police have an arsenal of legislation allowing for detention-without-trial. These powers fall into two main categories:

- the Public Safety Act of 1953, which provides for the declaration of states of emergency and which was first invoked after the events of Sharpeville in 1960; and
- security legislation such as the Internal Security Act, which allows for different types of detention: for interrogation (section 29), as a potential state witness (section 31) and preventive detention (sections 28 and 50). This form of security legislation dates from 1963, when section 17 of the General Laws Amendment Act of that year made provision for successive periods of 90-days detention.

The various bantustans (both 'independent' and 'self-governing') have similar legislation, and figures from these areas are included in the statistics presented below.

Table 1: Overview of detentions under security and emergency regulations

a) Detentions under security legislation

Years	Interro-gation	Witness	Preventive	Sub-total	Home-lands	Total
1963-1966	1 095	247	-	1 342	472	1 814
1967-1975	800	293	94	1 187	187	1 374
1976-1977	2 500	504	350	3 354	120	3 474
1978-1981	800	260	1 700	2 760	500	3 260
1982	107	100	3	210	83	293
1983	149	16	38	203	215	418
1984	339	47	191	577	532	1 109
1985	463	4	1 932	2 436	1 953	4 389
1986	477	143	3 512	4 132	520	4 652
1987	532	84	-	616	286	902
1988(6 months)	149	1	-	150	28	178
TOTAL	7 411	1 736	7 820	16 967	4 896	21 863

b) Detentions under emergency regulations

Year	Detail	No of detentions
1960	Partial state of emergency (29 March 1960 to 31 August 1960)	11 727 (official)
1985/86	Partial state of emergency (21 July 1985 to 7 March 1986)	7 996 (official)
1986/87	Total state of emergency (12 June 1986 to 11 June 1987)	25 000 (estimated)
1987/88	Total state of emergency (11 June 1987 to 10 June 1988)	5 000 (estimated)
1988/89	Total state of emergency (10 June 1988 to December 1988)	3 000 (estimated)
TOTAL		52 723 (estimated)

Detention statistics

Tables 2 and 3 below are compiled from the records of the Human Rights Commission, and show the statistical breakdown of detentions under the emergency regulations by area and target grouping. The figures cover the period January 1988 to December 1988. Many of the detentions fall into the category of 'unknown' and these are chiefly cases taken from official lists tabled in parliament, which give no information about detainees, apart from their names.

Table 2: Geographical distribution of 1988 emergency detentions

	No	% of known
PWV	1 393	45,1
Northern Transvaal	89	2,9
Natal	1 003	32,5
Orange Free State	45	1,5
Northern Cape	7	0,2
Eastern Cape	361	11,7
Western Cape	187	6,1
Unknown	622	
TOTAL	3 707	

In 1988 the main thrust of detentions was aimed at the Pietermaritzburg area; at the school-going, or potentially school-going, youth; and at members of civic organisations in the PWV area. The year saw an overall drop in the total number of detentions, but the detainee population experienced an increase in average time served as many key activists entered their third year of detention.

Many of the new detentions in the Pietermaritzburg area arose from particular events, most notably the October municipal elections. Nelson Mandela's 70th birthday celebrations and an anti-apartheid conference organised for September (and banned) also drew fire. New detentions were markedly lower in the Eastern Cape and Border areas where organisations in the democratic movement had suffered badly in the past. However, these regions are well-represented amongst the long-standing detainees.

Table 3: Targets of detention in 1988

	No	% of known
Students, scholars and teachers	528	46,5
Trade Unionists and workers	132	11,6
Community and political workers	439	38,6
Other (includes journalists, clergy, lawyers)	37	3,3
Unknown	2 571	
TOTAL	3 707	

The two largest categories of detainees were students and community workers, particularly in the PWV, where 445 and 287 respectively were detained. This reflects the ongoing education crisis - still very much on the boil with thousands of pupils being denied re-entry to schools - and the rent boycott, which remained unresolved throughout the year.

It is probable that a large proportion in the category 'unknown' are from Natal.

Other means of formal repression

Detentions are the state's most obvious attempt to disorganise its opposition. But apartheid's minions make use of a panoply of weapons to attack popular organisations and individuals. These include use of the courts, bannings, restrictions, listings and deportations.

Use of the courts

There has been a substantial escalation in the number of political trials in recent years. In the year ending June 1988, for example, 51 political trials involving 165 people were completed, with 80 convictions and 85 acquittals. A further 58 trials were in progress, involving 232 accused.[6] Charges ranged from treason, terrorism, membership and furthering the aims of banned organisations, sabotage, undergoing military training, harbouring guerillas, attending illegal gatherings, to murder and public violence, which at first glance are not political, yet closer investigation reveals them to be so.

Some of these trials involved prosecution of notable community leaders on tenuous charges of conspiracy with the African National Congress, and plots to overthrow the state by making townships ungovernable. The tenuousness of these charges has not saved some accused from conviction at the hands of enthusiastic judges, who have not been reluctant to pass punitive sentences.

The best known of these involves the 'Delmas trial' which tied up three senior UDF leaders and various religious and civic leaders in a trial which took three years to complete, and where the judgement alone filled 1 500 pages. Only five of the original 22 accused were actually jailed, receiving lengthy sentences. Pre-trial detention, the refusal of bail, and these sentences removed both the general secretary and publicity secretary of the UDF from the political arena.

An equally important trial ended in April, involving Moses Mayekiso, a leader of one of Cosatu's most powerful affiliated unions. Mayekiso was at the forefront of a union initiative to play a leading role in community politics. After a long treason trial, he and his co-accused were acquitted. However, the same procedures employed in the Delmas trial - detention and refusal of bail for most of the trial - removed them from political activity and leadership.

Flushed with the success of the Delmas trial, it appears the state is about to launch a similar treason trial in the Eastern Cape. It is likely to involve 30 Port Elizabeth activists, who had run an extremely effective consumer boycott in the area. This belated invocation of the judicial process is an exercise in deep cynicism, as most of the likely accused have been in detention since 12 June 1986. Some were also detained during the 1985 emergency.

Bannings and restrictions

After a lull of a few years, 1988 witnessed a spate of bannings (or restrictions, as the state euphemistically prefers to call them). These fall into three main categories: banning of organisations, of persons, and of gatherings.

In 1985, the Congress of South African Students (Cosas) became the first organisation to be banned by the South African government since 'Black October' of 1977. (Bantustan governments had been active during this period, however.) For the next 18 months, no bannings took place, as the state employed emergency powers to quell popular protest. But on 24 February 1988 the minister of law and order imposed a series of restrictions which crippled 33 organisations, ranging from Azapo, through many UDF structures, to one extreme right-wing group. The bantustan administrations of Transkei and Bophuthatswana each banned one organisation.

Table 4: Affiliations of restricted organisations

Black Consciousness	3
UDF	24
Trade Union (COSATU)	1
Right Wing (BBB)	1
Independent (CDD, DTU, NECC, ECC)	4
Bantustans	2
TOTAL	35

A chronology of the 1988 restrictions

February: In February 1988 a new mechanism was employed for banning. Seventeen organisations were restricted under state of emergency regulations. These restrictions have the effect of banning the organisations until the lifting of the emergency.

If the Internal Security Act had been invoked instead, the restricted organisations could have demanded that the minister supply a list of reasons for the banning. In addition, the banning orders could have been reviewed by the chief justice. By using the emergency regulations, the minister entirely circumvented the courts.

The organisations (mainly UDF affiliates, and including the

Figure 3: History of bannings of organisations since 1948[7]

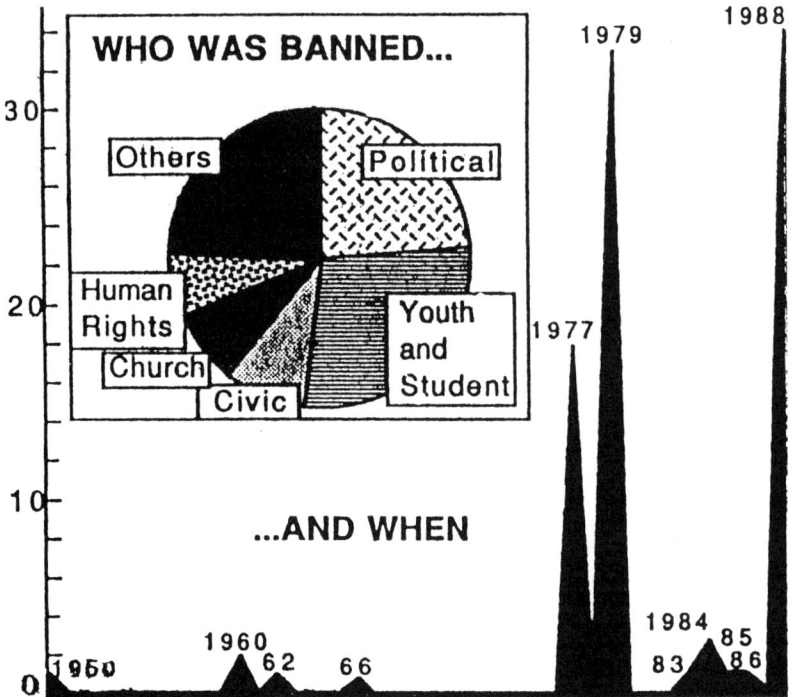

WHO WAS BANNED...

Others   Political

Human
Rights
Church
Civic   Youth
             and
             Student

...AND WHEN

1979   1988

1977

1984 85
83   86

1960
62   66

1950

UDF itself) which were prohibited from 'carrying on or performing
any activities or acts whatsoever' were:
- Azanian People's Organisation (Azapo)
- Azanian Youth Organisation (Azanyo)
- Cape Youth Congress (Cayco)
- Cradock Residents' Association (Cradora)
- Detainees' Parents' Support Committee (DPSC)
- Detainees' Support Committee (Descom)
- National Education Crisis Committee (NECC)
- National Education Union of South Africa (Neusa)
- Port Elizabeth Black Civic Organisation (Pebco)
- Release Mandela Committee (RMC)
- Soweto Civic Association (SCA)

- Soweto Youth Congress (Soyco)
- South African National Students' Congress (Sansco)
- South African Youth Congress (Sayco)
- United Democratic Front (UDF)
- Vaal Civic Association (VCA)
- Western Cape Civic Association

At the same time, a different set of restrictions were imposed on Cosatu, effectively prohibiting activities not confined to employment and work-place issues.

March: Following the February restrictions, the Committee for the Defence of Democracy was formed in the Cape. Six days after the launch, it was subjected to the same conditions as those 17 organisations restricted in February.

June: With the second annual reimposition of the state of emergency on 10 June 1988, the restrictions imposed earlier in the year on 18 organisations were renewed.

August: The 10 February 1988 coup attempt in Bophuthatswana led to the banning, at the beginning of August, of the opposition People's Progressive Party (PPP). Bophuthatswanan officials claimed that party leader Peter Rocky Malebane-Metsing was planning an operation from outside the country. Its object was the release of party members in jail on charges of high treason for involvement in the attempted coup. This ban removed all opposition from the Bophuthatswanan legislature.

After a long build-up of public threats and a concerted campaign in the government-controlled media, the End Conscription Campaign (ECC) was restricted under the same conditions as the 18 silenced organisations.

September: The Prisoners' Welfare Programme (Priwelpro) was declared an undesirable organisation in terms of the Public Security Act of the Transkei.

October: The Soweto Students' Congress (Sosco) and the Azanian Co-ordinating Committee (Azacco) were restricted under the emergency regulations. Azacco had been formed in February as a response to the restrictions on Azapo. Sosco was formed in 1985 after the banning of Cosas.

November: At the beginning of November, a further two organisations were restricted - the Port Elizabeth Youth Congress (Peyco) and the Transvaal Students' Congress (Trasco). Later in the month, restrictions were placed for the first time on a right-wing organisation, the extremist Blanke Bevrydingsbeweging (BBB). At the same time its leader, Johan Schabort, was also restricted.

December: During November the state had intimated that a further five organisations were being considered for restriction orders. In fact, a further eight were restricted during December, bringing the total to 34 organisations restricted in 1988.

On 8 December, the Black Students Society (BSS) at the University of the Witwatersrand and the Black Students Movement at Rhodes University, Grahamstown, were restricted.

On 13 December, two more student organisations were restricted: Mitchell's Plain Student Congress (MSC) and the Western Cape Students Council (WCSC).

This was followed on 29 December, when restrictions were imposed on the Democratic Teachers' Union, the National Detainees' Forum, Western Cape Students' Congress and the Western Cape Teachers' Union.

Table 5: History of the banning and restriction of organisations

Year	SA excluding 'independent' bantustans	Transkei	Ciskei	Bop.	Total
1950	1				1
1960	2				2
1962	1				1
1966	1				1
1977	18				18
1978		1		1	2
1979		33			33
1983			1		1
1984		3			3
1985	1				1
1986		1			1
1988	32	1		1	34
TOTAL	56	39	1	2	98

Table 5 above includes organisations restricted from carrying out any activities whatsoever. It excludes the 1988 restrictions on Cosatu. The banning of the military wings of the ANC and PAC in 1963 and the bannings of organisations that were considered to be the ANC and PAC under different names, are also excluded for clarity.

Some organisations have been banned in more than one area. The South African Allied Workers Union (Saawu), for example, was banned both in the Ciskei (1983) and the Transkei (1986). Cosas and the UDF have been banned in the Transkei as well as in the rest of the country.

Table 6 below roughly classifies organisations banned according to their areas of operation. As the table shows, political and student/youth organisations have been worst affected by bannings and restrictions. In the bantustans, the church has paid heavily for criticism of the 'homeland' governments.

Table 6: Nature of organisations affected

	SA excluding 'independent' bantustans	'Independent' (TBVC) bantustans
Political	11	12
Student and Youth	24	4
Civic	7	2
Church	-	8
Human Rights	4	3
Others	10	13
TOTAL	56	42

In 1976, the Suppression of Communism Act of 1950 was amended by the Internal Security Act. This Act consolidated all existing security legislation. Up to that time, five organisations had been banned under the Suppression of Communism Act and these bannings were now subject to the ISA. The 18 organisations banned in 1977, and Cosas, banned in 1985, were all subject to Internal Security Act provisions.

The regulations of the Public Safety Act, under which the state of emergency exists, allow for the restrictions that are now placed on 33 organisations (including Cosatu). The bantustans each have their own security legislation which allows for bannings.

Restrictions on persons

An emergency-related trend has been the restriction of many detainees as a condition of their release. Police served similar orders on many not detained, but whose political involvement they wished to restrain. In February 1989, at least 135 detainees had been served with restriction orders on their release. The figure has escalated considerably since the hunger strike (see below).

The restrictions tend to prohibit the restricted person from engaging in activities which occupied them before their detention. This effectively renders such people unemployed, and sometimes unemployable. Zwelakhe Sisulu is one such case, as is Raymond Suttner. Both are also house arrested and have to report twice daily to the police.

Special restrictions were also placed on two political figures during 1987/8: the ANC's Govan Mbeki and KwaNdebele's Paramount Chief David Mabusa Maboka.

Under the emergencies promulgated since 1985, the restriction or banning of individuals has been widely used to hamstring political figures who are politically embarrassing to detain.

Released detainees regard their restriction orders as a further form of imprisonment, which they are required to police themselves. Some orders are extremely harsh, even punitive. For example, a trade unionist from Mohlakeng, near Randfontein, was detained for over two years, then house arrested and restricted to that township. He worked in Germiston, 60 km away, and his employers were willing to re-employ him, but his order makes this impossible. He is now unemployed.

Another punitive aspect of restriction involves reporting to a police station on a regular basis.

In the case of two young West Rand activists, they have to report at 10 in the morning and 2 in the afternoon. This makes it impossible for them to find employment. They share the fate of a Natal south coast man who is restricted, unemployed, and must report to police stations a considerable distance from his home, costing him R7 per day in travelling expenses.

Table 7: Banning of persons under emergency regulations[8]

Date	Number
July 1985 to March 1986	68
12 June 1986 to 10 June 1987	82
11 June 1987 to 9 June 1988	43
10 June 1988 to 31 Dec. 1988	50
1 Jan 1989 to 31 March 1989	+/-400
TOTAL	(estimated) 643

Restrictions and bannings of gatherings

In the period under review, the minister of Law and Order invoked both the Internal Security Act and the state of emergency to ban or restrict gatherings. The ISA was used nine times in 1988, including the banning of five meetings which were to celebrate May Day.

The bantustan administrations were also active in this arena, with Transkei and Bophuthatswana in the forefront. In August 1988, the Bophuthatswanan administration refused permission for the National Seoposengwe Party to celebrate its 15th anniversary. All political meetings were subsequently banned, forcing the NSP to hold its celebrations outside the bantustan.

Table 8: Bannings and restrictions under the state of emergency

Date	Event	Number
Jan-March 1988		13
April	Wectu launch	1
June	16 and 25 June commemorations	3
July	Mandela birthday	2
August		4
September	Biko, Regina Mundi, Anti-Apartheid conference	3
October	Rally for Defence of Democracy	1
November		5
December	Zeph Mothopeng, Delmas Trial	2
TOTAL		34

A large number of funerals were also restricted. The police are of the opinion that these events can be emotionally charged, especially if the deceased is an activist who died in police custody, during police action or in politically significant circumstances. A more recent police method of handling such occasions is to place heavy restrictions on the number of people who may attend, who may speak, and how the coffin shall be conveyed to the graveyard. Almost all funerals of a political nature are restricted in this way.

Informal repression

A number of different activities which can be described as informal repression fall into two broad categories:
- those activities carried out by organisations and structures which fall unambiguously under state control, and which operate with full legal sanction; and
- activities which are clearly beyond the law, but which are pro-government or pro-apartheid, and are carried out by anonymous agents or organisations, perhaps linked to the state, or by surrogate or right-wing groups. These activities frequently step beyond the bounds of the law, and they are very seldom successfully investigated or the culprits punished.

During 1987, South African government 'reform' policies involved the decentralisation of repressive power to regional and local levels, administered by joint management centres (JMCs). The enforcement arm of the JMCs, the municipal police, have risen in prominence to take their place beside the security police and the defence force as a major means of political control in the black townships of South Africa.

Another aspect of informal repression is illegal, yet appears to enjoy, if not official sanction, then at least less-than-enthusiastic investigation. The activities detailed below are seldom brought to court, let alone successfully prosecuted. They are secretive, and often more violent than other forms of repression. They include attacks on individuals and property, such as the massive blast which destroyed Khotso House in Johannesburg, a building housing anti-apartheid organisations.

In the same vein, 'vigilante' groups (which we call the South African Contras) abound, attacking progressive groups and their members. There is a disturbing trend for these vigilantes to be recruited into the new municipal police units, and it is often difficult to recognise the difference between the two, save that the latter wear uniforms, and the arms they carry are officially sanctioned.

The year 1988 witnessed a steady tempo of kidnappings and assassinations of anti-apartheid activists. These took place both within South Africa's borders and in foreign countries. When cross-border raids are made, the perpetrators are often identifiable, since they are clearly South African military commandos, and the government does not deny responsibility.

There are, however, assassinations carried out by clandestine groups or death squads, referred to many years ago as the 'Z' squad - though their name has probably changed. In a recent murder trial involving drug squad detectives, one line of defence was that one accused knew of security policemen who were members of a political assassination squad operating in Swaziland, and portrayed his own activities as being in this vein.

The Paris assassination of the ANC civilian emissary to Europe, Dulcie September, is another case where there has been speculation of South African security involvement.

Legally-sanctioned organs of control and repression

Joint Management Centres (JMCs)

The JMCs represent the major method of local-level control in South Africa. They are a decentralised form of repression, reaching into townships, factories and rural communities.

The JMCs implement the dual aspects of government policy: reform and repression. Reform takes the form of providing services and infrastructure which are genuinely needed, and which may improve some aspects of township life.

Repression, which is exerted simultaneously, removes from circulation the leaders of the communities who articulate grievances and political aspirations. These leaders are either detained, forced into hiding (thus hampering their ability to organise politically), or subjected to campaigns of intimidation. The JMC strategy involves meeting some of the community's expressed demands without giving credit to those who raised public awareness of them.

The new municipal police

The main vehicle for repressive control under the JMCs is a new police force created to serve the needs of the unpopular black local authorities. These municipal police are poorly trained (mainly by means of crash courses of between three and six months), and a good number of them have been identified as township thugs or former bodyguards of councillors.

These 'kitskonstabels' (instant police), as they are known, have an appalling record of discipline, and are frequently accused of assault. In October 1987 the minister of Law and Order admitted that weapons issued to kitskonstabels had been used in at least 95 crimes, including murder, rape and armed robbery.

A Black Sash survey in the Eastern Cape revealed three major trends in municipal police activity: firstly, they are used to strengthen the power of community councillors; secondly, they act as auxiliaries to the police, especially security police; and thirdly, they have close links with vigilantes, often siding with them in confrontations with progressive community organisations.

Extra-legal and surrogate repression

The South African Contras

In the past 18 months, a trend first noted two years ago has continued. It marks a serious police rethink about control of the townships. In 1985, 66% of people who died in township unrest were killed by police. In 1987/8, the overwhelming majority of deaths are due to what government commentators refer to as 'black-on-black violence'.

By stressing this, the impression is created that the state and its security forces are playing a peacekeeping role, and that the black community is tearing itself apart.

This change in tactics emerged after the enormous political opprobrium which descended on the police after the Langa massacre (21 March 1985) in which 22 died and over 50 were wounded (the state paid R1,3-million in compensation), and the Mamelodi massacre (21 November 1985), in which 12 people were killed.

The change in police tactics took a number of forms: firstly, police became more proactive, banning meetings, rallies and funerals rather than breaking them up while in progress; secondly, they began controlling numbers who could attend, and laying down restrictions on conduct for the occasions; and thirdly, free rein was allowed to vigilantes and other 'third force' or 'contra' groups, who perform the same divisive and disruptive work which the police formerly undertook.

Deaths in township unrest due to police action have declined. But the total number of deaths has not declined accordingly. In fact, in some areas, such as Natal, they have increased. The victims of the violence are largely the same - members of organisations belonging to the broad democratic movement and trade unions, but the perpetrators of the killings have a new identity - they are black.

However, the new killers are generally identifiable in their support for broad aspects of apartheid policy. For example, the Ama-Afrika group in Port Elizabeth, led by the Rev Ebenezer Maqina, has conducted a concerted campaign, sometimes amounting to war, against UDF and Cosatu members. This is a form of 'surrogate' repression.

The most significant of such groups is the Inkatha organisation, based in Natal. It is tribalist in ideology, regards Natal as its fiefdom, and is intolerant of other organisations operating in its sphere of influence. Despite its mobilisation along ethnic (Zulu) lines, Inkatha has met with only mixed success. In particular, it is losing ground among youth, especially schoolchildren, the urbanised, and organised workers.

Over the last two years Inkatha members have been involved in one of the most sustained periods of brutality ever witnessed in the Natal region. Here, the methods of confrontation and violence are naked, in opposition to Inkatha's other face, that of moderation and negotiation, polished up for white and international consumption.

There are many other contra-type organisations, characterised mainly by their complete lack of support and substance, yet they receive both large funding and official media support, being interviewed and lionised by the state-controlled radio and television.

Intimidation, violence and assassination

The section above details organisations which enjoy a degree of official sanction. But there are a number of organisations and activities which fall beyond the law. These extra-legal activities have the effect

of intimidating, inhibiting, or even eliminating political opposition
to the government.

Attacks on individuals and property

An analysis of 51 right-wing attacks listed in *Jodac News* (no. 1,
1989), covering the period 1986 to September 1988, shows six attacks in 1986, 26 in 1987 and 18 in 1988. Of these 51 attacks, 13 were
perpetrated on private homes and property. In 16 incidents, trade
unions were amongst the victims, including Cosatu offices in Johannesburg, Durban, Cape Town, Nelspruit and Vereeniging. The
UDF, Namda, Descom, and the ECC were also targets, as were
church organisations, student organisations, newspapers, news agencies and publishing houses. Cric (Community Research and Information Centre) and Pena (Port Elizabeth News Agency) were both
attacked three times in two years. Table 9 below shows the nature of
these attacks.

Table 9. Analysis of right-wing attacks, 1986 to September 1988

Type of attack	No of attacks
Arson	15
Petrol and fire bombs	6
Bombs	10
Shots fired	1
Break-ins and burglaries	15
Vandalism	4

A large number of attacks on individuals and property have taken
place in the period reviewed. Their object is intimidation and disruption. It is true that a low-intensity war is being waged in South
Africa, with guerillas of the ANC engaging in 'armed propaganda' in
the form of bomb blasts in strategic places. These have brought
retaliation from pro-apartheid forces in the form of massive demolition blasts at Cosatu House (the Johannesburg-based head office of
South Africa's largest trade union federation), Community House in
Cape Town (which houses a large number of progressive community
organisations and trade unions), and Johannesburg's Khotso House,
bombed in August 1988, which housed the South African Council of
Churches, the Black Sash and many other church or anti-apartheid

organisations.

These massive attacks on property can be crippling to organisations. There is the physical loss of irreplaceable records, and the need to find new premises - a difficult task, as potential landlords quickly learn the intended lesson of the bomb blasts: it could be you next. The sophistication of the explosive devices, their size and expert placement, all point to a military or paramilitary connection.

However, there have also been bomb placements which appear to be the work of ultra-right-wing organisations, aimed at the black population, such as a bomb placed in September 1988 in the *Why Not* Discotheque in Hillbrow, Johannesburg, a venue supported by a black clientele. These bombs have an intimidatory effect, and appear to be meant as a warning to black South Africans moving into areas previously reserved for whites only.

'Mysterious' robberies of progressive organisations

The past 18 months have witnessed an escalation in robberies and break-ins at the offices and property of trade union and political organisations. This is particularly common in the Eastern Cape, notably in the towns of Port Elizabeth, East London and Grahamstown.

Frequently the only items stolen are important documents belonging to the organisations, leading the victims to suspect that security police are involved. On other occasions, valuable equipment, including personal computers (which store important organisational information), typewriters and the like have been stolen.

These intrusions have caused considerable disruption of work. The same organisations have also been subjected to arson attacks which have destroyed premises and property.

Dirty tricks campaigns

Progressive organisations have been subjected to numerous dirty tricks campaigns. The most common form this activity takes is the production of pamphlets or publicity material purporting to be from a progressive organisation, but which carries a disruptive message. The purpose of the dirty tricks campaigns is to create confusion and promote internecine argument and fighting. In Soweto the two main (but rival) anti-apartheid groups, UDF and Azapo, have been particularly singled out for this treatment.

The perpetrators of these campaigns have never been brought to justice, but events relating to the ECC are revealing. Here, three members of an army 'dirty tricks' unit had second thoughts about their activities, and told ECC of their plans. They were court-martialled, but a subsequent court action by ECC drew an army admission that it was directly involved. Thus, in at least this campaign, the involvement of military intelligence has been publicly demonstrated.

The two other intelligence agencies in South Africa - the security police and the National Intelligence Service (NIS), both take a keen interest in progressive organisations and their disruption.

Kidnappings and disappearances

Unsolved disappearances and kidnappings of activists continue. But this has recently been concentrated in the Transvaal, as opposed to its previous predominance in the Eastern Cape.

Kidnappings take two major forms: the abduction of activists from neighbouring countries, and local abductions within South Africa. In the latter, some of the 'disappeared' people have subsequently been discovered in detention, but others have never been found.

One difficulty in solving these cases involves the secrecy of security police activities, and the detention system. 'Disappeared' activists may be detained, in hiding, in exile, or dead.

This confusion works to the benefit of those who would disrupt political organisations, and is particularly welcomed by members of death squads, who are able to take advantage of the confusion caused by secrecy.

In one instance, Vusi Mashabane, from Duduza, was abducted, interrogated and tortured by a group of men who, it later transpired, were soldiers. Other examples include disappearances from a neighbouring state, Lesotho, and incidents where the disappeared person is never heard from again. Particularly alarming is the disappearance of Stanza Bopape, a prominent Pretoria activist from the Mamelodi Civic Association and employed by the Community Research and Information Centre (Cric). Police say he escaped from detention on 12 June 1988, but he has never reappeared. The already-suspicious circumstances of his disappearance are strengthened by the fact that, for some time after 12 June 1988, police continued to lead Bopape's family to believe he was still in detention.

Assassinations

Assassinations have the effect of controlling opposition when all other methods, such as detention or intimidation, have failed. It is very rare that such assassinations are solved. A recent reversal of this trend relates to the death of Eric Mntonga, a director of the Institute for a Democratic Alternative for South Africa (Idasa) in East London. Due to internal rivalries in the Ciskei security police, new facts about his death have emerged, and senior security police members are now being charged with his death, which occurred on 24 July 1987.

Table 10. Recent assassinations

Name	Organisation	Died	Perpetrator
Samuel Seliso Ndlovu	Sosco	02.09.87	unknown
Sicelo Dhlomo	DPSC, Sosco	24.10.87	unknown
Linda Brakvis	UDF	29.01.88	unknown
Pearl Tshabalala	UDF	10.02.88	unknown
Amos Boshomane	Seawu	25.02.88	unknown
Nomsa Nduna	mother of unionist	06.03.88	unknown
Michael Banda	Potwa	01.07.88	unknown
Sidney Msibi	ex-ANC	05.07.88	unknown

The above list does not begin to address the killings in Natal, where a sustained campaign is being waged against Cosatu and the UDF. Hundreds have died in this violence. The list above refers to individual killings of known activists.

Repression and resistance to detentions

On 24 February 1988, the minister of Law and Order imposed restrictions tantamount to banning orders upon 17 organisations. Prominent among these were the Detainees' Parents' Support Committee and Descom (Detainees' Support Committee), two of the major support organisations for detainees and their families.

These bannings caused considerable disruption of detainee

monitoring and support work. The work was, nevertheless, continued by existing groups such as the Black Sash and Detainees Education and Welfare (DEW). Later, the Human Rights Commission was launched at the behest of numerous human rights organisations, and it is continuing to monitor repression trends.

An *ad hoc* Hunger Strike Support Committee was formed in 1989 to meet the particular crisis of detainees on hunger strikes.

This remarkable form of resistance to detention-without-trial had previously been used by detainees in a limited way to make a point of protest or win a local concession. It had generally met with little success.

In 1989, however, 20 detainees began a hunger strike in Diepkloof and, in a carefully planned and disciplined action, more and more detainees joined the strike. The campaign then spread to other areas until it became a nation-wide event, garnering local and international publicity.

The state at first ignored the growing action, then poured scorn upon it, claiming it was a publicity stunt initiated from outside the prisons. However, they reckoned without the detainees, many of whom had reached a point of desperation after two or more years of detention with no end in sight. In the end, Minister Vlok's nerve cracked first, and of the 850-odd detainees held in February 1989, he had released all but 150 by the end of March.

Repression will intensify

This overview has attempted to draw together many of the means used by the state to neutralise or eradicate political activity opposed to apartheid.

The picture which emerges is a complex one, for the state by no means has complete control over the situation. Many organisations and individuals have had their activities severely curtailed and disrupted, but new and creative forms of resistance and organisation have evolved to cope with changing political climates. Nevertheless, state power in the form of repression has never been as strong in South Africa. Even the spectacularly successful hunger strike, which won major concessions from the government in the form of releases, was a desperate and last-resort form of brinkmanship, which could have backfired on its participants.

Repression exists because South Africa is burdened with a

government which is fundamentally undemocratic and unrepresentative. The ruling group thus has to govern without the consent of the majority of South Africans. The end of detention-without-trial is accordingly not imminent. On the contrary, repression is bound to intensify, in all its forms.

Notes

1   Charles Humana, *World Human Rights Guide*, Pan, London, 1987.
2   David Webster, 'Repression and the state of emergency', in G Moss and I Obery (eds), *South African Review 4*, Johannesburg, 1987.
3   Figure taken from *Human Rights Commission Fact Paper, FP1*, Johannesburg, 1988.
4   *Focus on Human Rights*, published by the Human Rights Commission and the *Weekly Mail*, 10.12.88.
5   *Human Rights Update*, 1(3), April-June 1988.
6   Figures drawn from issues of *Human Rights Update*, 1988.
7   Reproduced from *Human Rights Commission Fact Paper, FP2*, 1988.
8   Information drawn from *Human Rights Commission Fact Paper, FP3*, March 1989, and *Citizen*, 05.04.89.

People's War or Negotiation? African National Congress Strategies in the 1980s

Tom Lodge

The complexity of the African National Congress as an organisation is reflected in tensions between the two strands of its current intellectual evolution. On the one hand the concept of an insurrectionary people's war appeared to be influential in military planning from at least as early as 1983. On the other, ANC diplomatic spokesmen continued to stress the organisation's commitment to a negotiated settlement. Both themes can be understood within their specific historical contexts, but both are also informative about the long-term development of the organisation and the broader political movement of which it is a part.

'People's war'

In 1982, a series of articles in the *African Communist*, journal of the South African Communist Party, debated whether the time had come 'for arming the masses'. The publication of this debate seems to have accompanied a major strategic decision, reflected in a senior Umkhonto we Sizwe official's use of the term 'people's war' in *Sechaba*, a leading ANC publication, during 1983. This notion became central to the ANC's military planning from then on, and was used to define tasks related to the further unfolding of the armed struggle.

Within ANC thinking, 'people's war' involved a perceived need for greater numbers to participate in the war. By the close of 1982 ANC strategists believed that the growth in mass organisations - and particularly trade unions - had created 'enormous potential for the advance of the revolutionary movement'. The ANC believed it had attained - at least in the cities - the goals of 'armed propaganda': spectacular and technically sophisticated guerilla operations directed at demonstrating the state's vulnerability and encouraging popular political assertion.

The insurrectionary perspective

Events in the succeeding years helped confirm the ascendancy of an insurrectionary perspective among ANC planners. As a commission appointed by the National Executive Committee argued at the 1985 consultative conference, the ANC was 'never going to have a rear base with a friendly border' from which to conduct an orthodox campaign of guerilla warfare. Instead, the 'positive factor' in the ANC's future development was 'a people in political motion' - and moreover, unlike most guerilla wars, a people who would be mobilised most easily and formidably in the towns, not the countryside.[1]

Rural insurgency was important, not least in prompting the dispersal of enemy forces. But the decisive battles, both military and political, would remain in the cities and particularly within an organised and, ultimately, armed working class. Special attention, suggested a commission on cadre policy at the 1985 conference, 'needs to be given to drawing workers into Umkhonto we Sizwe':

> This is vital, both for the overall perspectives of people's war, as well as the possibility for a long-lasting national work stoppage backed by our oppressed communities supported by armed activities aimed at bringing the regime to its knees.[2]

The series of localised rebellions against black local authorities, which began with the Vaal uprising in September 1984, gave this scenario more force. The level, intensity and character of popular political assertion were quite unprecedented and, as the 1985-86 regional state of emergency demonstrated, well beyond the repressive capacity of the state's normal policing resources.

For some of the veterans of earlier township conflagrations who, since exile, had joined the ANC, these were exhilarating times. Alex Mashinini, writing in *Sechaba* of April 1986, perceived in the 'level of mass political and military participation...the possibility of revolution'. In his analysis, 'the existence of a large black working class...brought to the fore of the struggle the issue of insurrection...as it can be conceived in any highly industrialised country'. In the more immediate context, the masses had created in certain localities 'a peculiar form of dual power, not in the administrative but in the political sense'. Though the state was still intact, there were emerging 'organs of self government' which had the potential to evolve both as sources of 'alternate power' to the state and as 'organs of insurrection'. These bodies, Mashinini argued, should be 'strengthened, consolidated, and developed into "Revolutionary People's Committees"'.[3]

Mzala, a former law student at the University of Zululand, and a contemporary of Mashinini, took the argument a stage further; the ANC had no need to develop such committees, they already existed in the various township-based organisations, and in themselves represented, as had the Paris Commune, 'a new type of governability...ready to serve as organs of people's war and insurrection'. Such structures were 'clearly destined to be the embryos of our future democratic state'. In the people's communes, asserted Mzala, was a government of a new type. Such a perception was premised on the certainty that 'our strategy emphatically calls for the seizure of power in contrast to the reformist call for power-sharing'.[4]

At a more prosaic level, people's war represented a pragmatic military response to the ANC's logistical task of conducting insurgent operations over lengthy external lines of communication. Policy and practice from 1985 onwards emphasised the local recruitment and local training of guerilla auxiliaries who were to be equipped with grenades, limpet mines and automatic rifles. There was also a move towards the shortening of guerilla training courses in Umkhonto base camps as well as a movement away from ambitious and complicated operations such as the Sasol oil refinery attack. Activities requiring less co-ordination, less technical expertise, and smaller units, were now favoured.

With its first deployment of landmines in 1985, the ANC began to develop a military presence in rural areas, an important dimension of its post-1985 strategy of expanding the field of operations. The number of attacks would increase significantly from under 50 in 1984 to the present annual total of around 250-300 which has been maintained since the beginning of 1986 and which probably represents the limits of existing Umkhonto capacity.

Targets of Umkhonto military action have been broadened to include the government's political collaborators (and occasionally, unavoidably, members of their families), as well as supermarkets and other commercial premises. Many ANC leaders dislike hurting civilians - white or black - but only rarely do they disavow such attacks. As a senior Umkhonto official observed bitterly to a journalist in 1987: 'People have been making so much noise about some of these bombs that have gone off in supermarkets. It is true that we sent these people there. But the world is only concerned when whites die - when blacks die it is acceptable'.[5]

In fact, as was evident in 1988, many ANC notables are extremely uneasy with the argument advanced by Umkhonto's new chief of staff, Chris Hani, that terrorist action directed at whites might have positive political effects. There is a resigned acceptance that relatively superficially-trained rank-and-file combatants, in exercising considerable tactical discretion in the field, have often been disinclined to pay too much attention to the humanitarian conventions of the ANC's code of guerilla ethics. And there is, after all, no evidence to suggest that attacks on white civilians have had an adverse effect on the ANC's standing within its black constituency.

Towards an insurrectionary objective

But people's war does not merely represent the choice of a particular set of military tactics over others. It is a strategy which is directed at the accomplishment of a particular political objective: an insurrectionary seizure of power. To be sure, the meaning of the term 'seizure of power' is frequently debased through its usage by governments whose accession was relatively peaceful if not constitutional, but ANC theorists do not merely use the term rhetorically. As Mzala puts it: 'Our insistence on seizure of power...is not an idle proclamation'. [6]

A senior ANC office-bearer, interviewed in 1985, spoke of 'armed insurrection...a culmination of guerilla warfare' as the 'classic method of making revolution', citing both the Iranian and Bolshevik revolutions as suggestive alternative precedents for a South African insurrectionary change of regime.[7]

The strategy and tactics commission at the 1985 Kabwe conference forecast a situation in which the 'advanced combat formations' of Umkhonto we Sizwe would be engaged in tandem with 'the people in arms - the advanced active elements of the masses, prepared and charged by the vanguard formations'. The commission also suggested optimistically that groups might be joined by 'those elements of the enemy forces ready at the decisive moments to side with the revolutionary forces'.[8]

Demoralisation within the SADF ranks is accordingly a crucial consideration in the strategic situation projected by ANC insurrectionists.

Interpreting armed struggle: the insurrectionary vision

For many people within the ANC, there is an inseparable link in South Africa between national liberation and social emancipation. Within Umkhonto ranks during 1985, this implied a strategy which must lead to the revolutionary overthrow of the existing ruling class, and the complete dismantling and replacement of the state apparatus. For the ANC's advocates of insurrection, seizure of power implies this process. Cassius Mandla, writing in the November 1985 edition of *Sechaba*, noted that 'there seem to be different interpretations of the tactic of armed struggle'. 'If,' he went on

> our strategic objective of armed struggle is to demonstrate that we can make the enemy bleed and die, the purpose being to force him to a negotiating table or to force universal adult suffrage without racial discrimination, in practice we will wage war for freedom and victory. We are talking here of armed struggle as a pressure tactic and armed struggle as warfare. Which of the two is our revolution?[9]

Mandla's article contended that the first approach was outmoded; it was no longer the case that the ANC wanted 'the enemy to come to terms with us and recognise the legitimacy...of our cause'. In the new order of things

> the enemy has no role in the solution of our problems... There can be no going back to the practice of frittering away our energy in activities calculated to prize the case-hardened conscience of white oppressors to invite us to negotiations to bring about a dispensation acceptable to them and us. We are for a dispensation that excludes the enemy as a factor in its making.

Mandla presents the insurrectionary case in its most polarised form, and by implication his article is more critical of ANC leadership than the work of other contributors to *Sechaba*. But his views are a reasonable extension of the logic running through the insurrectionary argument. A group of theorists within the ANC's intellectual community in the mid-1980s argued that it was possible for the ANC's struggle to culminate in an insurrectionary transfer of power in which a revolutionary vanguard would triumph over a shattered state apparatus. In this process, 'the passage of political power from the racist minority to the democratic majority will ultimately be decided by material strength, and military force is the key element of that strength'.[10]

This view seemed to accord with authoritative ANC statements, which in 1985 interpreted the seizure of power as 'mass action involving revolutionary force, such as land occupation, factory occupation, people's control of the township in the face of constituted authority, etc'. As a rider to this, ANC planners observed that 'our capacity to play an effective role at this level will depend largely on our combat presence and the availability of minimum armouries of weapons'.[11] The transfer of power which is projected in such formulations is an accession through force of arms, not a negotiated change of regime.

This is one vision of the future which may animate ANC cadres. A reading of official documentation from the 1985 conference as well as the statements of ANC leaders suggests that its adherents extended at that time well beyond the confines of habitual contributors to *Sechaba*.

It is a vision which may have been especially compelling in 1985 and 1986 when the conflict in the townships was at its peak. But insurrectionists have to compete with an older and more socially conciliatory tradition in ANC thinking: the espousal of negotiated settlement and constitutional reconstruction. Ironically the events of 1985-1986 may have also supplied sustenance for the revitalisation of this tradition.

The negotiation option

The ANC's willingness to contemplate a negotiated accession to power runs as a counterpoint to the language of seizure and insurrection which is so central to a section of its polemical literature. A senior ANC leader was quoted in January 1986 as asserting that the ANC's main objective lay not 'in a military victory but to force Pretoria to the negotiating table'.[12] A little later another senior ANC executive member elaborated the argument:

> The call you make to people is the same, saying you are not prepared to negotiate, that you must intensify the offensive. But in the course of that

offensive it is clear that one of the most important things is breaking up the
power structure... Even the army will have its problems because it is a conscript army subject to all the pressures that the general public is subject to.
There are 'homeland' leaders also subject to pressures and we hope that more
of them will desert Botha's camp as well. Out of all this you will get a realign-
ment of forces. We are not talking of overthrowing the government but turning so many people against it that it would be forced to do what Ian Smith had
to do.[13]

In October 1987, the ANC's national executive produced a statement on negotiations which 'reaffirmed that the ANC and the
masses of our people as a whole are ready and willing to enter into
genuine negotiations provided they are aimed at the transformation
of our country into a united and non-racial democracy'. The document set a number of pre-conditions to be met before the ANC
could participate in such negotiations and for the first time
announced the ANC's acceptance 'that a new constitution could include an entrenched bill of rights to safeguard the rights of the
individual'.[14]

The constitutional guidelines

In mid-1988 the ANC published a set of constitutional proposals or
guidelines. These do not represent a repudiation of the ANC's
adopted programme, the Freedom Charter. Instead they are
presented as a translation of the broad principles of the Charter into
explicit constitutional proposals. They and the intellectual process
which helped to produce them deserve some detailed examination.

The document begins with a preamble. This stresses the requirement for 'corrective action' to guarantee an irreversible redistribution of wealth and 'opening of facilities' to all. Twenty-five clauses
follow, divided into sections concerned with the state, franchise,
national identity, a bill of rights and affirmative action, the economy,
land, workers, women, the family and international relations. The
clauses specify a non-racial form of sovereignty based on a universal
suffrage represented in a central legislature, executive, judiciary,
and administration.

To enhance efficient government and popular participation in
decision-making, powers should be delegated from central government to more localised administrative bodies. Judicial, police and
military organs of government, as with other sections of the

administration, should be popularly representative and democratic in their structure. The constitution should guarantee 'free linguistic and cultural development', the eradication of racial discrimination, the prohibition of racism, fascism, Nazism and ethnic or regional sectionalism; and subject to these constraints, protect freedom of association, thought, worship and the press.

Within the economic sphere, the guidelines prescribe a mixed economy in which 'the state shall have the right to determine the general context in which economic life takes place' and in which the private sector 'shall be obliged to co-operate with the state in realising the objectives of the Freedom Charter'. Apart from the public and private sectors, the economy should include co-operative and small-scale family enterprise. The state should direct a programme of land reform. Workers' trade union rights, including the right to strike, as well as the equality of women should be constitutionally entrenched.

The guidelines fall well short of a socialist reconstruction of South Africa. Significantly, the political provisions of the guidelines suggest a more radical degree of restructuring than do the prescriptions for the economy. Even so, the changes to the political system which are envisaged in the guidelines do not suggest a complete dismantling and reconstruction of the state apparatus, let alone the 'people's dictatorship' espoused by some of the more polemical *Sechaba* Jacobins. This emerges clearly from the proceedings of an internal seminar held by the ANC in March in Lusaka where a draft of the guidelines was reviewed.

Kadar Asmal, a former South African and founder member of the Irish Anti-Apartheid Movement, wrote a paper on electoral systems in which he situated the discussion firmly within the context of a multiplicity of parties competing for office. Tony O'Dowd (the treason trial lawyer between 1956 and 1961) drew up a programme of legal reforms which included the codification and retention of a substantial proportion of the existing body of legal doctrine. O'Dowd's paper also suggested that with respect to those courts which had a primarily criminal field of jurisdiction 'many present judges could probably be continued in office'.[15]

The economic principles spelled out in the guidelines are all-embracing but cryptic. They can be interpreted in a number of ways, but do not include the specific commitment of the Freedom Charter to nationalise the mines, the banks and monopoly industry. This need not be taken to signify the existence of a bourgeois predisposition within ANC leadership. It is the case, though, that some ANC and Communist Party authorities contend that a fairly long period of transition may be required before all the aims of the Charter can be realised.

In 1985, for example, 'Nyawuza' in an *African Communist* polemic directed at the Trotskyite 'Inqaba' group argued that

> the problem with people advocating 'socialism now' is that they expect those blacks who cannot read or write to run socialist industries and mines. The danger here is that we find ourselves depending on the expertise of the very forces we want to defeat: people who are against our socialist principles. The result would be economic chaos.[16]

A senior ANC official conceded in 1986 that 'for some while after apartheid falls there will undoubtedly be a mixed economy, implying a role for levels of non-monopoly private enterprise represented not only by the small racially oppressed black business sector but also by (white) managers and business people of goodwill'.[17] And when a senior ANC delegation met the South African business delegation at Mfuwe, Zambia, in September 1985, ANC leaders suggested that the nationalisation of 'monopoly' industry could take the form of a 51% state shareholding, which would still leave a substantial role for large privately-owned companies.[18]

Even non-communist leaders within the ANC favour an ultimate transition to socialism, but there is a growing perception within the leadership that the transition may be a gradual one accomplished, in words used by a senior ANC representative, through 'debate rather than in the streets'.

Even in the early stages of transition the social and economic role of the state would of course be pivotal, serving as the most effective guarantor of what Albie Sachs - a political lawyer in South Africa in the 1960s who, until his injury in a car bomb attack in 1988, had been involved in developing the Mozambican legal system - termed in his contribution at the Lusaka seminar 'second generation rights' or the broad socio-economic liberties which include the right to health, shelter, nutrition, and so forth. In the economic context, not only would the state control the public sector but it should also regulate activity in the private, co-operative and family domains.

The last of these probably refers to agriculture. A contributor, writing in a 1985 commemorative volume on the Freedom Charter, asserted 'that most peasant workers...are not yet class conscious enough or ready for the adoption of socialist solutions'.[19] And Sachs believes that the ANC in power would show some sympathy for the position of those 'white farmers with a deep attachment to and love of the land'.[20]

In Sachs and other ANC theorists' thinking, there is a conviction that ownership of productive assets should be limited to those with 'productive skills'. But he does suggest as a possible scenario that a post-apartheid South Africa might retain a group of landlords (for whom certain rights and obligations would have to be legislated), though presumably these would be required to perform an active managerial role as opposed to merely living off the land.

The ANC constitutional discussion is notable for its caution, degree of qualification, and eschewal of utopianism. In general, the constitutional restructuring envisaged by the ANC includes strong elements of adaptation. Theorists demonstrate considerable sensitivity to what they understand to be broad legal and politico-administrative traditions in South Africa and employ these to help justify many of their recommendations. They also draw on other models and experiences which range from the Anglo-Saxon and more generally Western bourgeois liberal heritage to Eastern European and Third World models of democratic practices.

This is hardly the political programme suggested by the implications of an insurrectionary conquest or seizure of power, and it is significant that the ANC itself linked the discussion of constitutional rights to the subject of a negotiated transfer of administration by including the reference to a bill of rights in its October 1987 statement endorsing negotiations.

This surely implies a future scenario in which, to cite the option rejected by Mandla in *Sechaba*, sections of 'the enemy (would) come to terms with us and recognise the legitimacy...of our cause'.[21]

Interpreting opposing discourses

The two opposed streams of discourse in the ANC have a much broader importance than their immediate location among small groups of exile intellectuals would suggest. The polemics in *Sechaba* are not primarily directed at an external academic readership; the journal is a principal source of political inspiration within the organisation and among its following, both inside and outside South

Africa. Here the ANC's strategic prescriptions are rendered into slogans and directives, which have great influence in the communal context in which its moral authority is rarely contested.

Both the discourses discussed here have been influenced by, and have had an impact on, wider political events. Supporters of insurrection were surely influenced by the mood of 'immediacy' which pervaded the ranks of teenage street fighters in 1985, as this passage by Mandla vividly testifies:

> A different will holds sway, a popular will... This is a moment of truth, a revolutionary flashpoint... We have not witnessed in living memory our country being a constantly boiling cauldron of unbroken ever escalating street battles... We are at an historical crossroads...a revolutionary mood pervades all ages and classes among the oppressed...the moment of revolution is now or never in our lifetime... Yesterday's approaches must be broken with...a month of sustained...armed action which may well prove to be the long awaited abracadabra for the dawn of freedom in South Africa.[22]

Such a view might well have had immediate implications for the way the ANC perceived politics should be conducted. In 1985, both in policy statements and polemics, ANC voices expressed criticism of the local-level negotiations which were occurring between civic organisations and state bodies. The fear was that they risked compromising what the ANC believed to be the insurrectionary character of popular organisation.[23]

It is possible, conversely, that the efforts at negotiating specific issues with the authorities - especially in the sphere of education - helped the ANC to begin elaborating its own conception of a restructured social order. Another important contribution to that process was the new source of pressure emanating from the politically and socially heterodox procession of delegations arriving in Lusaka to seek clarification of the ANC's policies.

A machiavellian interpretation of the ANC's intellectual dualism can, of course, be constructed. This would see either one or other discourse as a purely tactical ploy designed to win the loyalty or allay the fears of a particular constituency. References to businessmen or white liberals in certain *Sechaba* commentaries and internal Communist Party documents are sometimes less than flattering: they are portrayed as ultimately hostile groups with whom short-term alignment can be made for the expediency of isolating the regime rather than as potential constituents or long-term allies.[24]

Such rationalisations of united front policies may help to defend and justify the politics of social conciliation to revolutionary purists within the ANC. But they do not supply an adequate explanation for the present vigour of constitutional discourse within the movement. And participation in encounters with liberal and business elements may have reduced any element of cynicism present in initial ANC perceptions of the purpose of such events.

The elaboration of the constitutional guidelines was debated very thoroughly, and the outcome reflects a balance between the ANC's conservative and radical traditions. The organisation's sensitivity to legal and politico-administrative traditions represents a more profound commitment to social restraint than merely an elaborate exercise in public relations. The ANC's own historical consciousness includes a strong respect for tradition and continuity.

In a popular history of the organisation published recently by the editor of *Sechaba*, a major theme is 'that there is no historical justification for an artificial demarcation between the ANC of today and the early ANC'.[25] This sense of historical deference to the legacy of previous generations may help to explain the evolutionary and adaptive features of the ANC's own procedures - doubtless the influence of lawyers who have made such a contribution to the direction of the organisation since its inception. This is seen in the carefully hedged and qualified language of ANC public policy statements; in its painstakingly drafted internal constitutions; in facets of its behaviour - it remains the only insurgent organisation to be a signatory to the Geneva convention; and in its concern for precedents set by its own history.

Negotiated settlement or armed struggle

The persistent imbalance of military power between the ANC and the South African administration may strengthen the attractions of a negotiated settlement in which the strategic trajectory would designate armed struggle as a source of pressure rather than the agency of an overthrow of state power.

According to an ANC executive member, 'war is not going to be fought in a straight line... It must be seen to be going up and moving towards its climax... War must be war in South Africa'. But it appears likely that the war, as such, will be fought in a straight line for a long time to come. The ANC's capacity to expand the sphere of controlled military operations is likely to continue to be limited by the problems of commanding, supplying, and reinforcing its army over very long lines of communication. The removal this year of its training facilities from Angola to Tanzania and Ethiopia - a consequence of the Namibian settlement - serves to highlight this enduring constraint.

Nevertheless, the socially radical traditions represented by the discourse of insurrection cannot be neglected. These reflect the internal changes which have affected the ANC since its prohibition in 1960 and in particular its organisational transformation into a vanguard-style bureaucracy. But it is also an expression of an ideological tradition of social emancipation present in the ANC's history which goes back much further than 1960 and which in the last decade has drawn inspiration from the organised struggles of Africa's mightiest industrial working class.

There is a political tension within the ANC and it cannot be easily reconciled analytically. Nor should it be reduced to the detection of factions - real or imagined. For the tension transcends generational membership, political affiliation and social experience. It is an expression of the contradictions and anomalies which exist in the broader society from which the organisation has come and in which it continues to evolve, a society which can simultaneously combine features of racial autocracy, bourgeois liberalism and industrial democracy.

It is a dynamic tension which should be welcomed because it may well prove creative in producing a social order which can attain that elusive balance between social justice and political liberty which represents democracy.

Notes

1 'Commission on strategy and tactics', document presented at the Kabwe consultative conference, 1985, 15.
2 'Commission on cadre policy, political and ideological work', document presented at the Kabwe consultative conference, 1985, 8.
3 Alex Mashinini, 'Dual power and the creation of people's committees', *Sechaba*, April 1986.
4 Mzala, 'Building people's war', *Sechaba*, September 1986.
5 Text of an interview with Joe Modise, Lusaka, January 1987, *Byline Africa*, Sandton, South Africa.

6 Mzala, 'Building people's war', 12.
7 Interview, *Sechaba*, May 1985.
8 'Commission on strategy and tactics', 18.
9 Cassius Mandla, 'Let us move to all-out war', *Sechaba*, November 1986, 25.
10 Mzala, 'MK - building people's forces for combat and insurrection', *Sechaba*, December 1986, 19.
11 'Commission on strategy and tactics', 16.
12 *Financial Mail*, 17.01.86.
13 *The Observer*, London, 02.03.86.
14 'Statement of the national executive committee of the African National Congress on the question of negotiations', 09.10.87.
15 Tony O'Dowd, 'The legal system and the judiciary in South Africa today', Lusaka seminar, March 1988, 8.
16 Nyawuza, 'New "marxist" tendencies and the battle for ideas in South Africa', *African Communist*, 103, 1985, 58.
17 *Guardian Weekly*, 17.08.86.
18 Notes on a meeting between ANC and South African businessmen, Mfuwe, Zambia, September 1985.
19 *Selected Writings on the Freedom Charter*, London, 1985, 103.
20 Albie Sachs, 'Towards a bill of rights in a democratic South Africa', Lusaka seminar, March 1988, 10.
21 Mandla, 'Let us move to all-out war', 25.
22 Cassius Mandla, 'The moment of the revolution is now', *Sechaba, November 1985.*
23 See Alex Mashinini's criticism of 'elements who would like to pull these committees into unpopular deals with agents of the apartheid system', in 'Dual power and the creation of people's committees', 6.
24 See, for example, SACP politburo document on negotiations, released by the office of the state president, Cape Town, 12.06.86.
25 Francis Meli, *South Africa Belongs To Us: a history of the ANC*, Harare, 1988, ix.

Between Apartheid and Democracy: The South African Election of 1989

Rupert Taylor

When the National Party came to power in 1948, apartheid was initially advanced as 'a radical alternative to the principle demands of liberalism'.[1] Its basic underlying, yet unproven, premise was that because racial contact leads to friction and conflict, integration must not be encouraged. To this end the NP pushed through the Population Registration Act (1950), the Group Areas Act (1950) and the Reservation of Separate Amenities Act (1953). The terms 'separate development' and 'group rights' were advanced by the NP's leading intellectuals in an attempt to give racial practices a seemingly presentable moral and philosophical facade.

In the 1970s ever-growing contradictions, fed by changing economic and class forces, necessitated a rethink. A growing belief that apartheid must be reconciled to some liberal democratic principles took root. Moving away from Verwoerdian principles, concepts of pluralism and consociationalism were used to inform PW Botha's unfolding reform programme which sought - through a new language of legitimation[2] - to build a wider base of consensus for NP rule. The changes initiated by Botha centred on the 1979 labour reforms, the granting of freehold title for Africans in urban areas, the abolition of 'influx control' for those resident outside the 'independent homelands' and the new constitution of 1983 which established a tricameral parliament - with separate houses for whites, coloureds and Indians.

This reform programme was inherently limited because, as the Commonwealth Group of Eminent Persons' report noted, it had to 'be viewed against the background of a determination not to give up white control'.[3] The NP was unwilling to democratise state power and grant freedom of association. Consequently the tricameral parliament fell far short of meaningful reform.

There was, for example, 'no question of segmental autonomy as prescribed by consociation theorists'.[4] With both the House of Representatives (for coloureds) and the House of Delegates (for Indians) being boycotted by the United Democratic Front and National Forum, their functioning has rested more on co-option and corruption than consensus. Reform under Botha was thus a pragmatic and uncoordinated process guided by the need to contain the unfolding contradictions of apartheid, and did not seriously address root causes.

Nonetheless, the rhetoric of 'power sharing' and the pace of change led to far-reaching splits in Afrikanerdom. In March 1982, 16 MPs left the NP to join the Conservative Party under Dr Andries Treurnicht, which 'sees itself as the political torchbearer of Afrikaner nationalism in the South Africa of the 1980s'.[5] Committed to reversing the limited reforms and re-establishing the principle of partition for each South African group and race, the CP captured 22 seats (all in the Transvaal) in the 1987 election.

On the other hand, as a direct reflection of the inherently limited nature of the NP's reform programme, a state of emergency was imposed to check the process of black political mobilisation. Protests, which endeavoured to wreck government initiatives at local level by making the townships ungovernable, were met with force. The security forces ruthlessly destroyed structures of people's power through intimidation, restrictions and detentions. Innumerable organisations were restricted and between June 1986 and May 1987 around 25 000 people were detained.

Unable to justify repressive actions in terms of liberal democratic principles, NP policy was rendered more or less bankrupt. In the 1987 election the NP received a mandate for a proposed multi-racial statutory National Council, but no serious black leaders have been prepared to participate in that forum, and it has come to nothing. Given the lack of any fresh movement the NP faced a tough challenge in approaching the 1989 election; it had to reformulate a position somewhere between apartheid and democracy that would generate new hopes and expectations.

The election campaign

The early months of 1989 signalled the end of the PW Botha era. On 18 January Botha suffered a stroke and on 2 February stepped down as NP leader, forcing a snap leadership election. In the final round of voting the NP caucus elected FW de Klerk (69 votes) over Barend du Plessis (61 votes). Botha remained state president, but as friction developed within the NP leadership it was agreed on 6 April that an election would be held on 6 September and thereafter Botha would retire (under the constitution an election for all three houses had to be held by March 1990).

During the campaign, which stretched over five months, the CP demanded a return to Verwoerdian apartheid; the newly-formed Democratic Party argued for real movement towards Western democracy; and the NP sought to establish a new footing somewhere between the two. Meanwhile, outside the electoral process, the mass democratic movement (MDM) of the black majority advocated a boycott of the House of Representatives and House of Delegates and engaged in an anti-apartheid defiance campaign.

For the CP, the crisis in South Africa is due to the failure of the NP to adhere to the pure principles of apartheid. Therefore the solution is simple: a return to Verwoerdian apartheid. Harking back to the past, the CP manifesto stated that the white community has a non-negotiable right to self-determination and that a constitution for a 'white South Africa' with separate states for other groups must be introduced. Influx control would be reimposed and only white trade unions would be allowed in white South Africa.

With party activists strongly represented in local authorities, civic associations and voluntary bodies throughout the Transvaal and Orange Free State, the CP campaign focused on taking a strong grassroots approach. The NP was attacked for mismanaging the economy and 'selling out' to a future black majority government. Given the implementation of United Nations Security Council Resolution 435 the NP was also accused of 'selling out' Namibian whites. Although there was little attempt to mention just where 'white South Africa' would be located and how its day-to-day functioning could be maintained, CP head office talked of winning 60 seats.

The DP stated a commitment to abolishing apartheid and laying the institutional foundations of a democracy - 'we say South Africa is ready for Western democracy'.[6] The DP manifesto stood for one-person one-vote in a federal, not a unitary, system; 'South Africa should consist of a federation of states or provinces with each conducting its affairs as it chooses, providing it does not conflict with the constitution or Bill of Rights'. Whilst the DP argued that the new constitution would be arrived at as a result of negotiation between the leaders of all South Africa's communities it was also stated, without indicating how, that the constitution 'will protect our various cultures, religions and languages and will prevent majority domination while allowing every South African a vote of equal value'.

Under the joint leadership of Zach de Beer, Dennis Worrall and Wynand Malan, the DP hoped to mask internal differences and appeal to different constituencies. During the campaign the DP generated a new dynamism, and meetings drew large crowds. With a fresh party image the DP hoped for a broad support base; Worrall talked of the DP gaining 42 to 44 seats. Founded on 8 April 1989 (as a merger of the Progressive Federal Party, Independent Party and National Democratic Movement) the DP also hoped to break down the divide between parliamentary and extra-parliamentary forces opposed to apartheid. But its strategy of going against the MDM's boycott position and putting up a limited number of candidates for the House of Delegates did not succeed.

Rejecting both the CP's commitment to Verwoerdian apartheid and the DP's support for liberal democracy as being unworkable in the South African situation, the NP's campaign resuscitated the belief that a position somewhere between apartheid and democracy could be arrived at. Drawing on ideas circulating within the Afrikaner Broederbond for a number of years,[7] a Five-Year Action Plan charted this middle road. The Action Plan, advanced at the NP Federal Congress in June, stressed the need to negotiate a system of full political participation for all and a move away from rigid race classification.

It argued that various groups would have their own governing bodies with sole jurisdiction over 'own affairs' and that each group would participate in central government to decide, by consensus, on 'general affairs'. Whilst groups remained the basis for political participation, the plan spoke of dialogue and negotiation; it argued for a universal franchise and a bill of rights. This clearly attempted to give hope and promise of a future built on both the principles of apartheid and democracy; talk of a bill of rights, a just society and a commitment to a universal franchise sounded well and good to those wedded to liberal values whilst expression of a continuing commitment to 'group rights' catered to the values of those socialised in apartheid.

By paying lip-service to the foundations of both an apartheid and a democratic state the NP's plan could mean different things to different people. In many ways this best reflected the ambiguous attitudes of the typical NP voter. In this sense the campaign strongly resembled the NP's electoral strategy in the May 1987 election where the key slogan 'Reform Yes, Surrender No' gave a double-edged emphasis on reform/security that could also be interpreted ambiguously. 'To the right wing voters it (the NP) could promise that its concerns with reform would never jeopardise their security, and to moderate voters it could promise that its concern with security would never jeopardise reform'.[8]

Similarly, as in 1987, to mobilise and build electoral support the NP made full use of its control over the mass media. With a halo constructed by media image-building, De Klerk talked of being committed to new action for a great and just South Africa. On the state-controlled South African Broadcasting Corporation the NP engineered a 'least-of-evils' position between the CP and DP, one day countering the claims of the far-right, the next day the claims of the left. De Klerk's international visits to various European and African countries were fully portrayed to show the NP leader as a skilful statesman who would best negotiate South Africa's future.

Continuing personal differences did, however, impede the smooth functioning of the NP campaign. Matters came to a head over a planned trip by De Klerk and Foreign Affairs Minister Pik Botha to meet with Zambian President Kenneth Kaunda. Alleging that the visit had not been properly cleared, PW Botha announced his resignation as state president in a live television address on 14 August. This was after De Klerk and the cabinet had told him it was 'time to go'.

The NP campaign was also affected by the MDM's grassroots defiance campaign.[9] In the weeks preceding the election, protests were organised country-wide. Hospitals in the Transvaal, buses in Pretoria, schools in the Border area and beaches in the Cape and Natal were targeted. Whilst a few of these protests passed by peaceably, most were met with state repression. Defence Minister Magnus Malan and Law and Order Minister Adriaan Vlok strongly condemned the defiance campaign as an ANC and communist-inspired plot and police used tear-gas, stun grenades, batons and quirts to disperse crowds and enforce segregation.

On 19 August protesters who defied a police and army cordon to visit whites-only beaches in the Western Cape were whipped and arrested. Four days before the election a non-violent march attended by 2 000 people in central Cape Town was met with tear-gas, whippings, water cannons and purple dye; a quarter of those present were arrested. According to the Human Rights Commission at least 37 meetings were violently broken up by the security forces and over 1 500 people were detained in the month preceding the election. The NP was hard-put to balance these actions with talk of a new, great and just South Africa.

Election results

In the event, the NP was returned to power with 93 seats, losing 17 to the CP and 12 to the DP. One seat, Fauresmith, was an NP-CP tie. On the right, the CP did not live up to its promised threat, and failed to make serious inroads into the NP vote. Winning a total of 39 seats the CP managed to break out of its Transvaal heartland, taking six seats in the Free State and two in the Cape (see Table 1). These victories, however, involved very narrow margins, mostly under 500 votes, and the CP failed to take any seats in Natal.

In the Transvaal the CP gained nine seats, but allowing for the benefits that flowed to the CP through the collapse of the right-wing Herstigte Nasionale Party and a DP presence in some constituencies splitting the vote, this did not represent an unexpected advance. The gains were mainly restricted to solidly Afrikaans-speaking areas in which small farmers and blue-collar workers predominate. Outside of Pretoria, the NP held its established ground. In fact it increased its majority in some seats with high-profile candidates, notably Innesdal, Krugersdorp and Potchefstroom.

Although an additional 600 votes for the CP in each of nine marginal NP constituencies would have brought the prospect of a hung parliament very close, whereby the National Party would have failed to get an absolute majority of 84 seats, the threat of the CP was contained. All told, the CP did not capture much ground beyond the support given by Afrikaners in those social classes most severely hit by economic recession and disillusioned or frightened by the NP's limited reform measures. Only around 7% of the English-speaking community voted CP.

The DP ended up with a total of 33 seats, winning over many of those voters who had deserted the Progressive Federal Party and New Republic Party (both to the left of the NP) in 1987, but its level of support fell short of the 34 seats and 27% share of the vote captured by these parties in the 1981 election (see Table 2). It is too optimistic to argue, as Mark Swilling has done, that the DP was the 'real winner'.[10] Almost all of the seats the DP gained were either PFP or NRP in 1981. Mass support for the DP was largely confined to English-speaking upper-middle and middle-class urban areas of Cape Town, Johannesburg and Natal.

Table 1: Directly-elected MPs, 6 September 1989

Party	Cape	Natal	OFS	Transvaal	Total
National Party	42	10	7	34	93
Conservative Party	2	0	6	31	39
Democratic Party	12	10	0	11	33
Total	56	20	13	76	165*

* One seat a tie

Table 2: Directly-elected MPs, seats, and percentage of votes

Party	1981	(%)	1987	(%)	1989	(%)
National Party	131	(57,7)	123	(52,5)	93	(48,0)
Conservative Party			22	(26,4)	39	(31,2)
Progressive Federal Party	26	(19,0)	19	(14,0)		
New Republic Party	8	(7,6)	1	(2,0)		
Democratic Party					33	(20,4)
Herstigte Nasionale Party		(14,0)		(3,1)		(0,2)
Independents			1	(1,3)		
Total	165		166		165*	

* One seat a tie

Much of the NP's support base in urban areas, gained in the 1987 election, crumbled away. The new dynamism generated by the DP also appealed to young voters and those who boycotted the poll last time around. Yet the DP did not manage to win over significant numbers of NP supporters to capture a number of its targeted seats such as Durban Point, Helderberg, Stellenbosch and Waterkloof. And in the Border area, once the stronghold of the now defunct United Party, the DP did not do as well as expected. Less than 10% of the Afrikaans-speaking community voted DP.

Altogether, the NP lost, to the right and left combined, some 33 000 votes on its 1987 showing. Half of the seats lost by the NP were won by the other parties with majorities of less than 800 votes. Around 46% of the Afrikaans-speaking community and 50% of the English-speaking community voted NP; the NP's core support held firm.

To most white voters the election seemed to be a good representation of the democratic process. The NP, however, received a mandate from only 48% of white voters, representing just 6% of the total South African population of voting age. Other factors in the electoral system made the perversion of democracy worse - there were various underhand electoral tactics; certain constituencies were unjustly loaded in number of votes; and the first-past-the-post system denied proportional representation. Some votes were worth five times more than others; North Rand had 43 637 eligible voters whilst Prieska had only 8 148. Given demographic patterns, the Transvaal should have 89 seats instead of 76, and the Cape 46 instead of 56.

It is clear that the real challenge to NP rule lies with the MDM. Whilst the NP received just over a million votes the MDM organised national worker and student stayaways on 5 and 6 September in which over three million participated. Election day saw the largest stayaway in South African history; in urban areas there was a 75% stayaway. Protests in the Western Cape saw more than 20 people killed in battles on election night and hundreds injured. The excessive use of force by riot unit police moved Lt Gregory Rockman to inform the press that they 'stormed the kids like wild dogs... You could see the killer instinct in their eyes', and an ambulance attendant stated 'I saw about 15 bodies on top of each other in an ambulance. One of them was a child of about nine. They all looked so young'.[11] The *Weekly Mail's* headline on the election results was poignant; 'Nat: 93, Con: 39, Dem: 33, Hurt: 100, Dead: 23'.[12]

In fact over 70% of South Africa's population who should be voters were unable to cast a vote. Taken as a group, less than 50% of those permitted to vote for the House of Assembly, House of Representatives or House of Delegates did so. For the latter two there was a higher level of rejection than hitherto; less than one-fifth of registered voters went to the polls. One candidate in the election for the House of Delegates made history by getting no votes - he did not even vote for himself. In liberal democracies the ballot has a moral justification by virtue of a universal franchise and the open and genuine debate offered by all. In South Africa no such moral justification can be offered.

A vote for 'reform'?

Nonetheless the results indicated that the NP's campaign best articulated the deep longing for security and justice amongst white voters. The NP's position provided mental comfort: it offered consistency between existing feelings and beliefs, relieved voters' fears and justified a comfortable position. What was distinctive about the NP's campaign, however, was the way in which its promises lacked any concrete form: no detailed, consistent and fully worked out position between apartheid and democracy was presented.

The Five-Year Plan - with its emphasis on group-based decision-making with protection against 'domination' ensured by group consensus - provided more questions than answers. How is a group to be defined? And what happens if consensus is not reached? The central problem around developing a form of parallel group decision-making within a unitary South Africa is that it points towards a vertical caste system which - in the words of a past professor of sociology at the University of Stellenbosch - 'has no known precedent or model in the whole of man's history and is completely in contradiction to any principle of social organisation known to the social scientist'.[13] Emphasis on the new leadership of De Klerk directed attention away from such crucial issues towards questions of style and appearance.

In many areas, NP policies have increasingly become ones of ambiguity and improvisation. There are, for example, no substantive plans for re-organisation of the Reservation of Separate Amenities Act or the Group Areas Act. The NP has refused to take a hard and fast position on separate amenities; this is most evident in its strategy of non-involvement in Boksburg and other CP-controlled municipalities.[14] The NP refuses to repeal the Act but is prepared to let CP councils suffer defeat through peaceful boycott moves. Similarly the NP is not eager to repeal the Group Areas Act but does want it to be less strictly and uniformly enforced, as through the introduction of Free Settlement Areas.[15]

The resignation of Chris Heunis, past acting-state president and Minister for Constitutional Development and Planning, was instructive. Before the election Heunis argued for the need to allow forms of freedom of association through the creation of an 'open' or non-racial grouping. Even though the DP quickly pointed out the absurdity of a non-racial grouping alongside other racially-defined groupings, for the NP caucus Heunis had gone too far in spelling out a concrete position. In the face of party pressure he resigned.

As the NP did not commit itself to any specific proposals for reform in its campaign, it cannot simply be assumed that the NP vote was a vote for reform. It is incorrect to argue, as NP leader De Klerk has, that the election results reflected a new mood among the white electorate with over 70% voting for reform. This conveniently adds together the NP and DP votes - and then gives them a value. The DP's views on reform are far removed from the NP's (the DP is, for example, committed to a common voters' roll as opposed to the NP's talk of separate rolls on a group basis).

And it cannot, with any certainty, be said what the 1989 election was supposed to be about - and what people thought it was about. With support based mainly on tradition, party identification and political symbolism, most people had only the vaguest notions of what the issues were. The issue of reform was one of many factors that lay behind individual voting behaviour.

It can be argued that many people voted for the NP despite its position - confidential NP opinion polls have shown that 60% of white voters agree with the basic elements of CP policy. Furthermore, much support is more firmly cemented through the politics of patronage, which helps hold the party together. The high levels of government expenditure and the 'gravy train' place many under obligation to the government - almost 50% of the white labour force is in the public sector. The granting of extra subsidies to public servants for home loans prior to the election was an obvious move to secure and buy votes.

Observers who interpret the election as being a clear signal to press ahead with reform, who claim that the results augur well for 'real' negotiation, have failed to distinguish the rhetoric from the reality. At the end of the campaign all the NP was given was rather generalised support for its record in power.

The problem facing the NP is where does it go from here? The truth is that it is trapped. Whilst the Five-Year Plan has an underlying commitment to 'group rights' (guaranteeing separate residential areas and schools), to gain widespread credibility it has to go beyond this. This is because 'group rights' are virtually a euphemism for racial exclusiveness. For years 'group rights' have been used to explain why the NP is still committed to segregation.[16]

The new minister for Constitutional Development, Gerrit Viljoen, has recognised that 'the group concept will only be acceptable if the rigidity in the current structure of what constitutes a group is replaced by a far more flexible approach'.[17] To move forward the NP must look for new plausible group definitions that reconcile 'group rights' with individual rights. But this is like trying to square a circle. Attempts to construct a halfway house between apartheid and democracy will always elude the NP because the two are diametrically opposed.

There is no constitutional model that can be applied or developed to make adequate provision for both sets of rights within a unitary South Africa. To support the NP's position is to defend the indefensible. That is why, to speak in George Orwell's terms, so much political language consists of euphemism, question-begging and sheer cloudy vagueness.[18]

Events since the election have seen movement towards establishing negotiations with serious black leaders. Under an open leadership style the NP has allowed a series of media events around protest marches, the release of political prisoners and mass rallies. To date, however, the chief function of such manoeuvering has been symbolic rather than substantive. The core issues affecting white minority rule continue to remain in suspended animation. And yet the NP, having generated new hopes and expectations, cannot merely rest on instigating talks about talks and creating spectacle after spectacle.

For the NP progress beyond the present pre-negotiation phase demands that the Five-Year Action Plan be given concrete form. This will have to be either in terms of a continuing commitment to 'group rights' or to principles of liberal democracy. Given that the tenets of apartheid and democracy are inherently incompatible, whichever path is taken will inevitably lead to splits in the NP caucus and an erosion of NP support. This vulnerability of the NP has recalled memories of the old UP, the centre party which collapsed under pressure from the right and left; thus Frederik van Zyl Slabbert has written that the NP 'finds itself in the position of the old UP after the '74 election - consuming all its energy in finding a new identity with contradictory trends in its ranks'.[19] As these contradictions play themselves out the twilight of National Party rule will come ever closer.

Notes

1 Hermann Giliomee, 'Apartheid, Verligtheid and Liberalism', in J Butler et al, *Democratic Liberalism in South Africa: Its History and Prospect*, Cape Town, 1987.

2 Deborah Posel, 'The Language of Domination, 1978-1983', in Shula Marks and Stanley Trapido (eds), *The Politics of Race, Class and Nationalism in Twentieth Century South Africa*, London and New York, 1987.

3 The Commonwealth Group of Eminent Persons, *Mission to South Africa: The Commonwealth Report*, Harmondsworth, 1986, 42.

4 Hermann Giliomee and Lawrence Schlemmer, *From Apartheid to Nation-Building*, Cape Town, 1989, 130.

5 Simon Bekker and Janis Grobbelaar, 'Has the Conservative Party bandwagon slowed down?', *Indicator South Africa*, 16(1/2), Summer/Autumn, 1989, 11.

6 Zach de Beer, 'Unity and Prosperity in Reach', *Sunday Star*, 03.09.89, 18.

7 As indicated in the confidential Broederbond Working Document, 'Basic Political Preconditions for the Survival of the Afrikaner', 1986. Reproduced in Centre for Policy Studies, 'Negotiations Package', Johannesburg, 1989.

8 Frederik van Zyl Slabbert, 'Pricking the Bubble', *Leadership South Africa*, 6(3), 1987, 23.

9 Jo-Anne Collinge, 'Defiance: A Measure of Expectations', *Work in Progress*, 61, September/October 1989.

10 Mark Swilling, 'A Swing to the Left but the Real Victors are the Non-Voters', *Sunday Star*, 10.09.89, 12.

11 'Election Carnage', *New Nation*, 08.09.89,1.

12 *Weekly Mail*, 08.09.89, 1.

13 SP Cilliers, *Appeal to Reason*, Stellenbosch, 1971, 7-8.

14 Harry Mashabela and Monty Narsoo, *The Boksburg Boycott*, Johannesburg, 1989.

15 Lawrence Schlemmer, 'Racial Zoning: Problems of Policy Change', Johannesburg, 1989.

16 Craig Charney, 'Restructuring White Politics: The Transformations of the National Party', *South African Review One*, Johannesburg, 1983.

17 'Interview: Gerrit Viljoen', *Leadership South Africa*, 8(8), October 1989, 74.

18 George Orwell, 'Politics and the English Language', in *Inside the Whale and Other Essays*, Harmondsworth, 1981.

19 Frederik van Zyl Slabbert, 'The Bumpy Ride FW has Chosen', *Sunday Times*, 27.08.89, 23.

The Emergency State:
Its Structure, Power and Limits

Mark Swilling and Mark Phillips

> We cannot respond by blind defiance based on the view that the state is panic-stricken and all that is needed is a big push. Nor can we respond by saying that the state's power is unassailable - that it has no weaknesses, and that we should retreat into total inaction.
> Cosatu discussion paper, special congress, 1988.

This contribution assesses the strategies, structures and resources that the emergency state has deployed to fight its battles on the 'political terrain'.[1] The state is implementing a new set of strategies in the face of mass resistance to the failed 'total strategy' reforms. While capital and the popular classes have pursued a range of strategies to transform apartheid, the state has been able to mobilise enormous resources and co-ordinate ambitious policies in response to these challenges.

By 1988, the state had arrived at a point where political reform had ground to a halt and 'crisis management' had become its primary strategic concern. In the process, state power and executive decision-making became concentrated in the Office of the State President (OSP) which, in turn, was increasingly dependent on the national security management system (NSMS).

Recent analyses of state institutions have failed to note the extent to which this office became the epicentre of an authoritarian power structure that bound together, in the person of the state president, the complex security hierarchy, the state departments and the facade of consociational parliamentary government.[2] The OSP used, as its main instruments of power, the cabinet, cabinet committees, state security council (SSC) and NSMS as mechanisms for ensuring that the policies of the executive president were accepted and implemented.

State policies and actual outcomes rarely coincide. The state has to do battle on a terrain of ongoing struggles, conflicts and contradictions generated both by tensions between its own institutions and by challenges and pressures emanating from without. Any assessment of the shifting nature of state strategy and the structures it has erected to implement these strategies must take cognizance of this social context.

Re-organising state structures

The highly complex and disorganised decision-making structures of the pre-Botha state, a 1980 white paper on rationalisation argued, 'make it difficult for the central executive machinery to act swiftly and effectively to solve problems and crises'.[3] By restructuring state decision-making structures, this report argued, 'a more manageable machinery of government to meet new challenges and crises' should be created.[4]

The re-organisation of state decision-making structures was one aspect of what became known during the early 1980s as the 'total strategy' programme. Originating in the security establishment and the 'verligte' wing of the NP, this programme aimed to co-ordinate and centralise reform policies and security strategies while stream-lining and rationalising the bureaucracy. The reform policies of the early 1980s were supposed to restructure four basic social arenas: work place, city, regional political economies and the constitutional order.

Six components of the broadly defined executive branches of the state were re-organised between 1979 and the introduction of the new constitution in 1983:
1. the department of the prime minister was turned into the office of the prime minister (OPM) and given extended powers exercised through its various planning branches responsible for all key aspects of policy-making;
2. the National Intelligence Service (NIS) replaced the Bureau of State Security;
3. cabinet committees, working groups for cabinet committees and a cabinet secretariat were introduced;
4. the state departments were re-organised and reduced in number, and a new department, the Department of Constitutional Development and Planning, was created to manage the macro-planning of reform policies;
5. the Senate was abolished and a President's Council (PC) established to work out a new constitutional dispensation;

6. the state security council was turned into a powerful policy-making body and the embryo of a complex national, regional and local system of committees was set up underneath the SSC - the NSMS.[5]

This re-organisation of the central decision-making structures prepared the way for the highly centralised and constitutionally sanctioned authoritarianism of the executive presidency which was created by the 1983 constitution.

From total onslaught to total strategy

Proceeding from the assumption that the state was faced with a 'total onslaught' aimed at undermining all levels of society, the new PW Botha administration moved swiftly to implement a 'total strategy' with three goals:

- maintain state security;
- reform the political environment; and
- co-ordinate all state action.

'Total strategy' planners identified four areas of 'reform': the legalisation of black trade unions through the Wiehahn Commission; the recognition of the permanence of urban blacks in line with the recommendations of the Riekert Commission; the formulation and implementation of a new constitution that brought Indians and coloureds into parliament as junior partners; and the introduction of a new regional development policy to co-ordinate industrial development planning in accordance with regional boundaries rather than racially-determined constitutional divisions. This involved the division of South Africa into nine 'development' regions for the purpose of integrated 'development planning'.

The internal logic of 'total strategy' reforms was coherent and premised on a specific perception of South Africa's social reality. The Riekert Commission's assumption was that bantustans could be retained but that, in response to the unrest of 1976-77, 'urban blacks' must be recognised as permanent members of the cities and towns.

Once the Riekert Commission had conceded 'urban rights', a range of other complementary rights necessarily followed: rights to form trade unions, purchase property, sell labour on a 'free urban labour market' without a contract, and trade. The municipal franchise was seen as the ultimate embodiment of the new 'urban identity'. Blacks were given, for the first time ever, fully autonomous municipal institutions - the Black Local Authorities (BLAs) - with extensive urban powers (eg allocation of housing and trading sites).

The new urban policy created several critical problems that soon contributed to black protest. First, because it wanted to create a privileged elite of 'urban insiders' divorced from the poverty of the rural masses, it required an intensification of influx control - hence the proposed Orderly Movement and Settlement of Black Persons Bill of 1982. Second, the Black Local Authorities, as a form of local political self-government, were designed to be financially self-sufficient even though they were excluded by the Group Areas Act from access to rateable commercial and industrial property. Third, because the BLAs' only link to higher forms of representation was through the bantustans, the issue of the overall illegitimacy of state constitutional reform was exacerbated.

The constitutional reforms were premised on the consociational theory that 'group identities' in multi-cultural societies must be protected but that structures for 'co-determination' and 'joint decision-making' must also be created. The result for whites, coloureds and Indians was the creation of a new 'consociational contract'.

As far as the majority African population was concerned, the new regional development policy rested on the principle of 'economic interdependence and political independence' or what the Buthelezi Commission called the 'soft-borders approach'. Bantustans would proceed to independence and then enter into supra-state agreements to form a 'confederation of Southern African states'.

'Total strategy' may have left many fundamentals of apartheid intact - bantustans, influx control and constitutional exclusion of Africans from central government. But it did introduce significant modifications to some basic institutions of political society. These institutions regulated access to three social arenas: the city, the factory and government. In reality, however, the maximum the white minority was prepared to concede in the early 1980s fell far short of the minimum the black majority was prepared to accept.

'Total strategy' failed because its intentions were thwarted by a sustained period of black resistance. This exacerbated, highlighted and brought to the fore the key structural contradictions that original 'total strategy' reforms failed to address.

As far as urban policy was concerned, the attempt to drive a wedge between 'urban insiders' and 'rural outsiders' by intensifying influx control was challenged and made unviable by two social movements. The first and most important was the squatter struggles. Displaying desperate and relentless determination to escape grinding rural poverty and live in urban areas, squatter communities broke through the urban wall and illegally invaded land to secure their right to urban existence.

The labour movement also helped undo the 'rural-urban' divide. By organising migrant workers and urban proletarians into single industrial trade union organisations committed to joint wage demands, the intention to create two entirely separate labour markets - one urban and privileged and the other rural and cheap - was severely undercut.

The local government system that was supposed to bind the new urban system together was soon in ruins. Because the state insisted that councils should raise their own finance for development, councillors were forced to increase rent and service charges. Their low levels of legitimacy meant neglecting development altogether was not an option. They needed to demonstrate the benefits of participation. The increases, however, triggered a nation-wide popular rebellion that began in the Vaal townships in September 1984 and spread across the country. By mid-1985 most BLAs had collapsed because of mass resignations or because councillors had been killed by residents.

Fiscal unviability was not the only problem. The root cause of the depth of popular protest was that the local franchise was not tied to a programme for granting full political rights to blacks. This facilitated the emergence through the local community organisations of a national political organisation - the United Democratic Front - committed to the total dismantling of apartheid and the creation of a non-racial democracy.

The new labour dispensation failed to achieve its objectives. When the unions threatened to oppose the legislation because migrants were excluded from the right to join unions, the state backed down and extended the definition of 'employee' to include migrant workers.

The consociational contract foundered on the rocks of popular resistance expressed through the election boycotts of 1984.

The confederal schema also began to come apart at the seams.

The steady political and fiscal decline of model bantustans like Ciskei, Transkei and Bophuthatswana made it clear that separate development and the grand vision of a confederation of states was turning into a nightmare.

'Total strategy' inadvertently created spaces for the initiation and deepening of mass democratic mobilisation and organisation. By granting industrial rights, unions were spawned that challenged relations in the factories, cities and political society. Opening up the urban system generated urban movements that politicised civil society and destroyed the cornerstones of the Riekert policy framework. And constitutional reform provided the focus for national organisation and resistance on a scale not seen since the 1950s. All these manifestations of resistance and opposition short-circuited key state strategies and prepared the way for new ones.

The route to counter-revolutionary warfare

Intense struggles were fought within the state during the 1985-86 period over what was to replace the failed policies of 'total strategy'. In some ways these struggles reflected the influence of big business on state policies. These interests responded to the failure of reform, township unrest, the politicisation of the work place and economic recession by pressurising government in both public and private ways to move beyond the original parameters of the 'total strategy' programme. The campaign against influx control spearheaded by the Urban Foundation is the best example of this. The public tone of business politics in 1985-86 reflected the urgently-felt need in these quarters for policies that would deracialise the capitalist system.[6]

These struggles also reflected competing intra-state conceptions of how best to protect the generalised interests of the white minority.

The failure of the 'total strategy' reforms immediately created the need amongst state planners for the formulation of alternatives. But this did not take place in a vacuum. The state is not a unified monolith. Different branches of the state were interfacing with the black communities on different levels and interpreting the challenges from below in different ways. To understand state responses to the contradictory consequences of its own strategies, a picture of the balance of institutional interests within the state is required.

The three most important institutional forces within the state that had a say in the formulation of alternatives to 'total strategy' were the OSP, the Department of Constitutional Development and Planning (DCDP) and the security establishment. The DCDP was, until early 1986, the flagship of reformist policy-making. It was this department that formulated both the early 'total strategy' reform programmes and the 1985-86 extended reform policies that moved the state well beyond its initial reform goals. However, the security establishment was also emerging with strategies of its own to defuse the crisis. In the end, it was when the OSP came down decisively in favour of the security establishment's proposal that the state defend itself by means of 'counter-revolutionary warfare' methods, that the balance of institutional forces within the state shifted decisively against the DCDP.

The DCDP was created in 1982 under minister Chris Heunis. Its institutional foundations were the 'planning branches' that had been created in the office of the prime minister in 1980. Each 'branch' dealt with a specific field of state action: physical planning, economic policy, constitutional development, security planning, social planning and scientific planning. The transfer of these branches to the newly created DCDP in 1982 (except for the security planning branch which became the secretariat for the SSC), turned it into the engine room of government policy-making and strategic thinking. It was here that overall co-ordination of all the major state reform actions during the 1982-85 period took place - the new constitution, restructuring of local government, new regional development policies, the population development plan and urbanisation strategy.

Proceeding from the assumption that more concessions would satisfy black demands and legitimise state strategies, the political reformers in the DCDP responded to black rejection of reform by extending the reform programme. There were five critical moments in the extension of this programme:

- in November 1984, Heunis announced that Black Local Authorities were to be included into the proposed regional services councils;
- in May 1985, Stoffel van der Merwe published his NP pamphlet entitled '... and what about the Blacks?' which argued that bantustans were a failure;
- in September 1985, PW Botha announced citizenship would be restored to all Africans residing permanently in 'South Africa';

- in late 1985, PW Botha conceded that the tricameral parliament was not the final solution, merely a step in a process;
- in September 1985, the President's Council published its report on urbanisation which culminated in the December 1985 statement by Heunis that blacks would be given freehold property rights and in the 1986 Abolition of Influx Control Act.

All these extensions of the reform programme were *ad hoc* responses to popular pressure from political movements, business organisations, the international community, trade unions, squatters and township organisations, and the deepening sense of crisis. These shifts unintentionally undid existing policy positions without being coupled to a coherent set of alternatives.[7]

The extension of the reform programme during this period met with the strategic requirements of the 'counter-insurgency' framework that was still dominant at the time. In an address to the Institute of Strategic Studies in 1981, ex-police commissioner General Johan Coetzee isolated five key elements of the 'counter-insurgency' position:[8]

- 'a dynamic policy of change' to 'make the RSA a difficult target to pin down for assault';
- 'a clear political objective...to ensure a free, independent and united country';
- 'propaganda...directed at the counter-insurgents, the population and the insurgent';
- the avoidance of conflict and defusing of 'explosive situations';
- 'an effective information organisation and spy network'.

'Counter-insurgency', he said, 'is based on the cornerstone of information or intelligence'. Yet by mid-1986 it seemed to many in the security establishment - and elsewhere in the state - that the 'dynamic policy of change' embodied in the reformist policies pursued in the 1978 to 1986 period had failed to defuse conflict. Decimation of the spy network through attacks on informers, moreover, had undone the very cornerstone of 'counter-insurgency'.

To increasingly assertive security hardliners, the democratic opening which reform previously provided had to be closed and 'counter-insurgency', with its bias toward political solutions based on good intelligence, replaced with a far more thorough-going and co-ordinated programme that would crush the democratic opposition, 'counter-organise' the communities and upgrade socio-economic conditions.

The NSMS was activated to co-ordinate the implementation of this new strategy. Although established silently in the wings at the outset of the post-1978 period, the NSMS was not initially fully activated or turned into the nerve centre of state action that it became. Instead, it was created as a fallback in case the formal civilian structure collapsed or proved unable to govern the country on its own.

Hitting the emergency switch

During the heyday of 'total strategy' in the early 1980s, officials who staffed the NSMS, to quote one former state official, 'were just keeping the seats warm'.[9] The turning point came in May 1986 when PW Botha and the generals decided to scuttle the Eminent Persons' Group's proposals by bombing three capitals of Frontline states. Instead of negotiating with the black opposition - as Chris Heunis's department was beginning to do at the local level - the state moved to smash it by declaring a national state of emergency and activating the NSMS at all levels. In the words of one ex-security policeman, 'when the moment came, all we needed to do was to hit the switch'.[10]

'Hitting the switch' meant turning the NSMS shell into a crisis management machine. The pre-1986 organisation of the NSMS was, in the words of a reliable pro-government intelligence source, 'found to be lacking as far as proper communication and control channels were concerned, apart from the fact that it was designed for a totally different function and was not geared for the day-to-day running of the emergency'.[11]

The NSMS was structured as a separate arm of government under the auspices, direction and effective control of the OSP.

The office of the state president

The OSP became the epicentre of state power.[12] No major macro-policy decision was taken without the president's personal approval or direct involvement in its formulation. US political scientist Samuel Huntington's 1981 proposition in *Politikon* that the 'route from a limited uni-racial democracy to a broader multi-racial democracy could run through some form of autocracy',[13] became an accurate reflection of centralised executive power. The OSP managed to achieve this by subordinating key executive structures to itself, taking direct control of some executive powers and then rooting itself in the 'crisis management' functions of the NSMS (see Diagram 1).

This is the best method, deputy minister Leon Wessels argued, of dealing with 'the question of political power' because 'one of the most important principles of counter-revolutionary action is that unity of purpose within the framework of a co-ordinated plan is essential'.[14] This view was echoed in a Department of Military Intelligence (DMI) booklet which concluded its analysis of 'counter-revolution' with a quote from one-time commanding officer of British forces in Malaya: 'Any idea that the business of normal government and the business of the emergency are two separate entities must be killed for good and all. The two activities are completely and utterly inter-related'.[15]

The OSP involved a number of committees, advisors, secretariats and ministries dealing with public administration, public expenditure priorities, constitutional affairs, economic policy, socio-economic development ('welfare'), and propaganda and national security. The most important of these are the President's immediate advisors, the National Priorities Committee, Commission for Administration, Information Ministry, National Intelligence Service, Welfare Secretariat, Economic Advisory Council (EAC) and the National Joint Management Centre (NJMC).

Since 1986, welfare officials have been squeezed out of the cabinet committees and a welfare secretariat controlled directly by the OSP created to administer the functioning of the 'welfare' cabinet committees. This placed the macro-co-ordination of non-security policy in the hands of a secretariat directly controlled by the OSP. The cabinet became merely the political structure supposed to legitimise this decision-making process.

Up until 1986, the OSP co-ordinated the activities of the welfare departments and the security establishment via the National Co-ordinating Committee (NCC). After the declaration of the state of emergency, this overall co-ordinating function was transferred to the newly-activated NJMC that fell directly within the NSMS. When the NJMC took over this role, co-ordinating these two arms became a critical component of the 'crisis management' system. It also meant that the OSP became more directly involved in the day-to-day management of the emergency state by rooting itself in the NSMS.

The National Priorities Committee, originally established in

Diagram 1: The Office of the State President

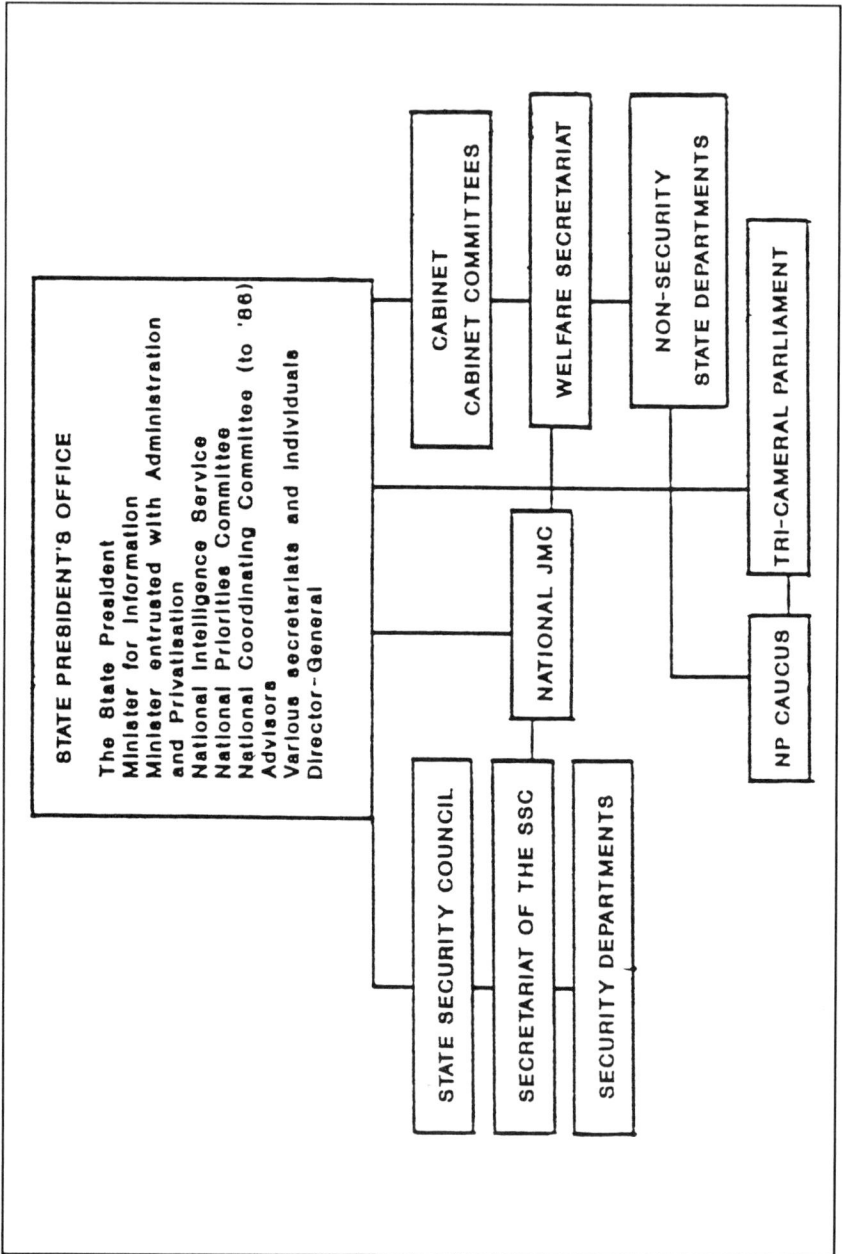

1984 to oversee the design of the budget, has taken over economic policy-making functions previously the responsibility of the Department of Finance, EAC and the Central Economic Advisory Service. The National Priorities Committees makes all the key decisions about how the state spends public money in the context of sanctions and deepening recession. In particular, it is this committee that balances the competing but complementary demands of defence and socio-economic upgrading.

The Commission for Administration (CFA) is an extremely powerful tool answerable to the president. Its 1 000 employees design organisational plans to ensure the effective implementation of executive decisions. The CFA is entitled to issue orders that ministers and directors-general are obliged to carry out. Although it handled 211 investigations into bureaucratic restructuring in 1987 alone, its major task at the moment is to execute the privatisation programme. The CFA falls under the minister in the OSP entrusted with administration and privatisation.

The NIS, with its 5 000 employees under director-general Neil Barnard, is responsible for intelligence interpretation and evaluation. The OSP relies heavily on the NIS for assessments and analysis of security-related issues and for the formulation of counter-strategies.

The Ministry for Information in the OSP under minister Stoffel van der Merwe regulates and controls the communication activities of all government departments. The Bureau for Information is involved at all levels in what one official called the 'war of words' - an essential component of the 'counter-revolutionary' strategy. The Bureau's involvement in the Communication Committees of the Joint Management Centres (JMCs) links the OSP directly to grassroots propaganda efforts. The Bureau claims that its publications, including 45 township newspapers, have a readership of 2,5-million.

Although the OSP has always been central to the new constitutional system, during 1987-88 its power and significance became more pronounced.

This was particularly apparent in the way the president unilaterally reconstituted the EAC in mid-1986. Prior to 1986, the EAC was the most important institutional link between organised business interests and the state. It included representatives of business organisations, academics, top officials and cabinet ministers and was chaired by Simon Brand of the Development Bank.

In mid-1986 PW Botha announced that most of the officials and academics were to be excluded from the EAC and that from then on he would personally appoint the business representatives. Those appointed were the leading lights of monopoly capital.

In business circles, there are different interpretations of the changes in the EAC. While some see it as evidence of a deepening alliance between big business and the state, a more widely held view is that the changes have resulted in a downgrading of the influence of organised capitalist interests. As far as the state is concerned, the EAC is one of a number of forums where it would like to consolidate an alliance with monopoly capital around common interests in deregulation, privatisation, sanctions busting, urban development and economic reform in general. This is part of the state's view that business has a responsibility to play a leading role in the economic and welfare components of the 'counter-revolutionary' effort while leaving security and constitutional matters to the state.

The OSP emerged as the lynchpin of key strategic thinking and action. By 1988 this office and its incumbent had become the most decisive decision-maker in the state - a level of power centralisation not uncommon in societies going through a violent interregnum.

Implementing 'counter-revolutionary warfare'

The implementation of the 'counter-revolutionary warfare' strategies through the NSMS involved four significant changes:
- the establishment of a new structure of centralised command under the direct purview of the OSP, ie the NJMC;
- the deployment of new security personnel in key positions in the system;
- a vast increase in the numbers of people drawn into the system at JMC, sub-JMC and mini-JMC level; and
- the imposition of a new theory of state action, ie 'counter-revolutionary warfare' in order to 'win-hearts-and-minds' ('WHAM').

The 'counter-revolutionary warfare' position replaced the 'counter-insurgency' ('COIN') perspective that former security police chief General Johan Coetzee propounded in the early 1980s. 'Counter-insurgency' was seen to have no answer to the unprecedented scale of the 1984 to 1986 uprising, or to the powerful local and international pressures for fundamental change. By mid-1986, the hardliners' advocacy of an alternative strategy designed to break the opposition and rebuild support bases for the state enabled them to force the scuttling of the EPG mission in May 1986 and the declaration of a permanent national state of emergency.

Unlike 'counter-insurgency', which concentrated on intelligence gathering while leaving politics to the civilians, 'counter-revolutionary war' involved the security establishment in countering the 'revolutionary strategy' at every point. It relegated the politics of reform to a matter of secondary importance, to be dealt with at some point in the future. The immediate task was to secure the security of the state; or as the DMI booklet put it: 'To defeat the revolutionaries and...regain the initiative'.[16]

The new 'counter-revolutionary' strategists - the so-called 'securocrats' - proceeded from three basic points of departure:

- law and order must be restored before reforms can be introduced;
- socio-economic development or what PW Botha called 'social reform' must precede political reform;
- constitutional development must begin at local level and proceed upwards to higher levels.

'Re-establishing law and order' involved the declaration of what effectively became a permanent state of emergency and the detention of around 30 000 people, the banning of 34 organisations, widespread vigilantism and the silencing of leaders and alternative newspapers.

A large-scale socio-economic upgrading programme has been launched in the most troublesome black townships. Thirty-four of the most volatile townships - the so-called 'oilspots' - have been earmarked for special attention. About 1 800 urban renewal projects are currently in progress[17] in approximately 200 townships countrywide, and R300-million will be spent by the 13 Transvaal RSCs.[18] Official sources have claimed that R34-billion will have to be spent on housing development over the next 20 years.[19]

PW Botha criticised sanctions on the grounds that they would deprive black townships of R9,5-billion for socio-economic upgrading over the next ten to fifteen years.[20] The Department of Development Planning spent R59-million on upgrading during 1986-7 and R57- million was budgeted for 1987-8. The Development Bank spent R2,4-billion on services and housing during 1984-7. Budgets for the three model 'oilspots' are Crossroads - R90-million; Alexandra - R95-million; and Mamelodi - R410-million (including housing).[21]

Sources of funding for this urban development programme include the Development Bank, RSCs, the Department of Constitutional Development and Planning, Provincial Administrations, South African Development Trust, South African Housing Trust, Urban Foundation, private sector companies, Department of Finance and local authorities. Reliable sources have confirmed that special funding for upgrading is also being channelled through SADF budgets. The deputy governor of the Reserve Bank, Jan Lombard, estimated in 1988 that R4,5-billion per annum is needed until the year 2000 to finance a large-scale 'settlement programme' for blacks. This, he argued, was R3,3-billion more than the R1,2-billion currently invested in this area by the public and private sectors.[22]

The logic of the 'counter-revolutionary warfare' strategy implied that only once security actions had run their course, grievances had been addressed and municipal government restored, would political reform again be on the agenda. But to be successful, the security-upgrade strategy had to fall under centralised command and be tightly co-ordinated. The chosen instrument was the NSMS (see Diagram 2). The most decisive institutional expression of their new ascendancy in the state was the establishment and activation of the National Joint Management Centre (NJMC) in the heart of the NSMS. Through the NJMC, the security establishment was entrusted for the first time with the co-ordination of both security and welfare decision-making functions.

The NJMC, chaired by deputy minister of Law and Order, Leon Wessels, was responsible for ongoing management of the state of emergency. Under the direct guidance of the OSP, the NJMC co-ordinates state action at all levels. The NJMC has a 'Nasionale Staatkundige, Ekonomiese en Maatskaplike Komitee' ('Semkom') to co-ordinate welfare policy and a 'Gesamentlike Sekuriteits Komitee' to co-ordinate security strategies.

Who's who in the SSC

The SSC is at the apex of the NSMS. It brings together the most important of the cabinet ministers and the country's security chiefs.

Diagram 2: The National Security Management System

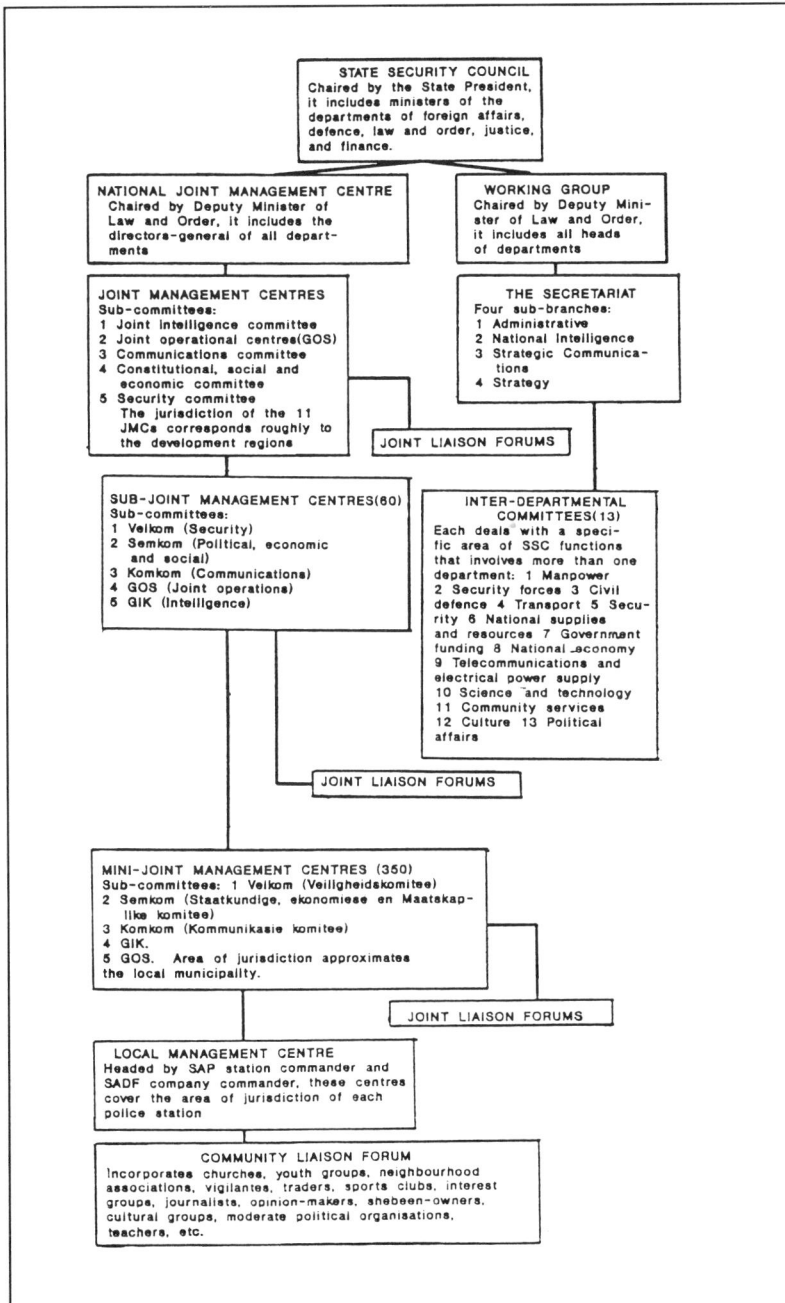

STATE SECURITY COUNCIL
Chaired by the State President,
it includes ministers of the
departments of foreign affairs,
defence, law and order, justice,
and finance.

NATIONAL JOINT MANAGEMENT CENTRE
Chaired by Deputy Minister of
Law and Order, it includes the
directors-general of all departments

WORKING GROUP
Chaired by Deputy Mini-
ster of Law and Order,
it includes all heads
of departments

JOINT MANAGEMENT CENTRES
Sub-committees:
1 Joint intelligence committee
2 Joint operational centres(GOS)
3 Communications committee
4 Constitutional, social and
  economic committee
5 Security committee
  The jurisdiction of the 11
  JMCs corresponds roughly to
  the development regions

THE SECRETARIAT
Four sub-branches:
1 Administrative
2 National Intelligence
3 Strategic Communica-
  tions
4 Strategy

JOINT LIAISON FORUMS

SUB-JOINT MANAGEMENT CENTRES(60)
Sub-committees:
1 Velkom (Security)
2 Semkom (Political, economic
  and social)
3 Komkom (Communications)
4 GOS (Joint operations)
5 GIK (Intelligence)

INTER-DEPARTMENTAL
COMMITTEES(13)
Each deals with a speci-
fic area of SSC functions
that involves more than one
department: 1 Manpower
2 Security forces 3 Civil
defence 4 Transport 5 Security 6 National supplies
and resources 7 Government
funding 8 National economy
9 Telecommunications and
electrical power supply
10 Science and technology
11 Community services
12 Culture 13 Political
affairs

JOINT LIAISON FORUMS

MINI-JOINT MANAGEMENT CENTRES (350)
Sub-committees: 1 Velkom (Veiligheidskomitee)
2 Semkom (Staatkundige, ekonomiese en Maatskaplike komitee)
3 Komkom (Kommunikasie komitee)
4 GIK.
5 GOS.   Area of jurisdiction approximates
the local municipality.

JOINT LIAISON FORUMS

LOCAL MANAGEMENT CENTRE
Headed by SAP station commander and
SADF company commander, these centres
cover the area of jurisdiction of each
police station

COMMUNITY LIAISON FORUM
Incorporates churches, youth groups, neighbourhood
associations, vigilantes, traders, sports clubs, interest
groups, journalists, opinion-makers, shebeen-owners,
cultural groups, moderate political organisations,
teachers, etc.

Its regular members are the state president (the chairman); the minister of Defence; the ministers and directors-general of the Departments of Law and Order, Foreign Affairs, Finance, Constitutional Development and Justice; and a few politically important cabinet ministers. The SSC is a macro-policy-making body and not in direct daily control of emergency management, and the OSP is more important than the SSC when it comes to the running of the NSMS.

The security officials on the SSC are the chief of the SADF (General Jannie Geldenhuys) and the chiefs of the army (Lt-General Kat Liebenberg), navy (Vice-Admiral Glyn Syndercombe), airforce (Lt-General Jan van Loggerenberg) and medical services (Lt-General Daniel Knobel); the directors of NIS and of military intelligence (Neil Barnard and Lt-General CJ van Tonder); the commissioner of police, Major-General Hennie de Witt and the chief of the security police, Major-General Johann van der Merwe; the director of security legislation, Andre Bosch, and the director-general of the state president's office, Dr Jannie Roux (*Africa Confidential*, 08.07.87).

Of the 23 permanent members of the SSC other than the state president, 12 are security officials. The SSC is also able to co-opt onto itself for the purpose of specific discussions any other cabinet minister, senior civil servant or even industrialists - such as the chairman of Armscor. It meets fortnightly, on a Monday, the day before cabinet, which meets on a Tuesday. It assesses, from a security perspective, all executive decisions of state and decides on matters as fundamental as the Angola-Namibia peace settlement and the 'invasion' of Bophuthatswana - a 'foreign country' - to rescue its president from a coup bid.

The SSC is backed up by the four legs of the NSMS. The work committee, which provides expert back-up to the SSC, consists of the heads of government departments represented on the SSC and the heads of the other three cabinet committees (the committees for social, economic and constitutional affairs, now subsumed under the welfare secretariat). Support to the work committee and the SSC is given by the secretariat of the SSC. The personnel of the secretariat are drawn from the following departments: SADF - 70%, NIS - 20% and Foreign Affairs - 10%.[23]

The work committee is responsible for intelligence interpretation, strategic communication, strategy formulation and administration of the security management system. In addition, there are 13 inter-departmental committees, created to ensure the co-ordination of departmental resources and activities.

As the definition of security has broadened to include the management of most problems and processes, so the official designation of the system itself is beginning to change to simply the National Management System. This suggests that the NSMS is no longer regarded as simply a temporary measure, but that this 'structure could become a permanent feature of government'.[24]

The real lifeblood of the NSMS is its fourth leg - the network of over 500 regional, district and local JMCs. The NJMC co-ordinates the activities of both the welfare and security secretariats above it, while attempting to ensure the smooth functioning of the hundreds of committees below it. The latter include:

- 11 JMCs which divide the country into management regions corresponding to SADF area commands and, roughly, to the nine economic development regions;
- approximately 60 sub-JMCs, with some having borders that correspond to the RSCs; and
- at the local level somewhere between 250 and 500 mini-JMCs and local management centres which co-ordinate state action in almost every settled community in the country.

These co-ordinating structures bring together military, police and civilian officials, usually under the chairmanship of a senior military or police officer. The functions of a JMC and the essence of the 'counter-revolutionary warfare' strategy are apparent in each JMC's committee structure.

Although there are frequently local variations in their precise make-up most, whether they are the NJMC, regional JMCs or local mini-JMCs, have four functional committees and a co-ordinating executive with representatives of each on it. These are the intelligence, security, and communications committees. In many ways, the functions of these committees mirror the concentration of power at the highest level in the OSP, which centralises a similar set of functions: co-ordination, intelligence, security, welfare and communications.

Each intelligence committee gathers and interprets the intelligence on which co-ordinated state activities are based. Staffed by representatives of military intelligence, the security police and the NIS, it seeks to promote unity of effort between these traditionally competing intelligence agencies. The security committee, the repressive arm of the NSMS, acts on the intelligence provided by the intelligence committee. Staffed by a combination of riot police officers, military officers, security branch officers and officers of the municipal police, kitskonstabels, and commando and civil defence units, it co-ordinates the implementation of security strategies laid down by the SSC.

The welfare committee, on the other hand, takes responsibility for co-ordinating the functions of the civilian administration. At the national level this includes the co-ordination of, for example, constitutional and labour issues. But at most levels welfare implies a particular concern with local upgrade programmes.

Its membership consists of officials of the various non-security state departments - roads, education, welfare, health, constitutional development, transport services, the Provinces' community services offices, etc. More than anything else the welfare committee tries to prioritise development projects, cut red tape and unblock bottlenecks.

The functions of the security committee and the welfare committee - the two essential arms of the system - encapsulate the overall intentions of state security strategists. These are, to use the words of Law and Order minister Adriaan Vlok, to 'take out', 'eliminate' or 'annihilate' activists on the one hand, while on the other 'addressing grievances'.[25]

The overall national security management strategy is sold to the public via the fourth of the JMC's committees, the communication committee (Komkom). The Komkoms are run by the Bureau for Information, though SADF communications operations and personnel and public relations personnel from other government departments do play some role. There is sometimes a clear overlap of personnel between SADF and Bureau officials. The Bureau's national head of planning and research, for example, is a senior military intelligence officer, Major-General PJ Groenewald (*Africa Confidential*, 08.07.87).

Komkom has three aims. Through the co-ordination of lectures, television programmes, pamphlets, newspaper articles and the publication of local newspapers it attempts to explain and justify actions against activists and opposition organisations. Komkom aims to ensure the maximum publicity for welfare-type projects and government-supporting 'counter-organisations'. Wherever possible credit for JMC co-ordinated programmes will be taken by 'civil bodies' such as BLAs. At the same time the sincerity of state reforms is widely publicised both inside and outside the country. The third apparent function of the Komkoms is the co-ordination of disinformation via a variety of media forms, pamphlets and township graffiti.[26]

The JMC system interfaces with the public informally through talks and lectures on the nature of the counter-revolutionary war given by JMC officials. The head of the strategic communication branch of the SSC secretariat once claimed to have given 'about a thousand lectures' on the system in a six-year period.

A more formal relationship between the public, the business community and the security system is being formed through a network of Joint Liaison Forums, Joint Liaison Committees and Defence Manpower Liaison Committees (Demalcoms). These allow the security establishment to market their strategies to influential members of the public and broaden their informal sources of information and support.[27]

The secretariat of the SSC is the only part of the NSMS which employs full-time staff. This is one of the major strengths of the system. At all other levels it relies exclusively on officials co-opted from the various state departments themselves. This means that decisions and recommendations of the NSMS at all levels are the decisions and recommendations of senior officials of these departments, rather than being policy imposed from the outside. Just as it is highly unlikely that the cabinet will override the carefully considered and security assessed decisions of the SSC, so it is equally difficult for any single government department to ignore plans formulated in the JMCs.

This review has tried to trace the evolution of the structures and strategies of the emergency state. Its responses to pressures within the white power bloc and challenges from below in the black communities have been twofold: it has changed strategic tack in search of new solutions; and it has adapted its power structures to manage and co-ordinate resources under emergency conditions. In the process, an authoritarian decision-making structure emerged with the state president at its epicentre and the security system operating as its most important executive arm.

The 'total strategy' package was adopted in the early 1980s

because it was both a set of means and a coherent conception of ends. The ends comprised a grand reformist vision of a stable multi-ethnic society coupled to a set of means whereby reform could be managed from above by carefully deploying the finely honed instruments of state power. In the end, 'total strategy' failed because the state proceeded from the assumption that it was neutral and above social conflict and could, therefore, define solutions as if the apartheid state form itself was not the cause of the problem.

The explosion of mass resistance from 1984 onwards not only led to the collapse of the pre-1984 reform policies, but directly challenged the apartheid state form itself. It was in direct response to this threat that the state of emergency was declared and the 'counter-revolutionary warfare' strategies implemented. However, these strategies are largely reactive means designed to defend state power.

It is now openly admitted by the state that short of 'surrender', it has no coherent conception of a blueprint or vision of how the fundamental political conflict can be resolved. Despite Law and Order minister Vlok's claim at the Natal NP Congress in 1988 that the state has a coherent 'master plan' for resolving this conflict (*Star*, 19.09.88), the emergency state and its structures are involved in little more than the business of long-term 'crisis management'.

Although the strategies of the emergency state are having profound social and political effects, the outcomes of state actions do not correspond to the intentions of policy-makers. Vlok's 'master plan' will more than likely founder on the rocks of racial oppression, limited resources and ongoing resistance.

The official position of the NP government remained, until recently, the insistence that, in PW Botha's words, 'as far as I am concerned, I am not considering even to discuss the possibility of black majority government in South Africa'.[28] Preventing this outcome remains, in essence, the purpose of 'counter-revolution' in South Africa. Yet in at least three instances elsewhere in the world where 'counter-revolutionary warfare' strategies have been used to counter essentially national struggles (Vietnam, Algeria and Malaya), the basic demands of nationalism had to be met. In South Africa, the complete absence of any constitutional resolution to the 'national question' will guarantee ongoing mass resistance to white minority rule.

At a time when the emergency state needs to finance rising

security expenditure and the expensive 'winning-hearts-and-minds' programme, there are increasing constraints on the national resource base. Sanctions, economic recession, political unrest and declining investment will exacerbate this structural limitation on the scope of state action. There is no evidence that the economy will achieve the five to seven percent growth rate that Reserve Bank deputy governor Jan Lombard argued is required to finance the socio-economic upgrading programmes.

Finally, the survival of mass opposition in the communities, schools and work places during the emergency era calls into question the confident claims by security officials that the state of emergency has succeeded in 'eliminating the revolutionaries'. Even they concede that there is 'still a revolutionary climate'.

Notes

1 This term is taken from S Nolutshungu, *Changing South Africa: Political Considerations*, Manchester, 1982. We are not concerned with the relationship between state and capital or with the structural question of the relations between capitalism and apartheid. Our concern is more descriptive and focused on specific state strategies that have emerged out of its attempts to resolve the fundamental political conflict between the white minority and the oppressed majority. Nolutshungu uses the concept of 'political terrain' to refer to those actions by the state that are not directly reducible to class interests. The interests of the white minority are reflected in their purest and most concentrated form in the racially exclusive practices of the state. This has been the underlying determinant of state action during a decade of unprecedented challenges from below by the black majority.

2 See M Swilling and M Phillips, 'The Politics of State Power in the 1980s', unpublished mimeo, Centre for Policy Studies, 1988; M Morris and V Padayachee, 'State Reform Policy in South Africa', *Transformation*, 7, 1988, 1-26; D Geldenhuys and H Kotze, 'Aspects of Political Decision-Making in South Africa', *Politikon*, (10)1, 1983; K Grundy, *The Militarisation of South African Politics*, Oxford, 1988.

3 Republic of South Africa, 'White Paper on the Rationalisation of the Public Service and Related Institutions', *Journal of Public Administration*, (15)2, June 1980, 49.

4 White Paper 1980, 63.

5 See J Selfe, 'The Total Onslaught and the Total Strategy', unpublished MA dissertation, University of Cape Town, 1987; and D Geldenhuys and H Kotze, *Politikon*, 1983.

6 See R Lee and F Buntman, 'The Business Sector and Policy Involvement', unpublished mimeo, Centre for Policy Studies, 1989; M Mann, 'The Giant Stirs: South African Business in the Age of Reform', in P Frankel et al, *State, Resistance and Change in South Africa*, Johannesburg, 1988.

7 This contrasts with the capital logic interpretations provided by Morris and Padayachee, 'State Reform Policy', 1988; and W Cobbett et al, 'A Critical Analysis of the South African State's Reform Strategies in the 1980s', in P Frankel et al, *State, Resistance and Change*, 1988.

8 See J Coetzee, 'Rewolusionere Oorlog en Teeninsurgensie', in M Hough, *Nasionale Veiligheid en Strategie met Spesifieke Verwysing na die RSA*, Pretoria, August 1981 (translated by Centre for Policy Studies staff).

9 Interview with ex-official of the Department of Constitutional Development and Planning, August 1986.

10 Quoted in *Newsweek*, June 1988.

11 Africa International Communications, *New Emergency Management*, Cape Town, 1986, 13.

12 The argument in this section is based on an extensive research project conducted by researchers at the Centre for Policy Studies during 1988/89. The findings will be published in the forthcoming annual policy review publication of the Centre. The references to primary documents and the approximately fifty interviews that were conducted are contained in unpublished drafts produced by M Swilling and M Phillips. See also B Pottinger, *The Imperial Presidency*, Johannesburg, 1989.

13 S Huntington, 'Reform and Stability in a Modernising Multi-Ethnic Society', *Politikon*, (8)2, December 1981, 20.

14 W Ebersohn, 'The Headmen', *Leadership SA*, (7)5, 1988, 21.

15 Department of Military Intelligence, *The Art of Counter-revolutionary War*, unpublished mimeo, 1987, 23.

16 *The Art of Counter-revolutionary War*, Department of Military Intelligence, 1987, 7.

17 South African Broadcasting Corporation, 'Comment', 17.06.88.

18 Research notes provided by Richard Humphries, Centre for Policy Studies.

19 South African Broadcasting Corporation, 'Comment', 17.06.88.

20 Speech by PW Botha to the Free State Congress of the NP.

21 Figures drawn from various editions of South African Institute of Race Relations, *Socio-Economic Update*, 1988.

22 J Lombard, 'Housing Finance and the National Economic Scenario', keynote address to conference on Finance, the Pathway to Housing, CSIR, 2 June 1988.

23 G Sheldon, 'Theoretical Perspectives on South African Foreign Policy Making', *Politikon*, (13)1, June 1986, 17.

24 Africa International Communications, *New Emergency Management*, 1986, 13.

25 M Schneider, 'Interview: Adriaan Vlok', *Leadership SA*, (6)1, 1987.

26 See proceedings of *End Conscription Committee vs Minister of Defence*, case number 88/2870 CPD; also *Africa Confidential*, 08.07.88.

27 See Kate Philip, in J Cock and L Nathan (eds), *The Militarisation of South African Society*, Cape Town, 1989.

28 Speech by PW Botha to the Natal Congress of the NP, August 1988.

Opposition Politics and Ideology in the Western Cape

Bill Nasson

The liberation struggles of South Africa's dominated classes represent a tangled complexity, rooted in the material circumstances of the local, the particular and the diverse. The consequent importance of 'keeping in mind the regional specifics of "political cultures" and longstanding local traditions' was recently stressed by Martin Murray, in his synthetic survey of popular insurrection in the 1980s.[1]

Discussion of political currents should not merely concentrate on the national features of resistance, but also on the peculiarities of each particular regional formation. No one area - Durban, the Western Cape or the Witwatersrand - can simply be held up as a mirror to reflect what is inevitably occurring within democratic forces elsewhere. South Africa is, after all, a mixed bag of historically distinct localities, with different traditions and different social forces. And, perhaps more than anywhere else, the Western Cape possesses an 'exceptionalist' political constituency and culture.

An adequate understanding of the nature and development of contemporary resistance politics must include an appreciation of some of the strong historical, institutional, ideological and personal relations which have imparted a discrete texture to the Western Cape political terrain. This means, above all, being sensitive to an historical situation.

Today, rancorous divisions between, for example, a Charterist-aligned Cape Areas Housing Action Committee (Cahac) and a Unity Movement-oriented Federation of Cape Civic Associations are not simply determined by the particular configuration of political choice and practice at any given moment. They are also crucially affected by the sulphurous inheritance of historically-derived traditions.

Perhaps only in the Western Cape can it be said that history, even when invisible to the more high-spirited political activists, always defines the inventory of left politics in some sense. There are fertilising historical traditions which carry an exceptional kind of human sensitivity, for local political memories are uncomfortably long.

The long-standing character of fractures between more-rooted marxist and left socialist or social democratic impulses in local resistance politics is equally important. For decades there have been many 'isms' contending for ascendancy within the various constituencies of the oppressed in the Western Cape.

The popular insurgency of the 1980s has certainly not reconciled these conflicting traditions. Indeed, they may harden. Since 1986, conditions that could support a sustained mass mobilisation have been increasingly eclipsed by intensified state coercion and control, and the elimination of leading dissident groupings and individuals. Bannings and proscriptions which have mainly targeted populist forces, affect different sections of the Western Cape democratic movement unevenly, furthering divisive rather than affirmative traditions of socialist and radical politics.

This makes it quite possible that established ideological rifts - between those still able to maintain a level of permeating organisational activity, like Unity Movement-derived groups, and those whose popular mobilisation is suppressed, like UDF-related groups - will become more persistent as traditional institutions and newer coalition groupings cut adrift from any common engagement.

In this respect the Western Cape experience crystallises the insistent need for a clear-headed view of the prevailing balance of forces in the present crisis. Above all, it should be a view which does not 'exaggerate the contradictions in the forces that sustain the status quo and downplay the contradictions in the resistance'.[2]

The Unity Movement inheritance

Local enthusiasm for Charterist politics in its various strains, especially since the mid-1980s, has run high. Its redoubtable populist discourse presides over the University of the Western Cape, where non-conformist radicals are now said to walk a sight carefully. But it would be wrong to assume that it has simply become a local hegemonic force.

Anybody who is at all familiar with Western Cape oppositional politics will know that at one fundamental level it is bound up with the history of the Unity Movement. And to consider the Unity Movement inheritance is to return once more to the theme of historical tradition. For it is manifestly within the revolutionary socialist tradition - created in the first instance by post-1917 developments within and outside Russia - where 'political currents often take their very definition from historically-derived divisions'.[3]

This is certainly true of the Trotskyist and semi-Trotskyist principles and influence which have animated the Unity Movement's bilious opposition to what it defined historically in classic policy documents as the 'pro-capitalist, anti-working class nature of ANC bourgeois social democracy', and the 'unprincipled Stalinism of the Communist Party'. This critical position is carried forward: towards the end of 1988 the *New Unity Movement Bulletin* hurled abuse at the negotiationist national convention 'scenario on which the ANC and the "left" theorists and political leaders base their wheeling and dealing with the multinationals and liberals in Lusaka'.[4]

All of this suggests that in the Western Cape a marxist presence - and an independent, minority socialist politics - continues to retain a distinct and visible constituency. If anything, it has become commonplace to draw attention to this socialist influence; even the recent Peter Berger and Bobby Godsell volume carries an astute contributor who has noted the 'long tradition of Trotskyism in the Cape'.[5]

The old Non-European Unity Movement has left a palpable 'Unity Movement' identity as a significant cerebral element in Western Cape political life and consciousness. And while not now, nor ever, a mass socialist phenomenon, the Unity Movement has come to exemplify a gritty, articulate and, at times, almost subterranean tendency which, borrowing a phrase from EP Thompson, we might fairly identify as 'a very substantial minority tradition'.[6]

Traditionally purist in doctrine and resting its policies on a theoretical analysis of liberation politics, the Unity Movement programme has long represented a taut, coherent and uncompromising Western Cape alternative not only to Congress Alliance/liberal coalitions but also to Pan Africanist and Black Consciousness forces.

The programme has an unambiguous class imprint - committed to the primary representation of proletarian interests in a socialist society. The Unity Movement is also unremittingly hostile towards what it sees as 'opportunistic' class coalition politics. This form of politics, it believes, undermines the building of exclusive working-class unity and diverts the oppressed into incremental negotiation with 'imperialism and the local ruling class and its agents'.[7]

The Unity Movement programme has a tough formulation of non-racialism. It denies the validity of any concept of racial/ethnic differentiation, or group/minority rights. An obvious consequence is virulent opposition to any 'multi-racial' liberation strategy - which gives the 'four nations' inscribed Freedom Charter programme such an unacceptable pedigree.

The Movement's position is underpinned by a principled policy of boycott and non-collaboration with ruling class agencies. This puts beyond the pale not merely the regime and its various apartheid institutions, but also a spectrum of ascribed associational interests which includes conservative and radical-liberal forces alike. In 1988 this encompassed organisations and individuals such as the NDM, Idasa, the Five Freedoms Forum, Tom Lodge, Steven Friedman, Mark Swilling and so on.

Finally, a significant feature of Unity Movement working political practice is a rather hydraulic version of marxist historical materialism.[8] Under conditions of accumulation which inevitably reduce the masses to ever-deepening immiseration, independent organisations, primed with a socialist consciousness and under working-class leadership, will rise up against their unbearable conditions and effect a transition to a society in which 'there will be no apartheid, no oppression and no exploitation'.[9]

In a period where popular alliance movements were hard hit by the compulsions and confiscations of a continuing state of emergency, the non-violent Unity Movement left was still able to provide a base for the Western Cape's radical socialist tradition.

Fragments consciously derived from that tradition - such as the Cape Action League (CAL), the New Unity Movement (NUM) and Students of Young Azania (Soya) - have sought to insert old Unity Movement organisational and theoretical formulations into the strategies of the liberation movement. And the earlier conspiratorial texture of Unity Movement social practice has continued to influence the form and content of this essentially close-knit, self-enclosed political tendency.

In 1988, tiny pedagogic societies, fellowships and study groups continued to sustain partisan political education through lectures and discussions. Topics ranged from 'Christian National Education'

with the Cape Town Cultural Society (a New Unity Movement affiliate), to the Russian Revolution - perhaps not the ideal anniversary year for this, but 1987 had already been poached by populist
organisations for a commemorative rally at the University of the
Western Cape.

Attempts at public forums by groupings such as New Unity
Movement and the Federation of Cape Civic Associations often fell
foul of state of emergency restrictions on assembly. But a customary
activist core of radical middle class, mainly coloured intelligentsia
was able to keep Unity Movement ideas in currency through
sporadic pamphleteering and the issuing of New Unity Movement
or CAL press statements in response to this or that issue.

The intermittent circulation of Federation civic newsletters, the
Educational Journal of the Teachers' League of South Africa, and
the astringent *New Unity Movement Bulletin* helped to outline
strategy and pose demands. These publications campaigned against
state stratagems like the October 1988 'dummy elections', and
poured vitriol over internal and regional negotiation politics, con-
temptuously dismissed as the 'Harare Safari', the 'Sellout in
Namibia', and the 'Angola Sellout'.

This kind of diffuse, critical format was by no means the only
recent representation of the Unity Movement's orientation. In Cape
Town in April 1988, the South African Council on Sport staged a
major sports festival (an 'Olympics of the Oppressed'). This af-
firmed its belief in the indissoluble link between non-racial sport
and liberation struggle, and its ideological inheritance of tight ad-
herence to a correct boycott and non-collaboration position.

At times, the Unity Movement was also able to mobilise a wide
social constituency around day-to-day issues. This recalled CAL's
involvement, during the early 1980s, in community struggles over
rent, housing, and the campaign against the 1984 tricameral
parliamentary elections. So in 1988 the Unity Movement tradition
again fused with a set of prickly local political conditions - group
areas in inner-city Cape Town. The pressures of local conditions
ensured that its pervasive conception of total emancipation became
interwoven with more particular and exclusive concerns.

Ever since the gutting of District Six, the few surviving educational institutions on the boundaries of this area have been
threatened with closure. These, especially Trafalgar High School
and Zonnebloem Teachers' Training College, are historic sites of

captive apprenticeship in Unity Movement ideology, and are thus invested with all manner of political, moral and social attachments by the local community. It is not surprising then, that educational, cultural and other organisational elements rallied to make a vociferous defence of their continued existence, through meetings, public statements, and even the self-assertion of fund-raising fetes.

Furthermore, recent District Six reconstruction initiatives by both state and capital also rubbed an especially raw nerve. Official disclosure in September 1988 that District Six might become a 'free settlement' area followed an earlier BP corporate proposal to rebuild the locality as an 'open suburb', a phoenix-like model for a post-apartheid society. Somewhat short on gratitude, a diverse amalgam of over 20 local bodies mobilised under the leadership of the Salt River-Woodstock-Walmer Estate Residents' Association, to constitute the Hands Off District Six Committee (Hods).

The starting point for Hods was opposition to 'the conception of creating an island of "normality" in our divided land' and the retention of an unreconstructed District Six as 'a symbol of the cruelty' perpetrated by apartheid in Cape Town. This perspective on group areas has since been subsumed into a larger set of demands for full and equal citizenship rights in all spheres.

Hods used the press to make official critical statements and in mid-1988 organised a significant one-day public conference on District Six. It drew on a wide range of community and academic resources and provided a base for the intended construction of a community-based District Six museum. This Unity Movement-imprinted civic initiative also travelled some considerable social and ideological distance by integrating fraternal support from the Anglican Archbishop of Cape Town and from UDF affiliates such as District Six Youth.[10]

Another fleeting convergence of constituencies took place around the 1987 'Out' campaign. This was a nominally single-issue action to try to block the construction of parliamentary houses in Walmer Estate for Labour Party ministers. Both Unity Movement and UDF-connected civic and other groupings in Salt River, Woodstock and Walmer Estate were keen on a marriage of campaigning convenience; a madcap plan to appropriate a section of the proposed building site as a 'people's park'. And a small but shrill protest march through suburban streets suggested that Movement elements were, for once, surprisingly amenable to practices from outside their own more recent political and cultural inheritance. But such common engagement was both transitory and extremely rare.

Far more representative of old political relations was the New Unity Movement response to an invitation to attend the (eventually banned) Cosatu-convened Anti-Apartheid Conference (which excluded CAL), to have been held in Cape Town in September 1988.

Calling its decision against attendance 'a painful one' in view of the recognition of 'the importance of trade unions in our liberatory struggle', the New Unity Movement nevertheless wrinkled its nose at the 'Cosatu call for unity'. It pointed to the presence of 'yesterday's oppressors who are paraded as today's revolutionaries', and argued that the accommodations of principle and strategy required to reconcile some bourgeois constituents, such as Idasa or the Black Sash, carried too crippling a political price. This price would involve 'opening the doors of the freedom movement to all sorts of political class enemies dressed up in different guises'.

The Movement stressed that what was at issue, among other substantive matters, was an incoherent and unconvincing conference representation of unity which failed 'to make clear the great difference between the interests of the workers and the capitalists'. The long-term consequence of such a coalitionist programme was that 'the long, bitter struggles of workers and the landless peasants will be sold down the river'.[11]

In a sense the debate within the Western Cape progressive movement has been about nothing but appropriate alliances for most of the contemporary period.

Charterist imagery

Popular political resistance can sometimes too easily be presented as a national phenomenon with local derivations. For many regions, the reality may instead be local movements with some fundamental national themes.

It is worth recalling the concerted Western Cape struggles of the earlier 1980s, pioneered from the barricades and culminating in the feverish militancy of the 1985 'rebellion' in Cape Town. For the Charterist or Congress movement, the later phase has been an engulfing state of emergency era when its actively assertive public presence has been confronted, and constrained or repelled.

At one level, the history of progressive struggle is simply a casualty list of local organisations or Western Cape branches of national movements stifled by state edict. Bannings fell predominantly - but not exclusively - on formations tied to the UDF. Apart from the Western Cape UDF itself, these groupings included the South African National Students' Congress (Sansco), Western Cape Civic Association (WCCA), Western Cape Students' Council (WCSC), Mitchell's Plain Student Congress (MPSC), the Western Cape Teachers' Union (Wectu), and the Democratic Teachers' Union (Detu).

In this period of very considerable losses, the model of sustained mass mobilisation lost its immediate hold. Yet despite a shrinking organisational base, the ideology and imagery of the Charterist tradition continued to occupy a vital part in the lexicon of progressive politics.

This is especially noteworthy as the Western Cape had no great reputation as an area of Congress influence prior to the troubles of the 1980s. As Colin Bundy has observed, in historical terms the rising graphic expression of ANC loyalties represented 'a real departure'.[12] And, as is well known, the outstanding feature of the recruits was their youth and schooling. An 'ANC-oriented political culture', to quote Jonathan Hyslop, 'became a strong feature of student movements in the Western Cape, which had previously been dominated by the Unity Movement, PAC and Black Consciousness'.[13]

During 1987 and 1988, the public accretion of Congress solidarities was contained but not displaced. One sharp episode involved the aftermath to the police shooting of the young activist, Ashley Kriel, in July 1987, in which the victim's funeral triggered a display of mass defiance. Western Cape ANC trials also provided a moment where the legacies of the immediate past were thoroughly worked over, prompting exhibitions of political sentiment which greatly displeased authority.

Ashley Forbes described the forging of his ANC loyalism in the crucible of the intense political affrays in Cape Town during 1984-85, declaring in September 1988 that he 'had no choice, no other way of expressing myself other than becoming a member of the ANC'. Of the municipal elections, Forbes stated,

We still have a state of emergency,...organisations are not allowed to express their grievances, the ANC is still banned, people in exile have not been allowed to return and not one of the demands of the Freedom Charter has been agreed to. As far as the ANC is concerned there should be a blanket boycott of the elections. I align myself with that call. I would call on the people of South Africa, all freedom-loving people of South Africa, to show their rejection of the October elections by refusing to take part. The mere fact that you are unable to call for such a boycott shows how undemocratic the elections are (*Argus*, 21.09.88).

Similarly, the 'Yengeni' trial of fourteen alleged Cape Peninsula ANC members charged with bombings and other offences, reaffirmed the radical intransigence of Congress loyalism into 1989.

Furthermore, if mass Charterism - taking its broadest possible definition - lacked the presumed practical relevance and immediacy which it clearly possessed in the earlier 1980s, its divergent constituencies continued to search for political space and ways of more mellowed organisational activity.

Survival, and the ability to provide a significant impetus, depended on combinations of many local circumstances, conditions and events. ANC ideological penetration could secure newly interested constituencies. Thus, in November 1988 the Thornhill Residents' Association was able to hold a report-back meeting in the conservative Indian suburb of Rylands on the Natal Indian Congress meeting with the ANC in Lusaka. Police harassment fortunately stopped at video cameras.

Similarly, white democrat constituency activists have helped to keep features of Charterist politics in public view, mainly through small events, workshops, and cultural gatherings. Attempts at larger public assembly involved a mostly frustrating tussle with state of emergency regulations.

Driven by imaginative slogans like 'Forward to Freedom! Forward to South Africa where the People shall Govern! Forward!', newly launched bodies such as the Mowbray Youth Congress, Gardens Youth Congress and the Cape Democrats were especially appealing to a mainly professional and intellectual middle class, alternative culture, white constituency.

More than this, however, in the 1988 municipal elections these organisations responded to the deeply ingrained boycott consensus within local black resistance politics by providing 'critical thinking' in the same direction. As the Cape Democrats put it, the understanding was that 'fundamental to the holding of free and fair elections is the right to vote or not to vote, to call upon to vote or to call upon not to vote; yet the Government has attempted to outlaw any call not to vote'.

The parliamentary re-emergence of the group areas issue in the second half of 1988, which coincided with what appeared to be a significant increase in police investigations of 'offenders' in some inner-city Cape Town areas, opened up possibilities for new single-issue community-based organisation and action.

Thus, there was the launch of a Woodstock Residents Against Group Areas: 'To Campaign Against Group Areas and the Amendment Bill; To Inform People of Issues Surrounding the GAA Bill; To Support Those Affected By The Bill; To Oppose All Forms of Forced Removals in the Country'. The familiar strategic presupposition was the construction of a popular alliance that would include local UDF components like the United Women's Congress but would also seek to generalise its basis and appeal around a pressing, day-to-day anti-Group Areas campaign.

But, by the end of the year, an essentially short-term, pragmatic movement was finding it difficult to sustain a constituency on the basis of collective missionary self-activity.

The 1988 October elections issue was undoubtedly the major public campaigning factor over this period. Despite increasingly-augmented state measures to sap the power of the boycott movement, the election factor provided a mobilising link between a great range of groupings. Typically, these ran from the Call of Islam and the Muslim Youth Movement to UWCO and Cahac affiliates in Cape Flats townships, from various school and tertiary education lobbies to other youth-based resistance bodies, such as the Bonteheuwel Youth Movement. Their tactics fused with an inherited set of boycott and non-collaboration political conditions, adding fresh emphasis.

The influence and impact of anti-election agitation was clearly variable. In townships or older suburbs where abstentionism is deeply entrenched, voter turnout was abysmally low. But in rural areas such as Paarl or Wellington, the Labour Party's administrative politics of patronage and compulsion in jobs, housing and services contributed to greater support for electoralism. But even here the popular base remained relatively small, with more militant boycotters causing 'candidates to complain of intimidation to prevent them standing for election' (*Paarl Post and Wellington Gazette*, 13.10.88).

Other political currents

In the Western Cape there are a number of other important constituencies of popular movement and experience. During 1987 and 1988, the established, mainly conservative Western Cape trade union movement showed no real evidence of transcending its historical limitations of generally moderate and compromising bureaucratic leadership, despite the creeping politicisation of influential segments such as GAWU. The highly sectional basis of union action was both ideologically and materially weakening, particularly where attempts were made to link workers' struggles explicitly to broader popular demands and struggles. Predictably, the Western Cape registered the lowest stayaway percentage in the 1988 Cosatu three-day work stoppage call. Undoubtedly a related factor would also have been the absence of heavy mass production in the region.

With the exception of occasional skirmishes between police and pupils over the display of ANC plumage or the use of 'subversive' literature (a generous category which even included the *Reader's Digest*) in classrooms, schools' resistance was not conducted as in those near-off days of militancy in 1984-85.

The return of generally more stable, if tense, conditions for pupils created the need to find a less-windy political conception and language in which to sustain support and solidarity in struggle.

For the UDF's mass African constituency - the third current - this period saw an undoubted retreat. By the end of 1986, the state was able to create the conditions for some political tidying up, particularly through the shredding of the Cape Youth Congress (Cayco) membrane of direct-action political mobilisation within squatter settlements in and around Crossroads.

Restabilisation was built mostly on the expulsion into hiding or into detention of 'comrades' activism. By 1988 certainly, things were sufficiently tidy in Crossroads for visits by both PW Botha and Adriaan Vlok. In the new settlement of Khayelitsha, populist elements were only able to maintain an abbreviated, low-key civic presence in public life.

Only in the marginalised and stubborn remains of the KTC squatter camp areas did intense internal struggles for local power and control continue between conservative and radical elements. In a municipal election declaration, one squatters' committee truculently made participation conditional upon prior material provision: 'If the government wants any of us to do anything, they must first upgrade our areas and give us decent houses'(*South*, 15-22.06.88). Tolerable housing became a key focus for the leaderships of competing interests.

A further trend which seems to have retained a faltering, intermittent, yet underlying persistence is Black Consciousness. It was a distinctive, if fairly isolationist, current in the Western Cape in the early 1970s. And it again picked up some local momentum from the late 1970s into the earlier 1980s, after the formation of the Azanian People's Organisation.

Black Consciousness's non-collaborationist stand, its anti-capitalist rhetoric and an orientation towards black working-class unity, produced the prospect of a National Forum (NF) cloning with amenable Unity Movement structures, like CAL. But the 1988 state onslaught against Azapo and other BC groupings ensured that the ideals and vision of black solidarity were more of an ideological than an organisational kind. This involved a genuine if uneven deposit in community politics, unified by sentiment and by new, fluttering, initiatives such as the Black Students' Society (BSS) at the University of Cape Town.

BSS was an expression of exclusiveness within a context of linked involvement, its wider situation being the populist school and tertiary education-based Western Cape Student Front. It is worth noting the only faintly echoing commemoration of the death of Steve Biko: a thinly distributed biographical flysheet underlined the subdued survival of the 'Black Consciousness Movement...an idea that will not die'.

A more steely link in the Africanist chain is the Pan Africanist Congress, which can justifiably lay historical claim to being an indigenous strain. Both the Cape peninsula and the rural Western Cape were areas of considerable PAC support in the 1960s. Since then, its radical Africanist consciousness, always with some residual regenerative potential, has lived on among loyal migrant adherents in townships like Langa and Guguletu.

Outwardly, the organisation has long ceased to be a local mobilising political force in any serious sense. Hence the general assumption that the PAC has been clubbed and smothered to virtual death. Yet, although much has to be guessed at because evidence is sparse and fugitive, survivalist stirrings have hinted that the PAC is still capable of some level of reconstitution.

For example, PAC perspectives were flashed in educational institutions through discreet pamphlet drops. At Mbekweni township in Paarl, a PAC consciousness emerged to colour some local constituency conflicts. And there was the grip - especially on Western Cape Islamic politics - of the 1988 political trial in which the state charged activists of the radical, shadowy Muslim group, Qibla, with furthering the aims of the PAC. As any reader of *Muslim News* would have noted, here was an indication of an intriguingly volatile constituency, binding together the separatist radicalisms of Pan-Africanism and post-Iranian, big-drum Islam.

Politics both fractious and complex

No discussion of Western Cape politics, however partial and abbreviated, can fail to reiterate the point that it has remained a fractious politics of quite Byzantine complexity.

The gesture and nuance of political identities and associations remain endowed with particularly personal meanings. And because they spring from a political culture more finely grained than perhaps anywhere else in the country, the ideologically-impregnated grammar of Movement or 'Congress' attitudes can infect the transactions of everyday social life.

Here, to talk of recapturing black social history to the wrong cultural or educational group audience is still to risk becoming a terminal victim of 'BC' or 'Pan-Africanism'.

Categorisation, be it as Congress adventurist (Western Cape Student Front), Movement dogmatist (New Unity Movement) or neo-liberal infiltrationist (Cape Democrats), is the very essence of those antithetical traditions and fragments which freeze political relations within the democratic movement.

Yet, equally, aside from generally fitful squatter camp violence, such as that between 'comrades' and 'vigilantes' at KTC, the Western Cape has been spared the ugly sectarian bloodletting that took so great a toll of UDF-Azapo in Soweto and Port Elizabeth in 1987, or UDF-Inkatha in the Natal/KwaZulu region.

This also means that derided coloured management committee members or 'Mancom quislings' and 'collaborators' for the most part need not fear physical attack. Instead, in the midst of a mostly abstentionist community, they live under the shadow of neighbourhood moral scrutiny, with its turned cheek and quiet contempt. This has been the ingrained social politics of the past four decades, a solidity of popular sentiment that 'all state structures must be delegitimated by the practice of boycotts'. For 'a tactic born of impotence',[14] to use RW Johnson's words, it has been conducted in the Western Cape with quite masterly effect. In Wynberg or Schotsche Kloof, a poll of around 18% in October 1988 was once again the ultimate check upon original sin.

Both popular front and united front traditions have exhibited a very marked set of strengths and weaknesses. Despite ever more restricted space, the former has still had the capability of rapidly putting together an assertive and strident mass political force, as witnessed even by the aborted Govan Mbeki reception in Cape Town at the end of 1987.

Papers like *Grassroots* have effectively articulated the language and concerns of mass agitational demands. But the increasingly enforced transition to a different tempo and social infrastructure, away from the whiff of tear-gas and the clatter of the SAP helicopter, revealed strategic weaknesses and organisational limitations.

Unity Movement radicalism has a settled form of disciplined politics. It has continued to encapsulate a healthy antipathy towards the composition and orientation of nationalist or progressive, 'entryist' white bourgeois politics, and has remained theoretically premised upon the self-emancipatory mission of the organised working class. But it is an organisationally tiny grouping. Fearful of being marginalised or even eliminated by a 'Congress' consensus which dislikes its political complexion, its very confinement within hard-nosed, inherited thinking and practices has been an unattractive and weakening factor.

Evidence from the immediate past bears this out: a tendency to cling to customary dogma and a focus upon the preservation of doctrinal and organisational integrity over a whole range of linked bases, from sport to education or health; a traditional propensity for sectarian polemic; and a reluctance to innovate theoretically and strategically around the relationship between autonomous components of the progressive movement.

If the past is a reliable guide to the future, then for Western Cape politics it will continue to be a question of understanding the complex aspirations, expectations and suspicions sustained and developed in autonomous popular struggles. As we survey this scenario, this is also assuredly the region in which the current 'new realism' or 'Great Participation Debate' dilemmas about boycott and non-collaborationist strategies are likely to face their most exacting test.

Notes

1 Martin Murray, *South Africa, Time of Agony, Time of Destiny: The Upsurge of Popular Protest*, London, 1987, 438.
2 Bill Freund, 'Review of *South African Review 4*', *Transformation*, 7, 1988, 96.
3 Gareth Stedman Jones, 'The Labour Party and Social Democracy', in Raphael Samuel and Gareth Stedman Jones (eds), *Culture, Ideology and Politics: Essays for Eric Hobsbawm*, London, 1982, 321.
4 *New Unity Movement Bulletin*, 2(3), 1988, 14.
5 Peter L Berger and Bobby Godsell (eds), *A Future South Africa: Visions, Strategies and Realities*, Pretoria and Cape Town, 1988, 214.
6 EP Thompson, *The Poverty of Theory and Other Essays*, London, 1978, 66.
7 *New Unity Movement Bulletin*, 2(4), 1988, 16.
8 For fuller discussion, see my 'The Unity Movement and its Historiographical Tradition', *Radical History Review*, 45, 1990.
9 *New Unity Movement Bulletin*, 2(4), 1988, 8.
10 See David Penkin, 'Building Developments in District Six: BPSA's Proposal', BA Honours dissertation, University of Cape Town, 1989, 33-38.
11 *New Unity Movement Bulletin*, 2(3), 1988, 16.
12 Colin Bundy, 'Street Sociology and Pavement Politics: Aspects of Youth and Student Resistance in Cape Town, 1985', *Journal of Southern African Studies*, 13(3), 1987, 323.
13 Jonathan Hyslop, 'School Student Movements and State Education Policy: 1982-87', in W Cobbett and R Cohen (eds), *Popular Struggles in South Africa*, London, 1988, 196.
14 RW Johnson, 'Via Mandela', *London Review of Books*, 11(1), 1989, 11.

Managing the Urban Crisis, 1986-1989: The Role of the National Management System

Andrew Boraine

Black urban townships have long been centres of resistance to apartheid controls: names like Sharpeville, Soweto, Alexandra, Crossroads and Mdantsane are in many ways synonymous with the history of resistance in South Africa.

The strongest challenge to urban apartheid came between 1984 and 1986, when township residents nation-wide refused to pay increasing rent and service charges levied by black local authorities which were perceived as unrepresentative and corrupt. Urban controls disappeared under the weight of combined mass resistance, and were gradually restored only through the declaration of a national state of emergency in 1986 and the resultant detentions, bannings and repression.

Since then, the state has attempted to regain control over black urban residential areas. The various institutions of the National Management System (formerly the National Security Management System) have played a key role in this process.[1]

Traditional urban apartheid

For many years, African residents in the urban areas were subjected to a number of well-defined urban apartheid principles and policies. The first of these regarded all African urban dwellers as temporary residents of white cities and towns, present only to provide labour, and then returning to reserve areas or bantustans. This policy was enforced primarily through influx controls (eg pass laws) and housing workers in single-sex hostels.

A consequence of this was that Africans were not entitled to political rights in 'white' South Africa. Prior to 1971, African urban residents were administered by white municipalities for which they could not vote, and subsequently by authoritarian bantu affairs administration boards which attempted to centralise all administrative and political controls over urban Africans under the Department of Bantu Affairs. Until 1977, the only forms of local government in the African townships were ethnically defined bodies with advisory powers.

Grievances of African residents were regarded as illegitimate, as they were not supposed to be in 'white' urban areas in the first place. Accordingly, all organisational and political activity in African townships was suppressed, and resistance was invariably met with repressive action such as bannings and detentions.

Urban Africans were also subjected to spatial forms of control. Residents were removed from the central areas of towns and cities and relocated in segregated dormitory townships. Townships were constructed to facilitate direct forms of social control. Most urban townships had single access roads and entrances, and were surrounded by 'buffer zones'. Police stations were located at township entrances and residents were subject to curfews and dawn raids.

African townships were also subjected to the principle of 'financial self-sufficiency' which was applied very ridgidly by the BAABs in the 1970s. Separate financial accounts were kept for 'white' and 'black' urban areas, with black townships being denied access to a revenue base including the commercial and industrial areas.

These policies reached their height in the 1960s when the state attempted not only to limit but reverse African urbanisation: this involved wide-ranging influx controls and forced removals, a freeze on land development and building of family houses in African townships, the removal of freehold rights, and promotion of the ideology of separate development and the concept of 'political representation' through tribal authorities in the bantustans.

Urban Africans of all classes were confined to overcrowded townships with inadequate housing and sub-standard or nonexistent social services such as electricity, water, roads, storm-water drainage, sewerage disposal, recreational areas, hospitals, creches and homes for the aged. In addition, urban Africans were burdened with high transport costs (townships usually being located far from work and shopping areas), inferior 'bantu education' and a system of local administration that governed without consent. In contrast, white urban residents have traditionally enjoyed an exceptionally high standard of living, with easy access to housing, high-quality services, and local-government institutions which respond to their needs.

These urban apartheid policies deeply politicised local government, education, housing, rents, service charges and transport. This led to the emergence of both spontaneous and organised forms of resistance within African townships - pass law evasions; squatter struggles; demonstrations, protests, deputations, marches and stayaways; boycotts of trains, buses, schools, businesses, rentals, service charges and local government institutions; and campaigns of defiance, civil disobedience and violence. It also led to the creation of different forms of mass organisation - student, youth, civic and women's groups; shop-steward local committees; and street and area committees.

Residential segregation promoted cross-class and inter-township unity amongst an increasingly differentiated African population, making collective mass mobilisation strategies in African townships highly effective and powerful.

Current urban policies

The challenge to, and gradual collapse of, traditional urban apartheid since 1976, has resulted in the evolution of new state urban policies over the past decade. These shifts were marked *inter alia* by the introduction of community councils and 99-year leasehold in 1977, the 1979 Riekert Commission, the 'Orderly Movement and Settlement of Black Persons' Bills of 1982, the President's Council report on urbanisation in 1985, and the white paper on urbanisation in 1986.

Central to current state strategy in the urban areas is the policy of 'orderly urbanisation'. This defines black urbanisation as both inevitable and economically desirable. The emphasis has shifted from controlling the influx of people into the cities to the establishment of formal and informal controls within the urban areas. These include supposedly racially-neutral slum, squatting and trespass legislation.

Current urban policies are still based on the premise of residential segregation, despite the increasing number of 'grey' areas. They differ from previous urban policies in the way in which they attempt economically to reintegrate the bantustans into metropolitan-centred planning regions. There is also an attempted reorganisation of urban space, with development plans promoting industrial and residential deconcentration to the edges of the cities through incentives, rather than the previous policy of industrial decentralisation to bantustan borders.

The central aim of current state strategy is the creation and maintenance of a variety of social, geographical and political divisions. This is undertaken through differentiated and segmented strategies in areas such as land allocation, black housing, provision of township services and facilities, urban management and financing, black unemployment, health and social welfare, and access to resources. Privatisation and deregulation are key mechanisms in this process of urban differentiation.

The identification and release of more land for black urban development is an important aspect of state policy. This marks the first time that urban land defined as 'white' in terms of the 1913 Land Act has been allocated for black development.

Land policy has involved both the allocation of more land to existing townships, mainly for the development of elite suburbs where the black middle classes can 'escape' the confines of poor working-class townships, and the allocation of land for peri-urban informal shack settlements.

Current state housing and urban development policies follow similar lines. The state maintains that the provision of housing is the responsibility of the private sector, while the cost of housing is the responsibility of the individual home-owner. Regulations affecting the involvement of private sector investment in the African townships have therefore been amended to promote this policy.

State-owned housing stock in the townships is for sale at subsidised rates to those who have been renting. This involves an attempt to relieve the state of its landlord role and depoliticise the housing issue. The state aims to cut housing subsidies and allow the provision of housing to be determined by the market. Those unable to afford a conventional house at market rates, or a self-build 'starter' house offered by the South African Housing Trust or the Urban Foundation's FHA Homes, are to house themselves on minimally-serviced sites on the fringes of urban areas. Housing standards are in the process of being deregulated to allow for the construction of differentiated forms of housing.

The state's privatisation policies also extend to the provision of services. African townships are meant to be self-financing through rates and taxes, and services like electricity, water, refuse removal and sewerage disposal are meant to be provided on an individual cost-recovery basis.

In recognition of the parlous financial position of most black local authorities, external sources of finance have been established in the form of Regional Services Councils (RSCs). This recognises the need for the partial transfer of resources between racially-defined municipal areas to finance the upgrading of existing services in the black townships. This ends the former policy of strict financial self-sufficiency applied to African townships. It also introduces the idea of metropolitan, as opposed to racially-exclusive, planning processes.

State urban policies are also concentrated on the revival of the system of black local authorities. The establishment of RSCs is an attempt to provide town councils in the townships with access to wider resources; town councils are now legally in charge of the development process and the allocation of land within the townships. There is also an emphasis on finding ways to end the rent and service charge boycotts which continue to deprive town councils of much of their income.

The intended consequence of these policies is to create divisions between middle-class home-owners in new elite suburbs and working-class residents of council housing and back-yard shacks, as well as a separation between formal black townships (the new insiders) and the burgeoning peri-urban shack settlements (the outsiders).

The National Management System in black urban townships

The NMS began exerting its power and influence during the nation-wide urban revolt of 1984-86, initially as a way of trying to restore 'law and order', and then as a longer-term mechanism for implementing new policies and controls to restabilise urban areas.

Although the NMS is a security structure par excellence, and is highly repressive, it should not be seen in terms of its security or repressive functions alone. It has come to form an integral part of current state urban strategies, particularly in maintaining control over urban black townships.

During 1986, the security forces identified 34 black townships as 'high-risk' security areas. These were defined as 'oilspots', where security forces have attempted to regain control through a combined process of repression and selective upgrading. The joint management centres (JMCs) established in each of these townships are identified as the means through which this process of restabilisation takes place.

The theory of low-intensity warfare, which underpins these security strategies, is based on the writings of various British, French and American military experts who observed anti-colonial struggles in Malaya and Algeria, and America's involvement in Vietnam. South African security officials, following these teachings, now place emphasis on economic upgrading and social restructuring, in addition to direct security action, as the way in which the black townships can be pacified and contained.

Many black material grievances are now acknowledged as legitimate. According to security officials, there is a need for direct or grassroots interventions in the form of upgrading to 'cut the ground' from under the feet of the 'revolutionaries' who are perceived as 'exploiting' black grievances. Known as 'winning-hearts-and-minds' or 'Wham', this strategy has already been applied for a number of years in Namibia and the South African bantustans.

The JMCs in the townships have identified a number of interconnected functions for themselves: co-ordination of security, welfare and finance; intelligence gathering and monitoring; re-establishment of black local government; communication of state policies; and specific 'hearts-and-minds' campaigns.

The JMC co-ordinates different security forces operating in the black townships. This is usually undertaken through a joint operations centre, which acts as the headquarters for the security forces.

The security committee usually comprises representatives from the police, security police, riot police, special police ('kitskonstabels'), municipal police, traffic police, SADF local command, SADF area defence units, military intelligence, and the National Intelligence Service (NIS).

The security committee, depending on the area of its operations, also co-ordinates or has contact with bantustan police and paramilitary forces, security personnel in private firms, the reserve police force (volunteers, often schoolchildren), the police reserve (former policemen), neighbourhood or crime watches in white suburbs, and the SAP Wachthuis, consisting of amateur radio operators.

Security action in the townships, or 'hard war' as it is known in security terminology, includes the establishment of roadblocks, and the conducting of patrols and house-to-house searches. A central security task is to monitor all forms of black township organisation, including political, civic, youth, trade union, shop-steward and women's structures, as well as all cultural, religious, sporting and business bodies. This is done through a joint intelligence committee (JIC). Security forces attempt to infiltrate these organisations to secure information about their activities. Mass democratic organisations have been identified as the main targets of JMC security action.

Security forces also engage in direct 'hearts-and-minds' activities, attempting to improve their image in the townships. In many cases, these take the form of providing sporting facilities for black schools.

The role of the JMCs is not restricted to security or repressive actions alone; a large part of their work is concerned with identifying areas of discontent in a township, hence the need for a wide-ranging intelligence network.

This information is then sent to the JMC's welfare or 'soft war' committee, which co-ordinates urban upgrading projects. These include the purchase of urban land for black development; differentiated housing schemes; the provision of infrastructure (roads, drainage, street lighting) and services (electricity, water, sewerage); the building of schools, sports stadiums and parks; and the improvement of transport networks. Altogether, several billion rands have been spent in the black urban areas over the past four years on over 1 800 projects.

Forms of finance for black townships are currently provided by various government departments, the RSCs, the National Housing Commission, the South African Housing Trust, the Development Bank of Southern Africa, and the provincial administrations. There are also investments and loans from the private sector, including the Urban Foundation's FHA Homes, housing and construction companies, urban development companies, company housing projects, and banks and building societies.

The welfare committee has the task of compiling a coherent and co-ordinated development plan that combines the responses of government departments with the investment programmes of the private sector.

Tapping the private sector

The major reason for poor housing and facilities in the black townships in the past has been the lack of finance, brought about by decades of state policy of forcing impoverished township residents to pay for their own development. A task of the JMC welfare committee therefore is to identify and co-ordinate different forms of revenue that can be used for upgrading programmes. The welfare committees maintain contact with a range of individuals, private companies and business organisations. Private developers are encouraged to invest in the township upgrading programmes, complementing the ability of the central government to make finance available.

In return, private companies have benefitted from the existence of the JMCs in a number of different ways. Many companies were confronted with disruptions to production in the form of strikes, stayaways, arson attacks and consumer boycotts during 1984-86. Consequently, many businessmen have lent direct or tacit support to the state of emergency and the re-establishment of 'law and order'.

Private companies and individuals also have a direct material interest in JMC programmes in the townships. Town planners, surveyors, civil, electrical and mechanical engineers, developers and builders have been awarded contracts and consultancies for a wide range of construction and upgrading work. The emphasis of the JMCs on encouraging urban development has opened up the possibility of new investment markets - private investment in urban land, installation of services, and construction of housing.

Numerous private companies, particularly multinationals, have channelled money into 'corporate social responsibility programmes' in the black townships, either as a form of advertising and public relations or more specifically to deflect international and local pressure to disinvest from South Africa. In many cases, the JMCs have attempted to utilise these funds for their own strategic ends.

The JMCs have identified existing political and financial weaknesses of the black local authorities and are attempting to re-establish the town councils as administrative structures. To this end, the JMCs assist in raising funds from a variety of sources for upgrading programmes conducted in the name of the black local authorities; they have also engaged in a systematic campaign to break the rent boycotts which have paralysed the councils for the past three years.

The JMCs seek to promote the councils by presenting them as responsible for various upgrading schemes; JMC officials write speeches and press releases for black councillors, run their election campaigns, and provide crash courses in public administration and finance. White officials, usually former bantu affairs board employees, now employed by the black councils, usually play a central role in these programmes.

The re-establishment of the system of black local authorities is part of the initial campaign of the JMCs to restore 'law and order'; however, the town councils are also meant to provide the basis for political representation of urban black South Africans at a later stage. A key JMC strategy aims to project councils as 'representative' of township residents. Because of this, the JMCs have a strict policy of 'no negotiations' with organisations outside of official government structures, and have often actively intervened to block local-level negotiations.

Restructuring township life

According to the perceptions of security officials, 'unrest' in black townships is caused either by 'agitators' manipulating township residents, or because there is 'poor communication' between the authorities and the residents. The JMC 'solution' to these two problems is to 'eliminate the agitators' on the one hand, and establish ways in which local authorities and the security forces can 'communicate' with township residents on the other.

Communication committees have been attached to all JMCs in the townships. These are run in conjunction with the Bureau for Information. The JMCs have established a large number of township newspapers, which claim a widespread black readership. They are also responsible for certain programmes on TV 2 and 3, and on radio, portraying life in the townships as 'back to normal', or outlining the results of urban upgrading. The committees also publish media on behalf of the town councils, and distribute comics, newsletters and pamphlets to township residents.

Under cover of the state of emergency, and the suppression of the democratic movement, JMCs have attempted to restructure the social basis of townships. They intervene directly to facilitate the establishment of different forms of organisation such as sports clubs, youth camps, fashion shows, drama groups and church choirs.

JMCs also attempt to draw non-government groupings and individuals from either inside or outside the townships onto JMC structures. This is done through the establishment of 'community liaison forums'.

The NMS structures, which operate from the mini-JMCs at a local level, to the state security council at a national level, have been able to co-ordinate a centralised yet localised response to the challenge from the democratic movement. Security officials regard previous state policies as having been fairly diffuse and random; the task of the NMS is to tie up the 'loose ends' of state strategies into a coherent whole.

The NMS attempts to avoid departmental bureaucracies which are perceived as having hampered swift and effective action in the past. To achieve this, the NMS has created parallel structures at a local, regional and national level, allowing for rapid intervention in the identification of problems and the implementation of 'solutions'.

Overall state strategy in the urban areas aims at creating or heightening divisions within or between black communities as a form of control. The NMS contributes directly to this process. NMS 'WHAM' strategy does not attempt to win the hearts or buy the support of all black people living in the townships. The upgrading strategy is deliberately selective, favouring a few at the expense of many. This is particularly evident in shack-settlement areas, where a selective or partial allocation of scarce resources by the state can turn one side of the community against the other.

The effect of the NMS programme, in certain areas at least, is to encourage the growth and co-option of various elites through the power of patronage, and at the same time neutralise groupings or individuals opposing the state.

In many cases, township upgrading and development programmes have exacerbated this process of division, with shack-dwellers being driven off land which is then 'developed' and allocated to middle class home-owners. Very few of the current development plans, whether they originate from the JMCs or private sector investment, cater for the urban poor and the unemployed. Where company housing schemes exist, they tend to favour higher-income employees; the financial policies of the banks and building societies make it difficult for low-income or unemployed workers to raise finance for a starter house or a serviced site.

Traditional urban apartheid policies tended to 'compress' various black classes into common localities. Current urban strategies are intended to reverse this process, opening up black communities along class lines. Many previous mass democratic strategies have been based on the premise of black unity. Whether this unity can still be assumed, given the processes outlined above, is unclear.

Current urban policies attempt to make Africans, and particularly members of the working class, carry the costs of urbanisation. State policies aim to create divisions between middle-class and working-class areas within townships; and between existing townships and informal shack-settlements on the fringes of urban areas, which act as 'catchment' areas for those marginalised within the urbanisation process - the unemployed, the unskilled and the unhoused.

The location of urban controls within the urban areas (between the urban core and urban periphery) is central to the policies of 'orderly urbanisation', rather than the maintenance of a division between 'insiders' in the urban areas and 'outsiders' in the rural areas, as traditional urban apartheid up until and including the Riekert Commission recommendations attempted.

Limitations of JMC upgrading programmes

NMS structures cannot function exactly as security officials intend. More often than not, state policies are replete with their own contradictions, and face resistance from below. Security officials have not been able to extend management structures to all black townships in South Africa, and many JMCs exist on paper only.

JMC involvement in urban development, and its control of the expenditure of millions of rands on projects in the black townships, are drawing criticism from a number of different sources, including state institutions and the private sector. Many urban upgrading projects, motivated by security rather than welfare concerns, were initiated without the necessary planning and research. Black local authorities, meant to be in charge of the development process, are too inexperienced, and in many cases too corrupt, to do so. Most black local authorities lack the necessary financial, managerial and institutional skills to manage the millions of rands worth of assets that have been invested in townships.

In addition, township upgrading programmes have had a limited effect because they are not linked to black political rights. The fact that economic and social upgrading is seen by many security officials as a substitute for political representation undermines much of their effectiveness.

The success of current state strategies in the townships relies heavily on reviving the system of black local authorities. Available evidence indicates that black town councils either still do not exist (many townships are run by white administrators), or are near collapse (as in Soweto). The results of the municipal elections in October 1988 show clearly that township residents have continued to withhold political support from the town councils. Financially, black town councils face massive debts due to the continuing rent boycotts, and in most cases are unable to meet their current running costs, let alone allocate resources for expansion and upgrading.

Many black councillors find themselves squeezed between township residents unable or unwilling to repay arrears and rents and service charges, and officials located within councils or on provincial administrations demanding that town councils start evictions and raise rent and service charges.

Even if residents began paying rent and service charges in full, the revenue generated would still be insufficient to meet the needs of the black townships. For townships have little access to commerce and industry which could provide a source of revenue. At the same time, townships have particularly large capital expenses, given the historical apartheid legacy of minimising the construction of housing and installation of infrastructure.

JMCs also face growing resistance in townships. There has been a slow but steady revival of civic and street committee structures in many areas, and the adoption of increasingly sophisticated strategies such as local-level negotiations with local and provincial authorities over rental arrears, transfer of housing to residents, upgrading of services, affordable service charges, and the demand for black townships to be included in a single tax base with neighbouring white municipalities.

JMCs and security officials no longer occupy the central position they enjoyed between 1986 and mid-1989. The decision by the new De Klerk administration to allow public demonstrations and release ANC leaders from prison suggests that different strategies are under investigation. Other state or parastate institutions such as provincial administrations, local authorities, Eskom and the Development Bank of Southern Africa have entered into a series of negotiations with a range of extra-parliamentary township organisations. This runs contrary to attempts by the JMCs to undermine and suppress community organisations, and create 'moderate' black organisations.

While the apparent eclipse of the National Security approach to urban management suggests that mass organisations within the townships will have a chance to regroup, the longer-term implications of 'orderly urbanisation' policies should not be minimised. Downgrading the role of JMCs in the upgrading of black townships does not mean that tensions and divisions caused by privatising urban land, and the provision of housing and services, will diminish.

Notes

1 This is a revised and updated version of a paper entitled 'Grassroots Interventions in South African Black Townships: New Trends in Policing and Security Management', presented at a conference *Towards Justice? Crime and State Control in South Africa*, Institute of Criminology, University of Cape Town, 4-6 May 1989. It draws on a wide range of published sources, as well as numerous published and unpublished case-studies of specific townships that have been conducted over the past two years by PLANACT, Johannesburg. While the ideas contained in this article are my own, they draw largely on discussions, seminars and projects within PLANACT, as well as discussions with colleagues within the Economic Trends group.

People's Courts and Popular Politics

Jeremy Seekings

The much-publicised emergence of 'people's courts' in 1985-86 gave rise to many varied and highly-polarised interpretations. For many participants and sympathetic observers of South Africa's township resistance, people's courts were one of the key 'organs of people's power', an innovative and positive achievement in the history of popular struggle. People's courts were seen to be important as a form of insurrectionary struggle, and as a precursor of future, post-apartheid democratic organisation.

In contrast, for the state and most of the press, people's courts were a barbaric instrument of intimidation and repression, used by township agitators to enforce the compliance of moderate township residents in unpopular consumer boycotts and other campaigns.

Remarkably little analysis has been undertaken on people's courts, and these contrasting interpretations remain largely untested. Suttner, in a paper written in the heady days of May 1986, celebrated people's courts, whilst warning of possible weaknesses, but his analysis relied on interviews with leading township activists.[1] Bapela, in a more recent paper, aimed to refute some of the wilder allegations concerning people's courts, in particular their responsibility for necklacing. But he presented almost no substantiating evidence.[2]

By far the most complete account and analysis of any courts is Burman and Scharf's research in Cape Town's townships.[3] Their study, based on very intensive interviewing, emphasises the diversity and complexity of issues surrounding township courts. The Cape Town courts varied in terms of their particular origins and their relations with different sections of the 'community'. These differences gave rise to contrasting court procedures and patterns of judgement and punishment.

This contribution is concerned with broad historical and sociological, but not criminological, questions: in what ways were people's courts innovative? Why did they emerge? What were they concerned with, and who was involved in them? What was the relationship between the courts and township politics?

There is some evidence on these questions, emerging from a number of trials - in state courts - which dealt with the issues of people's courts in the Pretoria-Witwatersrand-Vaal (PWV) region.[4] But trial evidence, even if sensitively used, can only provide an incomplete picture, and consequently this article no doubt misses subtleties of the kind picked up in Burman and Scharf's more detailed exploration of people's courts.

Defining people's courts

The category of 'people's courts' is widely embracing. The assumed general characteristics of such courts tend to be imputed from the context (political conflict) rather than from any features of the courts themselves. Many alleged 'people's courts' were in fact not described as such by their participants. An example is the supposed people's court in Kagiso, which was known to participants as the disciplinary committee.

The term 'people's court' can then broadly be used to refer to extra-state township courts which:
- met repeatedly (and possibly regularly);
- had a recognised structure (either formally or informally constituted);
- addressed civil and/or criminal charges in the township; and
- were broadly aligned with the 'radical' opposition.

This definition does not include 'courts' convened spontaneously, on a one-off basis, for example to 'try' consumer-boycott breakers on the spot at the entrances to townships.

In short, the existence of township courts reflected a form of struggle against the state, by removing issues from its jurisdiction into the extra-state township arena.

Historical antecedents

'People's courts' arise from a long tradition of extra-state township courts in South Africa. 'Informal dispute settlement and informal policing have been features of South African black townships and squatter areas since their establishment', as Scharf has pointed out.[5]

In several key respects, the PWV people's courts of the mid-1980s were broadly similar to their historical antecedents. They were 'extensions of existing and politically accepted practices, under the discipline of progressive organisations'.[6] This is largely a result of the chronic alienation of township residents from the state's courts. Wilsworth, for example, in a study of Grahamstown townships conducted in 1974-76, wrote that: 'If at all possible the police as well as various other official channels are kept out of conflict'.[7] And this avoidance of state legal institutions was characteristic of township politics: problems, wherever possible, were solved within the townships, by residents who were subject to a limited degree to popular sanction.

Before 1985, township courts in the PWV area tended to be run by township councillors or members of their ward committees. Many of these courts were known as *makgotla*. The term generally has pejorative connotations in the urban township context, derived from perceptions of the brutal activities of *makgotla* in Soweto especially.

Hund and Kotu-Rammopo, however, in a study of Mamelodi[8] apply the term *makgotla* to a wide range of courts with varying styles of operation and degrees of popular support. And even in Soweto, some *makgotla* enjoyed a measure of popular support, at least in the early 1970s. In some parts of Soweto as late as the early 1980s, there were other courts, often not known as *makgotla*, with some support.[9] The loose usage and broad applicability of the term *makgotla* has been mirrored more recently in the use of the term 'people's court'.

Soweto's *makgotla* developed during the period 1969-74. They seem to have begun in Meadowlands in response to a wave of break-ins, assaults and rapes. The *makgotla*, under the initial leadership of a Sinah 'Madipere' Makume, combined vigilante patrols with regular courts. These were primarily concerned with juvenile delinquency and family disputes. Madipere said, in an interview in 1976:

> I believe naughty children should be sjambokked. So if children under 18 are brought to us for disciplining, we sit as a *lekgotla* and thrash out the problem, giving the child (a chance) to defend himself. If the kid is found guilty, we let the parents lash him in front of us so the child cannot fight back. People with domestic squabbles come to us and we talk to them, but if their differences are too great, we refer them to the Bantu Commissioners. Robbers, attackers, murderers, rapists, and car thieves, we lash if they are under age, but if they

are 18 we send them to the police who sort them out (*World*, 11.03.76).

From the mid-1970s, Soweto's *makgotla* became increasingly coercion-based and deeply unpopular. The *makgotla* earned a reputation for arbitrary and barbaric practice. Their 'policing' became increasingly violent, and court sentencing became much more severe. In 1976, for example, a 19-year-old died from over 100 sjambokking injuries.

Furthermore, many *makgotla* were self-appointed, with few voluntary members and weak or no links with most township residents. Many of their activities seem to have involved brutalising residents to advance the interests of members of the *makgotla*. In short, they were often paramilitary rather than judicial institutions.

Few Soweto residents recognised the jurisdiction of these *makgotla*, and their operation depended more on their coercive capacity than any popular support or legitimacy.

There were courts in many other PWV townships.[10] As in Soweto, they were primarily concerned with civil and family disputes, especially disrespectful or delinquent youths, and were presided over by older residents, usually councillors or members of the ward committees.

In some of these courts, at least, trials were open, sentences were limited, and the objective was conciliatory rather than punitive. Hund and Kotu-Rammopo wrote of a court in Mamelodi's Ward 4:

> The emphasis is on the individual and not the alleged crime as such. The central aim of the Ward 4 court process is to prevent the breaking of relationships and to make it possible for parties to live together amicably in the future.[11]

When sentences were lenient, when judges were local and respected, and the objective of the court was conciliatory, the jurisdiction and judgements of the court were generally accepted. Mamelodi's Ward 4 court thus operated on the basis of broad adult participation and popular legitimacy.

But in the early 1980s, Mamelodi's Ward 4 court was increasingly overshadowed by courts run by the populist but hierarchical and authoritarian Vukani Vulimehlo People's Party (VVPP). The VVPP courts operated more through coercion and fear than assent.

In the years between 1974-84, the range of township courts in the PWV tended to be presided over by older, conservative leaders. And for most courts, the primary concerns were the maintenance of 'community', and in particular of a specific social order that they thought bound the 'community' together. Court leaders were often able to impose their own conception of order.

If these courts enjoyed popular support, it was because residents believed the kind of 'order' the courts imposed was functional to daily life. The courts solved pressing problems easily, without recourse to the state.

Other courts, which resorted to coercive discipline, aroused considerable discontent. *Makgotla* in Soweto (at least from the mid-1970s) or the VVPP in Mamelodi were not conciliatory, but divisive; and they relied more on the threat of considerable coercion than popular legitimacy.

The emergence of people's courts

During 1985, according to the state, over 400 people's courts were established in townships across South Africa. Police Brigadier Herman Stadler, in both the Mayekiso and Zwane trials, alleged that these courts were established as part of an ANC conspiracy to undermine the authority of the state.

During 1986-88 the state sought to convict many of the alleged organisers of people's courts on these grounds. Many of the accused have been convicted, but charges of treasonable conspiracy have not been upheld. The accused in *Mayekiso* were acquitted of all charges; the 'Alexandra Eight', some of whom were members of the Alexandra Youth Congress, were not convicted of treason, but of sedition in the *Zwane* case; the accused in *Ntshilele* were similarly convicted of sedition.

In both *Zwane* and *Ntshilele* the conviction of sedition rested on the respective judges finding that the accused had 'arrogated to themselves certain functions of the police and the judicial authority of the state' by conducting an anti-crime campaign and/or running a court. But in neither case was there evidence of conspiracy with the African National Congress or other local organisations, nor of hostile intent to subvert the authority of the state. And from a sociological perspective, the evidence for conviction even of sedition seems to be weak.

Many sympathetic observers have shared the state's view of

people's courts as revolutionary in intent. But the evidence presented in court points to different explanations for the rise and activities of these popular structures. Given the wide embrace of the category 'people's courts', this is not surprising. There does seem, nonetheless, to be a broad pattern.

Courts were a response to changing conditions in the townships. Widespread protests provided new problems or exacerbated existing problems - both for political organisations and residents. At the same time, the complete breakdown of relations with the police and the disruption or demise of the councillor-based courts of previous years left a vacuum. New institutions were needed. This explains why, when people's courts emerged, residents broadly supported or participated in them.

Particular groups had different motivations for establishing new courts. In some cases, township organisations formed courts to address specific organisational problems arising from either the general disorder or protest campaigns themselves. In other cases, new groups sought to establish themselves at the centre of township political life, and courts provided a good opportunity to do so.

From mid-1985 onwards, activists in radical civic organisations across the PWV region became increasingly concerned with the lack of discipline within their organisations or related to their campaigns. Consumer boycotts, in particular, were often accompanied by an unacceptable degree of coercive enforcement, impairing the organisations' popularity. In response to this problem, often attributed to 'comtsotsis', civic activists initiated anti-crime campaigns in several townships.

In Kagiso and Munsieville, on the West Rand, the people's courts were formed following an anti-crime campaign. The campaign itself was concerned with 'hooliganism and criminal behaviour perpetrated by people claiming to be our members enforcing and monitoring the (consumer) boycott' (*Star*, 18.12.85).

In practice, countering the boycott-enforcement problem involved a broader project: 'making the life of the people more bearable, because the problem of criminals was a real problem that affected everybody and people could not walk at night', according to Lawrence Ntlokoa, KRO's secretary.

These kinds of courts, initially established as a forum to discipline young people caught in the anti-crime campaign, emphasised re-education. Corporal punishment was limited - and subject to the approval of parents or other family members.

In Alexandra too, the establishment and activities of courts were linked to anti-crime campaigns. But the decisive factor in that township was the final breakdown of routine state policing and judicial activity during and after the 'Six-Day War' of February 1986.

During the Mayekiso trial, numerous state witnesses from Alexandra testified that, during early 1986, township residents who reported crimes to the police were told by the police themselves to take their problems to the 'comrades'. These witnesses also said violent vigilante activity in Alexandra was widely attributed to the police. Residents consequently saw the police as lacking both moral authority and practical utility.

Courts seem to have begun as advice offices, with residents such as Mike Beea (chairman of the Civic Association) and Sarah Mthembu (the mother of one of the first youths killed by municipal police) helping residents trace relatives who were missing after the 'Six-Day War'. These advice offices grew, as people brought a wider range of problems. Mthembu's house became home to members of the Alexandra Youth Congress. And according to a leading civic activist:

> The youth used to come there to console her... On a certain day...her sister-in-law came in the company of another woman. They said they had gone to the police station to report their husbands who had assaulted them... When they got to the Wynberg police station she said they were told that they must go to the comrades... After Sarah had seen this and the youth nearby, (she) said she is now going (to) open up to solve people's problems.[12]

Mthembu's lounge became a place to get advice and settle disputes, and became known as a 'people's court'. Younger residents later set up their own court in a shack in her yard. Subsequently, a number of courts were established by other youth groupings, partly in connection with their anti-crime campaigns.

The groups that established or ran courts stood to promote, indirectly, their importance in the township. The proliferation of courts in Alexandra - there were at least five - reflected not only the degree of breakdown of state policing and the court system, but also the variety of different groups vying for political leadership.[13]

In both Atteridgeville and Mamelodi, regularly-constituted people's courts were formed after malpractices or excesses occurred in more *ad hoc* courts run by the youth.

In Moroka (Soweto), a court was formed at the Inkanyezi Youth Centre by a group of elderly residents who were members of the Molapo-Moroka Civic Association, a branch of the Soweto Civic Association. A leading member - Ntshilele - was a former card-carrying member of the official *makgotla* movement. Their concerns were primarily local and social: 'People staying in our area had complaints... Domestic disputes, children who do not listen to their parents' advice, and if neighbours are in dispute'. The problem of children was particularly emphasised, as Ntshilele said during his trial: 'Die probleem is die kinders het baie stout geraak en ons het gedink ons moet 'n plan maak om hierdie kinders te berispe' (the problem is that the children had become very bad, and we thought we had to make a plan to discipline them).

Activities of people's courts

The state repeatedly alleged that people's courts were instruments of intimidation and the enforcement of unpopular campaigns, but failed to produce evidence to substantiate this view. Instead, court evidence generally showed that people's courts were used to sustain particular patterns of social rather than political 'order' in the townships, and also suggests that courts were often popularly supported.

More unpopular activities did not generally involve the enforcement of campaigns, but rather the enforcement of 'order' either through unpopular means (ie excessive sentencing) or by arbitrators whose jurisdiction was not recognised (ie the 'youth' in cases concerning 'adults').

Initial concerns in people's courts often involved organisational and campaign discipline, especially in relation to consumer boycotts. In practice, however, 'disorderly' activity around consumer boycotts was difficult to separate from general disorderliness in the township - particularly when the amorphous 'youth' were involved. Thus most courts became involved in the settlement of a range of civil disputes and social problems.

In Kagiso and Munsieville, residents urged the court to extend its role beyond cases arising out of the crime-prevention campaign. Older residents replaced younger ones as the court became involved in civil and family disputes. In one case, a local builder was accused of using too little cement when he built a *stoep* for a woman. He refused to repair it, but the court decided that it was his responsibility.

Many cases concerned family rows. In the *KRO* trial, witness E Motsowaye from Munsieville told the court:

> My mother once took a complaint to (the court) about my younger sister. (Four unarmed comrades) came to the house and called my mother and my younger sister. They then said to my mother, will you please tell us the story you told us before about your daughter, so that she must hear what you were complaining about. The complaint was that my sister had got a child and she is neglecting the child. She would go about in the township, leaving the child alone at home. Then (my sister) said yes, what my mother said was correct. She said the reason why she was doing this was because the father to that child resides at Kagiso and the father would come and take her along to Kagiso and bring her back late. They said to her, she should see to it that she does not do that again, because if she did that again, they would not come again to talk to her, they would come and discipline her and give her a hiding on her haunches.

Other non-activists gave similar accounts. According to a 42-year-old woman, the court was 'a committee which corrects people of their faults or of their mistakes... It appeared as though it was a good committee which was good for our children'. Ex-councillor Mabasa witnessed a court session where an old man was reprimanded for beating his wife, and warned that he would be 'dealt with' if he did it again.

In other townships, people's courts seem to have dealt with similar disputes. In Moroka, a woman who had a quarrel with her neighbour learned of the court's existence on the Johannesburg train:

> In the train I overheard certain women discussing about her child who was not obedient. I overheard this woman telling the other one that his (sic) boy was naughty so he (sic) took the boy to Inkanyezi (the court) and the people at Inkanyezi, they talked to this boy, now the boy is obedient, he even goes to school.

Many witnesses in the Ntshilele trial told of Moroka residents who took their problems to the people's court:
- a woman who accused the father of her children of not paying maintenance had heard about the court on the train;
- several women reported that their husbands assaulted them;
- one man accused his son of stealing his money;

- another accused his brothers of smashing his furniture and driving him out of his house.

In Orlando East (also in Soweto), the mother of a 16 year-old rape victim approached local street committee leaders after the police had refused to act. The parents of the alleged rapist wanted to sort the matter out locally, rather than go to the police. The accused admitted his guilt to his parents, who agreed with the victim's mother that he should be beaten and they should pay R350 in compensation. An agreement was even signed to this effect. The rapist's father was the first to beat him.[14]

When police raided a people's court in Atteridgeville it was dealing with a dispute over a pair of broken glasses. The plaintiff and the accused had been told to settle their differences amicably, but had allegedly later shouted abuse at each other in the street.[15]

Broadly similar activities seem to have characterised at least some of the courts in Alexandra. For example in the Mayekiso trial, Richard Mdakane testified that:

> The women were getting help when their men were no longer giving them money or supporting their children, and when they sleep out at places where they liked, the women used to go to the people's court and the people they find at the people's court were able to help them.

Other typical cases were inter-neighbour and parent-children quarrels, and disputes over child maintenance payments. One woman took her husband to court because he refused to have sex with her. But the Alexandra Youth Congress (Ayco) executive member presiding at the court said he was unable to deal with that case.

Court 'judgements' were generally conciliatory. Husbands were told not to assault their wives, or to make their maintenance payments, for example. Courts seem to have been particularly conservative when faced with marital strife, almost invariably telling the husband and wife to live together.

'Sentencing' in the courts was often limited to immediate needs. For example, maintenance must be paid (sometimes via the court). In several courts - including that in Moroka and some in Alexandra - corporal punishment was routine. Residents apparently approved sjambokking if it was limited to a few lashes.

State trials have produced little evidence of murder trials or death sentencing in people's courts. In Alexandra, one court investigated accusations concerning a murder charge, but did not resolve the issue. In Soshanguve, a court allegedly handled a murder case, and the 'defendant' claimed he was lashed 400 times. But the court members were acquitted of assault.[16] And members of a Tembisa people's court were acquitted of an alleged people's court necklacing.[17]

Courts and township politics

After a while, a number of courts became self-serving - particularly when the amorphous category of 'youth' in the township started imposing severe corporal punishment. These previously popular structures rapidly lost community support.

Scharf describes how, during the second half of 1985, the 'people's courts' of Guguletu in Cape Town lost support when they began to reach apparently arbitrary decisions, impose severe, and escalating, sentences of corporal punishment, and the 'youth' adjudicated 'adult' cases.

Alexandra was unusual in terms of the lawlessness of the 'youth' and the scale of the disruption arising from the bitter conflict. The courts' concerns became increasingly all-embracing, and their activities less popular.

Whilst the courts' anti-crime campaign had been popular, the escalating sentences imposed by the 'youth' (including on their elders) caused rising discontent. A number of witnesses in the Mayekiso trial stressed this.

Alexandra Action Committee minutes presented before the Mayekiso trial show that the committee condemned the use of corporal punishment in a court run by local youth, and sent one of its leaders to investigate. He reported that the youth who were controlling the people's court at 'number 31' were very aggressive and they did not like any person who criticised or condemned the lashing of people.

The AAC urged that re-education should be preferred to corporal punishment, and resolved that more problems be solved through the more broadly-based yard committees.

But the AAC hesitated before challenging the 'youth', because, as they later claimed,

> The youth that were participating in the people's court did not want to hear...any criticism about the people's court... The youth becomes uncon-

trollable at times. We realised that a certain bad element amongst the youth
might attack a person if he says such a thing.

But the problem intensified. As Mdakane told the *Mayekiso* trial:

(Some people) were saying that the people's court has helped in reducing
crime..., they reduced the *tsotsis*. But I remember there were people who were
against (the courts)... They were saying (that) in the people's court people are
being lashed... They are not so much against (the court), but they do not want
that there should be lashing.

In May 1986 a meeting of block committee representatives resolved
that the courts must be stopped, and AAC leaders were sent to
disband them, Mdakane told the *Mayekiso* trial court.

Discontent may have been focused on sentencing, but the age of
the court's members was clearly part of the problem. Many older
residents resented being tried by their juniors. Alexandra residents
also grew increasingly unhappy with what they saw as the 'arrogance'
of the 'youth'. In the main Alexandra people's court trial (Zwane
and others), the ages of the accused (at the time of trial) varied from
16 to 23. An older resident interviewed in May 1986 said that 'they
must fight the boere, not us'.[18]

Whilst the AAC and Ayco were aware of these problems, it
seems unlikely that they were 'solved' before the courts were suppressed by the state.

In Kagiso the youth appeared to recognise the problem. Initially,
the court was run by predominantly young civic activists. But they
resigned when people began to bring civil disputes involving older
residents to the court. The former chairman felt that he was too
young: 'As a young man I cannot handle an adult's affairs'. Bongani
Dlamini, in the KRO trial, said that the court was subsequently run
by older residents, many from the churches, and the chairman and
other leaders were in their mid-50s.[19]

The Moroka people's court was dominated by older residents -
including Ntshilele himself, born in 1926 - although younger residents also participated. But people who took their cases to the court
identified it as run by older people.

In August 1988 the head of the security police claimed 'the
almost total elimination of the so-called people's courts' (*Cit*,
25.08.88). In fact, most 'people's courts' seem to have been suppressed by the end of 1986, through the severe repression under the

nation-wide state of emergency in June 1986.

But suppression of people's courts may not end the tradition of extra-state township courts. The security police claim that township residents, freed from intimidation by necklacing radicals, will return to state courts. But even before the emergence of people's courts, many township residents seem to have had little faith in state courts. It is likely, therefore, that any lessening of state repression will be followed by the re-emergence of township courts, to meet a chronic need. And some will probably be presided over by relatively conservative residents.

But if the current concerns of progressive township activists with concrete local issues continue, some 'courts' will no doubt involve such activists and hence more radical conceptions of justice.

Continuities and discontinuities

People's courts, like the pre-1985 township courts, presided over civil and family disputes. Rarely, if ever, were they concerned with serious criminal cases. Their aim was generally conciliatory, and their sentences not severe. They were often popularly supported. However, also like their antecedents, people's courts sometimes became increasingly unpopular, when patterns of judgement and sentencing grew out of tune with what was popularly accepted.

The continuities between *makgotla* and people's courts are particularly striking in terms of how township residents saw them. In the Ntshilele trial, numerous state witnesses testified that they thought that the Inkanyezi Youth Centre court was a *lekgotla*. A 60-year-old man told the court that his impression was that 'these people were definitely a *lekgotla*', and the courts were similar in that they dealt in the main with local problems, rarely trying to deal with serious crime.

Despite some continuities, there were considerable differences between even the popular *makgotla* and the people's courts. These differences were due to the changing context of township politics.

In the period before 1984, township residents sought to redress their grievances through a variety of mechanisms that did not directly involve the state. Many of these involved local conservatives, who attracted popular support when they promoted popular causes. The more conservative elements often seem to have had greater organisational resources than more radical residents, and were relatively immune to state retaliation. These factors underlay some of the more popular *makgotla*.

After 1984 the political context changed, altering the character of participants in township politics. 'Popular conservatism' collapsed, at least temporarily, as involvement with the state could no longer be reconciled with popular support.

Members of progressive civic and youth organisations took over the roles previously played by conservatives. The motivations of these more progressive people were varied: sometimes they were altruistic, sometimes self-serving, and usually both.

Amidst a combination of polarisation and widespread radicalisation in township politics, conservative vigilante groups emerged. In the context of South African townships, 'vigilantism' has come to refer not to popular crime-prevention campaigns but to the sinister and violent paramilitary groups based around some township conservatives. And some *makgotla* in the period before 1983 seem to have been the antecedents not of people's courts but rather of these violent vigilante groups.

What was fundamentally innovative in the process of township dispute settlement in the mid-1980s, was the involvement of extra-state civic associations and youth groups. This convergence of township courts and radical township politics was a strengthening factor for these organisations. For while conservative township courts had been extra- state, people's courts were part of a broad national movement that sought to transform the state itself. And people's courts were integral to township-wide political organisation, in contrast to many of their conservative predecessors whose activities had been confined to dispute-settlement and policing.

It was the link between township judicial institutions and other radical extra-state structures that posed such a problem for the state. This was exacerbated by the fact that the courts constituted an important ideological as well as organisational development - although the extent of their ideological innovation should not be overstated.

By 1986, people's courts in the PWV had extended their judicial project much more extensively throughout townships than their conservative predecessors ever had. Most PWV people's courts were concerned with popular justice, 'popular' in the sense of being broadly supported as well as being outside of and alternative to the state. Scharf has described the Guguletu people's court in terms of the enforcement of a new 'people's morality that conformed to the political ideals of their liberatory projects'. Whilst similar concerns were articulated by many activists in the PWV, in most townships (excepting Alexandra) older residents seem to have established their control over people's courts, and their sensitivity to popular legitimacy seems to have curtailed the radicalism of the new 'people's morality'.

These factors combined so that, as Lodge writes:

> of all the manifestations of people's power, the efforts of local organisations to administer popularly acceptable forms of civil and criminal justice were the most challenging to the state's moral authority.[20]

State suppression of township courts probably reflected its recognition of these developments. Whilst *makgotla* before 1984 were often tacitly approved of by the state, people's courts were unacceptable. And the state's emphasis on necklacing, which its propaganda alleged took place through people's courts, perhaps also compelled it to take severe repressive action against township courts.

Ironically, association with township courts could also be a liability to civic associations. As the case of Alexandra shows, if a court became unpopular and was popularly associated with the civic organisation (whether rightly or wrongly), then the popularity of the latter was at risk.

PWV people's courts do not appear to have been primarily involved in the trial of 'informers and other enemies of the liberation struggle'. Trial evidence suggests that alleged informers were murdered without trial in a court, as in the much publicised case of Maki Skhosana in Duduza in July 1985. Such murders were committed on a more summary basis by groups of 'comrades'. Indeed, Boraine relates that a summary necklacing of an alleged informer was prevented in Mamelodi in February 1986, when community leaders urged the crowd to use people's courts.[21]

Instead, people's courts in the PWV seem to have been concerned with the resolution of everyday township problems. This convergence of political radicalism with everyday problems was probably a greater threat to the state than courts trying alleged 'sellouts' would have been.

Notes

This research was made possible through the kind assistance of lawyers involved in the cases. I am also grateful to Karen Jochelson and Nicoli Nattrass for their comments.

1 Raymond Suttner, 'People's Justice in South Africa Today', unpublished paper presented to a Sociology Department Seminar, University of the Witwatersrand, 5 May 1986.
2 MSW Bapela, 'The People's Court in a Customary Law Perspective', unpublished paper, University of South Africa.
3 Most of this research has, sadly, not yet been published. A brief account of the Guguletu/Nyanga East people's court can be found in Wilfried Scharf, 'People's Justice', *Sash*, March 1988. Related unpublished work is: Scharf, 'Images of punishment in the people's court of Cape Town', unpublished paper, Cape Town, May 1988; Sandra Burman and Wilfried Scharf, 'Informal Justice and People's Courts in a Changing South Africa', paper presented at the International Sociological Association Research Committee on the Sociology of Law Conference, Bologna, Italy, May 1988.
4 The principal trials referred to are: Krugersdorp Residents Organisation vs Minister of Law and Order and others (henceforth referenced as KRO); State vs Mayekiso and others (henceforth Mayekiso); State vs Zwane and others (henceforth Zwane); State vs Ntshilele and others (henceforth Ntshilele).
   The full references to these trials are:
   KRO - Supreme Court of South Africa, Witwatersrand Local Division, case no. 3464/86, before Mr Justice Goldstone;
   Mayekiso - Supreme Court of South Africa, Witwatersrand Local Division, case no. 115/87, before Mr Justice Van der Walt;
   Zwane - Supreme Court of South Africa, Witwatersrand Local Division, case no. 50/87, before Mr Justice Grosskopf;
   Ntshilele - Regional Court for the Regional Division of the Southern Transvaal, held at Johannesburg, case no. 41/3383/87, before HJ du Toit.
5 Scharf, 'Images of Punishment'; also see Burman and Scharf, 'Informal Justice'.
6 Andrew Boraine, 'Mamelodi: From Parks to People's Power', unpublished Honours dissertation, University of Cape Town, 1987, 129.
7 Mercia Joan Wilsworth, 'Strategies for Survival: Transcending the Culture of Poverty in a Black South African Township', Rhodes University, Institute for Social and Economic Research, occasional paper 24, 1980, 233.
8 John Hund and Malebo Kotu-Rammopo, 'Justice in a South African Township: the sociology of Makgotla', in *Comparative and International Law Journal of Southern Africa*, 16, 1983.
9 Interviews, residents of Chiawelo and Moroka, 1988-89.
10 Alexandra seems to have been something of an exception, with the local makgotla (or 'Thiza') active only until 1973-74 (*Zwane*: Zwane evidence). Thereafter, with the occasional exception of councillor-run courts, dispute settlement probably took place primarily at the level of the yard.

11 Hund and Kotu-Rammopo, 'Justice in a South African Township', 187.

12 Mdakane evidence, *Mayekiso*.

13 Pascal Damoyi, 'Dispute in Alexandra Township', *Work in Progress*, 44, Sept/Oct 1986; Mdakane evidence, *Mayekiso*.

14 The case went to a state court when the rapist laid charges of assault (he excluded his father). State vs SJ Mabuza and 4 others, in the Regional Court for the Southern Transvaal, held at Orlando Magistrate's Court, September 1987.

15 Glenn Moss, 'State moves against people's courts', *Work in Progress*, 46, February 1987, 38.

16 State vs Morris Ngobeni and 5 others, Pretoria Magistrate's Court; Moss, 'State moves against people's courts', 38.

17 State vs Vusimusi Mtimbane and Fani Mahlangu, Supreme Court, Witwatersrand Local Division.

18 Quoted in Karen Jochelson, 'Urban Crisis: State reform and popular reaction, a case-study of Alexandra', unpublished Honours dissertation, University of the Witwatersrand, February 1988, 154.

19 See also the ages of the court's leading members in the indictment for State vs Mary Ncube and 14 others (this case concerned the Kagiso and Munsieville 'people's courts'; it never came to trial).

20 Tom Lodge, 'The United Democratic Front: Leadership and Ideology', in J Brewer (ed), *Five Minutes to Midnight*, London, 1989, 220.

21 Boraine, 'Mamelodi', 131. He does not relate the outcome of the trial.

The Deregulation of Hawking: Rhetoric and Reality

Chris Rogerson

In a recent analysis Stephen Gelb and Darlene Miller argue that the question of deregulation has 'become one of the central themes in the debate about South Africa's economic future'.[1] Deregulation, the removal of certain rules and regulations governing various areas of the economy, was seized upon by the apartheid state as a key component of broader reform initiatives in the 1980s. The growing significance attached to deregulation derives from the state's implicit acceptance of the inevitability and desirability of future black urbanisation.

The government's 1986 adoption of 'orderly' or 'positive urbanisation' necessitated the introduction of certain economic reform measures (such as deregulation) designed 'to maximise the process for stability rather than attempt to undermine it'.[2]

Increasingly, strategies for deregulation and orderly urbanisation have been woven together in proposals for a development strategy of 'inward industrialisation', geared to generating economic growth through the production and supply of goods and services for a rapidly urbanising black population.[3]

Considerable controversy has been generated around issues of deregulation. Government, supporters of 'free markets' and many in business are lined up on one side, contending that deregulation can make a vital contribution to an expansion of the economy.[4] In particular, deregulation strategies linked to programmes for 'developing' the black small business and informal sector are viewed as a vehicle for reducing unemployment and deracialising the legal system. Proponents of deregulation go so far as to argue that it is the means 'whereby socio-economic apartheid can be dismantled' and fundamental rights restored to the black community.[5]

Set against these views are critics of deregulation, such as Cosatu and others in the labour movement, who perceive the programme as a direct assault on gains won by workers in past struggles. If the informal sector is to grow, unions suggest 'reconstruction' of the economy permitting the emergence of the 'real informal sector', namely the co-operative movement (*WM*, 15.12.88).

Cosatu sees economic deregulation as an attack on the immediate living standards of South African workers (*NN*, 23.12.87). But the threat deregulation strategies may pose to progressive political organisations through contributing to political stabilisation could pose a greater threat in the long term.[6] The apartheid state hopes that deregulation will facilitate the greater participation of black entrepreneurs in the economy, restoring their faith in 'capitalist structures' and giving them a stake in 'the system' they should defend.[7]

The debate on the implications of deregulation has been conducted so far largely with respect to the domain of manufacturing production. Yet, the greatest ramifications of deregulation will probably be felt outside of the industrial sphere in areas such as transport operations or informal trading. This article focuses on the rhetoric and reality which surround the contemporary deregulation of street trading or informal hawking in South African urban areas.

Hawking: volume, composition and characteristics

There are some 900 000 people engaged either on a permanent or casual basis in hawking or informal trading activities.[8] The quantitative significance of hawking activities is sharply exposed if compared to employment statistics for other economic sectors. The accuracy of official employment statistics for certain sectors is questionable, but the crude data disclose current recorded employment in manufacturing as 1,4-million, in agriculture 1,1-million, in mining 800 000 and in construction 600 000 persons.[9] When viewed against this synoptic picture of formal sector employment, the sheer volume of hawker activity places it ahead of construction and even employment in mining.

This static picture does not disclose the recent rapid escalation of informal vending operations which must be contrasted with the stagnation and demise of certain spheres of formal sector employment.[10] The position of informal hawking in South Africa thus appears to be following a similar trajectory to that in other African economies, namely one of rising significance amid a national economic climate of stagnation or decline.[11]

No data presently exist allowing for a finer breakdown on, for example, the composition, racial complexion or geographical distribution of the national hawker population. Nonetheless, a broad-brush portrait of informal trader communities can be gleaned from the findings of several studies conducted during the 1980s which investigated aspects of hawking in various parts of South Africa.[12] Most striking is the remarkable diversity within the sphere of hawking in relation to the manner in which it is organised, the nature of its participants and income-earning potential.

In terms of organisation, hawkers may be either independent sellers (especially prevalent in small-scale fruit and vegetable trading) or dependent/disguised wage-workers (common in clothing and newspaper selling) whose livelihood is inextricably linked to formal-sector retail or wholesaling concerns. Participation varies markedly in respect of different racial categories and gender composition. For example in Johannesburg the most lucrative hawking trades, such as flower-selling and fruit and vegetable hawking, have been dominated by members of the Indian community, with black hawkers relegated to the role of assistants.

A consistent picture emerges of male dominance in the higher income kinds of hawking (including clothing and the vending of live poultry), with women being confined to the lowest income echelons of hawker operations. In several kinds of hawking, notably the sale of live chickens, flower-selling and the vending of soft goods, income returns often can be quite substantial with many traders on the path to becoming petty capitalists. However, for the majority of South African hawkers - typically female, poorly educated and black - returns from hawking are meagre and merely a strategy for household survival.

In cities, returns above a bare minimum for household survival are made problematic by an apparatus of controls designed to contain, if not eliminate, street trading operations. In the perception of many South African local administrators, hawkers are a 'reminder of underdevelopment', 'aberrations' or 'social evils' to be purged.[13]

There appear to be four major sources of objection to the activities of hawkers in urban areas. Firstly, hawkers are victims of the historical importation of British and American planning traditions into South Africa and therefore in conflict with urban planners' notions of a 'tidy cityscape' or 'city beautiful'. Secondly, despite the fact that hawkers often have close ties to the formal sector, many of them fall prey to objections of 'unfair trading' practices from established business enterprise.

Health considerations are a third basis for discriminatory anti-hawker controls. But, as in settler-colonial societies elsewhere in Africa, it is the *perception* of health hazards more than the actual hazards themselves which conditioned the strengthening of negative hawker policies in South Africa.[14] Finally, the operations of hawkers have been prejudiced by the stigma of being in some way associated with a 'threat' to the existing political order. The most recent manifestation of authorities' fears about the informal sector generally is the paranoia stemming from the unfounded belief that hawkers are potential conduits for acts of urban terrorism.

Reform and deregulation

The changing reformist policy environment on deregulation of street hawking is an about-turn on long-established practices designed to suppress such activities. Ironically, 'activities previously seen as threatening to urban control are now presented as a means of resolving the urban crisis through creating employment'.[15] In an urban development study recently conducted for the Development Bank of Southern Africa, it was asserted that policies applied 'within individual urban centres will need to be fundamentally different to what they have been in the past'.

Moreover, it was admitted that the innovation or non-innovation of urban accommodationist policies 'can play a decisive role in determining whether poverty, unemployment or inequality will grow progressively worse, or whether the urban phenomenon can be turned into a vehicle of national development and personal betterment' in South Africa.[16]

The emphasis upon deregulation as a policy tool was expressed in the 1987 white paper on privatisation and deregulation which reiterated the imperative for 'encouraging entrepreneurship' and that 'the approach to regulation must therefore emphasise the promotion of economic activities and be less directed towards their control'.[17] A further call for deregulation was sounded by the Development Bank urban study which argued that 'in the immediate future, the very least that needs to be done is to lift restrictions on informal economic activities'.[18]

The question of hawker deregulation is inseparable from the

extent to which local authorities are willing to reshape the traditionally racially segregated form of South African urban areas. Of vital importance to hawkers is the opening up of the central areas of cities - the zones of greatest market potential - for increased trading opportunities by members of the poorest urban populations. A sharing of the cities and a freeing of trading opportunities for black hawkers necessitates, however, that local authorities suspend an impressive arsenal of controls which includes restrictions on licensing, geographical limitations on areas of permitted trade, requirements for constant mobility of hawkers (move-on regulations), requisite storage facilities and occasionally, dress regulations that demand the wearing of light-coloured coats.[19] Advocates of the deregulation programme aver that it does 'not necessarily mean the total repeal of all regulations': emphasis is rather placed upon attaining 'appropriate or minimum standards'.[20]

The uneven pattern of deregulation

During 1987 and 1988, a study to determine the state of local policy towards hawkers was undertaken in a cross-section of South African urban areas.[21] The most striking finding was the geographical unevenness of efforts to deregulate the sphere of hawking by local authorities.

Broadly speaking, two groups of urban centres may be recognised. There are those small number of authorities where a serious rethinking of the constraining effects of past planning has been pursued and evidence exists of steps being taken to permit new hawking opportunities for elements of the black urban and urbanising poor. In the second group are a larger number of centres where little change is in evidence from the historical pattern of repressing street trading activities. At best, within this second cluster of South African cities, it is possible to discern token steps towards hawker deregulation, albeit under circumstances of the maintenance of stringent official control.

In terms of creating a more favourable policy environment for hawkers, a somewhat unpromising picture of the state of 'reform' was disclosed. In only three centres - Johannesburg, Cape Town and Durban - was it possible to discern evidence of the introduction of significant measures for hawker deregulation and the opening of new economic niches and geographical space for the urban poor.

Hawker deregulation in Johannesburg and Durban since 1984 has seen a movement towards greater tolerance of street traders as an integral element of the urban retail environment. Momentum for deregulation in Johannesburg was given impetus by the formation in 1986 of the African Council of Hawkers and Informal Businesses (Achib), the leading representative association of South African hawkers. It was founded to represent and promote the interests of black street vendors, to seek relaxed licensing procedures, afford assistance with the provision of storage facilities and secure services for membership in terms of educational support, burial policies and life assurance cover. From its inception, Achib has been a significant actor, lobbying Johannesburg city authorities on behalf of black hawkers in a campaign to end the repression of hawking.

Progress of deregulation accompanied a decline in prosecutions of hawkers for minor offences as well as a considerable simplification in the bureaucratic morass which formerly surrounded hawker licence applications. The introduction in Durban of the so-called 'vending licence' has been hailed as something of a minor breakthrough, creating an 'appropriate' set of licensing requirements. As compared to the awesome provisions attached to obtaining hawker licences, the much-simplified procedures for a vending licence require little more from holders than avoidance of traffic obstruction and littering.

The whole approach adopted towards deregulation in Durban departs from the Johannesburg policy of 'controlled licensing' under carefully monitored, 'sanitised' conditions. It favours instead a drive towards the acceptance of a more loosely regulated and wider spectrum of hawker operations. Several 'immune zones' have been declared in the city where hawkers trade on a 'free-for-all' or 'first-come-first-served' basis, a situation markedly at variance with Johannesburg's approach of allocating particular stands or trading space to select individuals.

In Cape Town, while official attitudes and enforcement practices towards street traders have become more relaxed, the city still retains an archaic set of restrictive legislation including the mandatory wearing of 'clean white coats' by hawkers. Nonetheless, Cape Town city authorities are in the process of crystallising a new relaxed policy environment for hawker activities. Notwithstanding the statement that existing regulations represent 'an undesirable attempt to impose First World standards on what are primarily Third

World operations' the significant caveat is inserted that deregulation in Cape Town 'must not be pursued at the expense of public health and safety'.[22]

Even in these three modest urban 'spearheads' of reform, only small numbers of hawkers have been direct beneficiaries of policy change. In Johannesburg at the beginning of 1988 there were only 1 004 hawker licences, a figure little-changed from the total number of licences issued as far back as 1927 to a community of 'poor-white' hawkers. A watershed in hawker deregulation, however, appears to have been reached during 1988. By July, 7 000 hawker licences had been issued in Johannesburg, with a daily average of 50 new applications being processed. Together with the group of permit-holders who operate as agents of licensed vendors, roughly 14 000 persons were legally able to hawk in Johannesburg by mid-1988. A similar pattern of escalation in legal hawker activity also occurred in Durban and Cape Town.

Yet the fragility of these three authorities' new-found tolerance of street trading is readily apparent. During January 1988 mass raids in Johannesburg were conducted on street hawkers in consequence of 'unfair competition' complaints lodged by established business enterprise. And during August, Achib launched a civil disobedience campaign openly defying several 'unjust laws restricting hawking' such as a prohibition on trade within 100 metres of general dealers selling similar goods (*WM*, 26.08.89).

The geographical unevenness of hawker deregulation is creating problems of 'swamping' for those authorities attempting to accommodate the needs of informal traders. The changed dispensation concerning hawker licensing in Johannesburg, for example, as compared to the continuity of repression of black street traders throughout much of the Transvaal precipitated a growing stream of applications from beyond the city to hawk in Johannesburg.

The continuity of repression

Outside of the three largest urban centres of Johannesburg, Durban and Cape Town, a picture emerges of the persistence of a set of long-held 'colonial' and racist attitudes and prejudices concerning urban planning generally and informal trading in particular. The rhetoric of deregulating and promoting the informal sector, widely espoused at the national state level, has filtered down only partially, if at all, to the level of the local state throughout many parts of South Africa.

The emphasis on continuity and the maintenance of strict controls over hawker activities was evident both in the resistance of many urban authorities to 'reduced' standards for licensing and in defence of planning for the pristine, ordered city. Such attitudes were demonstrated particularly in politically conservative and/or white working-class urban centres, including Middelburg, Newcastle, Pretoria, Pietersburg, Stellenbosch, Vanderbijlpark, Vereeniging and Welkom. Nonetheless, in centres such as Bergville, Grahamstown, Nelspruit and Port Elizabeth certain steps, albeit small and tentative, have been made to improve conditions for trading by hawker communities.

A total reluctance to abandon anti-hawker practices was manifest, however, in situations where authorities either allowed only a token number of licensed traders within central urban space or alternatively geographically marginalised the community of permitted traders to unsatisfactory peripheral locales.

The repression of hawking was assured by a battery of legislative controls, requirements that licensees be 'itinerant', blanket prohibitions on black street traders and police raiding of unauthorised hawkers. Typically, Vanderbijlpark maintains such high licensing standards that no black hawker has ever managed to qualify for a licence to operate in areas which are ostensibly viewed as 'white' space. A parallel situation occurs in Newcastle, where only white and Indian hawkers are permitted in town, 'the big problem' of black hawking being held at bay by assiduous police actions.

In Welkom, municipal officials described black hawkers as 'vultures', and spoke proudly of the ways in which they had preserved the 'orderliness' of the city against the mess and clutter of hawkers. In like fashion, Stellenbosch municipal officials were concerned that licensed hawking be confined carefully to 'well brought up' and 'educated' coloureds or Indians. Further defence against black hawkers was furnished by regulations in Stellenbosch requiring a 'separate store room acceptable only if at least nine square metres in size, constructed of brick, plastered and ventilated'. Similarly, at Kimberley licensed hawker storage must have a 'dust proof ceiling' and at Uitenhage, stipulations require that hawkers, in addition to having 'approved storage', must wear distinctive 'arm bands' and 'registration badges'. Finally, at George, the former state president's home constituency, deregulation rhetoric notwithstanding, the town's restrictive licensing regime had experienced 'no change in the past decade'.

Underpinning this continuing repression of street trading is a legacy of past planning attitudes and prejudices. Concern for preserving the tidiness and orderliness of the city was manifest variously at Stellenbosch, Welkom and Kimberley. Also strongly apparent was the influence of the 'sanitation syndrome' and the equation of black hawkers with a potential public health hazard. This surfaced most openly in interviews with municipal officials at Vereeniging, where vigorous opposition was expressed towards deregulation. At Pietersburg, the blanket discouragement of black hawkers in the 'white' town was defended on grounds of a 'health hazard for whites' with officials averring: 'We (whites) are higher developed than they (blacks) are and can't take germs as they can'!

Health considerations also were high on the licensing agenda in Uitenhage where aspirant black hawkers were required to pass an X-ray examination. Finally, the prohibition and suppression of hawkers was rationalised occasionally in terms of the extraordinary 'logic' of apartheid. For example, Pietersburg officials stated that black hawkers should not be permitted to trade there because 'we don't allow whites to hawk in black areas' and 'Pietersburg is unique because there are no blacks in the area'.

In all these centres at best, black hawkers were allocated sites on the urban periphery (Nelspruit, Pietersburg) or a token number of central sites (Grahamstown, Kimberley) many of which were unsatisfactory from the perspective of hawkers. In some centres, the 'progress' of deregulation had advanced only to the point where, as in Stellenbosch, licensed hawker store-room requirements were 'relaxed' to allow 'smooth walls' instead of the former provision of two metres of tiling.

Even in South Africa's smaller metropolitan centres a bleak situation was revealed with respect to the plight of hawker communities. In East London until 1984 the authorities' intense anti-hawking sentiment went to the extraordinary length that a local by-law was proposed to make the act of purchasing from hawkers a punishable offence. Significant change in the direction of a more felicitous policy environment for hawkers occurred only in the wake of the effects of a crippling black consumer boycott launched against white stores in the city. Among the key demands of the consumer boycott committee was that East London authorities permit the operation of black hawker enterprise.

Since 1985 a package of limited concessions has been granted by the city fathers, permitting some 600 black hawkers to ply their wares around the city. Nonetheless, this concession was not without strings: East London hawkers may trade on a basis of 'walk and hawk' only, necessitating that, at all times, goods be carried either in hand or, more commonly, on the head. This restriction underscores the East London authorities' expressed intention to retain 'control' in the continuing struggle for the city.

The situation in the neighbouring coastal city of Port Elizabeth is no more promising from the viewpoint of struggling black hawkers. The city typifies the repression of hawking in South African cities as a result of objections to hawkers' 'unsightliness', 'unfair competition', 'threat to public hygiene' and 'obstruction to pedestrian and vehicular traffic'. Planning notions of a sanitised Western city underpinned the establishment of a strict licensing regime in Port Elizabeth whereby aspirant hawkers had to run a gauntlet of some 18 requirements laid down by the local health department. By 1985, however, the burden of economic recession and escalating levels of unemployment began to cause the rethinking of official policies towards hawkers.

Out of this reassessment emerged proposals to formalise and upgrade hawkers through providing 'acceptable' colourful carts which would be operated by 'respectable' individuals adorned in blue uniforms displaying the label 'CBD Hawker Association'. This scheme, however, accommodates a mere 22 black hawkers who currently are token beneficiaries of the limited deregulation and upgrading policies so far adopted in Port Elizabeth.

Finally, in Pretoria, despite a strong organisational presence of Achib, the city authorities have yielded little ground in struggles over trading space in the city. The local authority has retained a long-standing antipathy towards hawkers on the city streets. Notwithstanding the national state's 'reformist' initiatives, policies enacted by local state in Pretoria continue to be at odds with programmes to effect deregulation.

Hawker deregulation - rhetoric and reality

A fundamentally different type of planning for urban growth in

South Africa will necessitate casting aside several long-cherished attitudes which have been cornerstones of urban policy. The most important of these include the lingering belief in the impermanence of black settlement in the cities, the preservation of 'First World' standards in areas designated as 'white' urban space, and the racist assumption associating the presence of urban blacks with a threat to public health and order. The abandonment of such underpinnings of anti-urban planning, a 'decolonisation of the official mind', is a precondition for the enactment of positive strategies for assisting communities of hawkers.

Against the backdrop of a recessionary economic climate, the limited growth of formal sector employment opportunities and the current weakness of alternative means of black economic empowerment, for example through the building of co-operatives, such positive strategies are urgently needed by hawker communities throughout South Africa.[23]

A large residue still remains of the heritage of settler-colonial and racist attitudes which have traditionally conditioned urban policy and assured negative policy reactions towards the operations of informal traders. This results in the appearance of a substantial gulf between the rhetoric of ministerial speeches espousing the cause of 'deregulation' and 'assistance' for the informal sector, and the contemporary situation of local urban policy across South Africa.

The overriding picture is one of slow progress towards 'deregulation' with many local state officials preferring a situation of continuity rather than substantive change in hawker policy. This conclusion parallels Greenberg's interviews with state officials concerning attitudes towards the abandonment of influx control, namely that local officials held tightly to the core tenets in state ideology, retaining an enduring commitment to traditional anti-urban controls.[24]

Change in the context of hawker deregulation in South African urban areas is not, however, simply a consequence of actions taken 'from above'. The abandonment of outmoded anti-hawker policies may be precipitated by actions 'from below'. The event of greatest moment in terms of effecting change 'from below' is the recent formation of organised associations of hawkers. The consolidation and strengthening of organised hawker associations during the late 1980s is one of the most positive signs on the horizon of future

change. Yet, currently there does not exist one single national association to represent the interests of hawker communities
throughout South Africa. Research disclosed the operation of at
least ten small, uncoordinated hawker associations which function,
mostly in ineffectual fashion, on a local or regional level.

The establishment of a nationally co-ordinated association might
more effectively concentrate a set of pressures upon the imperative
for a 'decolonisation of the official mind' and the widespread implementation of a changed policy paradigm for hawkers.

Notes

1 S Gelb and D Miller, 'The spirit of small enterprise', *Work in Progress*, 56/57,
   1988.
2 M Morris and V Padayachee, 'State reform policy in South Africa',
   Transformation, 7, 1988, 7.
3 RL Lee, 'An economic analysis of the policy of "inward industrialisation"',
   Old Mutual Economic Monitor, January 1987; HS Geyer, 'On urbanisation in
   South Africa', *South African Journal of Economics*, 56, 1988.
4 See for example Small Business Development Corporation, *Annual Report*,
   1988, Johannesburg, 1988; N Green and R Lascaris, *Third World Destiny*,
   Cape Town, 1988; and T Rudman, *The Third World: South Africa's Hidden
   Wealth*, Cape Town, 1988.
5 L Tager, *Deregulation in the South African Context: the Dismantling of
   Socio-Economic Apartheid*, Johannesburg, Trust Bank for the Law Review
   Project, 1988, 2.
6 Gelb and Miller, 'The spirit of small enterprise', 70.
7 See D Booth, 'Job creation and political unrest in South Africa', *TransAfrica
   Forum*, 5(4), 1988.
8 The estimate derives from M Kirsten, 'A quantitative perspective on the
   informal sector in South Africa', *Development Southern Africa*, 5, 1988; and L
   Mavundla, 'Hawkers - an integral part of the economy', in *Deregulation - The
   City Councils and the Private Sector: First Conference of the African Council of
   Hawkers and Informal Businesses*, Johannesburg, 1988.
9 Republic of South Africa, *South African Labour Statistics*, Pretoria, Central
   Statistical Office, 1987.
10 See CM Rogerson, 'Recession and the informal sector in South Africa',
   Development Southern Africa, 5, 1988.
11 See RE Stren, 'The ruralisation of African cities: learning to live with
   poverty', in RE Stren and C Letemendia, *Coping With Rapid Urban Growth
   in Africa*, Montreal, 1986.
12 See P Wellings and M Sutcliffe, '"Developing" the urban informal sector in
   South Africa: the reformist paradigm and its fallacies', *Development and
   Change*, 15, 1984; R Tomaselli, 'On the peripheries of the defended space:
   hawkers in Johannesburg', in R Haines and G Buijs (eds), *The Struggle for*

Social and Economic Space: Urbanization in Twentieth Century South Africa, Durban, 1985; KSO Beavon and CM Rogerson, 'Aspects of hawking in the Johannesburg central business district', in CS Yadav (ed), *Slums, Urban Decline and Revitalisation*, New Delhi, 1987; and N Nattrass, 'Street trading in Transkei - a struggle against poverty, persecution and prosecution', *World Development*, 15, 1987.

13 CM Rogerson, 'The underdevelopment of the informal sector: street hawking in Johannesburg', *Urban Geography*, 9, 1988.

14 AD King, 'Exporting "planning": the colonial and neo-colonial Experience', *Urbanism Past and Present*, 5, 1977-78.

15 D Hindson, *Pass Laws and the Urban African Proletariat in South Africa*, Johannesburg, 1987, 92.

16 AM Thompson and SF Coetzee, *Urbanisation and Urban Growth in the SATBVC States*, Development Bank of Southern Africa Research Report 7, 1987, 137.

17 Republic of South Africa, *White Paper on Privatisation and Deregulation in the Republic of South Africa*, Pretoria, Government Printer, 1987, 13.

18 Thompson and Coetzee, *Urbanisation and Urban Growth*, 137-138.

19 See Small Business Development Corporation, *Street Vendors: A New Proposal for Adoption by Local, Provincial and National Authorities on Hawkers and Other Traders in Public Places*, Johannesburg, 1988.

20 Tager, *Deregulation in the South African Context*, 3.

21 These findings are discussed fully in CM Rogerson and DM Hart, *Hawkers in Urban South Africa: Contemporary Planning and Policy*, Johannesburg, Report to the Friedrich-Ebert Stiftung, Bonn, 1988.

22 City of Cape Town, *Street Trading: Policy Plan*, Cape Town, City Planner's Department, Technical Management Services, 1987, 6 and 2.

23 For a review on the current state of co-operative development, see G Jaffee (ed), *Building Worker Co-operatives in South Africa*, Johannesburg, 1988.

24 SB Greenberg, *Legitimating the Illegitimate*, Berkeley, 1987.

Privatisation and the Educational Politics of the New Right

Peter Kallaway

The 1980s have seen a fundamental shift in the debate about educational change in South Africa. Most critiques of apartheid education previously involved moral arguments emphasising the absence of conditions necessary for human rights and racial equality in education. Bantu Education was held to deprive pupils of fundamental educational rights because it aimed to prepare black children for a role as inferior citizens and workers, and indoctrinate them with 'christian nationalism' in keeping with the ideological programme of the National Party.

This critique of apartheid education argued that equal provision should be made for all citizens regardless of colour, in keeping with the aims of education set out in the UN Declaration of Human Rights.

The issue of the economic dysfunctionality of such policies was always vaguely present in the general critique, but it was not until the late 1960s that it began to assume a systematic character.

In keeping with the international trend towards manpower planning in educational policy, the focus of the critique of Bantu Education now fell upon the links between the lack of educational provision and the needs of industry in a period of rapid economic growth. Bantu Education was criticised because it failed to take account of the growing need for educated youth in the industrial labour market.

The early emphasis can to some extent be explained by the fact that debate on educational reform was entirely in the hands of educationists, churchmen and liberal and civic rights representatives between the 1940s and 1960s. It was South Africa's deviation from the policies of the UN Declaration of Human Rights on grounds of *racial* inequality in education that caught the imagination and attention of critics.

The continuities with and similarities to the totalitarian school-ing systems of Nazi Germany and fascist Italy were the dominant concern of liberal anti-apartheid lobbyists in South Africa. The in-terest of economists and establishment politicians only tended to be drawn to the area once the educational problem was seen as relating to the productivity of a suitably-trained workforce for industry.

Both trends have remained significant in the debate around educational reform since 1976, though educational policy is still popularly criticised more for its ethical/political implications than its economic ones. The people's education debate, for example, is resonant with the terminology of equality and democracy in the sphere of education as well as notions of 'conscientisation' and per-sonal liberation through education - following a path first trodden by liberals.

The main thrust of this policy is towards improving the quality of education offered, although the nature of the curriculum appears to remain conventional even if the content is Africanised. There is lit-tle attention to the economic functions of education beyond a bland assumption that all apartheid education - increasingly defined by radical organisations as capitalist education - is 'for domination'.

The basis of people's education appears to emanate from the Freedom Charter statement that 'the doors of culture and learning shall be opened'. In the era of 'reform' during the 1980s, the state it-self brought the rhetoric of equality into education. The state's desire to link educational planning more directly with the needs of production, and allow more scope for the private sector in the plan-ning process, is significant. Both the call for vocational education and the privatisation of education have been key themes in the reform debate.

Although this approach is not new to educational reformers in South Africa, the current tendency is important because it repre-sents an international trend in policy-making which is transforming educational systems worldwide.

The reform rhetoric of the privatisation lobby, whether in the context of the reformist politics of the De Lange Committee inves-tigation and the subsequent Education Commission, or in the con-text of private sector education reform lobbies (Syncom, the Free Market Foundation and the Anglo American 'think tank', for example), invokes an appeal to human rights and individual freedoms conventionally associated with the UN Declaration of Human Rights. But many reformers and radical educationists have not fully understood the shift in meaning which has occurred around these concepts during the last decade. It is an important shift because of the political implications of private sector education reform initiatives.

Whether the private sector's increasing involvement in educational reform will simply promote the old demands for an extension of formal/liberal education to all, or whether those demands will be superseded by other demands for an educational dispensation to suit the needs of modern capitalism, remains to be seen.

The rise of the new right

The emergence of the politics of the new right in Britain and the US, associated with the rise of Thatcherism and Reaganism, has frequently been represented as the politics of crisis, manifested in extremely high levels of unemployment, unprecedented levels of inflation, economic recession, a marked decline in balance of payments deficits, and an alarming growth in the public debt of most Western countries.

But these negative economic indicators are not *causes* but *symptoms* of the crisis. The actual cause of the crisis is the decline in the profitability of the largest industrial and commercial companies of the capitalist world, which have triggered the present crisis to recreate conditions for profits to rise again.

International capitalism has implemented new policies which amount to a stepping up of class warfare against the working classes and popular masses of the Third World. The primary aim of these policies has been to reduce the working class's level of individual and collective consumption by cutting back the percentage of profits allocated to wages. Public expenditure allocations for social welfare and education have also been slashed and the trend has been to privatise these services. Such policies, implemented in Britain, West Germany, France, Spain and the US, present an attack on the service side of the welfare state. They involve a monetarist economic policy that lays stress on market forces rather than social service.

This policy shift away from the welfare state reflects a move from Keynesian economics to monetarism and supply-side economics (ie competitiveness, anti-inflationary policy and growth at all costs, as against a concern for full employment and an equitable welfare distribution).

As Duncan Innes puts it, 'the "privatisation revolution" is in essence the spearhead of a major international campaign to recapture for the capitalist class many of the resources won over generations of struggle by working people'.[1] This has led to an 'implicit and explicit assault on the rights of citizenship' or a redefinition of the terms of citizenship itself, in particular in relation to the rights of workers, women and minorities, especially 'blacks' or 'immigrants'. What was a public commitment to the right to work under the Keynesian welfare state has been transformed under Thatcherite monetarism into the untrammelled right of citizens to sell their labour as individuals in the market place, and the expanded freedom of employers to determine terms and conditions of work. This is known as *deregulation* in South Africa.

It is essential to see questions of social policy, welfare and education as part of the wider programme of the new right to restructure society in terms of its model regarding what is appropriate to the 21st century.

These formulations have had considerable global impact during the 1980s in the form of the new international economic order - and there is some indication that they are being promoted in communist economies like the USSR and China as well as in the so-called 'free' economies of the West and the Third World.

It is not surprising, therefore, that the South African government has looked to its major trading partners and political allies, the UK and the US, for guidelines on how to adapt the political agendas of Thatcher and Reagan to the context of reforming apartheid. Much of the ethos of the privatisation revolution has also been taken up in the South African context during the latter half of the 1980s by key private sector elements.

Reform strategies in South Africa during this period have been strongly influenced by global policies. Big business and government are increasingly following the lead of the new right in the advanced capitalist nations with regard to the adoption of monetarist policies, the desire to restructure industry, production and labour relations. There is also an emphasis on reform with regard to social and educational policies as well as privatisation, vocationalism and racial reform of a particular kind.

Liberal democracy has come to be widely identified with the

new right in recent years, in the sense that those who support it are by definition opposed to social democracy and socialism. When liberal democrats advocate the rights of the individual, equality of opportunity, and the virtues of the free market, they want to claim that they do so in the name of a 'better future', 'rights for all in a democratic society', and 'the liberation of the individual'. But this appeal has to be understood in the context of the evolving synthesis between liberal-democratic ideology and the politics of the new right, which is in fundamental conflict with the values and program-mes of social democracy as expressed in the idea of the 'welfare state' or in documents such as the Freedom Charter.

These are complex issues which cannot simply be ignored or wished away. They represent aspects of a fundamental legacy which any future government will need to deal with. These initiatives need to be recognised for what they are: not just the manoeuverings of a dying racist regime, but structural aspects of the global ideology of capitalism at the end of the 20th century.

Privatisation, deregulation and reform

As Duncan Innes has pointed out, the reform rhetoric of the Freedom Charter (1955) and the Business Charter/Project Free Enterprise (1986) find common ground in their opposition to apart-heid, providing a platform for the negotiations at Lusaka (1987) and Dakar (1988).[2] Yet they differ fundamentally in their respective ad-herence to social democracy and free market capitalism.

The Freedom Charter reflects many aspects of a socialist or wel-fare state philosophy, emphasising the need for state intervention in the economy in the interests of equality, a redistribution of wealth, and secure social policy and educational provision for all. The Busi-ness Charter, on the other hand, is essentially about the virtues of individual ownership, private entrepreneurship, and the privatisa-tion and deregulation of the free market economy.

The government's white paper on privatisation and deregula-tion[3] marked the most dramatic policy shift away from government control of the reform process, to an acceptance of the general as-sumption of the new right in politics and economics. Locating South Africa firmly within the world economy, the report committed the state to a policy which amounted to 'the systematic transfer of ap-propriate functions, activities and property from the public to the private sector', 'the reduction of public sector involvement in the economy' and 'a minimum of state intervention and regulation'.[4]

The purpose was to ensure 'effective use of production factors', 'optimal functioning of market forces' and 'increasing the percentage of net fixed investment in the private sector'.[5]

The selling off of parastatals, the removal of government regulations on the private sector, and the privatisation of key state functions and services (such as housing, health, transport, social security) are all held to be essential to the revitalisation of the economy, which is said to be a pre-condition for 'constructive change' away from apartheid.

Innes points out that these arguments for the withdrawal of government from the economy and the deregulation of the market have brought strong support from big business and the dominant sections of the media, and they are widely held to make sound 'common sense'. Perhaps of equal significance is the fact that they have drawn little substantial criticism from popular political organisations or 'left' academics.

Yet the benefits of such policies are illusory for the majority of working people. Levels of monopoly, concentration and centralisation of capital within the South African economy are amongst the highest in the world. The domination of a few very large companies which have the resources, capital and technological capability to run such complex operations (Sasol, Sats, Iscor, Eskom, Foscor etc) means the only sector capable of buying up what the state divests is big business and the multinational corporations. Thus monopoly capitalism will benefit most from the new system because it is the only sector that commands the resources to buy up the lion's share of the industrial and service sector divested by the state.

Privatisation allegedly provides a way out of the inequalities and rigidities of race discrimination under apartheid, and in theory allows everyone (regardless of race) full participation in the free enterprise economy. Yet seen in the light of global experience, it fails to address the issue of political, social and economic equality in politically realistic terms.

Since the majority of blacks are poor or working class and do not have the capital/entrepreneurship/education to enable them to benefit from the sale of state assets to the public, privatisation offers no direct solution to demands for equality or the redistribution of power and wealth. Evidence shows that 'while privatisation may help to resolve aspects of the crisis for the private sector in South Africa, it will do so in a way necessarily detrimental to the interests of working people and the poor in general',[6] thus exacerbating political and economic conflict rather than defusing it.

In the field of social policy the envisaged reforms will undoubtedly make for greater efficiency, more expertise and the allocation of greater resources to those who are able to choose to use *private* transport, health care, welfare services and education - ie those who are wealthy enough to be able to buy these services in the market place. But evidence from Britain and the US indicates that such policies have the overall effect of running down the public provision of welfare and education services for the majority of people. What is efficient, productive and enhancing of individual choice is by the same definition detrimental to the public good in the area of social services and education.

Within liberal-democratic societies such divisions and inequalities reinforce traditional class barriers. In Britain and the US they have been shown to have considerable potential for social divisiveness in economies already stressed by recession and unemployment. In South Africa, class boundaries are reinforced by the politics and economics of exclusion in an historically racist state. There is hence little chance of convincing the mass of disenfranchised people that such changes would be to their benefit - even if a few select blacks are conspicuously incorporated into the 'new middle class'.

New right social policies are acutely divisive in the context of the industrialised nations where the legitimacy and hegemony of the capitalist order are strongly entrenched. In South Africa, where the legitimacy of the state, and increasingly of capitalism itself, is being challenged, and where no social welfare net exists to 'catch' the victims of recession and retrenchments which inevitably follow the introduction of the policy of privatisation, these policies offer very little hope for peaceful reform in the long term.

Privatisation and education

In South Africa there has been a massive resurgence of private sector education during the 1980s. The total subsidy to South Africa's 330 private and 23 state-aided or subsidised schools during the 1986-87 financial year was R11,2-million. In 1983 approximately 6%

of all white children attended private schools.[7] By 1987, 7 608 black students attended private schools.[8] Thus it appears that the National Party's previously hostile stance to private education has been blunted over the last two decades.

During the 1950s the conflict between state control and private education came to a head over the issue of church control of mission education for black South Africans. From the 1970s the state has systematically backed down on the issue of non-racialism in private education - especially in the face of the outright defiance of the 'open' Catholic schools which had admitted 2 500 black students out of a total intake of 25 000 by 1984.[9]

By the 1980s a significant number of new initiatives had emerged, modelled on Waterford in Swaziland and Maru la Pula in Botswana, which focused upon promoting progressive education within a non-racial atmosphere. With considerable support from progressive mining houses and multinational companies, reputable private school educationists also established a chain of 'new' private schools - St Barnabas, Woodmead, and Sacred Heart College, and the New Era School Trust (Nest) schools at Uthongathi (Natal) and near Johannesburg, for example.

Most of these acquired new campuses, where a strong attempt was made to get away from the conventional image of 'English' private schools while retaining the benefits of a caring and progressive educational background. Many also attempted to ensure that the curriculum reflected a less conventional perspective, incorporating a significant 'Africanist' bias while submitting to the conventional requirements of the Senior Certificate or JMB examination system. These schools are intended to provide a model of non-racial co-operation and curriculum innovation for the future society they claim to herald.

The state's contribution to this process, based on the De Lange proposals, was to recognise that subsidised private schools should have a large measure of freedom of choice and association within the bounds of broad national policy. It was recommended that they should have access to the support services of the second tier of government in the new constitution.

A policy of registration and subsidisation of private schools which came into operation in 1986 meant that private schools could open their doors to all races without government intervention while simultaneously receiving state financial support.

Private sector pressure was very significant in bringing about these changes. The establishment of the Industrial Fund for Assistance to Private Schools in 1959 was the first significant move by the private sector to support school education. At that stage the emphasis was upon the need to provide an alternative avenue of education for middle-class white English speakers who wished to avoid the CNE influence in state schools.

From the 1970s the pressure shifted to accommodate children of the 'new middle class' of blacks in schools outside of the township and away from the disruptions of boycotts and stayaways. In addition there was a need for an increasing number of blacks in managerial and professional positions in industry, and these initiatives helped to provide the source of such a supply.

The state did not play anything like the same role in promoting private education in South Africa as it did in Britain, but from the time of the De Lange investigation it gave increasing, if hesitant, support to private education in the name of the principles of free market choice and individual freedom - the basis of the report's philosophy.

This expansion of assistance to private schools was paralleled by the massive expansion of private 'cram colleges' and private sector 'in-house' educational and training programmes for employees - often aimed at upgrading existing formal qualifications or helping employees to obtain a senior certificate. These developments reflected many of the recommendations and initiatives of the Manpower Services Commission in Britain.

Private education and training of whatever kind, although demonstrating considerable growth during the 1980s, only catered for a tiny proportion of the school-going population in South Africa by the end of the decade, and the majority of the students who did attend were drawn from professional middle-class white families, despite sincere attempts on the part of the schools to create the impression that they are not elitist. Nest schools, for example, offer bursaries to ensure that there is an intake of black and white students, that they are representative of a broad class spectrum and that there is representation from rural as well as urban communities.

Yet there are very limited numbers of lower middle-class or working-class children at such schools. This sector is so small that it can hardly be argued that the new policy of acceptance by the government is aimed at saving the taxpayers money. It is quite simply the result of political and social pressure from a source which the government can no longer neglect, and a policy which is an inevitable consequence of the acceptance of increasing private sector influence in the field of social policy.

As the quality of state educational provision is affected by cutbacks necessitated by the redistribution of educational resources away from white communities, so the popularity of these schools might increase amongst those who can afford them. Faced by the prospects of sharp declines in funding from the state and the resultant cost of white schooling, many parents who were previously satisfied to send their children to government schools may be tempted to shift their support to the private school sector. For these reasons there has even been talk recently of a major Afrikaans private school initiative.

Privatisation and local control of education

Critiques of the 1981 De Lange report and the 1983 government white paper introduced a wide-ranging debate on the links between private sector engagement in education and its relation to reformist state policy. In South Africa it was not simply the nature of the relationship between state education and capitalism that was held up for critical analysis: the critique of De Lange was also aimed at the private sector's 'liberal' argument which 'fostered the image that the free enterprise system is the enemy of racial ascription'.[10]

Whereas the private school initiative originally related to educating small numbers of upper-middle class English-speaking children only, the thrust of private sector involvement in education changed significantly from the time of the Education Panel[11] reports which addressed the urgent need for a national educational system to suit the broader needs of industry.

The argument here was not that apartheid education or Bantu Education was problematic because it was education for political subservience on the grounds of race. Instead, attention was focused on apartheid education's failure to address the manpower needs of the industrial economy. The private sector thus needed to influence the authorities to meet those needs according to the principles of self-interest and economic growth.

But as Hartshorne points out, 'the realisation that South Africa could survive economically only by making full use of black skills in commerce and industry was slow in coming'.[12] It took a full decade of the emerging policy of *ad hoc* interventions (in the form of upgrading, skilling and technical training programmes) before the private sector began recognising that only by influencing the government to provide an adequate basic education for the whole population would the rest of these projects bear fruit.

Under these conditions 'the symbiotic relationship between the state and corporate capital has become increasingly apparent'. Since 1983 the extent of change has been considerable, so that 'the educational initiatives of the latter have come to complement those of the former'.[13]

Ever since the early 1970s employer organisations such as Assocom, FCI, AHI, and Seifsa have brought increasing pressure on the state to take a specific hand in developing a systematic programme for vocational education. This has led to an increase in the state's provision of such facilities with the financial assistance of the private sector. These include technical orientation centres, secondary technical schools, technical colleges and technikons, and comprehensive schools offering a variety of study directions including commercial and technical ones, or the direct provision of such facilities by the companies concerned with the aid of substantial tax concessions.

It was the De Lange report which provided the basis for this policy, drawing heavily upon the Manpower Services Commission experience in Britain. Despite the apparent rejection of many of the De Lange principles in the 1983 white paper, this aspect was rapidly converted into concrete policy. Even prior to the official acceptance of privatisation as a central government policy and the establishment of bureaucratic structures to implement such policies, the foundations for such an initiative were in place.

One of the major changes in the nature of private sector engagement in education since 1976 has been the increasing acceptance that attention to the direct instrumental needs of industry is an inadequate and short-term solution to the enormous problems facing employers and state. South Africa in the 1980s is realising that interventions which reflect too narrow a focus on 'skills shortages' or the need for technical education, only provide a short-term solution to problems of supply in industry.

There has been an increasing realisation that the entire

elementary education infrastructure requires upgrading if the minimum demands of the mass of the people and employers are to be met. 'There is now a much more overt initiative by the private sector to influence national educational policy and to be prepared to negotiate and exert pressure on government to achieve fundamental educational reforms'.[14]

It is important to recognise that South Africa is already well into a scenario aimed at a massive redistribution of social and educational resources in favour of the mass of the population. The era of free or virtually free, privileged state education for white children is rapidly drawing to a close. In terms of the new political dispensation being implemented through the tricameral parliament and privatisation, local communities will in future need to look after their own welfare in fields such as education, while firm control of the curriculum and the certification system will remain in the hands of the central government.

The recognition that massive resources have to be directed towards the basic education of Africans implies that resources might have to be drawn away from the public sector in general, and from white education in particular. This has provoked a backlash from conservative whites who do not have the funds to contribute to keeping up the standards of their local schools. The only communities which will be able to preserve their schools in their present form, or draw increased funding over the years for expansion and development, will be those in the middle or upper-middle class areas, and which have access to funders and the private sector.

The newly-acquired 'freedom' for the average township school will be spurious because there will be no easy alternative to the provision of state funds for working-class children. Hartshorne assumes that such changes will be progressive, and in so doing perpetuates the long tradition which assumes there is a benign and benevolent face to the private sector's relationship with popular education. That assumption is doubtful, especially in the light of the evidence regarding the relation between the global trajectory of new-right policy and its influence in the sphere of education.

If the policies and 'scenarios' of the ideologues of privatisation are seen in tandem with the De Lange proposals, and taken seriously as 'the voice of the private sector', the prognosis for the educational future of the majority of South Africans is indeed bleak.

To suggest that 'one of the few ways forward' as we attempt to grapple with the disastrous legacy of apartheid education lies in co-operation between 'the private sector' and 'the community', as Hartshorne does, is to neglect the fundamental conflict between the goals of each. The former seeks to implement a set of policies which will emphasise the goals of liberal democracy - freedom of the individual, the virtues of the free market, the needs of growth and productivity and the provision of education and training, a functional approach to the funding of institutes of learning, and the state's withdrawal from the provision of social policy and education.

On the other hand, the contrary demand, usually expressed more vaguely, and linked to the call for 'people's education', emphasises policies of social democracy characteristic of the welfare state and the Freedom Charter.

When these policies are seen against the background outlined above, they present a considerable array of intellectual, moral and strategic problems regarding the desirable trajectory of educational policy.

Some of the doubtful assumptions for the future have been set out here, and suggest that much greater critical attention needs to be paid to the role of the private sector in educational reform.

Notes

1 Duncan Innes, 'Privatisation: The Solution?', in Glenn Moss and Ingrid Obery (eds), *South African Review 4*, Johannesburg, 1987, 555.
2 Innes, 'Privatisation', 551.
3 Republic of South Africa, White Paper on *Privatisation and Deregulation in the Republic of South Africa*, Pretoria, 1987, 8.
4 Republic of South Africa, *Privatisation and Deregulation*, 8.
5 Republic of South Africa, *Privatisation and Deregulation*, 8-9.
6 Innes, 'Privatisation', 556.
7 AL Behr, *New Perspectives in South African Education*, Durban, 1984, 81.
8 Report in the *Cape Times*, 12.06.87, cited by Monica Bot, *Indicator*, 5(1), Spring 1987, 2.
9 Bot, *Indicator*, 5(1), 1987, 2.
10 J Davies, 'United States Foreign Policy and the Education of Black South Africans', *Comparative Education Review*, 29(2), 1985.
11 Education panel, *First Report: Education in South Africa*, Johannesburg, 1963; Ken Hartshorne, 'Private Sector Involvement in South Africa, 1960-86', in Peter Randall (ed), *Addressing Educational Crisis and Change: Relevant Private Sector Initiatives*, University of Witwatersrand, Johannesburg, 34.
12 Hartshorne, 'Private Sector Involvement', 34.
13 C de Wet, 'The Trend of the Anglo American Corporation's Initiatives in

Education in South Africa, 1955-87', unpublished B Ed course paper, University of Cape Town School of Education, 1988, 4.

14 Hartshorne, 'Private Sector Involvement', 47.

Section 2: The Frontline: Regional Policy in Southern Africa

Introduction

Glenn Moss

Six articles in this section of the *Review* graphically demonstrate the changing policies and balance of forces in the Southern African region. The period under review begins with South Africa treating the rest of the region as its back-yard, destabilising neighbours, mounting raids of both a conventional military and hit-and-run commando nature, and supporting surrogate forces in at least five other Southern African states. It ends with the formal decolonisation of Namibia, withdrawal of South African and Cuban forces from Angola, and regional peace initiatives jointly sponsored by the US and USSR.

This shift in the balance of forces has created a new regional conjuncture, in which changing relations between social forces and organised interests have altered the nature of regional politics.

Diana Cammack's carefully researched contribution on South African destabilisation of the region sets the background for these dramatic changes. Periodising South African destabilisation policies, she identifies six different phases, and in the process chronicles key events in Pretoria's bloody attempts to control and subjugate its neighbours.

The enormity of South Africa's war of destabilisation underlines the substantial nature of recent shifts which have limited Pretoria's capacity to dominate the region. Rob Davies and Thomas Ohlson argue that the military events associated with Cuito Cuanavale represent the turning point in regional relations: although South Africa did not suffer the massive military defeat that some would have us believe, Cuito Cuanavale revealed 'windows of vulnerability' in the SADF's military capacity. Ohlson provides a fascinating explanation of the technological background to these vulnerabilities, particularly South Africa's loss of air superiority in Angola. This was the single most important factor in South Africa's reassessment of military strategy in Angola and, ultimately, Namibia and the rest of the region.

But, as a number of the contributors to this section point out, military factors were only one element in the creation of a new regional context. South Africa's failing economy and growing need for external markets; white resistance to an increasing conscript death rate; international condemnation of the MNR's savage massacre of civilians and, by inference, South African support for the Mozambican 'bandit' movement; a changing international context in which the USSR and US were both prepared to sponsor peace initiatives, and put pressure on their allies to accept the compromises associated with restabilisation of the region; and Angola and Mozambique's desperate need to rebuild their war-torn economies, even at the expense of encouraging internal class differentiation associated with IMF and other Western development plans: these were other factors influencing the development of a new regional conjuncture.

Cammack rightly notes that, during the most recent phases of South African destabilisation, Pretoria has drawn distinctions between its neighbours. Swaziland - always more compliant to South African needs than Mozambique, for example - has not suffered anything like the civil war, famine, displacement and destruction associated with MNR activity. The necessity to draw distinctions in the analysis of South Africa's regional policy is as true for different time periods as it is for different countries. In his contribution, Bob Edgar analyses the changing nature of South African military and political relations with Lesotho after the 1986 coup which deposed Leabua Jonathan. South African precipitation of the coup, and subsequent support for the military government of Lekhanya, indicate policies vastly different to those pursued in regard to Mozambique, Angola and Zimbabwe.

As 1989 wound to a close, all eyes were on Namibia - Africa's last formal colony. In his contribution to this section of the *Review*, Brendan Seery documents the run-up to South Africa's acceptance of UN Security Council Resolution 435, and the arrival of Untag forces in Namibia. His analysis of those party to the internal Multi-

Party Conference and Transitional Government of National Unity demonstrates the sort of opposition Swapo is likely to face as a government. Indeed, the abortive attempt to infiltrate guerillas into Namibia on 1 April 1989, and the mounting pressure Swapo faces concerning detention and torture in its Angolan camps, suggest that Swapo may find governing and building national unity rather different from opposition and a national liberation struggle.

South African regional policy is changing. Not, as Davies rightly argues, because of the intentions of individuals, but because of the force of real events and processes. Events within South Africa itself are starting to mirror some of the dynamics associated with regional peace initiatives: the release of political prisoners and *de facto* unbanning of some restricted organisations indicate that the pressure on De Klerk's administration to start a negotiation process is almost as intense as it was on Botha to concede Namibian independence.

But South African policy in the region is still subject to a number of competing interests and factions. It would be naive to believe that the destabilisation policies associated with the state security council and factions within the SADF have disappeared just because South Africa has a new executive state president. Indeed, the institutional power base of De Klerk's new administration is unsure: while it does not rest on the state security council, national security management system and SADF in the way that PW Botha's government did, those institutional interests still wield enormous power. Reinstitution or continuation of destabilisation and armed intervention as a form of policy in Southern Africa cannot be ruled out at present, nor in the future. However, the context has changed, and it appears unlikely that South Africa will ever be able to treat the rest of Southern Africa in quite the same way as before.

South African Regional Policy Before and After Cuito Cuanavale

Robert Davies

'From now on the history of Africa will have to be written before and after Cuito Cuanavale'.[1]

The balance of forces in the Southern African region changed significantly during 1988. The SADF suffered a major setback at Cuito Cuanavale in Angola and more minor, but significant, reverses on other fronts of regional destabilisation. As a result of these and other factors, a new regional conjuncture began emerging in mid-1988.

The post-Cuito Cuanavale period followed a phase of escalating South African aggression and destabilisation against neighbouring states. In the period between January 1987 and the end of the first quarter of 1988

- the SADF became involved in probably its largest, and most costly, conventional military incursion into Angola;
- the SADF continued its covert war through MNR bandits in Mozambique and provided considerable support for a new offensive in the southern provinces of the country. Over 1 000 civilians were murdered between June 1987 and January 1988 in this assault, which included notorious massacres at Homoine (where 424 were killed on 18 July), Manjacaze (where 92 died on 10 August) and Taninga/Maluane (where 349 bus passengers were killed in two separate incidents during October and November 1987);
- MNR attacks were launched in eastern Zimbabwe with the apparent objective of opening up a new front of destabilisation in that country in addition to Matebeleland, where the South African-backed 'Super Zapu' continued to operate;
- the SADF conducted acknowledged raids against residences in Livingstone in Zambia on 25 April 1987, and the Botswana

capital, Gaborone, on 28 March 1988. There were also un-acknowledged attacks on residences in Gaborone (9 April 1987), Maputo (29 May 1987) and Lusaka (on numerous occasions) as well as attempted assassinations of ANC personnel in Swaziland (9 July 1987), Lesotho (March 1988), Maputo (7 April 1988) and Lusaka (on various occasions);

- on 2 January 1987, a 'suspected South African agent' threw a hand-grenade at Botswana President Masire during a visit to a residence at Ramotsha on the South African border, which had been attacked a few days earlier (*The Times*, 03.01.87);
- a few days before the SADF's April 1987 raid on Livingstone, President Kaunda of Zambia claimed that a coup plot had been uncovered, involving 'South African agents' and Zambian businessmen (*International Herald Tribune*, 16.04.87).

During February 1988, senior South African officials appeared to be signalling their intention to escalate this assault on the region even further. Speaking at the scene of a rocket attack near Messina on 19 February, Foreign Minister Pik Botha told the press, 'We will no longer urge or encourage (neighbouring states) to attend regional peace conferences. They can go their own way... The South African government has had enough. We reserve the right to act as we see fit' (*Rapport*, 21.02.88). Speaking at the same time and place, Defence Minister Magnus Malan said, 'Wherever the ANC is, we will eliminate it. If the Zimbabweans find themselves in between, I feel very sorry for them' (*Star*, 20.02.88).

Yet within a few months, this began changing. In the first week of May, South African negotiators travelled to London for the first of several rounds of talks on Angola and Namibia with officials from Angola, Cuba and the United States. These resulted in an agreement over the withdrawal of SADF troops from Angola (completed on 30 August 1988), followed by accords signed in Brazzaville and New York (on 13 and 22 December respectively) providing for Namibia to begin its transition to independence in accordance with United Nations Security Council Resolution 435 of 1978 on 1 April 1989. The two accords also provided for the withdrawal of Cuban forces from Angola in phases, to be completed by the end of June 1991.

A few weeks before the first round of Angola/Namibia negotiations in London, PW Botha received an envoy from President Chissano of Mozambique and agreed to initiate a process aimed at reviving the stalled Nkomati Accord.

The Joint Security Commission, established by the Nkomati Accord but suspended since September 1985 when evidence of continuing South African support for the MNR was discovered at Gorongosa, was reconstituted in July 1988. In September, South Africa, Mozambique and Portugal signed an agreement to act together to repair Cahora Bassa powerlines sabotaged by the MNR. In the same month, PW Botha paid official visits to Mozambique and Malawi and in the following month travelled to Zaire. Pretoria had now proclaimed as its ultimate goal of regional policy the convening of a multi-lateral regional security summit involving itself and neighbouring states.

The beginning of the second quarter of 1988 thus marked a turning point in the regional struggle. Pretoria's strategists appeared once again to be placing greater emphasis on diplomacy and economic action in the mix of 'incentives' and 'disincentives' through which they sought to advance regional policy objectives.

These developments raise a number of questions. What brought about this shift in Pretoria's stance towards its neighbours in the region? What was the real nature and extent of the change in the balance of forces which underlay it? What are the main objectives and instruments of its current policy? Is Pretoria really moving towards an abandonment of regional destabilisation? What implications do these changes have for the state's domestic strategy, and what potential implications could domestic developments have on Pretoria's regional policy?

In approaching these questions it is imperative to recognise at the outset that the decisive factors behind the shift did not operate at the level of the subjective intentions of decision-makers but at the level of the objective conditions within which decisions are made. Pretoria pulled back from the path of escalating regional aggression - along which its leading spokespersons and officials showed every intention of continuing - because setbacks were suffered in the region which made this course too costly in military, political, economic and diplomatic terms.

It is this which constitutes the distinctive element of the new regional conjuncture. Of course, Pretoria had suffered reverses in the region before. In previous phases it confronted situations where diplomatic and economic action failed to bring intended results. Nevertheless, in the past it could generally count on military action advancing at least short-term tactical, if not wider strategic, goals. What distinguishes the new, post-Cuito Cuanavale conjuncture is that Pretoria is now facing a situation in which it cannot always expect to get its own way through military action.

This has undoubtedly opened up new possibilities for regional states to make certain advances and gains. It also has some potential to affect the domestic balance of forces in South Africa in a way which enhances the position of the oppressed at the expense of the oppressors. At the same time, however, it is important to recognise that Pretoria has suffered setbacks and not decisive defeats. It still possesses a formidable military machine with considerable capacity to inflict damage on regional states thwarting its ambitions. South Africa also remains the dominant economic force in the sub-continent. These factors too have to be taken into account in assessing the limits and possibilities of the emerging regional conjuncture.

Escalating military aggression: 1987 to April 1988

The militaristic aggression and destabilisation which characterised 1987 and the first quarter of 1988 was a continuation of the cycle which began in mid-1985. This phase of escalating aggression and destabilisation followed Pretoria's failure to capitalise on the 1984 Nkomati Accord and Lusaka ceasefire with Angola.[2] Pretoria had hoped that these agreements, and particularly the Nkomati Accord, would spike the guns of the ANC and dampen the struggle inside South Africa; lead to a 'power-sharing' settlement with the MNR in Mozambique; and result in generalised regional acceptance of South Africa's dominance as the regional power and hence reduce Pretoria's international isolation.

By mid-1985, Pretoria had clearly failed to significantly advance these objectives. This was largely due to advances made by popular struggle inside South Africa which deepened the crisis of the apartheid system and state. In these circumstances Pretoria's strategists concluded that they had little to gain by projecting a severely-tarnished image of peaceful intentions towards its neighbours. They opted, instead, for an image of unassailable strength.

Destabilisation and aggression, never abandoned despite the signing of formal non-aggression treaties, escalated and became the main feature of South African regional policy. Important exceptions involved actual or potential allies - Swaziland, Malawi and post-coup Lesotho - where economic action and diplomacy featured prominently in attempts to build and strengthen ties.

The three main fronts of regional destabilisation continued to be the war in Angola, waged both through conventional military action in the south and support for Unita assaults throughout the country; the covert war in Mozambique conducted through MNR surrogate forces; and acknowledged or clandestine raids against anti-apartheid activists, particularly ANC members, in regional states.

There was also a lower-level destabilisation directed at Zimbabwe, where a new front in the east was opened up following the MNR's 'declaration of war' on the Mugabe government and the signing of an agreement between the MNR and Sithole's version of Zanu.

Although a more-or-less permanent SADF presence had been established in parts of southern Angola for years and several 'deep incursions' by SADF units were reported in early 1987, the decisive SADF involvement of the period began on 1 July 1987 with the launching of 'Operation Modular'. This was a response to the concerted attempt by Angolan forces to take the town of Mavinga in Cuando Cubango province.

'Operation Modular' was followed by 'Operation Hooper' in December 1987. The battles and troop movements which culminated in several abortive attempts to capture the town of Cuito Cuanavale are dealt with in Thomas Ohlson's contribution to this section. It is sufficient here to note that the SADF saw the presence of a large Angolan force at Cuito Cuanavale, some distance from its main supply lines, as an opportunity to inflict a decisive defeat on Angolan government (Fapla) forces in the south-east of the country.

At least 3 000 official SADF troops were involved, together with larger numbers of South West African Territory Force and Unita surrogate forces, in operations which saw the SADF deploying the most modern and sophisticated equipment available to it - modified Mirage aircraft, G-5 artillery weapons and, for the first time in Angola, Olifant tanks.

A *Work In Progress* correspondent has suggested that Pretoria's strategists hoped these operations would enable them to enlarge the SADF 'control zone' in southern Angola to embrace the whole of the area bordered by the Benguela railway line in the north and the Cubango river in the west. This area - more than a quarter of the country - would then have become a 'Savimbistan' administered by

Unita with South African support. This would have served as the basis for the proclamation of Unita as an 'alternative government' of Angola - a development which was confidently expected to lead to increased US involvement and support for Unita.[3]

On the Mozambican front the April 1987 MNR assault in the south followed the infliction of serious setbacks on the MNR bandits during the Frelimo government's offensive in the centre-north provinces at the end of 1986 and early 1987. The assault in the south involved groups of several hundred well-armed bandits infiltrating directly across the South African border. Their initial objective appeared to be the occupation of a corridor along Gaza and Inhambane provinces from the South African border to the sea, thereby thwarting the government offensive in the centre-north by cutting the country in two.

When this failed, they turned to attacks on civilian road and rail traffic to and from Maputo city and on inhabitants of small towns. The aim appeared twofold: to cut Maputo city from the rest of the country, and provoke an increased flow of displaced people into the city, overstretching its resources.

Although Pretoria repeatedly denied that it was involved in supporting this assault, evidence of continued SADF involvement with the MNR emerged from a number of sources. Several former bandits, either captured or surrendering under the government's amnesty programme, spoke of

- air and sea drops of supplies (see testimonies by two unnamed youths in *Noticias*, 22.09.87; by former MNR sector commander, Modesto Sixpence in *Noticias*, 10.02.88 and 14.05.88; by Fernando Tepo and Eduardo Manuel in *Noticias*, 23.03.88; by Paulo Oliveira in *Noticias*, 24.03.88 and by Ian Grey in *Noticias*, 28.03.88);
- SADF personnel operating with bandits in bases in Mozambique (see testimonies by Abilio Jangane in *Noticias*, 02.03.88 and by Luis Tomas in *Noticias*, 09.04.88);
- involvement in a special commando unit sent to plant bombs in Maputo and Matola (see testimony by South African passport holder George Alerson in *AIM dispatch*, 22.06.87).

Mozambican government communiques reported

- infiltrations of bandits across the South African border shortly before the Homoine massacre (*Guardian*, 15.06.87 and 22.07.87), the Manjacaze massacre (*Guardian*, 13.08.87; *AIM dispatch*, 18.08.87), and the Taninga/Maluane massacres (*Domingo*, 01.11.87);

- violations of Mozambican air space (*Noticias*, 16.11.87 and 24.11.87);
- operations conducted by units with specialist skills, and hit-and-run teams striking against railway bridges and powerlines and then returning to South Africa (*Noticias*, 17.10.87).

The remains of parachutes used to drop supplies to the bandits were also discovered in the vicinity of Homoine shortly before the massacre there (*AIM dispatch*, 29.06.87). Numerous observers pointed out that only South Africa had the motivation and capability to carry out air and sea drops to bandits in Mozambique.

As far as the raids and assassinations directed at ANC members and other opponents of apartheid are concerned, the period saw a change in the pattern 'from random slaughter to a sustained campaign; from individual covert hits to generalised policy'.[4] By the first quarter of 1988, this campaign had reached beyond the Southern African region with such incidents as the assassination of ANC representative Dulcie September in Paris and attempts on the life of Godfrey Motsepe, the ANC's representative in Brussels.

What continued to distinguish such attacks in Southern Africa, however, was the apparent indifference to casualties among innocent bystanders. Most victims in the incidents described above were not South African exiles but nationals of the countries concerned. This suggests that at least part of the motivation behind these attacks was to engender fears and insecurities in local communities where exiles were living.

The changing regional context

Pretoria's changed stance in the region was the product fundamentally, of reverses on all the main fronts of regional destabilisation described above. But the most decisive of these was that suffered by the SADF at Cuito Cuanavale. A Cuban publication[5] has aptly described Cuito Cuanavale as 'South Africa's Waterloo': it smashed the myth of SADF invincibility and revealed important windows of vulnerability.

The most immediately apparent set of vulnerabilities involved weapons and equipment. The South African Air Force found itself unable to penetrate Angolan and Cuban radar/missile defences in

Cuito Cuanavale. Its equipment was technologically inadequate, partly due to the effects of the arms embargo. Faced with modern Soviet equipment brought into the frontline by the Cubans, the SADF rapidly lost the air superiority it had previously counted on. In addition, the Angolan/Cuban air strike on the Calueque dam in June 1988 showed up important deficiencies in SADF air defences.

Cuito Cuanavale also revealed a major vulnerability in relation to personnel. A plan put forward to take the town by infantry invasion was rejected because it was estimated that it would cost the lives of 300 white conscripts. A loss of this magnitude in these circumstances was deemed unacceptable, thus highlighting a previously untested political constraint on the SADF.

In addition, Cuito Cuanavale pointed to economic vulnerabilities. It has been estimated that the war in Angola was already costing R4-billion a year by mid-1988 and any escalation would have put 'an intolerable burden...on an economy, already running out of steam' (*Sunday Star*, 04.07.88). An intensification of the conflict in southern Angola could have pushed Pretoria closer to the 'major fiscal crisis' some observers saw 'looming in 1989'.[6]

While Cuito Cuanavale represented the decisive setback which produced the new regional conjuncture, reverses were also suffered on the other main fronts of destabilisation. In the case of the bandit assault on Mozambique, government forces, supported by their Tanzanian and Zimbabwean allies, began achieving some military victories in late 1987 and early 1988. Although the overall security situation in the country remained far from satisfactory, the MNR was prevented from definitively cutting transport links to Maputo and isolating the capital.

By mid-1988, the security situation on main road and rail links to the capital had greatly improved and a new government offensive in the centre-north had succeeded in dislodging bandits from a number of bases they had occupied since 1986. In December 1987 the Mozambican government passed a law offering an amnesty to bandits surrendering to the authorities. By mid-1988, it had become evident that the military successes and the amnesty law had provoked considerable disarray within the ranks of MNR - both externally and on the ground in Mozambique.

By December 1988, around 3 000 former bandits had surrendered under the amnesty law; two former senior MNR officials, Paulo Oliveira and Chanjunja Chivaca João, had defected; Mateus

Lopes and João Ataide had been murdered in Malawi during December 1987 and former secretary-general, Evo Fernandes, was killed in Lisbon in April 1988; and clashes were reported between different bandit groups in Mozambique (*WM*, 08-14.07.88).

Pretoria's known involvement with the MNR also began threatening to increase its international isolation. The extreme brutality of MNR actions, highlighted by the well-publicised massacres at Homoine, Manjacaze and Taninga/Maluane, greatly undercut whatever international support the bandit assault might otherwise have mobilised. The technique of covert war was partly modelled on the 'Reagan doctrine' of supporting 'anti-communist' insurgencies against 'leftwing' governments. But even the US administration felt obliged to distance itself from the brutality of MNR assaults.

At an emergency aid donors' conference held in Maputo in April 1988, a US deputy assistant secretary of state, Roy Stacy, accused the MNR of 'one of the most brutal holocausts against ordinary human beings since World War Two' (*WM*, 05-12.05.88). A report commissioned by the US state department estimated that MNR had been responsible for the deaths of at least 100 000 civilians.[7] Written by Robert Gersony, the report was based on interviews with 200 informants in Mozambique and neighbouring countries.

Gersony's sources reported that they themselves had witnessed 600 murders - by shooting; knife, axe and bayonet killings; burning alive; beating to death; asphyxiation; and starvation and drowning.[8] Of those interviewed, 91% had 'very negative' and a further 5% 'somewhat negative' perceptions of the MNR, compared to 7% 'very negative' and 10% 'somewhat negative' perceptions of Frelimo government troops.[9]

While most official US and other Western public statements refrained from direct accusations against Pretoria, denunciations of armed banditry in Mozambique were accompanied by a less public diplomatic process in which South African officials were informed that continued involvement in such atrocities would result in increasing international isolation and could fuel pressure for sanctions (*Noticias*, 11.04.88; 12.05.88; *WM*, 05-12.05.88).

In hit-and-run raids, too, Pretoria suffered a number of setbacks. A network of suspected South African agents was uncovered after a bomb blast outside an alleged ANC residence in Bulawayo during

January 1988. Seventeen people were detained after this incident and three (Kevin Woods, Michael Smith and Phillip Conjwayo) later sentenced to death for murder. Two members of the SADF were convicted of assault and sentenced to ten years imprisonment and eight strokes of the cane after being captured during a bungled raid in Botswana in June 1988. Another alleged agent, Charles Dennis Beahan, was handed over by Botswana to Zimbabwe when he attempted to flee at about the same time.

An article in the *WM* (08-14.07.88) described the latter incidents as 'two minor SADF disasters which compounded the major one at Cuito Cuanavale'. The bungled raid in Botswana, in which a policeman was injured, reinforced calls in the US for South Africa to be declared a 'terrorist state' (*WM*, 24-30.06.88).

At the same time as these reverses, the stagnating economy, increased international aid to the region and the threat of sanctions began to strengthen the hand of those arguing that South African capital needed greater access to the region. This was, for example, the theme of a paper presented by the chairman of the Development Bank, Simon Brand, in July 1988.

Brand argued that a 'siege economy' response to economic stagnation, sanctions and disinvestment, involving the direction of large resources to capital-intensive, high-technology strategic investments, could impair South Africa's long-term economic development. Instead, he gave strong support to what has become known as the 'inward industrialisation' strategy, involving expanded production of manufactured consumer goods aimed at the 'middle and lower income markets'.

This pattern of industrialisation, argued Brand, would be 'the optimum strategy in terms of South Africa's relations with its neighbours'. It would provide a solid base for South Africa to expand its exports of manufactured commodities to Southern Africa. It would also probably lead to increased imports of raw materials from neighbouring countries and result in co-operation on infrastructural projects (*WM*, 15-21.07.88).

A number of corporations began formulating growth strategies based on greater access to the region. Eskom, for example, drew up plans for a regional electricity grid. This aims to enlarge the corporation's links with all SADCC member-states and Zaire, and eventually embrace all the 15 East and Southern African Preferential Trade Area members as well. It would directly challenge and compete with a number of SADCC projects aimed at reducing energy dependence on South Africa (*FM*, 29.04.88).

The view that greater involvement in the region might provide a more effective shield against sanctions than 'counter-sanctions' also began to gain ground.

By the end of the first quarter of 1988, it was clear that the military option in Angola had become extremely costly in military, political and economic terms. Pretoria's known involvement with MNR banditry in Mozambique was threatening to become costly diplomatically and ideologically. Raids against neighbouring states were going wrong. Agents and even official SADF members were being captured and this was threatening to increase South Africa's international isolation.

At the same time, objectives which seemed possible to advance through economic and diplomatic action began to assume increasing importance. A new situation was being created by changes in the regional balance of forces in which even those elements in the state most committed to escalating aggression and destabilisation were compelled to consider other options.

Pretoria's regional policy after Cuito Cuanavale

While the main factors precipitating Pretoria's apparent change of course in the region were setbacks, it is important to recognise that it did not suffer decisive defeats. As the chief Cuban negotiator, Jorge Risquet, has repeatedly pointed out, both sides in Angola could have continued fighting, albeit at an increasingly high cost in human and material terms.[10]

On other fronts too, impediments to escalating regional destabilisation were not overwhelming. It was not therefore a question of Pretoria simply surrendering, but of making some concessions while struggling to extract others on new terrains of regional struggle. The new regional conjuncture has indeed opened up possibilities for Pretoria to advance certain of its regional policy goals.

Western powers have considerable sympathy with many of Pretoria's long-standing economic and diplomatic regional ambitions. In addition, a number of the Frontline states are willing to trade economic and diplomatic openings to Pretoria in return for security guarantees. Thus, while the new context implies some losses for Pretoria - notably a withdrawal of the SADF from Angola;

independence for Namibia; and a scaling down of destabilisation against Mozambique and other Frontline states it has also opened up opportunities to advance other goals through economic and diplomatic action. Ideally, Pretoria's strategists would like to see the current phase resulting in

- a controlled or manipulated transition to independence in Namibia ensuring that an independent Namibia does not represent any serious challenge to South Africa's regional interests;
- South African retention of Walvis Bay;
- a non-aggression pact with an independent Namibia;
- a non-aggression agreement with Angola (included in the accord signed in New York on 22 December 1988);
- the exclusion of the ANC from Namibia, the closing of ANC camps in Angola (implicitly provided for in the 22 December accord), and the severe reduction, if not total elimination, of the ANC 'presence' elsewhere in the region;
- a negotiated settlement resulting in power sharing between the Angolan government and Unita, and possibly some form of negotiation between the Mozambican government and the MNR;
- greater access to regional markets for South African commodities, participation by South African capital in selected projects in the region, and the opening up of more 'facilities' in the region for sanctions busting;
- endorsement by regional states of the government's 'reform' programme;
- a reduction of international isolation based on acceptance of South Africa's 'constructive role' in the region.

These goals of current South African regional policy do not envisage only economic and diplomatic gains; there are also clear security objectives. These seek to accompany the abandonment of the 'forward defence' posture, implied by a withdrawal from Angola and Namibia, with a reinforcement and tightening of security demands in relation to other regional states.

All of this is wholly compatible with the approach defined in Pretoria's original strategy to build a 'Constellation of Southern African States' through promoting security and economic cooperation as a step towards political acceptance and even cooperation. What has changed is the rhetoric: the goal of establishing a constellation is no longer proclaimed as such, instead Pretoria's officials have spoken of their desire to convene a regional security conference with a number of regional states and of participating in the 'development' of the region in partnership with the West. This, plus the desire to see regional states endorsing its 'reformed apartheid' programme, aims at building legitimacy and gaining acceptance for South Africa as the 'regional power'.

However, while Pretoria has its regional policy ambitions, the Frontline states have their own agendas. In addition to the withdrawal of South Africa from Angola and independence for Namibia, they seek a generalised cessation of destabilisation to secure their independence and create conditions for economic development.

A number of Frontline states currently have policies that envisage conceding some economic and diplomatic openings available to South Africa with the aim of encouraging those forces favouring less bellicose options. But they seek simultaneously to ensure that such co-operation is on mutually beneficial terms. Moreover, the Frontline states have all reaffirmed their commitment to SADCC and hope that any reduction in the intensity of de-stabilisation will advance those SADCC programmes which aim at reducing dependence on South Africa.

The final outcome in terms of the balance of gains and losses remains to be seen. Whether in the end it is Pretoria or the independent regional states which gain most will be determined by the course of future struggles and the strategies and tactics followed by both sides. What seems certain is that the new phase of regional struggle will be both more complex and multifaceted than in the past, in the sense that it will take place on a number of terrains. It is also likely, at least on some of these terrains, to be characterised by compromises rather than clear-cut victories and defeats. It is unlikely in such a conjuncture that Pretoria will completely abandon destabilisation.

The present phase has been viewed by some observers as one of 'compromise' not only in terms of objectives but also in terms of conduct of regional policy (Shaun Johnson in *WM*,23-29.09.88). Perhaps, the most likely scenario is that there will be a partial pullback from destabilisation in some parts of the region. In these areas Pretoria will encourage involvement by South African capital with the objective of advancing strategic interests (including those related to sanctions busting) as well as promoting purely commercial gains. It will also attempt to use this involvement as a diplomatic lever to ward off sanctions and reduce international isolation. In other areas, however, it is likely that Pretoria will continue to be involved in destabilisation - although its involvement will probably be re-organised and concealed in new ways.

Domestic implications

Regional developments rarely occur in isolation from those on the domestic front of struggle, and the post-Cuito Cuanavale period is unlikely to be any exception.

Pretoria's rulers have expectations that the new regional context could reduce international pressure, including the threat of sanctions. They also hope that it might result in greater international and regional acceptance of the legitimacy of the regime's 'reformed apartheid' programme.

Redeployment of military resources inside South Africa could lead to improved 'border security'. Withdrawal from Angola is unlikely to lead to a reduction of defence spending as such. Most existing weapons development projects will still go ahead, with some given greater urgency by the vulnerabilities revealed in Angola. However, withdrawal from Angola and Namibia potentially releases resources which could be used to reinforce border control measures. A project to construct an electric fence along the Botswana border - costing R130 000 a kilometre - is being mooted and there are apparently plans to deploy more personnel in border areas. The SADF is said to believe that such measures could cut 'infiltrations' to 5% of their current levels (*FM*, 26.08.88).

Against this, the processes put in motion by Cuito Cuanavale could contribute to undermining domestic legitimacy of state policies further, and boost the morale of popular forces. The regime has largely failed in its attempts to portray the SADF as victors in Angola. It is widely recognised that Pretoria became involved in negotiations on south-western Africa because it suffered a military setback at Cuito Cuanavale.

As the process of implementing UN Security Council Resolution 435 in Namibia advances, the contrast between what is happening there and in South Africa can be expected to sharpen. In particular, when Namibia moves to elections under a one-person, one-vote system, the question why this cannot also apply in South

Africa will become more difficult to answer.

Finally, the spectacle of Swapo taking over government in Namibia might have some effect on the morale of the oppressed people of South Africa similar to when Frelimo took over in Mozambique.

In the short term, such developments could add to pressures for a reversion to militaristic options. If popular struggle becomes more intense and the regional conjuncture reinforces this struggle, the temptation to revert to intensified regional aggression could be enhanced. The post-Cuito Cuanavale period is likely to be fluid and evenvolatile, with an ever-present possibility of reversion to full-scale militaristic aggression. But critical windows of vulnerability in Pretoria's militarism have been exposed.

If Pretoria can be prevented from closing the existing windows of vulnerability; if the defensive capacity of the Frontline states is strengthened; and if sufficient international pressure can be maintained to make the costs of a reversion to full-scale destabilisation prohibitive, then current regional dynamics may influence events inside South Africa in ways strengthening anti-apartheid forces at the expense of the apartheid oppressors.

Notes

1 Fidel Castro, addressing the Non-Aligned Movement Ministerial Conference, Havana, May 1988.
2 Robert Davies, 'South African Regional Policy Post-Nkomati', in Glenn Moss and Ingrid Obery (eds), *South African Review 4*, Johannesburg, 1988.
3 Anon, 'Pretoria's newest homeland - 'Savimbistan'', *Work in Progress*, 54, June-July 1988.
4 Glenn Moss, 'Apartheid's Death Squads', *Work in Progress*, 53, April-May 1988, 25.
5 *Prisma*, 191, August 1988.
6 Howard Preece, 'Retreat from Angola would ease budget deficit', *Southern Africa Report*, 15.07.88.
7 R Gersony, 'Summary of Mozambican Refugee Accounts of Principally Conflict-Related Experience in Mozambique', report submitted to ambassador John Moore, Director, Bureau for Refugee Programs and Dr Chester A Crocker, Assistant Secretary of African Affairs by Robert Gersony, consultant to Bureau for Refugee Programs', Washington, Department of State, April 1988.
8 'Mozambican Refugee Accounts', 21.
9 'Mozambican Refugee Accounts', 23.
10 Anon, 'The Brazzaville deal-in-the-making', *Southscan*, 3(5), 28.09.88.

The Cuito Cuanavale Syndrome: Revealing SADF Vulnerabilities

Thomas Ohlson

By late 1988 it was clear that military realities in southern Angola had been the single most important factor forcing the South African government to the negotiating table in May. South Africa had also been forced to accept the implementation of UN Security Council Resolution 435 on Namibian independence.

During the last 25 years South Africa underwent a process of militarisation virtually unprecedented in the world. Political and military leaders in South African society are obsessed with national security, often narrowly defined as military security. The Botha government adopted the concept of a 'total onslaught', according to which all manifestations against apartheid - either inside or outside South Africa - were perceived as part of a Moscow-hatched conspiracy aimed at gaining control over Southern Africa's riches.

During the late 1970s and 1980s, Botha's government used this ideology to justify its militarisation and regional aggression along with its domestic oppression. The 'total strategy' adopted to counter this 'onslaught' aimed to keep the liberation struggle as far from South Africa as possible. The common goal for all representatives of that government - irrespective of which 'faction' they belonged to - was to remain in power, preserve white minority rule and safeguard South Africa's strategic and economic hegemony in Southern Africa.

Whenever a diplomatic process was seen as a potential threat to these objectives, Pretoria aborted the process, reneged on commitments, or resorted to violence. This provides reason for skepticism regarding the current regional peace process.

On the other hand, the overall strategic framework is heavily influenced by two new and powerful factors: first, there was a temporary, but important change in the balance of military power and, second, the United States and the Soviet Union have, for the first time, exerted fairly strong joint pressure for a negotiated solution.

Other possible factors - though not so easy to quantify - are the unhealthy state of the South African economy, and the significant changes in the balance of forces within the white minority.

Key events in the war with Angola

Although South Africa had troops in southern Angola earlier in 1987, heavy SADF involvement began in August/September of that year, when Angolan forces began a concerted effort to take the town of Mavinga in Cuando Cubango province. With Mavinga as a rear base, Angolan troops planned to attack the Unita headquarters in Jamba further to the south-east.

South African units, consisting of elements from the ethnically-composed 32 and 101 battalions, as well as regular white troops from the permanent force and units from the South West African Territory Force (SWATF), rushed to assist Unita. The Angolan advance was halted and the SADF launched a massive counterattack which forced Angolan troops to withdraw.

At Cuito Cuanavale the SADF/Unita counter-attack was stopped and a military deadlock ensued. Angolan troops dug themselves in, established firm defence lines and received increasing Cuban assistance, in particular air support. By January 1988, South Africa began to lose air superiority. An SADF analysis made then argued that it was possible to take Cuito, but that it would entail the loss of up to 300 white troops, along with some 2 000 SWATF and an unspecified, but large number of Unita troops.[1]

These substantial losses were deemed unavoidable by the SADF strategists, since the massive land assault required to take Cuito Cuanavale would not enjoy much air cover from the SAAF as a result of the introduction of advanced anti-aircraft missiles, radars and MiG-23/Su-22 fighters for the defence of the town.

The plan to take Cuito through an infantry assault was shelved. Instead Pretoria opted for a drawn-out artillery battle in which the SADF began to shell Cuito with G-5 and G-6 howitzers, Valkiri multiple rocket launchers and, when they could get within range, Olifant tanks.[2] But the Cuito defences held and the military deadlock around the town increasingly turned into a strategic disadvantage for Pretoria. Joint US-Soviet pressure for a negotiated solution was being exerted and, while a swift and successful strike on Cuito would have improved South Africa's position in any future negotiations, the saturation-type trench war could only serve to weaken such a position.

In a move to regain the initiative, Unita forces, supported by SADF elements, launched a March 1988 attack on the north of Cuando Cubango and the Bie and Moxico provinces. The tactical aim of this was to take the town of Cuemba, Bie province on the Benguela railway with its vital airstrip, and then move east to take Luena, the capital of Moxico province.

The underlying strategic goal was probably to establish a bridgehead for the proclamation of an alternative Unita government in the south-east corner of Angola.[3] Undoubtedly, Pretoria's negotiating position would have been strengthened, had this operation been successful. However, it remains unclear what role, if any, the US played in this context and to what extent US and South African interests converge on the Unita issue.[4]

After some initial successes, including the capture of Munhango - Unita leader Savimbi's birthplace - the attacks on Cuemba and Luena were repelled by Fapla and Cuban forces and the Unita/SADF troops were forced to retreat south towards Mavinga. At about the same time Cuban forces, joined by Fapla and Swapo troops, fanned out towards the south from the defensive line they had occupied for years along the 16th parallel in the south of Huila province. This move was supported by reinforcements from the sea (at Namibe) and by troops from the Benguela defensive line further north.

These rapid changes in the deployment of Cuban troops and equipment in the south apparently surprised the SADF. For example, in late June SADF forces were unable to repel a Cuban air attack on SADF positions near the Calueque dam inside Angola, but close to the Namibian border. In addition, SADF troops around Cuito Cuanavale risked becoming cut off from their exit route southwards along the Cuito river. From Pretoria's standpoint, the tactical and strategic situation had deteriorated considerably.

Two important South African tactical objectives, which had crystallised during the course of fighting, were not achieved: Cuito Cuanavale did not fall and neither did the South African and Unita forces manage to take the towns of Cuemba and Luena further to the north.

South Africa's overall strategic position was markedly worse than it had been prior to the initiation of large-scale fighting in the late summer of 1987. South Africa had lost the unchallenged air superiority it previously enjoyed in the region, which had major implications for the situation on the ground. Earlier, South Africa and Unita had relative control over the area, and could move about with near impunity in large parts of southern Angola. This was vital, since the area functioned as a buffer zone separating the Swapo camps and bases further to the north from the Namibian border. It also served as the main staging area for Unita's destabilisation war against the Angolan government.

The combined and interlinked effects of the loss of air superiority, the enhanced fighting capabilities of Fapla, and the introduction of more modern equipment and distinctly more offensive behaviour on the part of the Cuban forces, shrunk the South African control zone considerably. By early July it consisted of a small sector in the south-east only, delineated by Mavinga, the Cuito and Lombo rivers, and the Caprivi Strip.

The SADF image of invincibility was shattered, and Unita was left out on a limb. In the process, the SADF lost a considerable number of white troops which entailed a high political price. Vital and extremely costly major weapons systems, such as jet fighters, were also lost. These were the military realities leading to negotiations.

Military vulnerabilities

After its abortive 'Operation Savannah' invasion of Angola, immediately prior to the declaration of independence in 1975, the SADF identified a number of critical vulnerabilities in its force structure. These were underlined by subsequent SADF experiences in Angola in the second part of the 1970s, and led to the introduction of several major projects such as fighting vehicles and the Valkiri multiple rocket launcher. These weapons systems, which were fielded and used in southern Angola during 1987-88, opened up new 'windows of vulnerability' in the war.

There is an erroneous view that the SADF suffered decisive and irreversible defeats on the military battlefield and, broken and humiliated, sneaked out of the back door to lick its wounds while the politicians were left with the task of negotiating terms of surrender with the victorious Angolan and Cuban representatives.

The SADF did not suffer a decisive defeat. Neither did it

suddenly become militarily weak in an overall sense. The 'triumphalism' implied in the above viewpoint only serves to obscure the real significance, as well as the limitations, of the military setbacks suffered by the SADF. Under the circumstances, the SADF was fairly successful in southern Angola. It never had more than 9 000 troops inside Angola, although most sources claim the number was closer to 3 000. This is only a fraction of the troops the SADF could theoretically field, yet by all accounts they performed extremely well against the numerically superior Fapla and Cuban troops.

For the first time since its participation in World War II, the SADF was confronted with a full-sized, well-equipped and battle-hardened conventional force. Despite making full use of the sophisticated weapons and equipment at its disposal, the SADF failed to defeat the Angolan forces. The swift strike tactic failed, leaving the SADF to face a classical conventional war with large-scale battles, fronts and logistical routes. This highlighted specific vulnerabilities in the SADF military machine - weaknesses which otherwise would not have been so apparent.

Two separate sets of vulnerabilities can be distinguished. The first concerns what can be broadly termed 'human and personnel factors'. Most important here is the high loss sensitivity with respect to white troops. By its own account, the SADF lost more than 50 white soldiers in Angola - substantially more according to Luanda. It seems clear that the idea of an infantry assault on Cuito Cuanavale was rejected by Pretoria's politicians, mainly since it entailed the risk of losing up to 300 whites. The fact that this casualty level was deemed unacceptable is an important indicator of just how sensitive South Africa is to white losses.

As young conscripts returned from the front in plastic bags and as information from the international press filtered through, public opinion became more critical of the war. In early August 1988, 143 white conscripts publicly refused to accept an SADF call-up order (*WM*, 05-11.08.88). An opinion poll conducted by the South African Institute for International Affairs showed that 57% of white South African adults believed the Botha government should negotiate directly with Swapo on Namibia.

Even among low-income white households - the main political support base of the Conservative Party - support for such negotiations exceeded 50%. Moreover, three out of four white adults opposed increased military spending.[5] A wide range of factors influence the results of such opinion polls, but experiences from other wars suggest that public opinion is heavily influenced by negative casualty figures.

To avoid major white losses in future large-scale fighting, South Africa could either attempt to recruit foreign mercenaries on a much larger scale than is currently the case or integrate more blacks into the SADF. The first alternative is very costly and unlikely to produce a sufficient number of recruits. The second has already proven to be problematical. The series of mutiny-like protests among black soldiers in Namibia during late 1987 illustrated the problem of giving blacks the task of defending the apartheid system (see, for example, *WM*, 20-26.11.87). The sensitivity to white casualties is a major window of vulnerability, difficult to close both in the short and long term.

The other set of vulnerabilities has to do with weapons and equipment, especially the loss of air superiority. South Africa possesses a fixed number of relatively modern or modernised Mirage fighters. Due to the mandatory UN arms embargo, enforced since 1977, South Africa cannot openly acquire new fighters from abroad, neither has it yet mastered the technology to manufacture the Mirage-3, the Mirage F-1 or any other modern fighter aircraft. The South African Air Force is thus extremely loss-sensitive when it comes to its most sophisticated jet fighters.

South Africa lacks other essential components in modern warfare, such as dedicated attack helicopters. The country's air defences are vulnerable as access to modern anti-aircraft missiles and up-to-date radar systems is limited. Airborne long-range surveillance is a crucial aspect of modern warfare and South Africa has for years tried to find a replacement for its obsolete Shackleton surveillance planes. Similarly, the country's access to the most modern C3I (command, control, communications and intelligence) systems, is severely limited.

Neither is South Africa up to date with technology related to electronic countermeasures, counter-countermeasures and electronic warfare (ECM, ECCM, EW) vital for aircraft protection and for airborne attacks on defensive missile sites and other radar and missile-protected targets, such as armour concentrations, road or rail junctions, power stations and bridges.

Most of these factors can be explained in the light of the arms embargo. Despite many loopholes, the embargo has had an effect. And although these weaknesses were known to South African strategists and the upper echelons of the military, the practical implications were not experienced until the Angolans and the Cubans introduced modern Soviet equipment into the frontline.

Taking the seemingly insoluble problem of white manpower vulnerability into account, South Africa has no option but to continue its attempts at circumventing the arms embargo and aim for a marked qualitative edge over its opponents, much as the USA is doing relative to the Soviet Union, or Israel relative to the Arab states. South Africa will also aim at maintaining its role as the quantitatively superior military power in the region. This is particularly important with regard to land warfare, where mobility and armour protection of infantry soldiers will continue to receive maximum attention.

Establishing new balances

The process leading up to the signing of the Brazzaville Protocol on 13 December and its subsequent ratification in New York a week later was somewhat paradoxical: initially, South African politicians ordered an invasion of Angola to save Unita and at the same time strengthen South Africa's position in any future talks. The military tried to achieve a quick defeat of the Cuban and Angolan forces, but failed. The main concern of the military then switched to getting out of Angola. Sections of the military argued that it now became more important than ever to maintain Namibia as a buffer. The politicians, on the other hand, found themselves caught up in a quagmire of military setbacks and external diplomatic pressures aimed at Namibian independence.

This affected the internal balance of forces within the ruling white minority. The business community, both English and Afrikaner-based, favoured a solution which would further peace and stabilise the regional security situation, thus enhancing possibilities of expanding economic activities in the region. Foreign Affairs officials and certain senior military officers, with a good grasp of the political implications on the global level arising from continued militaristic regional behaviour, were also probably in favour of a negotiated settlement, but wary of compromising South Africa's national security.[6]

Sections of the military and security apparatuses saw Namibian independence as a 'sell-out' signifying a marked deterioration of South Africa's security position. Those who took the final decisions on South Africa's positions in the negotiating process thus had to balance the two regional objectives of 'apartheid safety' and 'apartheid profitability', whilst weighing the impact of the alternatives at hand on the unstable international political scene.

At the heart of the matter is military security for Angola and South Africa. It was not only a question of 'peace with honour', but the far more important 'peace with security'. With this basic requirement as the point of departure, Pretoria, Luanda and Havana entered the negotiating process.

Angola and Cuba were in the better bargaining position due in no small part to the military realities described. Nevertheless, in seeking to identify the maximum diplomatic bargain attainable - while simultaneously rushing to improve its military capability and keeping the option open of aborting the process at any given time - South Africa may have been helped by the mediatory role of the United States.

It is interesting to advance some speculations about the guarantees Pretoria sought in exchange for Namibian independence. Quite substantial guarantees were clearly a prerequisite if the militarists were to be convinced to forego the militarist option. The following list of guarantees constitutes a 'best-case' scenario from the point of view of the South African government:

- a non-aggression pact between South Africa and Namibia, modelled on the Nkomati Accord between South Africa and Mozambique;
- total banning of the ANC from Namibia;
- retention of Walvis Bay;
- a negotiated settlement and some form of power-sharing between the MPLA government and Unita in Angola;
- a non-aggression pact with Angola or some similar agreement which clearly defines and makes binding the mutual commitment in the 14-point plan adopted in New York in July;
- the closure of the ANC bases in Angola;
- an end to the US disinvestment campaign;
- an end to the US calls for comprehensive economic sanctions against South Africa;
- a commitment on the part of the US to channel, whenever

possible, its regional investment through South Africa;
- a lifting of both the embargo on the export of arms and related material to South Africa and the embargo against arms imports from South Africa. In exchange for this, South Africa may offer to sign the Nuclear Non-Proliferation Treaty (NPT).

This hypothetical Pretoria 'wish list' implies a remoulding of the regional setting that would be difficult to accept for the Frontline states and the SADCC member-states.

The final outcome of the peace process in south-western Africa and the nature of Namibian independence will depend on basic military realities, the relative strength and bargaining position of the parties and the relative importance and dynamic of the economic and political factors in the wider strategic context. However, given the level of militarisation of South African society and the predominance of security concerns in any major political decision, goodwill or astute diplomacy alone will not be enough.

At the heart of the matter is power - military power, in particular - and the concept of security.

Notes

1   *Southscan*, 2(33), 4 May 1988; 'Pretoria's newest homeland - "Savimbistan"', *Work In Progress*, 54, June/July 1988, 7.
2   However, it appears that the SADF later tried to take Cuito in armour attacks on at least two occasions, in March and in May. See *Southscan*, 2 (33), 4 May 1988; *Africa Confidential*, 15 July 1988. The chief Cuban negotiator in the peace talks, Jorge Risquet, said the SADF tried to take Cuito 'about five times'.
3   *Southscan*, 2 (34), 11 May 1988; *Africa Confidential*, 15 July 1988; 'Pretoria's newest homeland - "Savimbistan"', *Work In Progress*, 54, June/July 1988.
4   The circumstances around the northbound Unita offensive are unclear. Some observers see it as an idea hatched in Pretoria and supported by Savimbi. The final aim, in this interpretation, was the proclamation of an alternative Angolan government - or a 'Savimbistan' - in the vast area approximately comprising Bie, Moxico and Cuando Cubango provinces. It was also hoped to get US assistance via the nearby Joint US-Zairean military exercise 'Operation Flintlock', taking place in the second half of April and early May. The other main interpretation is that all northbound Unita movements resulted from US attempts to wean Unita away from its South African backers, dismantle Jamba and move the Unita headquarters to Quimbele in northern Angola. This would make it easier for the US to channel support via Zaire and, at the same time, effectively put Unita out of South African reach. It is conceivable that both interpretations are correct: the first may have been valid up to the Unita failure to take Cuemba and

Luena, and the second thereafter. For a discussion, see *Africa Confidential*, 27 May and 15 July 1988; *Southscan*, 2(34), 11.05.88; and 'Pretoria's newest homeland - "Savimbistan"', *Work In Progress*, 54, June/July 1988.

5   *Southscan*, 2 (46), 10 August 1988.
6   On the possibilities for South African manoeuvres with respect to Resolution 435, see 'The October municipal elections in South Africa in the light of the Angola/Namibia negotiations', *Southern Africa Dossier*, 39, UEM, Maputo, September 1988.

South Africa's War of Destabilisation

Diana Cammack

'When there is war, you don't have time to plough. You are running here and there, in the mountains, and under the rocks'. This is the lesson learned by victims of South Africa's war of destabilisation, which has wreaked havoc throughout Southern Africa.

The low-intensity war in the region associated with destabilisation reflects the militarisation of South Africa's decision-making bodies, especially the state security council and the executive presidency.

Destabilisation refers to the most aggressive aspects of regional policy adopted by the Botha government during the late 1970s and 1980s. It aims to maintain neighbouring states in a position of economic dependence - not solely for economic reasons, but also to undermine the initiatives of the Southern African Development Co-ordination Conference (SADCC) and make it more difficult for its nine member-states to advocate or implement sanctions against South Africa. Destabilisation was also devised in response to the strengthening liberation struggle within South Africa, and was particularly aimed against countries hosting ANC political and military personnel.

Implementing regional policy

The implementation of South Africa's regional policy since 1978 can be divided into six phases.[1] The first ran from 1978 until mid-1980, from Botha's assumption of power until the creation of SADCC following Robert Mugabe's election victory in pre-independence Zimbabwe. During this period, PW Botha promoted the 'constellation of states' as the solution to Southern Africa's economic woes and as a means of combating what Pretoria saw as a 'marxist threat' common to the whole region.

The second phase ran until the end of 1981 and was marked by an increase in South African sponsorship of dissident groups in neighbouring states and the escalation of SADF commando raids. The war in Angola was re-opened with 'Operation Smokeshell' while Pretoria adopted a policy aimed at disrupting the economic life of Zimbabwe. A South African agent blew up the ammunition depot at the Inkomo Barracks outside Harare, while others planted bombs at Zanu (PF) headquarters, killing seven and injuring 124, but missing the intended target of Mugabe and his cabinet. Other agents were more successful when, on their second attempt, they killed Joe Gqabi, ANC representative to Zimbabwe.

Commandos were also busy in this period. They raided the Lobito oil terminal in Angola and a year later attacked the Luanda oil refinery, causing over US$12-million in damage and $24-million in lost revenue. In Mozambique they raided a suburb of Maputo, killing 13 ANC members and a local passer-by. On several occasions they attacked the Beira Corridor, economic lifeline of Zimbabwe, which carries a railway, a road and an oil pipeline from Beira to Zimbabwe; and in November 1981, they destroyed marker buoys in Beira harbour in order to disrupt trade.

There were several incidents on the Botswana border, a portion of south-western Zambia was occupied, and two ANC members were killed and a South African refugee kidnapped in Swaziland. A number of 'economic techniques of coercion' - withdrawing badly-needed railway locomotives and threatening to cancel the Preferential Trade Agreement with Zimbabwe - were used. During this phase Botha approved, and the National Intelligence Service funded, Mike Hoare's abortive attempt to overthrow the Seychelles government. The SADF, according to Hoare, provided over half his men (*Guardian*, 14.07.86).

This phase gave way to the third as Pretoria focused on the removal of ANC members from neighbouring states and displayed more selectivity in its targets, leading commentators to believe it had begun to differentiate between its neighbours. In this period, Swaziland and Malawi, as potential collaborators, were offered economic assistance or land to distance themselves from SADCC initiatives and halt assistance to the ANC. The secret Swaziland Agreement of mid-February 1982 was Pretoria's first major success, when King Sobhuza II agreed 'not to allow any activities within (Swaziland)...directed towards the commission of any act which involves a threat to use force against (South Africa)'.[2]

Governments unwilling to compromise were more ruthlessly attacked. The SADF raided Angola almost weekly, while commandos remained active in Mozambique. In the 'month of the hawks', December 1982, several came ashore at Beira and successfully placed limpet mines on two oil-storage tanks - the same plan less successfully followed by Captain Wynand du Toit's raiders at the Molongo oil complex in Cabinda, Angola, during May 1985. The Beira depot was badly damaged, causing Zimbabwe severe petrol shortages for several months.

There were two further attacks on Mozambique within the year: a dozen SAAF jets strafed two suburbs with fragmentation rockets in May 1983, killing six people. Several months later commandos, including Du Toit, planted a bomb at the ANC office in Maputo, injuring five.

In July 1982 saboteurs planted incendiary devices at Zimbabwe's Thornhill air base near Gweru, destroying over a dozen aircraft and wiping out Zimbabwe's strike and jet interception capabilities. Less than a month later an SADF unit was intercepted by the Zimbabwean army after it had crossed the border. Three white South Africans, whose presence in Zimbabwe was never adequately explained, were killed in the ensuing shoot-out.

South Africa also turned its attention to Lesotho where Chief Leabua Jonathan was becoming increasingly outspoken in his condemnation of apartheid, and continued to declare his government's intention to receive South African refugees. Pretoria supported the dissident Lesotho Liberation Army (LLA), which attacked the kingdom's infrastructure and government officials, while South African commandos targeted South African refugees. On 9 December 1982 over 100 South African commandos crossed the border and attacked a dozen homes in Maseru, killing 30 South Africans and 12 Basotho. Some of the victims had been ANC members, though Western diplomats stated there was no evidence to suggest that their homes had been used as military bases.

By 1983 South Africa was being shaken by serious internal problems: the nation's GDP was in decline, the rand had collapsed on world markets, and interest rates and inflation were soaring. The war in Angola was not going well for the SADF, while in Mozambique, economic chaos had deprived South African businessmen of markets. Both the US and European governments wanted an end to attacks by South Africa and its surrogates: they were harmful to Western economic interests and could provoke further Soviet involvement in the region.

Finally, dissent among South Africa's establishment became public, as 'doves' worried aloud about the impact destabilisation was having on the economy and Pretoria's political credibility. The aggressive third phase ended early in 1984.

'Pax Pretoriana'

South Africa and Angola signed the Lusaka Agreement in February 1984, with the Nkomati Accord being signed shortly thereafter. The previously-secret Swaziland Agreement was made public shortly after Nkomati. Botha's reward was a tour of Western Europe at the behest of the British prime minister - the first such invitation in over 20 years.

Liberals in South Africa lamented the weakness of the domestic economy, which made it difficult for the nation 'to demonstrate what South Africa can do for the rest of the region'.[3] A constellation of states was mooted once again, with Botha alluding to it at the signing of the Nkomati Accord.

Pretoria now styled itself the regional peacemaker and shortly after signing the accord, sponsored talks between Frelimo and the MNR. Discussions took place over several weeks. But with Frelimo aiming to 'liquidate banditry', the MNR hoping to secure several provincial governorships and four major cabinet posts (including defence), and Pretoria seeking to 'dilute the hardline part of Frelimo', no compromise was reached.

Instead, on 3 October 1984 the 'Pretoria Declaration' was announced. It advocated an end to hostilities inside Mozambique, but offered no real means of halting the war. Ceasefire talks continued for a week, but were halted by a phone call - rumoured to have been from either an MNR sponsor in Lisbon or a disgruntled SADF leader - to the MNR's representative in Pretoria. With the breakdown in ceasefire talks, Pretoria's role as peacemaker ended.

A public relations peace

By then it was already clear that Pretoria's peace had been a public relations exercise only. While conservatives in the West continued to insist that Botha was anxious to keep the peace, those in the region were painfully aware of the continued SADF aid to the MNR.

Documents found in August 1985 at the MNR's Casa Banana base in Gorongosa provided proof of this support, revealing that even as negotiations leading to the accord were underway, Pretoria had provided the Mozambican dissidents with a 'team of South Africans... to train (MNR) soldiers...(in) conventional warfare' and with enough supplies to last for at least six months, and possibly as long as two years.[4]

Diary entries discovered by Frelimo soldiers at Gorongosa told how the South African military intended to 'continue to give support (to the MNR) without the consent of our politicians in massive numbers so as to win the war' (entry dated 23 February 1984). The diary, kept by the 'private secretary' to the 'president' of the MNR, also noted that three weeks before Botha and Machel met at the Nkomati River, two MNR members were sent 'with authorisation of...General Magnus Malan...to (be) trained in ultra-secret communications between Pretoria and the MNR'. The diary continues: 'The general will ensure resupply even after the agreement...especially ammunition and radio transmitters' (24 February 1984).

On 14 February the diarist noted that the MNR's base in South Africa was to be maintained: he accompanied 'Colonel Vaniker' (now Brig-Gen Charles van Niekerk) to Louis Trichardt 'to reconnoitre the new camp where we are going to live'. Meanwhile, some MNR soldiers were sent north to Malawi where they could more easily strike at the productive central and northern regions of Mozambique, and implement their plan to drive to the sea, cutting Mozambique in two.

After the accord was signed, President Machel - in keeping with the agreement, and to the amazement and frustration of many radicals in the West - began to expel ANC personnel from Maputo. Meanwhile, according to the Gorongosa diary, the SADF extended the MNR's airstrip at Gorongosa, and provided the movement with communications equipment, arms and medical supplies.

MNR leaders were brought to South Africa by submarine, car and aeroplane, and on at least three different occasions, Louis Nel, then Pik Botha's deputy in the department of Foreign Affairs, flew into Gorongosa for talks with the MNR. The diary confirmed the view of many cynics of the Nkomati Accord: that South Africa would never abide by any 'agreement of good neighbourliness and non-aggression'. Pretoria had been helping the MNR - set up originally by Rhodesian Central Intelligence in 1976 - since Botha replaced Vorster as prime minister in 1978.

In 1980, more than R1-million was pledged to the MNR, but when it became obvious that majority rule was coming to Zimbabwe, the SADF took over full control of the MNR. The shift of personnel - begun in March 1980 and undertaken with the knowledge of the staff of the Rhodesian governor, Lord Soames[5] - started by flying staff of the MNR's 'Voz da Africa Livre' radio station south in a SAAF C-130 transport plane. Soon after, Rhodesian SAS personnel drove seven MNR vehicles to the Voortrekkerhoogte military base, and then a second airlift took members of the organisation to Phalaborwa.

A new base was established near the Kruger National Park, and Pretoria began to choose targets which suited its needs: Mozambican railways, roads, bridges, and the oil pipeline that served Zimbabwe. Four years later the Gorongosa diary demonstrated how little the targets had changed in the intervening years: 'railways, Cahora Bassa, co-operantes (foreign aid workers) and other targets of an economic nature, SADCC' (entry 24 February 1984).

The signing of the accord solved neither Mozambique's nor South Africa's problems. The Maputo government did not gain the breathing space it hoped the agreement would bring as South Africa and Malawi continued to assist the MNR. And while the ANC was driven out of Maputo as Pretoria desired, the movement did not lose strength in the long term. Frelimo, while making some economic concessions to the West, was not 'diluted' as many had hoped, but retained its socialist vision under Machel.

The international kudos South Africa had achieved for its 'peace drive' was short-lived and its pariah status soon re-established through the disclosure of the Gorongosa documents, the emergence of the debt crisis, the internal unrest and state of emergency which undermined the National Party's 'reform programme'.

Raids and retaliations

The implementation of Pretoria's regional policy entered its fifth phase in early 1985 - coinciding with Reagan's re-election as US president.

By the first anniversary of Nkomati, it was widely accepted that Pretoria had not honoured the accord. Direct military support for the MNR may have been stepped down to the passing of secret information,[6] but the additional and open support Malawi was providing the dissidents more than made up for this. The war in Mozambique did not end, but escalated as it shifted north. Meanwhile, new dissident intrusions into Zimbabwe were reported, and the LLA, which had been inactive for months, re-emerged in Maseru.

The facade of good-neighbourliness suffered a fatal blow in May 1985 when SADF Captain Wynand du Toit was captured by Angolans in Cabinda during a mission to destroy the Molongo oil complex. His capture and the worldwide condemnation following the raid did not keep Pretoria from planning a second major operation, this time in Botswana. In the Gaborone raid of 14 June 1985, outdated information led commandos to the homes of innocent local residents.

An August 1983 raid on Namaacha was apparently aimed at the ANC. But because the commandos' information was so dated, they killed two Mozambicans and an aid worker instead. The same may be said about the raid on the ANC information office in central Harare in May 1986.

The South African government generally claims its raids are directed at ANC 'operation headquarters', or 'control centres'. But in the vast majority of cases, private homes have been destroyed and civilians killed. After an SAAF attack on Maputo in May 1983, the SADF claimed that 41 'ANC terrorists' had been killed, six ANC bases and a missile battery destroyed. But foreign journalists documented the death of six people - five locals (three at a jam factory), and one ANC member.

After the 19 May 1986 raids on Botswana, Zambia and Zimbabwe, the SADF reported it had hit an 'ANC installation' at the Makeni Refugee Transit Camp, 15 miles from Lusaka. But according to the Zambian Commissioner for Refugees,

at the time of the attack, there were no South Africans at the centre. Even if there had been, they would have been refugees and not combatants, with no affiliation to any political party or movement. The South Africans who are received here are only civilians.[7]

During March 1988, the SADF 'got the wrong man' in a raid on

Gaborone, according to neighbours. The commandos killed and burnt four sleeping people, one of whom was supposedly ANC military commander, Solomon Molefe. However, neighbours claimed he was Charles Mokoena, a man they knew well.[8] During the June 1985 raid on Gaborone the 12 'key' people whom the SADF killed were all civilians, according to an American church worker.[9] While four *may* have had some connection with the ANC, none was a likely candidate for combatant training.

The SADF claims its attacks are aimed at the ANC only, and not meant to disturb locals or the relationship between Pretoria and neighbouring governments. After the June 1985 raid on Gaborone, an SADF spokesman claimed that South African commandos had tried to avoid engaging members of the Botswana Defence Force: they were, according to this statement, actually helping the BDF do its job, as it was unable to close the alleged ANC infiltration routes through Botswana (*FM*, 28.06.85).

Raids are often retaliations for ANC sabotage in South Africa. The May 1983 raid on Maputo was a reprisal for the Pretoria car bomb explosion which killed 18 and injured 40. Raids in recent years have often followed border landmine explosions.

Pretoria has set out to prove it will not be bullied. Three raids into Mozambique in 1981 were all near the port of Beira which serves Malawi. The raids - six weeks prior to the SADCC meeting in Blantyre - involved a warning to President Banda to distance himself from the SADCC programme.

A 1988 car bomb in the Harare suburb of Avondale, which injured 17 people, exploded on the eve of the Commonwealth conference in Canada, where the issue of sanctions against South Africa was expected to top the agenda. More recently, the National Party's need to demonstrate its uncompromising stand in the face of black rule led Pik Botha to threaten neighbouring states about the 'consequences of colluding' with the ANC. This threat was soon followed by a car bomb in Gaborone which killed three and injured three - all Botswana citizens.

Another aspect of this fifth phase was the increased activity of 'Z squads' - groups of specially-trained killers who target ANC leaders. In keeping with Defence Minister Magnus Malan's promise to 'eliminate' the ANC 'wherever' it is, squads have attacked a number of notable ANC members in the last two years. Among them were Dulcie September, ANC representative in Paris, who was shot dead in March 1988, and Albie Sachs, an ANC lawyer and lecturer, who was seriously injured in a car bomb attack in Maputo in April 1988. ANC Secretary-General Alfred Nzo escaped an assassination attempt in Lusaka during January 1988, as did Godfrey Motsepe, the ANC representative in Belgium, the following month. In March 1988 ANC guerilla Mzizi Maqekeza was shot dead in a Maseru hospital.

The death of Mozambican President Samora Machel has also been seen as part of the offensive fifth phase. While the Margo Commision found that the air crash which killed Machel and 33 others in October 1986 was the result of pilot error, few in SADCC believe this to be the case. Critics of Margo's findings argue that a false beacon was set up inside South Africa, luring the plane off course from Maputo and into a hillside near Nelspruit.

Questions about the plane crash remain: why, if the plane went down at 9-20 pm, was the first message to Maputo sent many hours later and the first of the injured hospitalised only 11 hours after the crash? Why, when the plane deviated off course, did South African air traffic control - which admits it monitored the flight on radar for hundreds of miles - not tell the pilot he was off course? Why did South Africa refuse to take part in an international inquiry, while denying the Soviet Union (as builder of the plane) and Mozambique a full role in the Margo inquiry, inviting them only to cross-examine witnesses at the public hearing?

Why has South Africa refused to take part in Mozambique's on-going investigation? Why were South African security forces on full alert the night before the crash? Why, according to Paulo Oliveira, the MNR's former chief representative in Europe, did South Africa tell the MNR in Lisbon to 'stand by...to claim responsibility for downing the plane' and later tell them 'not to worry about the matter'?

Rumours compound the belief that the full story has yet to be told: the morning of the day Machel was killed, reports already circulated through the Frontline capitals stating that the 'assassination of the Mozambican leader appears to be on the minds of the South African generals' (*Sunday Mail*, Harare, 19.10.86). Rumours since then speak of an overnight camp near the crash site, with a tent large enough to house electronic gear, which had disappeared the next day before representatives of the Maputo government arrived at the scene. Finally, eye-witnesses report that police who arrived at the crash site spent their time collecting documents which were then photocopied. The originals were later strewn around the site to make it appear as though they had been blown by the wind. This being the case, what else was tampered with, and why?

'Thump and talk'

Pretoria has several tactical aims in its regional policy, such as forcing neighbouring states to expel South African political refugees or recognise the bantustans as independent. But its overriding goal is to maintain hegemony throughout the region. This means reducing the threats posed by the SADCC, the ANC, and to a lesser extent, the PAC. In an effort to achieve this, Pretoria has, since 1982, pressured governments to sign accords or failing that, come to some sort of 'working agreement' with the SADF. Pretoria's strategy of 'thump and talk' has had some notable successes.

The Swaziland agreement assures close co-operation between the Swazi state and the South African police, and though South Africa's actions have ruffled Swazi diplomatic feathers on occasions, Mbabane has done little to prevent the shoot-outs and kidnappings carried out by South African agents in the kingdom since the early 1980s.

Botswana, while refusing to sign an accord or acknowledge that the ANC transits the country, has had no option but to enforce new legislation aimed at controlling radical elements and halting the stockpiling of arms.

The coup in Lesotho marked a turning point in South Africa-Lesotho relations. Where before Chief Jonathan made regular statements against apartheid, the present government is, to use the words of Major-General Metsing Lekhanya, 'very keen' to 'explore all avenues of mutual understanding and development'. Where once South African refugees were allowed to live, work and study in the kingdom, now political refugees are flown north out of Maseru as soon as they arrive. Where once the progress of the Highlands Water Scheme was hampered by the volatile relationship between the two governments, the R4-billion project is now well underway.

A further step in the promotion of 'good relations' involved the opening of the South African trade mission in Maseru, and King Moshoeshoe II's visit to South Africa in late 1988 - the first state visit by a black leader in nearly 20 years. While the full story of

South Africa's role in the fall of the Jonathan government has yet to be told, it is clear that Pretoria has benefitted from the new spirit of pragmatism which resulted.

Talking peace

It was events in Angola, more specifically those near Cuito Cuanavale, which drew South Africa's aggressive fifth phase to a premature close. Several thousand South African troops were trapped by a combined force of Cuban-Angolan soldiers, giving the SADF no possibility of escape without unacceptable losses. Pretoria was thus forced to come to terms with the new reality in southern Angola: it no longer had domination over the area's air space, neither could it move in and out of the country as it had done for years.

But South Africa's reasons for talking peace with enemies were not confined to the mounting military costs in Angola, or even to the continuing financial drain of Namibia. They were linked to the vociferous rejection of further Angolan excursions voiced by the liberal (and increasingly, by the mainstream) Afrikaner community; the inability to create a credible internal settlement in Namibia; and the economic crisis confronting the administration. With a foreign debt continuous at US\$22,6-billion, a rising inflation rate, stagnating growth rate, a decline in the gold price and dwindling gold reserves, as well as the crippling impact of sanctions, Pretoria could no longer afford the war.

The international setting had changed, and both the US and USSR were prepared to back peaceful intentions in this region as well as other trouble spots throughout the world. Angola was ready to adopt a more pragmatic approach to reconstruction of its war-torn economy, and was expected to join the IMF and World Bank once the peace process was underway.

Meanwhile, South Africa's support for the MNR had become a focus for discussion in Western capitals and in the mainstream Western press when Robert Gersony's report for the US State Department, on conditions within Mozambique, was made public in April 1988. In it he outlined the MNR's reign of terror against civilians and, while he was circumspect in linking South Africa with the MNR, newsmen and senior officials in Washington were not. Thus, there was a 'convergence of interests', using Chester

Crocker's phrase, which ushered in the newest phase of South Africa's regional strategy.

As astonishing as it is to outsiders, the South African government presents itself as the benefactor of the sub-continent. It presents South Africa as the 'mainstay' of the entire region and claims to accept the 'role and responsibility for co-operation with its neighbours'.

Central to this co-operation is South Africa's 'transport diplomacy', as outlined by Kobus Loubser, former general manager of Sats. Transport, he said in September 1979, is one of the 'strongest links between countries' and a 'prerequisite for many other forms of co-operation'. Throughout the 1980s Pretoria has represented itself as a benefactor because it ships several million tons of goods through and from the Republic to SADCC states, and because some six thousand Sats railway trucks may be in one or another of the neighbouring states at any one time. Sats ports and railways are, according to Pik Botha, Southern Africa's 'lifeline'.

Pretoria encourages trade with Southern Africa because it is good for business: the Republic needs markets for its manufactured goods, and Africa provides them. For instance, in 1984 over 40% of South Africa's machinery, a third of its chemicals and a quarter of its transport equipment went to African buyers outside the Southern African Customs Union. Trade that passes through Sats ports and Sats railway lines generated a quarter of a billion US dollars in 1982 alone.

Transport dependence ensures that neighbours are in no position to close borders with, or advocate sanctions against, the Republic. Yet Pretoria has, at times, cut rail traffic to Zimbabwe, Botswana, Lesotho, Zambia and Mozambique. At critical moments, it has withdrawn locomotives and wagons 'on loan' from Sats, causing shortages of fuel, goods and fertilisers. Seven years ago Deon Geldenhuys of the Rand Afrikaans University suggested that one of the ways 'in which the Republic could use its economic relations in Southern Africa for non-economic purposes' was by 'limiting or prohibiting the use of South Africa's railways and harbour facilities ...by manipulating the availability of railway trucks and berthing facilities in harbours, or harsher measures such as imposing surcharges on goods transported'.[10]

In August 1986 Botha, upset by Kaunda's and Mugabe's pro-sanctions speeches, ordered a 'statistical survey' of all Zimbabwean and Zambian trade as well as the licensing of Zambian imports. These measures, according to a Johannesburg source, were meant to have a 'bit of a nuisance value', which they did, as lorries waited at the South African border for up to 30 hours while the 'survey' lasted.

In 1987 a different form of 'transport diplomacy' was used against Botswana and Zimbabwe when Pretoria tried to force SADCC railway crews to obtain Bophuthatswana visas before entering the bantustan - a move which would, in effect, have implied recognition of Bophuthatswana's 'independent' status. The scheme backfired when Botswana built a turn-around point for trains 9,6 kilometres north of the border, and told the South African crews to collect the wagons there.

The January 1986 blockade of Lesotho resulted in panic buying in Maseru and precipitated the coup. Pik Botha's threats, uttered later that year, to intercept US grain shipments to northerly neighbours were taken seriously in the region, as was the threat by Lt-Gen van den Berg that 'at the stroke of a pen, we (South Africa) could cut off the import and export facilities we allow them (Zimbabweans) and bring them to their knees in weeks, if not days'.[11] Not surprisingly, one of the SADCC's primary goals is to reduce transport dependence on South Africa.

Sabotage of Mozambique's railway routes is one of the chief causes accounting for this dependence. Since August 1983 the Limpopo railway line through Chicualacuala to Maputo has been virtually closed due to MNR sabotage, while sabotage on the Beira line, which began in earnest in mid-1982, has slowed traffic considerably. As a result, at independence, Zimbabwe's trade through Mozambique was nil because the border between the two nations had been closed for four years in line with Machel's adherence to the UN's call for sanctions against Rhodesia.

By 1983 Zimbabwe had managed to send half of its trade through Mozambique. A year later, due to sabotage of the railways, traffic had dropped to a third and by the end of 1985, less than 10% of Zimbabwe's trade transited Mozambique. The rest went through South Africa.

The cost to the Zimbabwean economy has been enormous, as is evident when comparing tariff charges on, say, tobacco sent from Harare to the coast in 1986: Z$806/tonne to Maputo via the Limpopo railway, compared to Z$1366,80/tonne to Durban via Beit

Bridge.[12] No wonder the Beira Corridor Group - a collection of businessmen raising money to renovate the railway and port - went to Pretoria to explain to the state security council that the 'motivation of the private sector in re-establishing these routes is one of survival. We've got to reduce the costs of getting our exports to the sea'.[13]

Efforts are thus underway to renew the Beira, Nacala and Limpopo lines through Mozambique, improve the functioning of the Tazara railway from Zambia to Tanzania and re-open the Benguela line from Zambia through Angola.

An armoury of dissident groups

The major weapon in Pretoria's destabilisation armoury is its support for dissident groups in neighbouring states. At least six years ago Malan outlined South Africa's policy: Pretoria would combat its enemies, 'even if it means we will have to support anti-communist movements...and allow them to act from our territory'. Since the enunciation of the 'Reagan Doctrine' - which justified US support for anti-communist guerillas anywhere in the world - Pretoria has tried to legitimise its support for right-wing dissidents in much the same terms.

Over the years five groups have received South Africa's material aid and logistical support: the LLA in Lesotho, 'Super Zapu' in Zimbabwe, the MNR in Mozambique, Unita in Angola and the Mashala gang in Zambia. The most active of these have been 'Super Zapu', the MNR and Unita.

South African support for the MNR and 'Super Zapu' has taken several forms: the provision of arms, ammunition, mines and rocket launchers, communications equipment, medicines, logistical support in the form of training, intelligence gathering, rear bases inside South Africa, construction of air fields, and transport to and from South Africa.

The MNR's radio station, 'Voz da Africa Livre' (known in Mozambique as 'Radio Quizumba' or Radio Hyena) was beamed from the Transvaal prior to the signing of the Nkomati Accord. The Zimbabwean dissidents' 'Radio Truth' has been beamed from the same location since 1983. While South Africa repeatedly denied it was responsible for the broadcasts, on 25 November 1983 the station at Meyerton mixed up the tapes: the 7 am vernacular broadcast of 'Radio Truth' began with the introductory music of 'Voz da Africa Livre' while the 'Voz da Africa Livre' programme began with the introductory music of 'Radio Truth'.[14]

South African involvement with Zimbabwean dissidents has also been well documented. For instance, according to 'A Chronicle of Dissidence in Zimbabwe',

> Joseph Dube crossed the border into Botswana in early 1983 and went to Francistown where he stayed with a group of Zimbabwean recruits for military training in South Africa. While in Francistown, the recruits were addressed by a Mr Callaway, a member of the South African Defence Force, who told them that South Africa was ready to train and arm them to topple the Zimbabwean government.[15]

In February 1983, a dossier on South African involvement with 'Super Zapu' was presented to Maj-Gen H Roux of the South African Chief of Staff (Intelligence). The dossier contained evidence of the provision of weapons, including assault rifles of Bulgarian manufacture and Soviet RPG-7 rocket launchers. Rumanian ammunition manufactured in 1980 - after the last Zapu arms shipment, which arrived in Africa in September 1979 - was also provided to the dissidents.

South African involvement in Zimbabwe has not been confined to helping 'Super Zapu', as a recent trial in Harare of three South African agents demonstrated. The sophisticated attempt of August 1988 by other agents working in Zimbabwe to rescue five men facing trial in Harare indicated the lengths to which Pretoria was willing to go to keep its involvement secret, especially in the run-up to the American election. Recent trials of South African agents in Botswana and Zambia have provided further proof of the extent of South Africa's use of agents inside SADCC countries.

Equally well documented is the logistical and military support given to the MNR over the years. The Gorongosa diary spoke of 'AK 47 ammunition' as well as 'material for urban guerrilla warfare (such as) time bombs and timing devices'. In July 1988, as Piet Koornhof assured the American public in his 'Letter from South Africa' that there was 'new co-operation' between Mozambique and South Africa, the MNR, with SADF assistance, was active in the south of Maputo province. Indeed, the South African authorities had to ask the Mozambican government to return the body of a soldier and material it had captured.

Earlier in the year, according to Chanjunja Joao, a former MNR official in Lisbon, new communications equipment, including a fax machine, was delivered to the MNR by Brig-Gen van Niekerk to improve contact between the MNR office in Portugal and its rear bases. In September 1988 PW Botha stated that South Africa would no longer aid the MNR, and a month later, Pik Botha reiterated the pledge. Yet evidence provided by the ex-commander of the MNR women's section indicates that the South African military was still active at Gorongosa as late as October 1988.

During June 1989 Lt-Gen AJ Liebenberg, chief of the army, told the press that neither private groups or individuals, nor the SADF, were supplying the MNR with logistical or material support. Yet many, especially in the Frontline states, believe that MNR is still supported from within South Africa.

Little concrete evidence has emerged from within South Africa recently to substantiate this. The collection and dissemination of such information is difficult, sometimes illegal. And refugees, aid workers and residents of the Eastern Transvaal, who may be in a position to provide details of support, are frightened to speak openly about what is often rumoured in the area.

These rumours allege that SADF assistance to MNR continues in the form of weapons delivered by helicopter inside Mozambique; that known MNR activists are given sanctuary in South Africa; that refugees are still being recruited for MNR within South African prisons; and that co-ordination of operations and assistance to MNR has been 'privatised', with Portuguese-speaking ex-colonials now farming in the Eastern Transvaal taking a leading role.

Paying the price of destabilisation

The cost to the region of South Africa's destabilisation has been enormous, both in economic and human terms. The amount in war damages, output losses, additional defence expenditures and lost economic growth reached over US$15-billion between 1980 and 1986, and was estimated to be in the same order of magnitude as a year's economic production of the entire population of SADCC countries.[16]

The human cost is also staggering in terms of deaths - upward of 150 000 in Mozambique alone - displacement of people, numbers of refugees, child mortality rates, the amount of food and industrial goods produced, the destruction of the environment and wildlife, the number of amputees and the decline in inoculations and health care.

Now, though, there is talk of peace, and there are signs of a new mood. But there are pitfalls which could plunge the region back into war. Questions arise over whether South Africa will really stop its assistance to Unita, and whether continued US support for Unita helps the peace process in Southern Africa.

South African retention of Walvis Bay may impact negatively on the Namibian peace process. To the east, Cahora Bassa is to be renovated and its pylons repaired, but there have been further attacks along the electricity lines since that agreement was reached. Continuing MNR attacks on villages along the borders of both Zambia and Zimbabwe leave those countries wondering if there is really to be peace.

Finally, there is apartheid, the major stumbling block to peace. Dr Chiepe, minister of External Affairs in Botswana, put it eloquently in 1986:

> The sole source of tension is the apartheid policy... Until apartheid is dismantled lock, stock and barrel, the confusion which this vile system generates inside South Africa will continue to spill over across the borders and threaten the peace and stability of the region (*Herald*, Harare, 27.01.86).

Notes

1 The first five phases were discussed in Rob Davies and Dan O'Meara, 'Total Strategy in Southern Africa: an analysis of South African regional strategy since 1978', *Journal of Southern African Studies*, 11(2), April 1985.

2 Copies of the letters between King Sobhuza II, PW Botha and Mabandla Fred Dlamini, which constitute the Swaziland Agreement, are published as Appendix 4 to David Martin and Phyllis Johnson (eds), *Destructive Engagement: Southern Africa at War*, Zimbabwe Publishing House, 1986.

3 Interview with John Barratt, South African Institute of International Affairs, 28 August 1984.

4 Mozambique Ministry of Information, *The Gorongosa Documents*, 2 vols, Maputo, 1986.

5 David Martin and Phyllis Johnson, 'Mozambique: To Nkomati and Beyond', in Martin and Johnson (eds), *Destructive Engagement*, 14-15.

6 This view is taken by Eddie Cross of the Beira Corridor Group. See *Cape Times*, 10 December 1986.

7 UNHCR, *Refugees*, August 1986, 28.

8 *Africa News*, 18 April 1988, citing the *Sunday Star*, Johannesburg.

9   Michael Appleby, then regional director of Church World Service, 'A Profile of the Victims of the South African Raid on Gaborone, Botswana', unpublished mimeo, 23 June 1985.

10 For a discussion of this point and several others made by Geldenhuys, see Joseph Hanlon, *Beggar Your Neighbours: apartheid power in Southern Africa,* London, 1986, 30-31.

11 *Southscan*, 29 March 1988.

12 Southern African Research and Documentation Centre, Harare, in association with Third World Foundation for Social and Economic Studies, *South Africa Imposes Sanctions Against Neighbours*, published for the eighth summit of the Non-Alligned Movement, Zimbabwe, September 1986.

13 Eddie Cross interview, *Cape Times*, 10 December 1987.

14 David Martin and Phyllis Johnson, 'Zimbabwe: Apartheid's Dilemma', in Martin and Johnson, *Destructive Engagement*, 64.

15 Ministry of Information, Posts and Telecommunications, Zimbabwe Government, *A Chronicle of Dissidence'*, August 1984, 25, 34-35.

16 Reginald H Green and Carol B Thompson, 'Political Economies in Conflict: SADCC, South Africa and Sanctions', in Martin and Johnson , *Destructive Engagement*, 271-73; and Reginald Herbold Green, 'Cutting off the Flowers: The Macro-economic and Human Cost of War to Southern Africans', unpublished mimeo, Maputo, November 1986.

Mozambique after Machel

Jeremy Grest

Mozambican president Samora Machel died in October 1986, during a phase of intense South African pressure on his country. This precipitated the gravest military, economic and social crisis Mozambique had faced since independence.

Frelimo weathered Machel's death without any power struggle for succession, demonstrating a high degree of leadership cohesion. The central committee's election of Joaquim Chissano as party president and head of state did not lead to major policy changes. It consolidated trends, already visible under Machel, towards a more collective style of leadership, decentralisation of administration, and economic reform.

War and emergency

The MNR's war against Frelimo and its effects on the entire population and the economy remain the dominant feature of life in Mozambique. South African destabilisation in the area intensified after the Nkomati Accord of 1984. Malawi became a major springboard for MNR strategy, developed and co-ordinated by South African military intelligence. This aimed to cut Mozambique in two and permanently occupy parts of Tete and Zambezia provinces.

Military pressure from the MNR was accompanied by an intensified campaign of economic sanctions, including a South African ban on migrant labour recruitment from Mozambique. The Mozambican army, backed by Zimbabwean and Tanzanian troops, succeeded in containing the MNR thrust into the central provinces in late 1986 and early 1987. In January 1987 an MNR drive towards Pebane and the coast was blocked. During February and March 1987 the recapture of MNR-held towns in Zambezia, Tete and Sofala began. As the Mozambican forces consolidated their position in the centre-north and the MNR lost ground, small MNR units began to move south. Simultaneously, direct South African infiltration into Gaza and Maputo provinces increased.[1]

The focus of the war shifted south as the MNR attempted to capture and hold territory in Gaza province and isolate Maputo from the rest of the country. When this failed after Mozambican counter-attacks in mid-1987, the MNR began attacking poorly-defended smaller towns in large numbers, massacring inhabitants. In July 1987 several hundred MNR troops overran Homoine in Inhambane province and systematically slaughtered everyone they found. Over 400 people died in the attack which the MNR disclaimed when the extent of the atrocity became known.

In a similar incident during August 1987, an estimated 600 MNR rebels attacked Manjacaze in Gaza, which was defended by nine Frelimo soldiers. More than 90 inhabitants of the town died before the MNR, which had lost ten combatants, was repelled. A captured MNR fighter, present at the camp from which the Homoine attack was launched, alleged that South Africa continued to supply the MNR by helicopter.[2]

In June 1987 the war spread beyond Mozambique's borders. For the first time the MNR attacked a town in north-eastern Zimbabwe, stealing food and blankets and murdering more than ten inhabitants. The MNR had declared war on Zimbabwe in 1986, after Zimbabwe began supplying military aid to Mozambique. The attacks which followed in the eastern border region of Zimbabwe may have been aimed at drawing Zimbabwean forces out of Mozambique where up to 12 000 troops were aiding the Frelimo offensive in the central provinces.[3] Zimbabwe's commitment to Mozambican security, however, remained firm.

Zambia was also affected by the war. From late 1987 border villages were attacked, residents abducted and cattle and maize taken back into Mozambique.[4]

Towards the end of 1987 the MNR began focusing on Maputo's communication links, aiming to isolate the capital. Convoys of buses and lorries were attacked, vehicles burned and passengers killed. In October 1987, 80 vehicles were destroyed and over 270 people died near Taninga, north of Maputo. Passengers on trains to Maputo were massacred.[5] Small towns and rural communities in the vicinity of Maputo were attacked and their infrastructures destroyed. Inhabitants were forced to take refuge in the capital, straining its already limited resources. Although the MNR failed to cut Maputo off entirely, the disruption was serious and costly.

Fighting in southern Manica, northern Tete and southern Inhambane continued during the first part of 1988, with the MNR
concentrating on assassination of local leaders and destruction of infrastructure. The MNR remained active in Niassa, Nampula and
Zambezia as well, but its control over these areas was limited.[6]

Between 1982 and 1986 over 500 000 people lost their lives as a
direct or indirect result of MNR activity.[7] The MNR invasion of
Zambezia in 1986 and the Mozambican army's counter-offensive
created a massive refugee population in Malawi. By May 1987 an estimated 235 000 peasants from Mozambique's most productive
farming areas were living as refugees in Malawi. An early 1989 estimate put the figure at over 600 000.[8]

The MNR's Gaza offensive sent refugees streaming across the
South African border into Gazankulu, while the war in the central
provinces forced peasants into Zimbabwe. A World Council of
Churches study published in June 1988 estimated that the total
number of displaced Mozambicans in neighbouring states exceeded
one million. Over half that total were in Malawi, 200 000 in South
Africa and the rest in Tanzania, Zambia, Zimbabwe and Swaziland.[9]
These figures do not take account of the number of internally-
displaced Mozambicans, estimated at 1,6-million in January 1988.[10]

The MNR invasion of Zambezia and Tete disrupted food
production and placed 3,5-million people at risk of starvation in the
rural areas, while the normal food requirements of the 2,8-million
urban dwellers also became dependent on imports, making a total of
over 6-million people - 40% of the population - reliant on relief
supplies for their survival.[11]

A 1987 UNICEF report on Mozambique and Angola gave these
two countries the highest infant mortality rate in the world -
between 325 and 375 deaths per 1 000 for children under five years
old. War and destabilisation was estimated to have caused 84 000
child deaths in 1986 alone, with a total of 320 000 child deaths between 1981 and 1986.[12] The MNR has also abducted children to
serve in the war.[13] A Ministry of Health report on the impact of
South African destabilisation states that over two million people
had lost access to health care by the end of 1986 as a result of the
closure of over 700 health posts.[14] The war has closed down 2 629
schools, destroying half the primary-school network and depriving
about 500 000 pupils of education. Four hundred teachers have
been killed, abducted or mutilated and 22 secondary schools cater-

ing for 8 000 pupils destroyed.[15]

By late 1986 the overwhelming problems of war and famine had forced the Mozambican government to petition the UN for relief aid. A new national structure, the National Executive Committee for the Emergency, was created in Mozambique to co-ordinate relief work between ministries and donors, and oversee the work of the Natural Disasters Office.[16]

An emergency airlift of supplies to Sofala province began shortly afterwards. Total aid pledged by 100 governments in 1987 was approximately US$276-million.[17] In 1988 a further $270-million was pledged towards Mozambique's emergency appeal for 1988-89. The bulk of aid pledged in 1988 - $183,4-million - was for food. Logistical support for transport of supplies amounted to $76-million.[18]

In November 1988 Mozambique was still facing a food deficit of almost 300 000 tons of grain for the period up to 30 April 1989. Since it lacks the foreign currency to buy food, Mozambique will remain reliant on aid to make up the shortfall. Frelimo military gains in several districts in Tete, Zambezia and Sofala during 1988 necessitated further urgent relief work. Returning refugees in the border provinces placed a further burden on the receiving areas which, in most cases, lack the resources to deal with them.[19]

Short-term relief aid has in some instances simply created further dependence. Government strategy for displaced people is to feed them, provide land, agricultural tools and seed, and encourage self-sufficiency. The government has also argued for a more flexible approach from donors to allow it to allocate resources according to changing circumstances.

The international community has responded positively to Mozambique's plight, and assistance has saved many thousands of lives. But the government has become increasingly dependent upon international goodwill for its survival, and thereby increasingly open to pressures on internal policy - both from other governments and from the host of Non-Governmental Organisations and agencies involved in the emergency relief programme.

Malawi, South Africa and the MNR

Relations with Malawi improved gradually following the creation of a joint security commission in December 1986, which aimed at eliminating armed banditry in Mozambique. Shortly before his death, Machel accused Malawi of allowing its territory to be used as a rear base for the MNR and threatened to place missiles on the border.

South Africa attempted to counter Mozambican allegations of continued Malawian support for the MNR with claims that Mozambique and Zimbabwe were planning to topple Banda's Malawian government.[20] But by March 1987 relations between Mozambique and Malawi had improved to the extent that Banda was prepared to send troops to protect the Nacala port corridor.

Several official meetings aimed at improving security and other forms of co-operation took place in 1988, culminating in Chissano's July state visit to Malawi. Mozambique's strategy of co-operation and dialogue has thus been relatively successful in neutralising the MNR threat posed from within Malawi.

Relations with South Africa reached an all-time low following the Machel plane crash. Both Mozambique and the Soviet Union rejected the South African Margo Commission's conclusion that the crash was caused by pilot error and negligence, insisting that a decoy beacon had caused the crash and demanding further investigations to establish its location and identity.

Mozambique continued to accuse South Africa of sustained support for the MNR, often backed by corroborating evidence from MNR personnel who had surrendered or been captured. South Africa denied these charges, countering that Frelimo had allowed the re-establishment of ANC guerilla command networks in Maputo. South Africa's agressive posture was summed up by President Botha on a visit to France in November 1986 when he said:

> Unless Mozambique co-operates with South Africa, it has no future. It will down the drain. If they co-operate with us then they have a future.[21]

A ban on migrant labour was announced in October 1986 as part of the pressure South Africa was mounting against Mozambique. This ban, which threatened the livelihood of approximately 61 000 mineworkers and 30 000 Eastern Transvaal farmworkers and their families, was a further blow to an already disrupted economy. Pressure from employers led to a modification of the ban. No more novices were to be recruited for the mines, while in the farming districts Mozambican workers were allowed to register for one-year contracts only. The ban was lifted in November 1988 and recruitment resumed, but a potent weapon remains in South Africa's hands for future use.[22]

In May 1987 South African commandos killed three people in raids on four separate targets in Maputo, leaving behind South African-registered vehicles and escaping by sea. This was the first raid of its kind since the Nkomati Accord. In June the South African minister of Defence, Magnus Malan, publicly suggested that the SADF had plans to step up support for the MNR.[23]

Following the Taninga convoy attack of October 1987, Frelimo's political bureau accused the South African government of complicity in the massacres which, it argued, had become the MNR's principal tactic. Opening a new air base at Louis Trichardt during October, Malan accused Mozambique of allowing the ANC to launch armed attacks from its territory, and warned that South Africa would not allow 'Soviet surrogates' to be given 'a loose rein indefinitely'. There were suggestions at the time that South African military intelligence was planning to re-activate the Comoros as a rear base for the MNR in the north of Mozambique, following Malawi's withdrawal of support for MNR activities.[24]

PW Botha's call in January 1988 for a regional security conference was met with scepticism in Maputo, where it was felt that Pretoria had consistently disregarded the terms of the existing non-aggression pact between the two countries. Pretoria's position on talks hardened soon after, reportedly because of Mozambique's refusal to participate and its continued assertion of South African responsibility for Machel's death.

Foreign Minister Botha warned that his government reserved the right to act as it saw fit.[25] In April Albie Sachs, a leading ANC legal adviser employed by the Mozambican Ministry of Justice, was critically injured by a car bomb. Mozambique immediately blamed South African security services, especially in the light of assassination attempts on leading ANC members in Botswana, Brussels and Paris at around the same time.[26]

In April 1988, Mozambique launched a new diplomatic initiative to revive the 1984 Nkomati Accord. Jacinto Veloso, the minister of Co-operation, delivered a message from Chissano to PW Botha in Cape Town, proposing a full meeting of the joint liaison committee created in August 1987. This was seen as a step towards reviving the joint security commission.

This body had ceased functioning at Mozambique's insistence

following the publication of the Gorongosa documents in 1985, which provided evidence of continued South African control of the MNR. The subsequent joint security commission meeting held in Pretoria during July paved the way for the two heads of state to meet at Songo at the Cahora Bassa power station in September.

The change in international and regional forces was among a number of factors which combined to improve official relations between Mozambique and South Africa. Both the Soviet Union and the US had shifted towards greater co-operation in the solution of regional conflicts, including Southern Africa, and their sponsorship of talks on Namibia and Angola created a new climate in the region. Mozambique had played a part in broking relations between Angola and South Africa and had received briefings from both senior Soviet and American personnel. And the outcome of the battle for Cuito Cuanavale in 1988 weakened those in the South African state apparatus favouring military solutions and strengthened those favouring diplomacy.

In addition, the need to counter the effects of sanctions and the realisation that South African industry is no longer internationally competitive have driven capital interests in South Africa to look more carefully at neighbouring states for trade and investment.[27]

The decisive swing in international opinion against the MNR, following the April 1988 release of Robert Gersony's report for the US State Department in which he documented the systematic brutality inflicted on Mozambicans by the MNR, was also a major factor in the changing regional dynamics.[28]

The MNR has been in a weakened state for some time, with increasing evidence of divisions in its external leadership. The March 1988 defection of Paulo Oliveira, MNR spokesman for Western Europe between 1981 and 1987, provided detailed insights of intense factionalism between what he described as the PretoriaLisbon and Washington-Bonn axes of the organisation. He also provided evidence of continued South African involvement in the MNR until at least January 1988.[29]

In April, former MNR secretary-general Evo Fernandes, expelled in 1986, was murdered in Lisbon. His widow accused the Mozambican security services of involvement, a claim denied by Mozambican Foreign Minister Pascoal Mocumbi. Oliveira suggested Fernandes could have been killed by the American faction of the MNR, or by the South Africans, or as a result of conflicts within his own group.[30]

In November 1988 Chanjunja Chivaca João, former head of the MNR department of Mobilisation in Portugal, handed himself over to Mozambican authorities under an amnesty offer, and gave details of the organisation's attempts to rehabilitate its international image following the Gersony report. He also described continued MNR liaison with South African military intelligence, and regional and ethnic splits within the MNR.[31]

In December 1988, Chissano rebutted speculation that Mozambique had been involved in secret meetings to set up negotiations with the MNR. The government insisted it would not negotiate with a grouping that is led from outside and has no internal social base. The only concession it offered the MNR was an amnesty to those who surrendered.

The amnesty, valid for 1988, and subsequently extended to 1989, met with some success in neutralising the MNR. Around 300 former fighters surrendered in 1988, and a greater number are believed to have abandoned the MNR and merged with the displaced peasantry.[32] Reasons given for surrendering were hunger, disillusionment with MNR methods and a feeling that they were not going to win the war.

Tripartite negotiations between Mozambique, Portugal and South Africa during 1988 led to an agreement over proposals to repair power lines from Cahora Bassa.[33] The cost was estimated at nearly R14-million in December 1988 with about 1 400 pylons having been destroyed by sabotage (*EPH*, 20.12.88). South Africa has, meanwhile, provided R10-million for non-lethal equipment to support Mozambican security forces guarding the line (*NM*, 29.11.88).

The seriousness of Pretoria's renewed commitment to co-operation with Mozambique will be measured in large part by the flow of electricity from Cahorra Bassa to South Africa. It is possible that MNR operations will now be directed away from areas of strategic importance to South Africa and the organisation will operate in other areas of the country and neighbouring states like Zimbabwe.

Subsequent to the Songo meeting a new South African trade mission building opened in Maputo, and the Afrikaanse Sakekamer invited businessmen to investigate business opportunities in Mozambique (*NM*, 08.11.88).

A joint development and co-operation committee formed after the Songo summit held its first meeting in December 1988. Relations between the two states are at present at a confidence-building stage, with the Mozambicans waiting for the South African government to prove its bona fides - especially regarding pledges to cease support for the MNR. The Mozambicans believe there are deep splits within the South African state over Mozambican policy. They aim to widen these splits and strengthen those favouring co-operation, including those in the capital sector.

Economic rehabilitation and social change

The introduction of the Economic Rehabilitation Programme (PRE) in January 1987 was one of the most important steps taken by the Chissano administration. The three-year crash programme has been adopted in an attempt to arrest economic collapse and restore 1981 production levels by 1990. Agriculture, especially the peasant family sector, takes priority in the scheme which encourages a shift to rural production and rehabilitates industries serving the agricultural sector. The PRE also seeks to undercut the massive black market, give a more realistic value to the currency and remove administrative blocks to the development of productive enterprises.[34]

The PRE has profound implications for the future of Mozambican society. Announcing the programme, the government indicated there would be dramatic and radical changes in exchange rate, budgetary, credit, monetary, pricing and wage policies. This shift in orientation has to be understood in the light of an effective declaration of bankruptcy to Western creditors, the signature of the Nkomati Accord, and the negotiations with the IMF and World Bank in 1985 and 1986 which preceded the PRE.

Currency devaluation has been used as a major instrument of economic policy. The metical has fell in value from its official 1987 level of MT40 to the US dollar to around MT620 in mid-1989, with the administration committed to continued adjustments as a method of restricting demand and managing the balance of payments problem. At the same time, the informal market rate dropped from between MT1 500 and MT2 000 to a rate of around MT1 200 per US dollar - a shift from over 40 to 'only' twice the official rate.[35] Rapid inflation was the immediate result of this, the official estimate for 1987 being around 160%.

Sharp rises in the official prices of a long list of commodities and services came into effect in 1987, and further increases were announced in 1988. The state has withdrawn subsidies on food in the official ration system. Increases in April 1988 ranged from over 770% for a kilogram of rice to 150% for a litre of cooking oil, while official price controls have been removed on a growing list of commodities in line with PRE strategy.[36] New minimum wage rates have also been announced, but they only partially compensate for the rises in the cost of food and urban accommodation which came into effect in June 1988.

The PRE is generally credited with having halted the downward economic spiral of 1981-85. But it is not a neutral instrument in social terms and the costs have been largely borne by the urban poor. Shops in the urban centres now carry a greater range of commodities than before - when many were bare - but the prices demanded are out of range of many urban dwellers. The officially-controlled foodbasket, which only covers about half the calorie requirements of an adult, costs more than the minimum wage of 12 500 MT per month and it is estimated that as much as half of available rationed goods are left unsold.[37]

The necessity of introducing or raising charges for a range of social services was discussed in the people's assembly during August 1988. Primary school enrolment rates are dropping below the 1986-87 level of 47% as parents keep children out of school for lack of money to pay for books and equipment, and normal clinic and hospital attendance rates have fallen rapidly in some areas.[38]

The Frelimo Party conference of July 1988 discussed the possibility of opening housing, schools and health care to private interests - a substantial step backwards, as nationalised services are seen as an irreversible gain of the revolution. Frelimo has arrived at this point because of progressive deterioration of services and state inability to finance reconstruction needed. Social services have been caught between the budgetary demands of the war and the PRE emphasis on increasing the level of capital formation through state investment.

New avenues for accumulation are being opened by the PRE, especially for traders who take advantage of the relaxation of controls on small-scale imports, travel to South Africa and return with goods to sell in Maputo. Profit levels are high and the process shows signs of consolidating the strong social and economic position of traders and transporters already in evidence before the introduction of the PRE. Rural trade is becoming increasingly privatised too, as the parastatal Agricom gradually withdraws from the retail sector. This process is hampered by the war and absence of suitable rural traders, whose numbers currently stand at around 2 500 - half the pre-independence level.

In agriculture the policy reforms set in motion by the fourth congress predate the PRE. Restructuring of major state farms and distribution of land to peasant and private farmers have gone ahead in most important agricultural areas. The World Bank has estimated that the area cultivated by state farms halved between 1984 and 1986 - from 123 000 to 67 000 hectares.[39] Reforms have led to an increase in the quantity of produce now available in local markets. But they also fuel differentiation in the countryside and foster class alliances at variance with the popular political structures Frelimo has been building since independence.[40]

Industrial output has increased, due largely to the availability of foreign funds for raw materials and spare parts. However, the growth rate of 18% in 1987 fell to just over 5% in 1988 because of problems with energy supplies, imported raw materials and spare parts.[41] While the greater availability of inputs has enabled many enterprises to increase production, there have been problems caused by PRE price increases which have lowered sales and created accumulations of stocks.[42] Manufacturers are now calling for import controls to protect local industries from foreign competition.[43]

Mozambique's GDP grew at a rate of between 4% and 4,6% in 1988 - less than the 6% planned. Agricultural production grew by only 2,6%, while the transport sector's growth was 4%. However, there was a 9,5% decrease in international rail traffic, which is an important foreign exchange earner. Mozambique's foreign debt now stands at around US$565-million. Defence and security will account for 40% of the total budget, whilst education and health care expenditure will be 9,9% and 5,4% respectively.[44]

Mozambique's dependence on external assistance seems likely to grow in the short term. The Office for the Promotion of Foreign Investment continues its efforts to attract foreign capital to Mozambique, offering attractive conditions to would-be investors. The idea of private enterprise has assumed an air of legitimacy in both official and popular discourse, and private farmers and other entrepreneurs have begun to assume a greater public role in their communities. Many public officials whose private economic activities benefit from the contacts and resources which their public positions provide are themselves becoming involved in entrepreneurial positions.[45]

Aeprimo is a new association of private sector companies established in May 1988, which is open only to privately-owned Mozambican firms. With 227 members, Aeprimo intends to be a conduit for a substantial part of the Western - especially US - aid destined for the private sector. Foreign exchange controls were relaxed in June 1988, allowing anyone to open an account with the Central Bank. Previously only foreigners and Mozambicans who could prove they had legitimately received foreign currency could do so.[46] This measure aims to attract foreign currency into the banking system and is effectively an amnesty for those who accumulated currency through illegal transactions.

SADCC transport corridors

Mozambique's ports of Maputo, Beira and Nacala provide strategic corridors to the landlocked SADCC (Southern African Development Co-ordination Conference) states of Zimbabwe, Zambia and Malawi as well as to South Africa. The Beira corridor project was established as the SADCC's priority in 1986, when a comprehensive ten-year rehabilitation programme with a $614,6-million budget was published. As the political climate in the region changed, both Western capital and aid donors began to look more carefully at investing in SADCC projects both as a means of supporting the Frontline states and as an alternative to sanctions against South Africa.

The Mozambican government created the Beira Corridor Authority to manage the rehabilitation of the port after successful donors' conferences had raised 90% of the capital necessary for phase one of the project. Traffic through Beira in 1987 increased by just over 31% - about half of this being Zimbabwean.[47] But Zimbabwean capital has been slow to switch to Beira and there is some mistrust between Mozambican authorities and Zimbabwean businessmen. However, the Beira Corridor Group (BCG) is strongly committed to the use of the port as it will save Zimbabwean business valuable foreign currency. Mozambique, in turn, needs the port revenues for its economic reform programme.

Since 1983, Zimbabwe has provided protection for the vital oil pipeline from Beira. In 1985, as calls for sanctions against South Africa and the threat of counter-sanctions grew, it increased its military commitment to Mozambique. The Zimbabwean army has between 7 000 and 10 000 troops patrolling the corridor and engaging in offensives against the MNR. This has reduced their attacks to hit-and-run raids which cause disruption, but offer no real threat to the continued functioning of the line. The BCG is encouraging a number of agricultural schemes as part of a strategy to repopulate the countryside and improve the security situation.[48]

The Maputo port transport system is the most important of SADCC's five transport routes, and is the only port in the region which could dramatically increase its throughput with minimal rehabilitation, given security on the main access routes. It is also the only port in the SADCC network which South African capital needs for its own exports and imports, being the most viable harbour for the Northern and Eastern Transvaal. Maputo is also the port closest to Swaziland and southern Zimbabwe, offers a wider range of handling facilities than Beira and crosses easier terrain than the Beira-Mutare line.

The Limpopo line also connects Maputo to the Zambia, Zaire and Botswana rail systems through a link with Bulawayo. At its operational peak in 1974, Maputo handled over ten million tonnes of cargo and was Africa's second largest port after Durban. The closure of the link to Rhodesia after Mozambican independence lost the port two million tonnes a year. Traffic increased briefly after Zimbabwean independence but by 1984 the route was permanently closed by MNR attacks. Swaziland traffic fell by half as a result of attacks on the Goba route, and South African sanctions on the port reduced flows from peaks in the 1970s of six to seven million tonnes to under 500 000 tonnes during 1987.

Since the inception of the PRE, Maputo authorities have embarked on an intensive campaign to promote the port. There has also been speculation that Maputo is being viewed as a sanctions-busting route for South African capital.[49]

Phase one of the rehabilitation of the Limpopo line was completed in 1988 and in March traffic moved down the complete route for the first time in four years. Phase two of rehabilitation started in 1989 with funding from USAID, West Germany,

Botswana, Portugal and Canada. Security for the route remains a key long-term problem. The Zimbabwe and Mozambican armies have been co-operating in Gaza province to keep the line clear, while Britain has provided non-lethal military assistance.

The Nacala port transport system is the third of Mozambique's vital transport links to the interior. Nacala is a natural deep-water port servicing Malawi. As the most highly-subscribed of all the SADCC transport projects it is newly refurbished, but desperately short of cargo. Security problems along the line have delayed rehabilitation work.

Combined Mozambican, Tanzanian and Zimbabwean forces launched offensives in the north in 1988, recapturing Milanje, a town on Malawi's eastern border with Mozambique held by the MNR since 1985. But Tanzanian troops were withdrawn in December 1988. No reasons were given for this, and official statements referred to a successful completion of their task. The costs to Tanzania and disapproval of Mozambique's rapprochement with South Africa in September 1988 may have been factors in the Tanzanian decision to withdraw. Although Frelimo seems to have gained a better hold on the towns in the north, the countryside still provides sanctuary for the MNR, making Nacala's future security uncertain.

Party, people and policies

Frelimo's fifth congress was held in July 1989, having been postponed for a year due to the war situation. Major events since the last congress included the Nkomati Accord of 1984, Mozambique's accession to the IMF in 1985 and the PRE of 1987.

The party has been badly weakened by the effects of the war. Its organisational structures at the base have lost coherence in large areas of the country and Frelimo has suffered a substantial loss of legitimacy through its agricultural policy failures. The build-up to the congress provided Frelimo with an opportunity to remobilise party structures whilst the congress itself provided an index of how democratic Frelimo remains and which forces within the society and party are setting the policy agenda for the next five years.

Theses debated at the congress were released for public discussion in November 1988 and provided some insights into the preoccupations of party leadership at this important juncture. Seven theses dealt with the party; national unity; social organisations;

national defence and security; economic scientific and technical development; social questions; and international relations.

The thesis on the party stressed its origins as a broad front organisation during the national liberation struggle, and its present status as a party of the entire people rather than the instrument of a particular social class or stratum. At the same time the vanguard nature of the party was emphasised, and socialism reaffirmed as the ideological base - 'synthesized and creatively applied to the national reality'.[50] Marxism-Leninism was not mentioned as official ideology and more stress was laid on bourgeois-democratic concepts such as well-being, progress, justice, liberty and equality than on class-struggle - a notion which seems to have dropped out of official pronouncements.

The second thesis, on national unity, argued that the national liberation struggle began the process of nation-building and that this was now under threat from a war which has undermined the social fabric and erased moral values. The call for unity undoubtedly reflects grave concern that social fragmentation is far advanced in Mozambique and that regional, ethnic and racial sentiments are on the upsurge.

PRE, which has exacerbated the hardships experienced by large numbers of Mozambicans, especially urban dwellers, has added to the war's negative effects on national unity. Placing national unity above all else has become a defensive strategy aimed at preventing the total dissolution of 'the nation' and at rallying support for the current programmes of the leadership.

The thesis on social organisations described the mass democratic organisations as having important patriotic, civic, economic, social and cultural tasks to perform as well as linking the party to the people. The fourth congress had noted that some sectors of the people had been marginalised in the process of mobilising the population into mass organisations. The question of Frelimo veterans in particular has been problematic as many were sidelined after independence and several leading personnel were dismissed in the re-organisation of the armed forces. Former combatants finally obtained their own association - 13 years after independence.

The mass democratic organisations have not been highly successful in harnessing the creative potential of the Mozambican people. Whilst affirming the right of association as a precondition for the development of democracy, Frelimo has nonetheless tended to view non-party organisations as conduits for the party line. Consequently they exercise little organisational autonomy in practice and their effectiveness as independent voices of the people has been limited.

The thesis on economic, scientific and technical development argued that the social and economic directives of the third and fourth congresses remained valid in general terms, even though the principal tasks laid down by the third congress and the national plan were not realised. The main reason given for this was stagnation brought on by external aggression, although other factors such as natural disasters and administrative errors played a part too. The thesis saw no contradiction between the PRE currently being pursued and Frelimo's development goals laid down earlier.

It remains unclear whether the PRE is regarded as a necessary but temporary expedient on the long road to socialism or whether it marks the beginning of a decisive shift towards private accumulation. Congress documents do not provide answers to this question.

The congress was held at a time when Frelimo has been at a crossroads. There are implicit contradictions in the theses between a party which seeks to be the party of all the people and remain a vanguard at the same time, and between notions of socialism and the implications and effect of the PRE. There is currently a fundamental reshaping of Frelimo's accumulation strategy. Alongside this, new class alliances are being forged which reflect the new priorities.[51]

The broad front strategy which won Frelimo independence from Portugal is being reasserted in an attempt to win back the social forces alienated by Frelimo's failure to build socialism in one country. It has been argued that this process merely reflects the replacement of 'symbolic socialism' with 'symbolic reform', and that the state's capacity to implement an alternative strategy will be equally limited by the same factors that prevented it achieving its first set of goals.[52] This underestimates the real effects of the new forces for accumulation which have been set in motion by the IMF/PRE strategy.

The combined impact of a change in state strategy and large injections of foreign funding must have a significant impact on the internal balance of forces. Given the weakness of the state, its ability to control experiments with 'market socialism' and keep private sector growth within clearly defined bounds is questionable. The growth of an indigenous entrepreneurial stratum under these conditions and the creation of new alliances with the state contain the possibility that the long-term objectives of socialism will be progressively abandoned without ever ceasing to be the official goal of the party.

The period since Samora Machel's death has been a testing time for Mozambicans. The military threat has been contained but the overall situation remains grave. The destruction and dislocation caused by the war have been almost incalculable and have had the effect of increasing dependence on the international community for survival.

The IMF/PRE strategy adopted by Frelimo has staved off total economic collapse but the full costs in social terms, and in terms of Frelimo's socialist project have yet to be assessed. The indications are that the period under review marks an important watershed.

At a regional level the Mozambican government has adopted a strategy of renewed dialogue with Pretoria. This is based on a clearer understanding of the limits and possibilities of such a relationship, and is informed by an analysis of the regime which seeks to exploit openings provided by its internal divisions. The regional conjuncture after Cuito Cuanavale is more favourable for peace and reconstruction than previously. But whether Frelimo is able to take full advantage of it, having squandered much of its political goodwill internally, and having heavily mortgaged the future to the IMF, remains to be seen.

Notes

1   Economic Intelligence Unit, *Country Report*, 4, 1987, 18.
2   AIM, *Mozambique News*, 135, October 1987, 10-11.
3   Diana Cammack, 'Mozambique Briefing', *Review of African Political Economy*, 40, December 1987, 73.
4   Economic Intelligence Unit, *Country Report*, 1, 1988, 23.
5   *Country Report*, 1, 1988, 18-19.
6   *Country Report*, 2, 1988, 16.
7   *Mozambique News*, 136, November 1987, 1.
8   AIM, *Mozambique File*, 150, January 1989.
9   *Country Report*, 3, 1988, 22.
10 *Mozambique News*, 138, January 1988.
11 *Country Report*, 2, January 1988, 17.
12 UNICEF, *Children on the Front Line: the impact of apartheid, destabilisation and warfare on children in Southern Africa*, 1987.
13 *Mozambique News*, 135, October 1987, 10-11.

14 Ministry of Health, People's Republic of Mozambique, *The Impact on Health in Mozambique and South African Destabilisation*, 1987.

15 *Tempo*, 934, 4 September 1988.

16 *Mozambique News*, 130, May 1987.

17 *Mozambique News*, 138, January 1988.

18 *Tempo*, 917, 8 May 1988.

19 *Mozambique File*, 149, November 1988.

20 *Africa Confidential*, 27(23).

21 *Country Report*, 1, 1987, 24.

22 *Tempo*, 953, 15 January 1989.

23 *Country Report*, 3, 1988, 21-22.

24 *Country Report*, 3, 1988, 21-22.

25 *Country Report*, 2, 1988, 18.

26 *Tempo*, 914, 17 April 1988.

27 For details of recent South African business interest, see *Weekly Mail*, 13 January 1989 and 10 February 1989.

28 R Gersony, *Summary of Mozambican Refugee Accounts of Principally Conflict-Related Experience in Mozambique*, 1988.

29 *Mozambique News*, 141, April 1988.

30 *Mozambique News*, 142, May 1988.

31 *Tempo*, 948, 11 December 1988.

32 *Mozambique File*, 150, January 1989.

33 *Country Report*, 3, 1988, 20.

34 Economic Intelligence Unit, *Country Profile*, 1988-89, 10.

35 Swedish International Development Authority, *Country Report: Mozambique*, SIDA planning secretariat, Stockholm, 1988, 16.

36 *Country Report*, 2, 1988, 21.

37 *Country Report: Mozambique*, 20.

38 *Country Report: Mozambique*, 20.

39 *Country Report: Mozambique*, 24.

40 O Roesch, 'Rural Mozambique in the Baixo Limpopo', *Review of African Political Economy*, 41, 1988.

41 *Mozambique File*, 150, January 1989.

42 *Country Report*, 3, 1988, 30.

43 *Country Report*, 4, 1988, 25.

44 *Mozambique File*, 150, January 1989.

45 Roesch, 'Rural Mozambique', 89.

46 *Country Report*, 3, 1988, 26.

47 Colleen Lowe Morna, *The SADCC Ports Handbook*, London, 1988, 22.

48 Morna, *SADCC Ports*, 26.

49 Morna, *SADCC Ports*, 8.

50 *Tempo*, 948, 11 December 1988.

51 For a development of this argument see V Hermele, *Land Struggles and Social Differentiation in Southern Mozambique*, Scandinavian Institute of African Studies, Uppsala, 1988.

52 M Ottaway, 'Mozambique: from symbolic socialism to symbolic reform', *Journal of Modern African Studies*, 26(2), 1988, 211-226.

Security Council Resolution 435 and the Namibian Independence Process

Brendan Seery

'Independence or death' was the rallying cry for supporters of the South West Africa People's Organisation during the days of South African rule over the desert territory of Namibia.

But on the day when the end of the freedom road was in sight - 1 April 1989 - it looked tragically as though 'independence and death' was more appropriate.

In the streets of the Namibian capital, Windhoek, hundreds of Swapo supporters gathered to welcome the start of the United Nations Security Council Resolution 435 peace and independence plan with singing, dancing and ululating. From the townships of Katutura and Khomasdal, a joyful snake of thousands of Swapo faithful wound its way towards the city centre. The presence of scores of riot policemen - armed with tear-gas, rubber bullets and quirts, and determined to stop the crowd from reaching the town's centre - failed to dampen the spirit of the people, even when ordered to disperse.

The arrival of the United Nations' Special Representative for Namibia, Martti Ahtisaari, on 31 March, confirmed for many that South Africa was finally serious about giving up what had effectively been its fifth province since 1915.

South African Foreign Affairs Minister Pik Botha left no doubts about that when he addressed a closed meeting of seconded civil servants on 1 April: they should be proud of what South Africa had given the territory, he said, and the South African withdrawal was not a sell-out of the whites. Botha made it plain Pretoria had no moral right to continue controlling Namibia.

Even as Botha was talking to civil servants in the austere and imposing Windhoek headquarters of the SWA Territory Force, the first clash took place in northern Namibia between an apparently lightly-armed patrol of the SWA Police and a group of about 80 Swapo insurgents, many carrying heavy weapons such as SAM-7 anti-aircraft missiles and RPG-7 rocket launchers.

That clash was only the start of a series of bloody battles the likes of which had not been seen before in the bush war between the South Africans and Swapo, which had started in August 1966. The South African foreign minister threatened to kick the UN out of Namibia unless it did something to 'bring Swapo to its senses'. His condemnation was echoed by British Prime Minister Margaret Thatcher, who was in Windhoek to visit British troops in the UN peacekeeping group and give the nation-in-embryo her blessing.

The South Africans eagerly pressurised Ahtisaari into releasing a number of their army battalions from confinement-to-base, making it clear to the UN man that whether or not he agreed, this would happen.

Botha and his spokesmen were also quick to exploit the opportunity for propaganda purposes. Few of the hundreds of foreign correspondents in the territory for the run-of-the-mill 'last colony gets its freedom' story had much sympathy for Swapo. Western countries were almost unanimous in their condemnation of Swapo's violation of the UN peace plan and its various, and related, accords.

By the end of April, the situation was still in flux, but close to 350 people had died, more than 315 of them Swapo fighters. The affair was one of the more grisly introductions to independence on a continent which has seen more than its fair share of strife.

Towards the unthinkable

Only a year previously, the situation was the last anyone could have foreseen. April 1988 in Windhoek was a grim time for those who hoped for independence, an end to the bush war and South African rule.

That month, the big guns of the South African National Party establishment flew to Windhoek to lay down the law. President PW Botha, Defence Minister Magnus Malan, Pik Botha, Finance Minister Barend du Plessis, and Education Minister Gerrit Viljoen came with the clear message that Pretoria was still calling the shots.

President Botha, in his best self-assured and finger-wagging form, indicated a more 'hands-on' approach by South Africa to Namibia. Administrator-General Louis Pienaar was given added powers to direct the pace and directions of constitutional change - in other words, steering constitutional matters down the 'own affairs, general affairs' path the NP favoured for South Africa itself. President Botha additionally warned of a crackdown on opponents of the system - especially newspapers - leading many political observers in Namibia to fear that South Africa's harsh state of emergency legislation was about to be exported across the Orange River.

The 1988 visit of President Botha and company to Windhoek should, however, be placed in perspective.

Many political analysts have considered that, decades ago, South Africa developed a 'two-track' policy towards Namibia. While its diplomats negotiated and wrangled with the international community about the territory's independence, Pretoria went ahead with its own constitutional and social experiments there. Analysts theorised that whichever track appeared to be the most promising at the time for Pretoria would be pursued most vigorously. Although few believed South Africa had actually crossed the psychological hurdle of 'giving away' Namibia, it was felt that the decision-makers in Pretoria's Union Buildings would do so if they believed the benefits for the National Party government outweighed the disadvantages.

Thus, although South Africa accepted the Resolution 435 plan in 1978, it seldom felt under sufficient pressure to allow it to be implemented. The peace plan collapsed in late 1978 and early 1979, and Pretoria was content to let matters lie as they were.

The entry of the United States into the arena, with the Reagan administration's policy of 'constructive engagement', did nothing to convince the South Africans they should be getting out of Namibia. Indeed, Pik Botha and his colleagues seized gladly on the principle of 'Cuban linkage' first mooted in 1981 by the Americans, concerned about the strategic implications of a massive Cuban - and Soviet - presence in Africa.

Washington and Pretoria were brought closer by their support for Savimbi's Unita rebels, and an unspoken hope between them that they could help reverse their loss in the Angolan civil war in 1975-76, when the Soviet and Cuban-backed MPLA was thrust into power in Luanda.

The mid-1980s was, therefore, a time when Resolution 435 was low on the South African agenda for Namibia.

The search for proteges

In the absence of significant international pressure for Namibian independence, South Africa continued to use the territory for its own
constitutional tinkering. 'Internal' policy in Namibia was based on
two legs. One involved cultivating a moderate alternative to Swapo
(for early on, South Africa's policy-makers realised, their public
statements notwithstanding, that Swapo would be the party they
would have to beat in Namibia). The other entailed using Namibia
as a 'test-bed' for reform-type legislation intended for use at home.
The cardinal rule, though, was that no matter the arrangement with
internal politicians, no matter how much power they were given,
Pretoria always ultimately called the shots.

Thus after the experiments with the Turnhalle and Minister's
Council - which ended when the Namibian politicians grew too
'cheeky' for South African liking - Pretoria continued to look for
proteges inside Namibia.

It found them in the six-party Multi-Party Conference (MPC), a
multi-racial collection covering the spectrum from slightly left-of-
centre to right of South Africa's Nationalists. The MPC fancied
itself as Namibia's 'Rainbow coalition' and hoped to appeal to a
broad range of ideological sensibilities.

The MPC was made up of
- the Democratic Turnhalle Alliance (DTA), led by veteran ex-
  Nat Dirk Mudge, and itself a collection of 11 parties largely
  operating along ethnic lines. It advocated a middle-of-the-road
  social democracy, with one-man, one-vote being implicit, along
  with the removal of apartheid;
- the Rehoboth Free Democratic Party (RFDP): led by Hans
  Diergaardt, the party represented the clannish and independent
  mixed-race Basters of Rehoboth, 80 km south of Windhoek, who
  have lived in Namibia since 1870, 20 years before German
  colonists arrived. The RFDP was almost xenophobic, to the
  point where it often allied itself with the proponents of apartheid in its efforts to avoid seeing its identity subsumed in a
  unitary state;
- the National Party of South West Africa (NP-SWA), led by
  Kosie Pretorius and unashamedly in favour of separate development, although couching it in the less-offensive term of 'protection of group rights'. Generally, the NP-SWA came to be

regarded as to the right of its sister party in South Africa, so much so that its members had more in common with the AWB and the Conservative Party than with PW Botha's Nats;

- the South West Africa National Union (Swanu): led by Moses Katjiuongua, the party was the 'right' faction of the oldest black political organisation in the territory. Swanu split in two over participation in the MPC, which leftists in the party considered a South African front organisation. Nevertheless, Swanu (MPC) promoted a socialist-leaning ideology;
- Swapo-Democrats: led by Swapo defector Andreas Shipanga, the party likewise adopted a left-of-centre ideology, although not as openly socialist-inclined as Swanu (MPC). Shipanga appeared to have little support;
- the Labour Party: led initially by Dawid Bezuidenhout, later by Reggie Diergaardt (no relation to Hans), the party was supported mainly by coloureds. It was strongly opposed to apartheid, yet with a moderate approach similar to that of Swanu (MPC) and Swapo-D.

The MPC parties officially formed their alliance in 1984, producing shortly afterwards a 'Declaration of Fundamental Rights and Objectives'. They visited a number of African countries, including Senegal and the Ivory Coast during 1984, in their search for support for their moderate ways. Later that year, they attended the abortive Lusaka Conference, after which there seemed to be further support for the notion that 435 was dead.

Neither national nor united

In 1985, the MPC six approached the South Africans with what amounted to a request for power. This fitted in well with Pretoria's 'internal' policy track at the time, and afforded South Africa the opportunity to convey the impression it was letting Namibians be the masters of their own destinies.

Thus, on 17 June 1985 - in the depth of economic woes at home - State President Botha flew to Windhoek to install the MPC coalition in office as a government. Styled as the 'transitional government of national unity' (TGNU), the coalition was installed with great ceremony - helped by the bussing in of thousands of DTA supporters. The Windhoek politicians were given considerable autonomy in running the day-to-day affairs of the territory, but

Pretoria retained control over the important portfolios of defence and foreign affairs.

After the collapse of the Lusaka Conference and the apparent lack of interest in the UN settlement plan, the installation of the TGNU struck some political commentators as an attempt by South Africa to defy world opinion and declare 'UDI' for Namibia.

Even the most bullish of South African government experts conceded at the time of the appointment of the TGNU that it would not attract much more than 50% of the Namibian electorate. Foreign Affairs Department analysts were more forthright: they said in private that they hoped the six MPC parties would be able to garner enough support to deprive Swapo of the whip-hand two-thirds majority it needed to dictate a constitution in the constituent assembly envisaged under the UN plan. The two-thirds majority stipulation was agreed by all parties to the dispute in 1982, as an effective codicil to the original Resolution 435 plan. The 1982 agreement also included a stipulation that any future constitution would have a bill of fundamental rights, guaranteeing freedoms of association, speech, religion and property.

A cynical South African foreign affairs expert described the TGNU, shortly after its inauguration, in terms similar to those used by an historian to describe the Holy Roman Empire (which was neither Holy, Roman, nor an Empire). The TGNU, said the South African, was not a government, it was not national and it was 'certainly not unified'.

Over the next three-and-a-half years, his comments were proved right in all respects, including the implied one that the government was 'transitional'.

The years of the TGNU were largely characterised by squabbling among its members, principally over ideological questions or, more crudely, what to do (if anything) about apartheid, which remained in force in the territory.

The wrangling was typified in the work of the constitutional council, which sat for two-and-a-half years under the chairmanship of retired South African supreme court judge Victor Hiemstra. Hundreds of hours of debate, more than two million rands in expenses and a number of overseas study trips later, the council had still failed to reach agreement on a constitution. The body's members - largely nominated by the TGNU parties (Swapo refused to take part) - blustered on and on about the protection of individual rights versus the protection of 'group rights'. They were, in effect, debating the future of separate development which, though not in its Verwoerdian guise, nevertheless dominated Namibian society.

Under South African government proclamation AG 8 of 1980, the administration of Namibia was divided up into 11 separate, ethnically-based systems. Each population group had its own legislative assembly, with elected representatives, and its own civil servants catering for its own needs.

The system was, because of its duplication, and inherent corruption and inefficiency, ruinously expensive. But, it effectively preserved white privilege and the 'kingdom' of the Basters at Rehoboth, and therefore was defended tooth and nail by both the NP-SWA and the RFDP.

It was not in South Africa's interests to have a unitary state, working on the principles of one-person, one-vote and equality for all, set up on its borders and on territory it still controlled. Thus, the NP-SWA and RFDP found themselves with a powerful ally in Big Brother in Pretoria, even though they were constantly out-voted by their partners on the TGNU.

The lessons of the right-wingers' clout within the TGNU were not lost on many ordinary Namibians who saw in it confirmation of their feeling that the Windhoek administration was nothing but a South African front.

Many of the genuinely well-meaning attempts by the left-leaning groups within the TGNU to change society thus came to nothing. Education Minister Andrew Matjila (a DTA man) foolishly predicted in the last quarter of 1986 that all schools would open to all races by the start of the first term in 1987. The right wing quickly put paid to that notion and, it is interesting to note, schools were still segregated almost two months *after* the start of the UN settlement plan in 1989.

The failure to tackle apartheid vigorously and do anything about pressing social problems meant the TGNU made little progress in extending its grassroots support. By appearing to do South Africa's dirty work on some occasions (such as the issuing of a certificate of indemnity for four SADF soldiers accused of killing a civilian) it probably even lost support.

The trip by President Botha to Windhoek in April 1988 was another nail in the coffin of popular support for the TGNU. It suggested, once again, that the Windhoek politicians were powerless to do anything which did not meet with Pretoria's approval. That the coalition was expendable and irrelevant as far as South Africa was concerned, only became apparent as the events of 1988 unfolded.

The South Africans had hardly left Windhoek after their whistle-stop lecture tour, when events elsewhere started to force their hand on Namibia. After President Botha's uncompromising stance and threat of a crackdown on dissent, anti-apartheid groups in Namibia despaired of ever seeing their freedom. Activists openly said they believed Namibia would not become independent until South Africa itself had first been liberated.

A turning point at Cuito Cuanavale

Whatever the battle for Cuito Cuanavale was or was not, it represented a turning point in the history of Southern Africa.

In the battles around the town of Cuito Cuanavale and the near-by Lomba River - which stretched from late 1987 well into 1988 - large numbers of South African forces were thrown in to help save Unita from possible annihilation. The rebels of Jonas Savimbi had never before come closer to having to defend their stronghold at Jamba. And the combined Cuban and Angolan forces had never before inflicted such heavy casualties on the SADF-Unita allies.

Which side was the outright winner in these battles of southern Angola is best left to military historians able to sort through the claims, counter-claims and propaganda. Angola's army, Fapla, and its Cuban allies claimed to have routed the South Africans and Unita, while the SADF claimed it had inflicted terrible casualties on the other side.

There is some doubt whether the SADF was forced to flee from the battles. But Cuito Cuanavale itself remained in the hands of the Angolans - although South Africa's commander said they never in-tended to take the town. The SADF - regarded by many white South Africans as Africa's 'superpower' and therefore virtually invincible - had failed to defeat a 'black' army. That in itself was a defeat.

Cuito Cuanavale had many important lessons for both the military 'hawks' and the policy-makers in Pretoria. It showed that the Angolans and Cubans had improved considerably in their fight-ing ability. They were now apparently being backed to the hilt by the Soviet Union, and the appearance of sophisticated armaments - such as jet fighters, anti-aircraft systems and tanks - suggested that the balance of power was swinging away from South Africa.

One important by-product of Cuito Cuanavale was the growing confidence of the Angolan and Cuban forces, and a determination to show they would no longer allow the SADF to treat southern Angola as its back yard.

This new confidence began to show itself in a worrying way for South Africa shortly after President Botha's April trip to Windhoek, when crack Cuban units began moving south, through western Angola, towards the Namibia border.

The South Africans initially tried to make light of the threat. One South African Sunday newspaper quoted military sources as saying the Cubans, in moving towards the Namibian border, were making the same tactical and strategic mistakes Adolf Hitler had made in 1941 in ordering the disastrous invasion of the USSR. The theory advanced in the article was that the Cubans had stretched their lines of supply and communication too far and would be extremely vulnerable to guerilla attack by Unita during the rainy season. This naively ignored the fact that the Cubans were linked to their rear supply areas by good tarred roads, and were building good all-weather airstrips. It also ignored the reality that there were few Unita groups operating permanently in western Angola.

Less than a fortnight after that article appeared, 12 white South African troops died in a bombing raid by an Angolan MiG jet fighter on the Namibia-Angola border town of Ruacana. The SADF claimed the bomb had been an unlucky hit, as it had been aimed at installations at the Ruacana hydro-electric scheme. Nevertheless, the attack did come after continued South African artillery shelling of Cuban positions at the Calueque water reservoir 20 km inside Angola.

The deaths at Ruacana looked like the start of one of the South African establishment's worst nightmares. Cubans were massing on the borders of Namibia, there was evidence that Swapo units were present and ready to make use of the Cuban 'umbrella of steel' to launch massive infiltrations or even a conventional campaign into northern Namibia.

South African military analysts reportedly arrived at the conclusion that it was possible to beat the Cubans, but that the fighting would be costly - with more than 300 white regular and national service deaths, and as many as 2 000 black soldiers being lost.

These casualties were unacceptably high for the political leaders, who would have been aware of a growing feeling among South African whites that Angola was becoming South Africa's 'Vietnam'. Even right-wingers, while pledging to fight to the bitter end for their own 'grondgebied' were saying publicly that the SADF should not be in Angola.

The military, strategic and international diplomatic realities saw the South Africans forced to the negotiating table.

Earlier in 1988, the US assistant secretary of state for African affairs, Chester Crocker, had begun another one of his seemingly endless rounds of diplomatic shuttling. But this time he began to find attentive listeners for his proposals. He put to Pretoria and Luanda the idea of the 'buffer state' in the form of an independent Namibia. For Luanda, Namibia would act as a buffer against the SADF. For Pretoria, Namibia would act as a buffer against the perceived communist 'total onslaught' (bearing in mind the South Africans well knew they could apply tremendous economic pressure on any Namibian government if they so wished).

The South Africans were forced to back down from their original stance that there could be no implementation of Resolution 435 until all the estimated 50 000 Cubans had left Angola. Instead, they obtained a commitment to a phased Cuban withdrawal, so that all would have left Angola within 27 months of the final signing of the peace agreements. Pretoria had to be contented with half the troops having returned to Havana by the date of the 435 election in Namibia. The remaining troops would, in addition, be some distance from the Namibian border.

On the plus side for the South Africans, they extracted a commitment from the Angolans and Cubans that Swapo's fighters would be restricted to bases in Angola north of the 16th parallel on the day of the implementation of the peace plan. This would also be the day on which a ceasefire between South Africa and Swapo would officially come into effect.

In Namibia, people of different political persuasions looked on the proceedings with some cynicism, having seen a number of other peace efforts come to nothing. Even after South Africa, Angola and Cuba signed the Geneva protocols in August, there was an air of unreality in Windhoek. Right-wingers believed South Africa would never sell them out, while the anti-apartheid lobby kept questioning South African sincerity and raising the possibility of a double-cross.

With the Geneva signings, South Africa and Swapo entered into a 'gentlemen's agreement' which led to a *de facto* ceasefire coming into effect in northern Namibia on 1 September. The South Africans, though, would not sign a formal peace treaty with Swapo, because they considered Swapo actions in Namibia since 1966 as 'terrorism' rather than a conventional war.

Towards independence

The peace was shattered almost immediately on 1 September by a bomb blast which devastated the Continental Hotel in Windhoek, leaving two dead and six injured. Each side accused the other of planting the device. The South Africans claimed the bomb could have been the work of either disgruntled Swapo cadres or of members of the organisation who had not received orders from their superiors to cease hostilities. Swapo counter-claimed that the bomb had been the work of the security service, pointing out that the Continental was a popular multi-racial drinking venue.

Yet, despite claims from the South Africans that Swapo had violated the peace understanding by engaging in acts of sabotage in the Ovambo area, the situation in the war zones changed markedly. So much so, in fact, that the dusk-to-dawn curfew, which had been imposed in the late 1970s, was lifted.

But 435 fever had yet to catch on among the general Namibian population, despite a visit from a UN technical team in October to prepare for the implementation of the peace plan. On 22 December, though, when the peace plans and accords were finally ratified by South Africa, Angola and Cuba at a ceremony in New York, even the most skeptical were forced to realise that, this time, South Africa was not playing games.

The TGNU, which quite correctly felt it had been left out of events, was the most let down by the peace process. As early as 1985, the Windhoek administration had been warned by Administrator-General Pienaar that Resolution 435 might come 'like a thief in the night'. Many of the TGNU members were stunned at just how quickly the independence reality did arrive. Those who had long trumpeted 435 as the only solution could not very well object, although they voiced their concern about UN impartiality in view of the world body's decree that Swapo was the 'sole and authentic representative of the Namibian people'.

Some government members were understandably apprehensive

about the prospect of having to go to the electorate, and give up their status, to say nothing of their R120 000-a-year jobs and life-styles.

Dirk Mudge, finance minister in the TGNU, echoed the fears of the business sector when he warned that too long a transition to independence could seriously affect confidence and investment. He also warned ominously that South Africa would drastically slash its budgetary contributions to the Windhoek exchequer, forcing 'belt tightening'.

Mudge was one of the most angry of the TGNU leaders when South Africa dealt the government its final slap in the face at the end of February 1989. The Windhoek politicians had offered to step down from office on 28 February, a month before the 1 April implementation date for the UN peace plan, to allow Pienaar to take over full control of the administration. The altruistic offer by the TGNU was not good enough for South Africa, which effectively tossed the TGNU members out onto the street by ordering their demise in a Government Gazette proclamation. That act was final evidence that the three-year TGNU experiment had meant little to South Africa.

Meanwhile, realists in Namibia began to consider Swapo seriously as the future government of the country. The organisation's economic policies, in particular, came under scrutiny. Despite its image as a marxist organisation, many analysts agreed that Swapo would probably adopt a pragmatic approach to both the economic and political life of the new state.

A number of prominent local businessmen risked their reputations by flying to Zambia to talk to Swapo, and some came away optimistic, even if they were not singing the praises of their hosts.

Political figures like Reggie Diergaardt of the Labour Party and Chief Justus Garoeb of the Damara Council, also flew to Lusaka and were cordially received by Swapo, sparking some speculation about possible alliances and election pacts come the time of 435.

An indication that old enmities are seldom buried came, though, when Peter Kalangula - leader of the Christian Democratic Action for Social Justice Party and head of the Ovambo Administration - had his invitation to visit Swapo withdrawn at the last moment. Kalangula had, apparently, still not been forgiven for his sin of joining the DTA in the 1970s - even though he fell out loudly and publicly with Mudge and had severed his association with the organisation many years previously. The row over Kalangula brought a critical reaction from veteran liberal politician Bryan O'Linn, chairman of the Namibia Peace Plan 435 study and contact group, who had originally invited Kalangula with the approval of Swapo.

As 1988 wound on into 1989, and the UN security council and general assembly finally approved budgets to enable the 4 500-strong peacekeeping force and its civilian and police components to be put into position in Namibia, there was still no sign of unease in the white community, nor signs of impending flight to South Africa. In the unsettled interim phase, many adopted a wait-and-see attitude, and hedged their bets by moving their money out of Namibia.

Swapo officials were fond of saying they did not believe there would be a white exodus from Namibia, pointing out that whites did not leave Zimbabwe in droves after Zanu (PF) took power. Those arguments did not, however, take into account the fact that Namibian whites - the bulk of whom are of Afrikaner extraction - would face no problems in liquidating their assets and moving them to another country, nor would they face problems of culture-shock of the sort that faced white Rhodesians moving to South Africa.

White businessmen, though, welcomed the arrival of the UN's Transition Assistance Group (Untag), seeing high profits for the taking from the free-spending foreigners. Prices rose significantly (particularly in the area of accommodation and rent) and locals began to feel the pinch.

The overall situation was calm and fairly optimistic on 1 April.

The violence which ripped northern Namibia apart will again best be analysed by historians when they can sort through claim and counter-claim, accusation and counter-accusation. But what is certain is that there will be few innocent parties, apart, of course, from the long-suffering inhabitants of the north.

Swapo must prove that it was not fully aware of, and hence bound by, the Geneva protocols - particularly in regard to the position of its forces north of the 16th parallel on the day the ceasefire came into effect. The organisation's conflicting and evasive statements - initially denying the fighters clashing with the South Africans had infiltrated Namibia, and then admitting they had, but only to hand themselves to the UN - did it no credit at all.

Even the organisation's allies must have asked why Swapo

thought it necessary to infiltrate the 1 500-plus troops it did, given that it would probably win a majority-rule election hands-down.

The South Africans were far from being innocent injured parties, either. The violent and often brutal reaction of the security force units (including many members of the officially-disbanded 'Koevoet' counter-insurgency unit) did much to deflect criticism away from Swapo. Claims that security forces were guilty of brutalities and atrocities, and of executing wounded Swapo members were not adequately refuted by the South African authorities.

Although the UN was, as it critics correctly charged, not in a position to step in and stop the fighting when it broke out, the situation was no fault of its own. The delay in approving the budget, and the decision to trim the UN peacekeeping force from 7 500 to 4 500 had meant the Resolution 435 timetable was some way behind schedule on 1 April. But, inextricably linked as it was to the timetable for the withdrawal of Cuban troops from Angola, the plan could not be postponed in Namibia. Nevertheless, Ahtisaari could still be criticised for allowing revenge-minded security force units free rein.

Yet the whole independence and peace process in Namibia and Southern Africa as a whole was too important to all parties - especially to South Africa - to allow it to fail. Despite all the threats from Pretoria about kicking the UN out of Namibia and halting the procedure, the plan continued.

South Africa's rulers had items on their agenda of far more importance than holding on to Namibia.

After the Coup:
South Africa's Relations with
Lesotho

Robert Edgar

After the military coup of 20 January 1986 deposed the government of Leabua Jonathan, South Africa moved to consolidate its influence in Lesotho. Although the new government did not totally conform to the one Pretoria desired, South Africa has deftly wielded its influence to exploit rivalries within the military, and between the military and the monarchy. By the end of 1989 Lesotho had adopted an accommodating line towards South Africa, exhibiting little of the independence displayed during the latter years of Jonathan's rule.

The post-coup government

Mounting South African economic and political pressure contributed to the crisis which led to the overthrow of Jonathan's regime.[1] Following the coup, the new government's first priority was to consolidate its power base and smooth relations with South Africa in an attempt to end its destabilising campaign.

There was little disagreement in Lesotho government circles over the need to do this, but there have been sharp differences over how far Lesotho has to go in placating Pretoria. Contrasting approaches towards South Africa reflect an evolving rivalry between the two leading figures in the government - Major-General Justin Lekhanya, head of the Royal Lesotho Defence Force (RLDF), and King Moshoeshoe II.

Lekhanya has favoured a close, almost collaborative relationship with South Africa. Moshoeshoe, more sensitive to African international perceptions, has aimed for a pragmatic relationship in which Lesotho retains room to conduct an independent foreign policy.

This split reflects the post-coup balance of power between the military and monarchy. Legislative Order Number 1, which theoretically gave considerable executive and legislative power to the king, also created a military council of six high-ranking RLDF officers who are to advise the king 'for the peace, order and good government of Lesotho...'. The king is empowered to appoint individuals to a council of ministers,[2] but the RLDF has placed several officers, including Foreign Affairs Minister Thaabe Letsie, in ministerial posts.

Complicating the situation is the fact that the king has allies in the military, such as Lt-Col Sekhobe Letsie who masterminded the coup and pulled a reluctant Lekhanya along with him. But the king's lack of a strong power base in both the military and the country leaves him unable to dictate to the army. Lekhanya's following outside of the military is also limited, adding incentive for him to rely on Pretoria for support, and seek its backing against Moshoeshoe.

The result of this rivalry is that there are often two policies on South African relations, with Lekhanya's 'unofficial' policy often - though not always - more attuned to Pretoria's needs than Moshoeshoe's official policy.

Pretoria has taken advantage of the divisions within the Lesotho government and assiduously courted individuals sympathetic to an 'accommodationist' line. These include Lekhanya, Tom Thabane, permanent secretary to the military council, and Evaristus Sekhonyana, the only carry-over minister from Jonathan's regime. The latter had been removed as foreign minister in 1984 after advocating the signing of a non-aggression pact with South Africa.

Pretoria has especially cemented relations with the RLDF. The provision of substantial funding towards a R2-million military hospital at Makoanyane military barracks in Maseru was one material expression of this. The SADF seconded 60 soldiers to the project from September 1987 to mid-1988. SADF head, General JJ Geldenhuys, accompanied RLDF commander Brigadier BM Lerotholi and other high-ranking Lesotho military officers to the opening of the hospital in July 1988. But no members of the military council attended, suggesting that there are limits to how much co-operation can be displayed openly. Geldenhuys marked the occasion by stressing the co-operative bonds that existed between the two military forces:

> Our two defence forces have raised our hands in greeting, we have clasped our hands to seal our mutual respect and good neighbourliness. We have put our hands to work constructively - and today we reap the fruits of our efforts.

Since the ceremony, SADF soldiers have not been seen in Maseru, but independent observers have confirmed that South African-manned Casspirs have been sighted in the Mafeteng district. A formal non-aggression pact may not have been signed, but all components of such an agreement are in place.

South African influence with the Lesotho government extends beyond the military sphere. After the coup, the Lesotho government agreed to the expulsion of a number of ANC and PAC refugees, removing what was once a safe haven for South African exiles. Since then, the number of South Africans passing through the refugee intake centre in Maseru has slowed to a trickle, with refugees understanding that they have to move on quickly.

ANC vulnerability in Lesotho was driven home in March 1988 when an ANC member, Mzizi Maqekeza, was shot dead by an unidentified gunman in a Maseru hospital, where he was recuperating from a bullet wound suffered in an incident with the RLDF a month earlier. Maqekeza sustained the wound when RLDF soldiers, who had stopped and searched him and two other ANC members, found a pistol on them. The soldiers lined the three up and shot them. Thandwefika Radebe, a National University of Lesotho law student, was killed outright and Maqekeza was admitted to the Queen Elizabeth Hospital. Before he could carry out his plans to go to Zimbabwe for further treatment, Maqekeza was killed.

South Africa also extended its influence in Lesotho when it opened a trade mission in Maseru in June 1987 which, like those in Mozambique, Swaziland and Zimbabwe, acts virtually as an embassy.

The Highlands Water Treaty

South Africa and Lesotho concluded the Highlands Water Treaty in October 1986. The scheme had been under discussion for three decades, but erratic relations between the Jonathan regime and Pretoria disrupted negotiations. And in the 1980s, Pretoria complicated matters by linking the treaty to Lesotho's signing of a security accord. Following the coup, Lekhanya and PW Botha discussed the Highlands scheme at a meeting in March 1986 where they were able to get the negotiations back on track.

According to the terms of the treaty, a newly-created Lesotho Highlands Development Authority (LHDA) will construct a series of six dams on the Upper Senqu (Orange) River over a 30-year period. Much of the surplus water will be diverted northwards to the heavily industrialised Pretoria-Witwatersrand-Vaal region, alleviating a critical shortage there. The first deliveries of water are scheduled to start flowing into the Vaal Dam south of Johannesburg in 1995. In addition Lesotho, now dependent on South Africa for almost all of its electricity, will be able to generate its own electricity supplies through hydro-electric power.

The scheme will generate annual revenues for Lesotho of over R100-million - the equivalent of a quarter to a third of Lesotho's current GDP. But there will be relatively little spin-off from the massive construction for Lesotho businesses or workers. Only about 14% of the dam's budget will be spent in Lesotho and about 2 500 Lesotho workers will be hired for work. Most of the skilled technicians will come from outside Lesotho and South African firms will supply most of the construction material and machinery.

Attracting international financing for the scheme has been a tricky problem since South Africa is its only customer and because it has been having its own debt problems in recent years. The LHDA will be responsible for raising loans of R7,7-billion for the dams and hydro-electric power scheme from the World Bank, EEC, export credit agencies, and donor countries. South African banks - with their government's guarantee - will raise the finance and loans for costs related to water transfer. Because of the political connotations related to making loans to South Africa, a trust fund is to be set up in the UK which 'will receive debt service payments from South Africa and channel them on a *pari passu* basis to all lenders' (*Financial Times*, London, 13.04.89). Thus, the US Export Import Bank, which is prevented from involvement in South African projects, will be able to make loans to the Highlands Water Scheme.

An immaculate deception: the pope's visit

Since the coup, Lesotho has gone to great lengths to improve relations with South Africa, but South Africa still takes every opportunity to impress on its neighbour that it is the dominant regional actor. No event better illustrated this than Pope John Paul II's Lesotho visit over 14-16 September 1988.

The pope was on a tour of five Southern African states - Zimbabwe, Botswana, Swaziland, Mozambique, and Lesotho. Both the Lesotho government and the catholic church in Lesotho - which had budgeted R5-million for the visit - had high expectations of the occasion. The pope was due to hold a series of masses, including a beatification ceremony for Father Joseph Gerard, a 19th century French priest.

Since 40% of Lesotho's citizens are catholic, and because South Africa had been pointedly excluded from the pope's schedule, an enthusiastic response was anticipated, and half a million catholics were expected to converge from Lesotho and South Africa. However, the visit, aimed at ignoring the South African regime and putting the spotlight on Lesotho, turned into a humiliating display of Lesotho's dependence on South Africa.

The elaborate preparations for Pope John Paul II's visit began to unravel on the eve of his arrival as four armed men claiming to represent the Lesotho Liberation Army (LLA) hijacked a bus carrying 70 pilgrims three hours after it had left Qachas Nek. After crashing through several roadblocks on the way to Maseru, they ordered the bus driver to take them to the British High Commission where, after being denied entrance, they parked outside the gates.

Their demands were not clear, but several notes sent out indicated that they wanted to have an audience with various figures, including King Moshoeshoe, the British High Commissioner, Pope John Paul II, and two Basotholand Congress Party (BCP) figures - Godfrey Kolisang, a Maseru advocate, and Chief Maroala Molapo. They also wanted to negotiate an end to military rule.

The pope was scheduled to arrive from Botswana the following day. He was briefed on the hijacking crisis, but decided not to alter his itinerary. However, as his plane flew over the Lesotho-South Africa border, it was decided that the overcast, rainy weather would make it difficult to land since two radio beacons necessary for a descent were broken. It was later announced that the plane had also developed engine problems. After ascertaining that airports at Maputo in Mozambique and Matsapa in Swaziland were also shut down by stormy weather, the Zimbabwean pilot, who had never before flown into Lesotho, took the plane to Johannesburg, landing at 11:13 in the morning.

Questions were later raised about the necessity of this diversion. Another Boeing 707 had been able to land at the airport less than 30 minutes after the Pope's plane tried to do so. And it was noted the flight could have been diverted to Bloemfontein - little over an hour's drive from Maseru - rather than Johannesburg (*West Africa*, 28.11-04.12.88).

The pope's landing in Johannesburg was an ironic twist of fate since the Vatican had turned down a South African government request to have a mass said at a technical stop-over at Jan Smuts Airport. But almost as soon as the plane was diverted, Pretoria treated it as manna from heaven and went into action. Foreign Minister Pik Botha rushed to Jan Smuts to receive the pope and after briefing him on the hostage situation, his entourage was sent on its way, shepherded by six Ford Sierras jammed with South African security men who not only squired the pope through South Africa but crossed the Lesotho border and accompanied him 30 km to Roma for an evening mass.

Waiting expectantly for the pope at the border were Lesotho's Military Council and senior government officials. Conspicuously absent was King Moshoeshoe, who could see a farce developing. He did not personally greet Pope John Paul until his airport departure several days later.

Twenty-five minutes after the papal entourage passed within 300 yards of the hijacking scene, the hostage drama came to an end. Lesotho troops had been keeping the hijacked bus under watch, but Lekhanya unilaterally called on a special SAP commando unit to storm the bus. The unit set off tear-gas under the bus and aimed a high-intensity spotlight to blind the hijackers while sharpshooters trained their sights on them. When the attack commenced, the hijackers tried to crash the bus through the High Commission's gates, but were stopped by shooting which lasted 15 minutes.

In the fighting several hostages were killed and 11 injured, while three of the hijackers were killed outright. The fourth tried to escape by mixing with the other hostages, but he was captured, taken away, and then inexplicably killed, raising the suspicion that there was a lot more behind the episode than those involved were willing to reveal. Immediately after the storming of the bus, a representative of the Lesotho government, Tom Thabane, denied that South African troops had been called in, but since any number of eye-witnesses (including a large international press contingent) could refute his claim, he later amended his story to say a 'combined South African and Lesotho force' had attacked the bus.

The hijackers claimed they were members of the LLA, but Ntsu Mokhehle, head of the main branch of the LLA, quickly distanced himself from the hijacking and denounced it as 'very despicable' and 'mean and treacherous'. For a number of years Mokhehle's LLA had Pretoria's backing in its attempts to overthrow the Jonathan regime, but Pretoria had never been enthusiastic about actually allowing Mokhehle to depose Jonathan. Now that Pretoria had a more pliable regime to work with, there was no need to sponsor Mokhehle further.

It is not surprising, then, that Mokhehle took up a Lesotho government offer of reconciliation. Following meetings outside Lesotho with Sekhobe Letsie in 1986 and Lekhanya in 1987, Mokhehle returned to Maseru with a delegation of BCP officials and LLA commanders in May 1988. After face-to-face negotiations with Lesotho government officials, an agreement was reached to allow BCP exiles to return, and by year's end, several hundred had come home. Mokhehle himself settled in Maseru in March 1989.

It was therefore not in Mokhehle's interest to disrupt the papal visit. However, the LLA had been riven with splits, and several factions had operated from the Transkei under the sponsorship of the Transkei army when led by former Selous Scout commander Ron Reid Daly. Since none of the hijackers survived, it may never be known whether they were operating on their own or under instruction from some branch of South African security.

This may have been in the mind of *Guardian* correspondent David Beresford, when he wrote that the papal visit 'could almost have been choreographed by the Bureau for Information' (*Guardian*, 16.09.88). Certainly the Bureau for Information took maximum advantage of the events by unleashing a stream of propaganda. SABC Radio made much capital out of the pope's stop-over in Johannesburg, claiming it highlighted South Africa's indispensability to a region in crisis. On 16 September it commented that

> The smooth and efficient way in which arrangements were made for the pope to travel to Lesotho...bore testimony to diplomatic skill, competent organisation - and the magnitude of an advanced infrastructure in an otherwise desolate African landscape.

Pope John Paul made a ritualistic statement disapproving apartheid but tempered it by criticising those who took the route of violence and protest to change the system. In a swipe at Allan Boesak and Bishop Tutu, the SABC observed on 20 September that 'Pope John Paul's interpretation...differs sharply from that of political priests whose interpretation of the role of the church causes confrontation and polarisation in society'.

Not everyone derived the same satisfaction from the papal visit. Among the big losers were South African business interests which had poured millions of rands into setting up food and souvenir stands for pilgrims. In the end, few showed up - the largest crowd was 10 000 for the pope's beatification mass for Father Gerard at the Maseru racetrack - and the businessmen took a beating and had to give away or sell their goods at huge discounts. One group expecting to gross R6-million employed a staff of 160 people - and netted less than R500 for their efforts. 'Now we are sitting here with 1 000 cases of Coke, 30 tons of maize, 42 tons of frozen chicken and tons of boerewors', lamented one entrepreneur (*Star*, 16.09.88).

Ongoing tensions between monarchy and military

Lekhanya's independent dealings with Pretoria had frustrated the king's attempts to place his own 'stamp of authority' on Lesotho's foreign policy.[3] Following the humiliating papal visit, Moshoeshoe arranged for a face-to-face meeting with Botha on 25 October. Moshoeshoe's direct flight from a state visit to Nigeria to the South African meeting infuriated the Nigerian government. However, the meeting had been set up that way to prevent the king's opponents in Lesotho from blocking it (*West Africa*, 28.11-04.12.88).

The meeting was intended to provide the king with access to the South African government; to dispel the notion that he was an anti-Pretoria radical; and to stress that the proper relationship between the Republic of South Africa and Lesotho should be based on mutual non-interference and the recognition and acceptance of differences between the two countries. The king's memorandum to Botha was a careful statement reviewing the interlinking histories of the two countries. It observed that even in times of conflict and disagreement during the 19th century, Afrikaner and Basotho leaders had found occasion to respect each other's sovereignty despite different viewpoints.

Tensions between factions in government were also evident when *The Mirror*, a Lesotho weekly, disclosed allegations of corruption. *The Mirror*, founded in mid-1988 by South African refugee Johnny Maseko, alleged that Finance Minister Evaristus Sekhonyana had, as Leabua Jonathan's Finance Minister, made it possible for Benco, an Italian firm operating in Lesotho, to transfer several million rands into a Swiss bank account. Benco later left

Lesotho after being implicated in a series of swindles and foreign exchange contraventions. *The Mirror* also charged Sekhonyana with padding his expense account claims for foreign travel. Sekhonyana denied all the charges and in late October brought a defamation suit against Maseko in the Lesotho High Court. Towards the end of November, Maseko was arrested and in mid-December deported, going first to Nairobi and then returning to South Africa. The Lesotho government used this to brand him an 'agent provocateur'.

The Maseko charges clearly riled Sekhonyana and the Lekhanya faction since it was evident that documents relied on for the exposè had come from disenchanted government officials who wished to embarrass and depose Sekhonyana.

Corruption charges also figure in a case involving a Taiwanese businessman, Vincent Lai, who gave Lekhanya a 20% share in his company for a meagre investment. Lai's partnership brought immediate dividends as a short time later the Lesotho government informed Guiseppe Florio, an Italian businessman with extensive holdings in Lesotho, that he was operating a stone quarry without a mining license. Lai was promptly granted a license for the quarry, boosting his company's worth (and the value of Lekhanya's share) enormously. Florio was then deported from Lesotho.

Lekhanya has also come under fire for covering up his involvement in a shooting. On the night of 23 December 1988, he shot and killed an Agricultural College student who, according to Lekhanya, was attempting to rape a woman. Immediately after the incident, Lekhanya's bodyguard admitted to the shooting, but the facts contradicted his confession and pointed to Lekhanya's culpability. After he revealed his involvement to top government officials, members of the Military Council asked him to step down while the incident was investigated. Lekhanya refused. In early October 1989, Lekhanya was let off the hook when a magistrate convening a judicial inquest accepted Lekhanya's version of the shooting and ruled that the killing was justifiable homicide.

It remains to be seen how much damage the shooting affair and the corruption charges have done to Lekhanya's position. Undoubtedly the incidents have given fresh life to the factional rivalries within the Lesotho government and have served as reminders that while the Lekhanya faction is in the ascendancy for the time being, the situation remains fluid and difficult to predict.

What is more certain, is that whatever the Lesotho government's composition is in coming years, its range of options for dealing with Pretoria will continue to be limited. This reality no Lesotho government can escape.

Notes

1   See Robert Edgar, 'The Lesotho Coup of 1986', in Glenn Moss and Ingrid Obery (eds), *South African Review 4*, Johannesburg, 1987, for an analysis of the 1986 coup.
2   Several ministers have been singled out for South African opprobrium because of past political associations. One was Khalaki Sello, minister of Law, Constitutional and Parliamentary Affairs, an ANC member before he came to Lesotho. Sello was dismissed from the Council of Ministers in 1988, but this was unrelated to South African pressure or divisions in the Lesotho government.
3   Lekhanya's excessive reliance on South Africa may have given rise to an erroneous story in *The Sowetan* (13.09.88) which reported that, after a state of emergency was declared on 24 August, Lekhanya had been suspended for three weeks and was reinstated through Pretoria's intervention.

Section 3: Labour

Introduction:
Labour at the Crossroads

Jean Leger and Eddie Webster

During the late 1980s, the labour movement reached an important crossroad. Relations between capital and labour had become increasingly polarised. Unions frequently resorted to strike action in support of demands, in some cases resulting in incidents of violence against fellow workers. Simultaneously, capital and the state moved to contain the power of labour, most significantly through the Labour Relations Amendment Act. Yet the labour movement has survived this offensive. Indeed, in response to the banning and repression of political organisations, it has taken on a leading role in internal resistance to apartheid.

Following the 1973 Durban strikes, elements of capital accepted, and even argued for, a new industrial relations dispensation. They were committed to limited forms of collective bargaining over issues like dismissal procedures and wages.

The 1979 Wiehahn reforms granted unions a degree of space to organise. While management prerogatives were challenged in intense skirmishes at individual firms, these battles generally remained localised. The labour movement, not yet sufficiently organised to act across whole industries or sectors, was usually forced to accept piecemeal improvements in wages and working conditions. Even the living-wage campaign for wages above poverty levels remained largely rhetorical. With strikes costing a negligible proportion of total workdays each year, capital did not feel the need to act decisively to curtail the labour movement.

But the protracted, and at times violent, mining and public

sector strikes of 1987 were something of a turning point. Managements' narrow conception of negotiation as a means of communication, and bargaining as being over wage increases to compensate for inflation, was shattered. As Bennett shows, the number of workdays lost soared in 1987, especially in the drawn-out trials of strength in the mining and public service industries. For the state, the protracted struggles of railway and post office workers destroyed the naive hope that collective bargaining would be limited to the private sector. The realities and implications of the workers' struggle for a qualitatively better industrial dispensation became readily apparent.

In response, both management and the state embarked on a concerted attempt to contain the union movement and reassert managerial prerogatives. On the mines this took several forms: the mass dismissal of at least 40 000 workers during the August 1987 strike, tight restrictions on union access and activities, and new security controls around mine compounds.

The state played an increasingly repressive role, trying to contain labour's growing involvement in wider political issues. Cosatu's headquarters were blown up, and many regional union offices suffered arson attacks. An openly hostile attitude towards Cosatu developed in 1987. This culminated in the extensive restrictions served on Cosatu and its allies in the UDF in February 1988. Increasingly, police have actively intervened in union meetings, videotaping proceedings and attempting to intimidate union members by a massive armed presence.

In the legal arena Benjamin shows how the changes introduced by the Labour Relations Amendment Act in 1988 threaten to undermine the gains made by the unions since 1979. The industrial court, in particular, has become a matter of controversy inside the labour movement, with some unionists discouraging its use. Instead of drawing the union movement into the industrial relations system, the Labour Relations Act has itself become a cause of industrial action.

In the short term, the strategies of capital and the state appeared to contain the labour movement. Bennett gives details of how in 1988 the number of workdays lost due to strike actions declined drastically. But in the longer term, the union movement has maintained its cohesiveness. While Bennett argues that state repression led to less frequent use of the stayaway tactic, major nation-wide stayaways have continued on a larger scale than ever before.

The labour movement no longer sees the stayaway as a one-off event. Stayaways are increasingly conceived of as a tactic in a broader campaign, as well as reflecting the wider concerns of the range of organisations involved in these calls. The highly successful September 1989 stayaway is a case in point. It was not only a ground-swell protest against the parliamentary election, but part of the campaign against the Labour Relations Amendment Act, a campaign of ongoing resistance which included negotiations, overtime bans, protest marches and consumer boycotts.

Maree shows evidence of a long-term project by management to establish a favourable collective bargaining environment. His examination of the food industry demonstrates how the changing balance of power in that industry has led Barlows to attempt to decentralise collective bargaining to plant level. Similar strategies have emerged in the paper industry with Barlows management refusing to join the industrial council. Barlows' rationale is that wages at individual locations should be dictated by local conditions, rather than having uniform conditions of employment across the corporation. In a scenario where unions are able to organise the workers of a company nationally, decentralised bargaining would reduce the wage bill significantly as wages outside the major centres are much lower. Decentralised bargaining would also stretch the union's limited resources as each set of negotiations would have to be repeated at each subsidiary rather than conducted for the company as a whole.

The unions' widening role beyond work place issues emerged clearly in the field of housing. Since the early 1970s, the state has increasingly neglected the growing need to provide accommodation for workers, leading to gross overcrowding of the existing housing stock and the mushrooming of squatter settlements. In a climate where the state is withdrawing from its limited social responsibilities through privatisation and deregulation, the contributions by Cobbett and Crush show how labour has begun to challenge managerial and private sector initiatives.

A major stumbling block for workers remains the high cost of housing, even in the case of employer-subsidised schemes. Unless creative means of financing housing can be developed, costs will restrict housing to the highest paid strata of the black workforce. The involvement of labour in the proposals put forward by the

Soweto People's Delegation on the rent boycott in Soweto is an important development. Central to these proposals is that Soweto should form part of a joint tax base with Johannesburg so that its citizens may rightfully share in the wealth they have helped create.

In the area of health policy and safety, Bachman et al show how the state is attempting to use privatisation and deregulation to reduce its financial contribution to health care and pre-empt union organisation in this area. Management prerogatives in the work place are being strengthened by the introduction of a system of management regulation of safety and health which largely replaces the role of the state safety and health inspectorates.

An important implication of these new initiatives by capital and the state in the fields of health as well as housing is the possibility of increasing stratification of the black working class. A housing campaign has the potential to worsen divisions between the employed - particularly in better paid sectors - and the unemployed. This could result in the creation of a new, privately housed elite drawn from better paid members of the workforce. Cobbett argues that unions, aware of the attractions of housing for sections of their membership, have been devising ways of ensuring that the benefit of their collective bargaining power is extended to embrace other, weaker sections of the working class.

Side by side with labour's widening role has been a more proactive approach from management to win greater co-operation on the shop floor. Maller focuses on employee share ownership schemes and argues that their limitations ensure that they are unlikely to win much union support. They do not provide workers with any greater control over the work place or any significant income. Above all they are rejected because they are seen as attempts to undermine the role of the trade union.

Labour has successfully forged a central place for itself in South African liberation politics. The political traditions that have shaped debates in the past persist, but as Fine and Webster argue, a strategic compromise is emerging that transcends these divisions. This growing unity within and between the two main union federations, Cosatu and Nactu, was expressed in the ongoing negotiations around the Labour Relations Amendment Act.

International pressure intensified during this period and, as the case study by Adler shows, so did divestment. About 130 US companies remain in South Africa after the withdrawal of 160 US subsidiaries over the past six years. Significantly, structural weaknesses in the South African economy began to influence debates in the white community. For the first time under Nationalist rule, economic issues played an important role in a parliamentary election (1989), pushing a growing number of voters either to the right or to the realisation that some negotiated solution involving the exiled political organisations is necessary. Labour's role within these negotiations remains to be clarified. Significantly, the ANC consulted with Cosatu in June 1989 on its proposed negotiating package before presenting it at the OAU conference in August and the Non-Aligned Movement meeting in September.

The kind of union movement that emerges, as well as the system of industrial relations that is presently being struggled over, will provide the framework for the relations between capital and labour well into the future. The struggle to define the future is taking place within institutions inherited from the past. The labour movement is clearly at the crossroads: capital and the state could succeed in containing it, or labour could continue playing a leading role in internal resistance to apartheid. The implications of unbanning the political movements for labour's role in the negotiation process remain to be seen.

Transcending Traditions: Trade Unions and Political Unity

Alan Fine and Eddie Webster

The relationship between trade unions and the struggle for national liberation has always been intensely discussed in South Africa. But with trade unions emerging as a leading force in the national liberation movement, these debates have now moved to centre stage.

There are at least three political traditions within the labour movement which structure the differing perspectives on its relationship to the national liberation struggle.[1] Although a strategic compromise transcending these traditions is emerging, they continue to shape debates over the role of the trade union movement in national liberation.

Three political traditions

During the 1950s the ANC, in alliance with Sactu, established its leadership amongst the oppressed classes. This was achieved through mobilising the oppressed - across class lines - around the demands of the Freedom Charter.

Sactu's participation in the Congress Alliance facilitated the rapid growth of trade unions in certain regions. But it also brought Sactu into direct conflict with the state and the organisation felt the full force of repression in the 1960s. By 1964 it had ceased public activities in South Africa and now operates in exile.

In the late 1970s this 'national-democratic tradition' re-emerged in the labour movement with the advent of general unions, particularly the South African Allied Workers Union, which followed in the tradition of Sactu. These 'community unions' argued that workers' struggle in factories and townships was indivisible, and that unions had an obligation to take up community issues. It was 'economism' and 'workerism' for unions to restrict activities to factory struggles.

This 'national-democratic tradition' involved a view that South Africa could not be understood in simple class terms. Social reality was based on a 'colonialism of a special type', necessitating national-democratic rather than class struggle as the appropriate strategic response. This meant a multi-class alliance under the leadership of the ANC, drawing on all sectors of the oppressed black masses and sympathetic whites, and aiming to establish a 'national democracy'.[2]

Supporters of this view differed over whether it was necessary to pass through a national-democratic stage before a socialist stage (the two-stage notion of revolution), or whether the national and class struggle take place coterminously. Most now argue the second position, stressing that the struggle for national liberation is part of the struggle for socialism.

The multi-class strategy of these unions influenced their approach to organisation. During 1983 many affiliated to the UDF and became increasingly involved in actions over rents, transport and local elections. This often resulted in neglect of shop-floor organisation in favour of premature confrontation with management and the state, leaving these unions exposed to the full force of economic recession and state attack.

During the 1970s an alternative political tradition developed in the union movement. The 'shop-floor' unions that first emerged in 1973 eschewed political action outside production. They believed it was important to avoid the path taken by Sactu in the 1950s, arguing that its close identification with the Congress Alliance and its campaigns was the cause of its demise in the 1960s.

Rejecting the 'community unions' as 'populist', the shop-floor unions developed a cautious policy towards involvement in broader political struggles. These unions (in particular those which were to affiliate to Fosatu) emphasised instead the building of democratic shop-floor structures around the principle of worker control, accountability and mandating of worker representatives. They saw this as the basis for developing a working-class leadership in the factories.

These unions argued that this strategy involved the best means of survival in the face of state repression, and of building strong industrial unions democratically controlled by workers. Some within this political tradition supported the creation of a mass-based working-class party as an alternative to the South African Communist Party.

The picture of unions and politics in South Africa has often been incomplete through neglect of a third political tradition - that of black consciousness. Its origins lie partly in the Africanist ideology articulated by the PAC, which broke away from the ANC in 1959 because of the latter's 'multi-racial' definition of the nation. The American black power movement was equally important in forming this political tradition in South Africa.

Black consciousness has similarities to the national-democratic position, in that it holds that racial oppression is a manifestation of national oppression. However, its concept of the nation is narrower in that it excludes whites. In addition, its emphasis on racial structures and identities virtually excludes class relations from its analysis. This has often given rise to voluntaristic and romantic forms of organisation and mobilisation.

By the late 1970s a class analysis had been introduced into the black consciousness discourse, whereby class was defined in racial terms. Associated loosely with Nactu in the trade union movement, the black consciousness tradition is distinguished from other traditions by its emphasis on 'black leadership' of the trade unions and its opposition to 'non-racialism' in favour of a policy of 'anti- racism'. In practice this has meant opposition to white intellectual leadership in the trade union movement. This led to this position's withdrawal and exclusion from the unity talks, which culminated in the December 1985 formation of Cosatu.

An ideologically more flexible position - labelled Africanist and sympathetic to the exiled PAC - has recently emerged in Nactu on the issue of whites in trade union leadership positions. The significance of this shift - from a black consciousness to an Africanist position - is not yet clear.

From Fosatu to Cosatu

The rapid mobilisation by 'community unions' in the early 1980s, the emergence of community organisations overtly associated with the national-democratic tradition, and the growing interest in political involvement amongst rank-and-file leadership, forced Fosatu to respond to criticism that it was isolating itself from wider political struggle.

A 1982 keynote speech by Fosatu general secretary Joe Foster reiterated suspicion of political action which was not worker-controlled but populist in character. Foster proposed an alternative political direction to the national-democratic tradition, based on the development of a 'workers' movement' under worker control. Fosatu officially began to take up non-factory issues, for example opposition by the Katlehong shop-stewards local to the destruction of shacks on the East Rand, and rejection of the tricameral parliamentary elections. But these efforts remained local and partial.

In the absence of one united national federation, there was no unified strategy in response to growing local pressure for political involvement. More importantly, although Foster had provided some general guidelines, his speech did not explain how - in practice - unions were to relate to the deeply-rooted national-democratic tradition.

Finally, it was the process of intensified struggle, where workers confronted the crisis in the townships daily, rather than in refined debate over the problems of 'populism', which marked the decisive break with political abstentionism. Fosatu entered into joint action with student and civic organisations in the 1984 November Transvaal stayaway. This was made possible both by the overlapping membership of these organisations and the irresistible pressure from union members demanding action in the face of rising rents, transport costs, Bantu Education and the repressive local government system.

According to Jay Naidoo, general secretary of Cosatu, the November 1984 stayaway

> symbolised in a real way the first political intervention by organised labour on such a large scale since the militant actions of workers in the 1950s. It further laid the basis for a developing alliance between students, youth and the worker-parents and gave the political leadership of the workers the confidence to assert the leading role of the working class in the broader struggle of our people.[3]

Trade unions' growing willingness to take solidarity action with one another began with the stoppage over the death in detention of union organiser Neil Aggett. This, together with joint actions with community groups in the aftermath of the stayaway, culminated in the December 1985 launch of Cosatu.

The new federation brought together unions from all three

political traditions described above: the well-organised industrial unions drawn from the shop-floor tradition; the general unions drawn from the national-democratic tradition; and the National Union of Mineworkers (NUM), having recently broken from the black consciousness tradition.

Cosatu now faced the difficult task of blending these diverse political traditions into a working-class political project. This challenge was captured by Cyril Ramaphosa, general secretary of NUM, in his opening speech to the inaugural congress of the federation. Cosatu, he said, would take an active role in national politics in alliance with other progressive organisations, but such an alliance would be on terms favourable to the working class.

A strategic compromise

The first six months of Cosatu's existence coincided with an uprising that began in August 1984 when the SADF occupied Vaal townships. The first state of emergency, declared in July 1985, did not contain this uprising. Indeed, the highpoint of township resistance came at the end of February 1986. Militant youth - known as com-tsotsis - went on the rampage in Alexandra township in what became known as the 'six-day war'.

Between September 1984 and May 1986, there were 12 local regional stayaways, but no national action. Then, in May 1986 over one-and-a-half million workers stayed away from work in a demand for May Day as a public holiday. This transition from localised to national stayaway action reflects the general political transformation which was taking place over the period from 1984 to 1986. It was widely believed among township activists that apartheid was about to crumble and that the establishment of 'people's power' was a prelude to a situation of 'dual power'.[4]

In her case study of Alexandra township, Karen Jochelson describes how

> local government did appear to have collapsed and had left a power vacuum. But this did not mean the township was ready for dual power. ...1986 did not involve a situation in which the state was unable to rule and in which alternative organs of government were effectively challenging the status quo. The state, with its untouched centralised power structure, still had its military forces firmly behind it and was able to repress township resistance with brute force.[5]

Responding to the euphoric mood, a Cosatu delegation visited Lusaka in February 1986 to speak to the ANC. A joint communique was issued in which the independence of Cosatu was acknowledged while the federation committed itself to the struggle for a non-racial South Africa under the leadership of the ANC. This clear identification with the national-democratic tradition brought the legacy of division in the labour movement to the surface again.

Critics within the shop-floor tradition believed that the Cosatu leadership had acted without a proper mandate in visiting Lusaka and should have made it clearer that Cosatu would struggle 'independently under its own leadership' within the broad alliance. By implying that Cosatu was 'operating under the leadership of the ANC', critics argued, a spirit of sectarianism could develop towards alternative political traditions. Furthermore, they feared some unions might feel it was no longer necessary to obtain proper mandates resulting in actions such as the failed stayaway called by Cosatu for 14 July 1986.

Some within the shop-floor tradition went further and argued that the national-democratic tradition stood in absolute contradiction to working-class politics. Organisational style and political content were such, it was argued, that any involvement of the working class in such politics must lead to the surrendering of trade union independence and with it the abandonment of working-class politics.

Alliance politics which stressed the 'people' over the working class failed to prepare workers for socialism. Unions which became embroiled in populist campaigns would lose their organisational independence because they would be unable to control 'people's' organisations of national democracy.

While these tactical, strategic and theoretical differences still persist inside Cosatu, they began to narrow in the first half of 1987 as Cosatu wove together what Jon Lewis has called 'a strategic compromise'.[6] One example of this was the political resolution adopted by the metal union, Numsa, at its May 1987 foundation congress. Rather than challenging symbols of the national- democratic tradition, Numsa endorsed the Freedom Charter as 'a good foundation stone on which to start building our working-class programme' - thus attempting to imprint on the Freedom Charter the strategy of the shop-floor tradition.

Commenting on the Numsa political resolution, and that of NUM which adopted the Freedom Charter at its annual congress two months earlier, Jay Naidoo argued that 'these resolutions together reflect the direction that workers are actually moving in struggle and the main debate that is going on within Cosatu. The struggle for socialism is already unfolding within the struggle for national liberation'.[7]

To suggest, as Alex Callinicos does, that this compromise is a form of 'capitulation to populism',[8] is both class reductionist and fails to grasp the specificity of the political terrain in South Africa. As Karl von Holdt has written:

> Nationalism is not simply a negative force. It is a very potent and deep political force that has been generated in resistance to colonisation and national oppression... Any serious struggle has to situate itself within this deeper popular current. Nor is this route quicker or easier. There quite simply is no other route: it is prescribed by the conditions of South African struggle.[9]

At the time of Cosatu's second national conference in July 1987, it appeared that the competing political traditions had sharply resurfaced, particularly after a telegram message to the conference from Sactu which stated that socialism should not be prioritised at this stage. Divisions within Cosatu appeared intense at the conference, with unions like Numsa believing they had received a raw deal, their positions being robustly overridden by the NUM-led bloc. Indeed, in his report to the 1989 Numsa congress, general secretary Moses Mayekiso confirmed that: 'Following the second Cosatu congress in 1987, Numsa members felt very pessimistic about Numsa's role in Cosatu. As a result, Numsa members in the regions played a less active role in Cosatu for some time'.

But with hindsight it could be suggested that was when the strategic compromise began to be forged, allowing Mayekiso to add that, by the special congress in May 1988, 'Numsa proposals were highly influential. Numsa has from the beginning demanded the right of free speech and open debate. Cosatu, especially at the national level, has changed substantially and has become more open. The sectarianism and suspicion which existed at the launch of Cosatu have now been very largely dispelled'.

Even though different meanings were being attached to the Freedom Charter, Cosatu adopted the charter as 'a guiding document which reflects the views and aspirations of the majority of the oppressed and exploited in our struggle against national oppression and economic exploitation'. However, this NUM-sponsored resolution also noted that Cosatu saw these struggles as 'complementary to each other and part of an uninterrupted struggle for total liberation'.[10]

There is, on this argument, no conflict between the struggles for national liberation and socialism. In adopting this approach, the dominant position in Cosatu was rejecting any chronological two-stage theory of change in favour of a view that the struggle for national liberation was part of the struggle for socialism.

At grassroots level this strategic compromise meant that shop-steward locals, very much an innovation associated with the shop-floor tradition, were opened up to community organisations either on a consultative basis or more permanently, as active participants in the locals.

Criticism of some of the affiliates for not having mandates and taking decisions as individual activists and 'not as worker representatives for their own factory' did emerge. But by 1988 the principle of 'disciplined alliances' with UDF affiliates had taken root in some Cosatu locals.[11] However, differences in the locals persist on the precise form these alliances should take.

Cosatu and the state

Conflict between the state and Cosatu reached a climax during 1987. In April of that year, police shot four striking Sarhwu members during a protest march in Johannesburg. A week later four 'scabs' were murdered after being interrogated at Cosatu House. The SABC mounted a vicious propaganda campaign against Cosatu, presenting its head office as a 'house of torture'. On 5-6 May a successful nation-wide stayaway against the white general election was held. Early the following morning Cosatu House was bombed. During this period, cabinet ministers made threatening statements about Cosatu's role in politics and 'labour unrest', warning that legislation would be introduced to 'clip their wings'.

During August, South Africa experienced its largest wage strike ever, when between 220 000 and 350 000 mineworkers went on strike for three weeks. The strike was br ken when 50 000 striking workers were dismissed, subsequently returning to work without significant wage increases. A month later the Labour Relations Amendment Bill was tabled with the declared intention of 'swinging the balance of power back towards management'.

NUM's set-back in the 1987 miners' strike, as well as the state's successful disorganisation of the democratic movement after the implementation of the second state of emergency in June 1986, strengthened the hand of those who argued for greater grassroots organisation. This surfaced clearly in a paper distributed in the name of the Cosatu secretariat at a special congress called in May 1988 to discuss protest action following the February 1988 banning of 17 organisations and restrictions placed on Cosatu.

The secretariat's report was Cosatu's first public reassessment of its high profile strategies and tactics of the 1984-1986 period, and of its new role as internal leader of the democratic movement.[12] The second state of emergency, the secretariat acknowledged, had been successful in disrupting the democratic movement. The emergency, including the restrictions imposed on Cosatu and the UDF, had weakened leadership, disrupted organisation and restricted Cosatu's ability to launch major campaigns.

Furthermore, the report said, the state had not acted out of panic. Rather, the repressive actions highlighted by the emergency and the introduction of the Labour Relations Amendment Act were just one aspect of a carefully formulated three-pronged strategy. The second facet involved the promotion of 'moderates' as 'true' political leaders, and the upgrading of some black townships. This required the silencing of the democratic movement. The more sophisticated sectors of the ruling class recognised that a resolution of the political crisis required negotiations with the democratic movement, including the ANC. But their intention was, thirdly, for state and capital to negotiate from a position of strength and on their own terms. The future strategy of Cosatu and the democratic movement, the paper argued, would have to take account of these factors.

The state's strategy, it noted, was not free of severe weaknesses. Among these were that it did not meet people's demands for true democracy and access to the country's wealth. Attempts to set up structures designed to bypass these demands would simply not work. Given the state's military priorities, there were insufficient funds for an adequate upgrading programme in the townships, and there were splits in the ruling bloc - particularly to the National Party's right.

But the democratic movement and Cosatu in particular faced their own weaknesses, argued the secretariat:

We have concentrated our energies on a narrow front. High profile political campaigns and mass organisation making clear political demands have pushed the democratic movement into the position of the leading opposition force in the country... (But) we have not done enough grassroots organisational work.

The democratic movement had been too slow to change its strategies in the face of the changing circumstances typified by the state of emergency. One reason for this within Cosatu, the paper argued, had been poor participation by affiliates in the federation's structures and difficulties in conducting open political discussion: 'Instead of arguing a position openly, delegates sometimes label other comrades' - the implication clearly being that there was a derogatory stereotyping of particular political positions. The paper said a code of democratic conduct was required, involving patient listening to all views, arguing positions, having mandates and all carrying forward the majority decision.

Cosatu structures, the secretariat continued, could provide a key focus in building organisation and mobilisation in the labour field. The UDF's banning meant that although Cosatu was not and could not be a political organisation, it and local UDF affiliates had to take up a more direct political role. This required unity and a common political direction and for this reason Cosatu had adopted the Freedom Charter at its 1987 congress. Community and youth organisations, the paper argued, could not lead themselves to a new society. But they could become strengthened through winning short-term changes through campaigns and struggle.

The nature and breadth of alliances

While the theme of unity and alliances was discussed at the 1987 congress, it became a major focus of debate within Cosatu from the time of the special May congress. The fact that these debates began to occur in a more open way suggested that Cosatu as a whole had taken note of the secretariat's plea for greater freedom of speech within the organisation. The questions surrounding alliances with Nactu, the black consciousness movement, black business organisations, the most enlightened sections of white capital and the liberal opposition parliamentary parties had become key areas for debate within Cosatu.

But there were limits to alliances. The relationship with the

South African Co-ordinating Committee on Labour Affairs, for example, a loosely knit confederation of nine major employer associations with whom extensive negotiations on the LRAA were held, fell into a different category, and there was no question that this represented an alliance.

It was on the breadth of the alliance that differences between representatives of the shop-floor tradition and those of the national tradition emerged. As the debate continued both sides shifted their positions so as to produce greater degrees of consensus.

Unions of the shop-floor tradition accepted symbols of the democratic movement such as the Freedom Charter as a basic if inadequate platform for working-class demands. In much the same manner, they now accepted the need for alliances with organisational representatives of that movement. However, they also identified serious shortcomings in these organisations which needed to be corrected, similar to those expressed in the Cosatu secretariat's paper.

In particular, unions of the shop-floor tradition expressed misgivings about relationships between the leadership and rank-and-file in many organisations of the democratic movement. They argued that while the progressive organisations had effectively mobilised people, their organisational structures were weak, and their leaders vulnerable to state attacks. Once leaders were elected, they often became unaccountable to those who had elected them. Disciplined organisations able to decide the correct time to advance or retreat were needed, rather than indulgence in 'kamikaze-style politics'. There was, in addition, excessive use of inexperienced 'children' in these organisations, who were unable to develop clear political strategies.

The shop-floor unions also indulged in some self-criticism.

By concentrating too narrowly on shop-floor issues some conceded they had failed to win over other interests, particularly the youth.

The principle of building alliances with groups comprising the democratic movement was thus no longer in dispute. Union reaction to the invitation list for an anti-apartheid conference planned for September 1988, but banned, demonstrates this well. Most unions in the shop-floor tradition accepted, after initial reluctance, black business representation at the conference. This demonstrated a further shift away from their previous insistence on an exclusively working-class alliance. Nevertheless, these same unions still questioned the wisdom of continuing relationships which began developing in 1988 between the democratic movement and representatives of the most enlightened sectors of business and the liberal political opposition. In a policy document at its 1988 congress Numsa stated

> All attempts to broaden the alliance by including political representatives of capital, homeland opposition parties, and all other forces outside of the mass democratic movement only serve to confuse, weaken and distract organised workers from our principal task of building the mass united front of working class forces within the democratic movement.

Unity and black consciousness

During debate on this proposed anti-apartheid conference, representatives at the Cosatu special congress expressed hostility towards the black consciousness movement. The NUM-led group initially insisted that the united front to be created by the conference should comprise only organisations sympathetic to the Freedom Charter. The Numsa/CWIU resolution, on the other hand, called for the establishment of a broad front of all working-class organisations, and organisations of the 'oppressed and exploited', aimed at ending 'apartheid and capitalism'. There was, it argued, a dire need for as broad a unity as possible to respond to the state's repressive measures aimed at political opposition, including the labour movement.

Some in the shop-floor tradition also hoped that the entry of black consciousness unions into the alliance would dilute the ideological dominance charterists had held in Cosatu's first 18 months. This would increase the spectrum of political thought in the organisation.

A TGWU compromise proposing a conference of a broad range of organisations (implicitly including those from the black consciousness tradition), convened jointly by Cosatu and its democratic movement allies, was eventually accepted.

The nature of this compromise meant that no group could claim an outright victory in the debate. But forcing a compromise from the NUM-led group on black consciousness participation was a clear reflection of the shift in the balance of power which had occurred in Cosatu. NUM was in a weakened state after 50 000 dismissals during the August 1987 strike and the subsequent disorganisation of work-place structures. Numsa, the largest union in the shop-floor tradition, had in contrast been growing at a rapid rate. NUM's numerical strength had decreased; its previously reliable ally, Fawu, was divided; and there was a growing need for greater unity. For the first time since the formation of Cosatu the NUM-led group - allied to the national-democratic tradition - had failed to dominate in a major policy decision.

But the decision that Cosatu and organisations of the democratic movement would convene the conference caused black consciousness groups to see themselves as unequal partners in the venture. Azapo turned down the invitation in the belief that it would be a second-string player in an event dominated by its ideological opponents. Nactu, which had co-operated with Cosatu during the 6-8 June 'peaceful protest', was saved from having to make a divisive decision on attendance by the banning of the conference.

Interestingly, internal representatives of the democratic movement had, at the time, adopted a more inflexible position on black consciousness than the ANC. Only a week before the special congress, a Nactu delegation met with the ANC in Harare and a joint communique from them accepted that allegiance to the Freedom Charter was not a prerequisite for participation in a united front.

By the beginning of 1989, resistance within Cosatu to co-operation with Nactu had ceased to be significant, and by March arrangements had been made for a joint Nactu-initiated worker summit. But by that stage obstacles to a closer unity were coming from Nactu. From the time of its August 1988 conference, Nactu had been dominated by an Africanist bloc. Initially there were no signs that this would hinder its developing relationship with Cosatu, and the Africanist approach was more flexible on dealing with non-racial organisations than the black consciousness approach.

The factor which led to Nactu's formal withdrawal from the summit was political. In a statement explaining the non-attendance of most Nactu unions at the 4-5 March worker summit, Nactu assistant general secretary Cunningham Ngcukana explained that the unity question had important political implications, and more time was needed to consider these. Other Nactu leaders elaborated, saying the Africanists had come to the realisation that Nactu was the only remaining internal Africanist organisation able to operate lawfully. They were reluctant to risk diluting this one source of strength - which would certainly occur if unity developed with the more powerful Cosatu. The question of formal unity between the two remains a difficult issue for the dominant groupings within Nactu. Nevertheless, once the rumblings over the March summit blew over, Nactu as a whole allied itself with the planned campaign against the Labour Relations Act and attended the August summit to assess the campaign's progress and plan further action.

Relations with the liberal community

Debate over the nature of the alliance also involved relations with liberal political parties and enlightened representatives of the business community. One parliamentary party, the NDM - which subsequently merged with the PFP to form the Democratic Party - was invited to the anti-apartheid conference. In the second half of the year, Cosatu for the first time met officially with the PFP. The forces behind this meeting with the PFP were NUM and Fawu, leading unions in the national-democratic tradition. Together with its allies in the democratic movement, Cosatu also held a two-day meeting at Broederstroom with the newly-constituted Consultative Business Movement, comprising 40 leading businessmen, mainly liberal in outlook.

Relations and contact of this nature fit comfortably into a broad strategy aiming 'to isolate the apartheid regime and strengthen and broaden the unity of anti-apartheid forces'.[13]

The 'Broederstroom encounter', according to the official record of the gathering, occurred after nearly 18 months of sometimes secretive meetings and discussions to develop the relationship and elementary trust that made the venture possible.[14] However, some Cosatu affiliates felt that the encounter, and the earlier meeting with the PFP, had gone ahead without sufficient discussion within the organisation.

Issues raised at the meeting included the municipal elections, labour legislation, militarisation, group areas and related legislation, the state of emergency and repression, and business involvement in the national security system.

The CBM agenda addressed issues of concern to the democratic movement in general, and steered clear of those Cosatu and UDF policies which troubled the business community - on sanctions and nationalisation of industry, for example. It remains to be seen whether the CBM, essentially a group of enlightened individuals associated with large corporations, is able to develop a power-base strong enough to allow it to transform liberal sentiments into deeds.

Some unions from the shop-floor tradition have argued that a premature relationship with the liberal community can have dire consequences for the struggle for socialism: liberal capitalists have their own agendas of which some are clear, and others hidden. Those who ignore these agenda's influence could undermine any socialist project. Before considering any kind of alliance with liberal groups it is necessary, on this argument, to consolidate the working-class alliance, and for the working class to develop clear ideas about its own strategies for change.

The liberal agenda, the argument suggests, includes strategies for the preservation and extension of capitalism. A broad front strategy with these interests, as it is presently developing, will at best lead to a mild form of social democracy where the interests of the capitalist class remain dominant. This, it is argued, is not the same as a mixed economy where working-class interests are dominant.

The nature of the post-apartheid economy has not been widely addressed by the labour movement and its allies, despite its centrality to the debate on alliance politics. Unions within the shop-floor tradition have favoured a decentralised form of socialism, with workers participating in work-place decision-making. But this view, and the ANC's apparent softening on nationalisation, and commitment to a mixed economy with public, private, co-operative and small-scale family sectors, still require a great deal of clarification.[15] And the worldwide trend away from strongly centralised economics has barely been discussed in South African debates.

National actions and negotiations

After the unprecedented success of the June 1988 three-day nation-wide stayaway, held in protest against the Labour Relations Amendment Bill, there was broad consensus that negotiations with Saccola should continue. This strategy - to attempt to get Saccola to put pressure on government to withdraw the Labour Bill - cut across the political divide and the only opposition to it came from small groups on the extremes of the political spectrum. The majority view was that frequent and regular mobilisation against the bill was not feasible. Despite the success of the June stayaway, stayaways should be used carefully and strategically. Given this, and the potential harmful consequences of the legislation, any tactic which may help and which has no apparent disadvantages should be employed.

Negotiations with Saccola were halted when the bill was promulgated in September 1988. This failure in negotiations created skepticism about any future negotiations with Saccola, especially regarding a proposal that the labour movement draft its own version of the Labour Relations Amendment Act for discussion at Saccola level. Nevertheless, the period of negotiations in the ten weeks following the stayaway had one positive consequence: while the labour movement was involved in serious negotiations with business this prevented the state from acting over the stayaway.

The special Cosatu congress debate on the stayaway, or peaceful protest as it was called, brought to the fore an area of tactical difference between the unions representing two different traditions in the organisation. There was consensus that the bill and the February restrictions called for a high-profile protest. But differences emerged over the proposed length of the 'protest' action. The shopfloor tradition unions urged a cautious approach, suggesting the action should be held over two days, Thursday 16 June as a *de facto* public holiday anyway, and the Friday of that week. Other proposals ranged from three to five days, excluding 16 June.

The congress eventually opted for a three-day action. Its success suggests that on this occasion the leadership of Numsa and its allies was over-cautious. But some of the unions which fought hardest for an extended protest - including NUM and the public sector unions - failed to mobilise their members. This may result in a more strategic approach to mass mobilisation campaigns in the future.

Local negotiations and grassroots alliances

Growing union involvement in negotiating local living conditions has developed side-by-side with the building of broad alliances over national political issues. For example, Uitenhage-based unions participated in negotiations over the supply of electricity to Kwa-Nobuhle, where most of the union members live. The union did not oppose the removal of electricity supply from the local authority and its transferral to a newly-established private company jointly owned by local companies and Eskom.

The growing involvement of unions in housing is more central. A Numsa paper produced in February 1989 posed some complex questions raised by widespread membership demands for decent housing. Conventional private home-ownership is presently the only option available in company housing schemes. A housing campaign thus has the potential to worsen divisions between the employed - particularly in better-paid sectors - and the unemployed. This could result in the creation of a new privately-housed elite drawn from better-paid members of the workforce.

A second problem in a housing campaign involves the direct negotiations with black local authorities and RSCs. For housing provision to be dealt with as more than a technical matter, community organisations capable of dealing with the issue would have to be revived. The Numsa paper proposes the establishment of democratically-controlled housing advice offices in each township, handling individual home-ownership problems and developing local housing campaigns. This opens up the possibility for alliances between unions and community organisations at grassroots level.

The clearest examples of grassroots alliances between unions and the community involve the NUM and Numsa-sponsored co-operatives established for unemployed union members. Initiated by union members dismissed during strikes, they provide limited job opportunities and are run on co-operative principles. Whether such projects are viable, what their aims should be, how they should be structured and their relationship to trade unions are matters of on-going debate in the labour movement.

The political differences that divided Cosatu in its first eighteen months of existence narrowed and in some cases were buried in the face of the state onslaught during 1987 and 1988. A strategic compromise was forged on how transition to a post-apartheid society could take place through a broad anti-apartheid alliance based on grassroots organisation. The debate, however, remains unresolved on how broad this alliance should be.

The necessity for unity forced a tactical and strategic compromise, but it has not removed the differences underlying the competing political traditions. Implicit in this broad-front conception of unity is the view that class contradictions are secondary to the national democratic struggle. In a post-apartheid society these class contradictions are likely to come sharply to the fore as the new state embarks upon the task of national development. Trade unions which emphasise their representative role by struggling to defend and improve members' working and living conditions could easily be seen as opponents of the new state's attempt at national development.

Attempts to limit the role of trade unions in post-colonial Africa have often been justified on the grounds that trade unions represent only a tiny fraction of the labour force. Compared to the mass of unemployed living below subsistence, workers organised in trade unions are regarded as a privileged labour aristocracy. Recent research on the labour market points towards a growing stratification of the black labour force, in particular the emergence of a skilled stratum of African workers. This creates the real possibility of an increasingly-divided African workforce in the 1990s.[16]

None of the political traditions specified in this contribution have adequately addressed this issue. In trying to confront this, the Numsa paper on housing reveals the complexities and real dilemmas that unions face in trying to accommodate the desires of members for improved living conditions without further entrenching the trend towards a stratified workforce.

If the debate over the nature of alliances is to confront these dilemmas, it has to define the nature of socialism as well as the role unions could play in the transition to an alternative economy. A necessary starting point is to examine the changing social structure of the African labour force. Stratification is taking place. This means unions face a choice: should they prioritise the immediate interest of their members, or do they set as a central aim the building of a social movement of working people as a whole. None of the existing union traditions give pointers as to the choice. But developments within the labour movement indicate that these traditions are being transcended by the objective conditions of struggle - and organised labour is forging a central place for itself in South African liberation politics.

Notes

1 The notion of 'tradition' is used here as a means of understanding the way unionists use the past as a resource to strengthen, maintain or challenge different political strategies.
2 Jon Lewis, 'Politics and trade unions', *The Independent Trade Union Guide*, Durban, 1988, 133.
3 Jay Naidoo, 'Building people's power', in *The crisis,* speeches by Cosatu

office-bearers, Cosatu, December 1986.

4 Anon, 'The historical significance of three days of national protest, June 6, 7, 8', *Phambile*, special supplement, 1988.

5 Karen Jochelson, 'People's power and state reform in Alexandra', *Work in Progress*, 56/57, November/December, 1988, 13.

6 Jon Lewis, 'Politics and trade unions', *The Independent Trade Union Guide*, Durban, 1988, 133.

7 'Jay Naidoo on Cosatu', (interview), *South African Labour Bulletin*, 12(5), July 1987.

8 Alex Callinicos, *South Africa Between Reform and Revolution*, Bookmarks, 1988.

9 Karl von Holdt, 'The political significance of Cosatu: a response to Plaut', *Transformation*, 5, 1987, 95.

10 Yunis Carrim, 'Cosatu: towards disciplined alliances', *Work in Progress*, 49, September 1987

11 Jabu Matiko, 'Cosatu locals', *South African Labour Bulletin*, 12(8), October 1987, 44.

12 'Cosatu secretariat report', Cosatu special congress, 1988.

13 Karl von Holdt, 'June 1988: the three day stay-away against the new Labour Bill', *South African Labour Bulletin*, 13(6), 62-63.

14 Max du Preez, Gavin Evans and Rosemary Grealy, *The Broederstroom Encounter*, Consultative Business Movement, 1988.

15 David Niddrie, 'Building on the Freedom Charter', *Work in Progress*, 53, April/May 1988.

16 Owen Crankshaw, 'The racial and occupational division of labour in South Africa, 1969-1985', paper presented to the Labour Studies Workshop, University of Witwatersrand, 1987.

Re-establishing Managerial Power? Changing Patterns in Labour Law

Paul Benjamin

Debate, dispute and protest over changes to the Labour Relations Act marked much of 1988. Neither the massive stayaway of June 1988 nor unprecedented talks between the independent trade union federations and capital were able to stop minister of Manpower Pietie du Plessis from enacting amendments to the act. By the beginning of 1989, trade unions and employers were adjusting to the changed rules of industrial relations, and the industrial court was using its new powers to intervene in collective disputes.

The new act, which took effect from 1 September 1988, contains the most extensive changes to labour law made since the amendments of 1979 following the first report of the Wiehahn Commission. The most significant of these are:

- alterations to the definition of an unfair labour practice, including changes to the law of dismissal and retrenchment;
- restrictions on the right to strike and on activities in support of strikes;
- changes to the powers of the industrial court, and the establishment of a labour appeal court;
- revision of the procedures for referring disputes to industrial councils and conciliation boards;
- provisions to facilitate continued operation of racially-exclusive trade unions;
- restrictions on trade union immunity against claims for damages arising out of strikes.

These changes have been criticised for poor drafting and lack of clarity. The act is now contradictory, ambiguous and at times incomprehensible.[1] This makes comment on the law difficult, and its potential impact will not be known until clauses have been interpreted by the courts.

This article examines some of the most significant changes to the act, discusses their likely impact on employment relations and looks at cases in which the industrial court has used its new powers.

The industrial court

The industrial court has been the central institution for implementing recommendations of the Wiehahn report. It was established in 1979 to develop a code of sound employment practice through its wide 'unfair labour practice' powers. The court has laid down guidelines for employment practice with considerable success and imagination in the areas of individual dismissal and retrenchment, with less consistency on the dismissal of strikers and, until recently, with a lack of perception on collective bargaining issues.

The extent of the court's unfair labour practice powers was not matched by ease of access for litigants wanting the court's assistance. For the court could not rule on unfair labour practices on an urgent basis. Delays were made worse by the large number of cases, and lawyers' dilatory tactics. By the time the court ruled on an unfair labour practice the dispute out of which it arose was often over or had become academic.

The court's power to grant urgent relief was limited to situations in which there had been a breach of statute, and it became chiefly a forum for employers to interdict illegal strike action. But judgement in such a case could still set a precedent for future disputes.

Despite these limitations, the court has had a major impact on patterns of industrial action. Since its inception, a decreasing proportion of strikes have resulted from dismissal disputes.[2] These have been referred to the industrial court or private arbitration for resolution. Dismissal cases - whether involving applications for temporary reinstatement pending conciliation or for permanent reinstatement - are the court's chief activity.

In addition, a major proportion of strikes on wages and working conditions are lawful, staged after the completion of the procedures in the Labour Relations Act.[3] This development is a consequence of the significant protections against dismissal the court has afforded to participants in lawful strikes.

Unfair labour practice

In restructuring the court's powers, the state revealed an ambivalent attitude to its performance. Changing the definition of an unfair labour practice implied that the court had intruded too far into the realm of managerial prerogative. In addition to these changing definitions, the court will now be expected to restrict strikes the state considers unacceptable.

The court's unfair labour practice jurisdiction is severely proscribed in the new amendments. The wide discretion of the old definition has been replaced by a lengthy schedule which lists particular actions as unfair labour practices. The court retains its previous discretion only on matters not covered in the schedule.

Strikes and lock-outs are no longer excluded from the ambit of the unfair labour practice, and the court is now empowered to use its unfair labour practice powers on an urgent basis. It can grant a temporary order restraining an employer, trade union or employees from committing an unfair labour practice. This makes the court a forum available for use in the course of labour disputes, and it has already been called upon to stop threatened strikes, lock-outs and overtime bans on the basis of unfairness.

The restructuring of the industrial court's power indicates tensions within the ruling class. The redefinition of unfair labour practice is a response by the minister of manpower and his department to pressure from more conservative sectors of capital. And in the key area of retrenchment, the new unfair labour practice definition is a political response to what conservative capital views as excessive incursion into managerial prerogative by the industrial court.

Trade union policies and the court

These changes reverse developments that have their origins in the policies of the independent trade unions of the middle and late 1970s. Arbitrary dismissal, disciplinary action and retrenchment provoked the majority of strikes in the 1970s. The desire of employees for protection against this arbitrary exercise of managerial power was a major reason for the recruitment success of the independent trade unions.[4]

The legal environment placed no check on the exercise of managerial power, and unions thus looked to negotiated collective bargaining agreements to limit this power. These agreements included procedures to be followed before dismissing workers, or prevented employers from dismissing workers without good cause. Retrenchment procedures negotiated in the late 1970s and early 1980s required employers to consult with trade unions over the need to retrench, and to consider ways of avoiding or reducing the extent of a retrenchment. Companies were required to select employees for retrenchment on objective and verifiable criteria, most importantly in terms of length of service as contained in the 'last-in first-out' (LIFO) principle.

Despite the rapid growth of the independent trade unions, only a small proportion of the country's workforce had the protection of negotiated dismissal or retrenchment procedures. These agreements were concluded primarily at establishments of the more enlightened sectors of capital or at factories where trade unions were able to mount intensive recruiting drives. Many important sectors of the economy were not unionised, employees in smallish factories were not recruited and overtly anti-union employers were often able to resist unionisation.

In its early years, the industrial court laid down guidelines on retrenchment and individual dismissals. On retrenchment, it took its cue from the procedures that had been negotiated by independent trade unions. On dismissals, the court went beyond what had been achieved by the unions and accepted recommendations of the International Labour Organisation as part of South African law. It ruled that a dismissal had to be both procedurally and substantively fair: an employee had to receive a fair hearing before dismissal and could only be dismissed for an adequate reason. The court imposed unprecedented restrictions on managerial power.

By the middle of the 1980s, industrial court guidelines on these issues were clear and widely known, and the court was sufficiently confident to disregard employer protestations that they were ignorant of the guidelines. A process of 'universalisation' had taken place, whereby the obligations voluntarily accepted by some employers had been extended by the court to all. This removed an argument often used by employers for refusing to enter into collective agreements with trade unions, namely that these would create economic disadvantages relative to competitors.

During this period, the independent trade unions dominated the industrial court, using it to protect their members from arbitrary managerial action. Employers often contended that the court had no power to challenge employer conduct that was lawful and complied with applicable statutes and contracts. This argument was repeatedly rejected by the industrial court, the supreme court and the appellate division for a number of reasons.

Firstly, it was the clear intention of the Wiehahn Commission and the 1979 legislation that the court's jurisdiction was to be based on fairness and not lawfulness.

And secondly, the court would have ruled itself out of existence had it found in favour of employers on this point. Without a jurisdiction based on fairness the court would not have attracted the employee litigants necessary for its continued operation. As a court designed to establish a code of employment practice, its primary function was to sit in judgement on employer conduct. This type of litigation had to be initiated by employees or trade unions willing to participate in court proceedings.

Changes to retrenchment law

The state has presented the 1988 changes to the Labour Relations Act as codification of the law on dismissal and retrenchment. A code usually summarises guidelines of the court's leading judgements. But this is not so in this case. The drafters of the Labour Relations Amendment Act have used the guise of 'codification' to reverse industrial court inroads made into employer prerogatives. The extent of this is shown by differences between the court's pre-amendment guidelines on retrenchment, and the new amendments to the act.

The industrial court established procedures to be followed before a retrenchment took place. An employer wishing to retrench workers had to give notice to a representative (majority) trade union of its intention to retrench; consult with the union over the need to retrench (ie whether the retrenchment could be avoided); and consult over the extent of the retrenchment.

In the past, consultations over the need to retrench have enabled unions to make creative proposals aimed at avoiding, or minimising the extent of retrenchments. The changes in the law, although ambiguously worded, may relieve employers of having to consult over the need to retrench. Their obligation may now be limited to talks with a recognised trade union over the dismissals that will result from a retrenchment. These talks are often a formality, leaving trade unions without the means to influence the number of employees retrenched.

Criteria for selecting employees for retrenchment have been drastically changed. The industrial court had ruled that criteria had to be objective and verifiable, and LIFO had become the major standard. This recognised employees' rights of seniority and required an employer to retrench employees with the shortest service. In contrast, an employer may now select candidates for retrenchment on any 'reasonable' basis, including ability, capacity, productivity and conduct of employees as well as the employer's operational requirements.

The court's objective criteria could be tested: it is easy to check whether seniority principles have been applied in a retrenchment. But there is no similar test of an employee's ability or conduct; these criteria allow employers greater latitude in retrenchment selection. This change could herald a return to the pre-industrial court era when a retrenchment could only be challenged if a trade union could prove that there had been overt victimisation of its active members.

In the late 1970s and early 1980s retrenchments were a major cause of strikes. Workers believed that employers used retrenchments to fire union activists and others they did not want to employ. Objective selection criteria were adopted because these helped to remove employees' perceptions of arbitrariness and created an increased sense of security of employment. The return of subjective criteria in retrenchment could lead to a resurgence of strike action over such disputes.

Prohibition of unacceptable strikes

State changes on fair retrenchment procedure are a deliberate intervention to reverse developments pioneered in the industrial court. Despite this apparent vote of 'no-confidence' in the court, the state has now entrusted it with preventing the occurrence of 'unacceptable' strikes.[5]

The impact of this will depend on how the court uses its new powers. Already both trade unions and employers have asked the industrial court to prevent the exercise of economic power in industrial disputes. Previously interdicts could only be obtained against illegal industrial action. This could be done in either the supreme or industrial court. Interdicts against strikes are often granted on an urgent basis, often without the union having a chance to present its case. Courts agree to deal with those cases on this basis because of the economic losses an employer will suffer as a result of a strike.

But now the court has the powers to interdict strikes because they are unfair, rather than illegal. This may be done even where all dispute procedures in an agreement or in the Labour Relations Act have been followed and the industrial action is lawful. The role of the industrial court is thus being changed to one of reducing the incidence of industrial action by prohibiting 'unacceptable' strikes.

The approach of the industrial court has resulted in an increase in disputes being channelled through statutory conciliation procedures. Here, the legality of industrial action is not affected by the reasonableness of the demand. A strike for an excessive wage increase or a lock-out to effect an unfair change in conditions of employment can nevertheless be legal. But the court's new powers are an invitation for it to inquire into the nature of the demand when asked to interdict industrial action. The court may now prohibit strikes and lock-outs because they are in support of demands that it considers unacceptable, even if full negotiation procedure has been adhered to.

Urgent unfair labour practice cases

Recent disputes and litigation in the hotel and metal industries have highlighted the problems in the court's new jurisdiction. In late 1988, a major hotel group and a trade union were in dispute over whether employees should have 1 May and 16 June as paid holidays.[6] Negotiations had taken place without settlement. The dispute had been channelled through the statutory conciliation procedures and both sides were in a position to stage lawful industrial action.

The hotel group, fearing strike action during the holiday season, implemented a lock-out. The union instituted urgent industrial court proceedings, arguing that the lock-out was unfair. The court issued an order prohibiting the company from locking out its employees, and preventing the union from calling strike action on any issue until a subsequent court case to clarify provisions in the recognition agreement. As this case could not be held until 1989, the judgement gave the employer a trouble-free Christmas and weakened the union's bargaining hand. Although there was no settlement, neither party could stage industrial action.

This type of court-imposed stalemate undermines sound industrial relations. The court's role is not to restrain industrial action staged after full collective bargaining. This judgement misconstrues the proper role of the court in collective bargaining disputes.[7] The court should ensure that collective bargaining relationships are established and prevent precipitate industrial action. This involves, for instance, ordering a recalcitrant employer to recognise a representative trade union, or interdicting a trade union from calling a strike prior to attempts to resolve the dispute. The court has no concern with the exercise of economic power after the completion of full negotiations, or where the industrial action is functional to collective bargaining.

BTR Dunlop v Numsa,[8] a case involving the dismissal of a leading shop steward, shows a different use of the court's new powers. The union demanded that the company either reinstate the dismissed worker or agree to refer the dispute to independent arbitration. The latter demand was included because of union members' lack of faith in the industrial court and because arbitration offered a quicker method of resolving the dispute. The company refused to accede to either demand and the union threatened a lawful strike. The industrial court granted an interdict against the strike because it would involve the unfair use of economic power.

The industrial court is thus likely to rule that any strike over a dismissal will be an unfair labour practice. The reasoning is that, as the industrial court can test the fairness of dismissals, trade unions should refer dismissal disputes to the court. Industrial action pressuring an employer to reinstate a dismissed worker may be ruled unacceptable because there is an alternative method - a court hearing - to resolve the dispute.

New unfair labour practices

In addition to giving the industrial court the power to interdict unfair strikes, the new law limits the type of strikes that can be staged. Sympathy strikes and repeat strikes will always be unfair labour practices. A strike will be classified as a sympathy strike if the employer concerned is not directly involved in the dispute giving rise to the strike. Although there have been few sympathy strikes in recent years, the concentration of capital in South Africa will create circumstances where sympathy action is the only effective means of pressurising an employer.

The prohibition on repeat strikes is extremely ambiguous. A strike will be unfair if it concerns a dispute similar to one that has caused industrial action in the past year and involves the same parties. There is an obvious difficulty in determining whether one dispute is similar to another. In addition, trade unions involved in a strike may not be willing to suspend the strike and resume negotiations because a later resumption of the strike could be considered a repeat strike.

Any act in support of a boycott of any product or service will now be an unfair labour practice. Boycotts are currently outlawed by the state of emergency regulations; the new Labour Relations Act ensures that boycotts will remain unlawful even if the emergency is lifted. This will restrict the ability of trade unions to draw on community support during industrial disputes.

Other unfair labour practices include the unfair suspension of an employee; the unfair alteration of conditions of employment; the breach of any agreement or any provision of the Labour Relations Act; any act of intimidation in an employment context; and any interference with employees' rights to associate or not to associate.

It is now an unfair labour practice to discriminate on grounds of race, creed or sex. At the same time the act contains provisions that facilitate the continued operation of racially-exclusive unions. This provision will protect the position of minority whites-only unions and ensure that they are not forced out of existence by the continued growth of non-racial unions.

The attack on independent trade unions

One of the central tenets of the independent trade union movement's approach to collective bargaining has been the concept of majoritarianism. This entails that, in any bargaining unit, a single trade union representing the majority of the workforce should be recognised by the employer. Once the trade union has proved its majority support, the employer must recognise it as the sole collective bargaining agent of the employees concerned. Minority trade unions may represent their members on issues such as discipline and grievances but have no right to participate in negotiations on collective issues such as wages and conditions of employment.

This system allows for uniformity in treatment of trade unions by the employer (only one trade union can have a majority at any time); and prevents an employer from having to bargain with a number of trade unions. Many of the country's major employers have accepted this approach and signed collective agreements implementing its principles. The new act makes it an unfair labour practice for a union to demand that an employer recognise it as the sole collective bargaining representative of its employees, or to do anything to prevent the employer dealing with minority unions.

This attacks a system of collective bargaining accepted and successfully applied by many of the country's major trade unions and employers. The statute is meant to provide a framework for the conduct of labour relations. It is formally neutral between trade unions and employers and provides structures for the regulation of industrial relations. But this new provision undermines the position of one of the participants in the industrial relations arena. It prevents employers from being compelled, either by industrial action or court order, to enter collective bargaining relationships that apply majoritarian principles. Disputes about the appropriateness of majoritarian trade unionism are removed from the jurisdiction of the industrial court. It will be an unfair labour practice for a trade union to call a strike with the aim of compelling an employer to sign an agreement incorporating a majoritarian approach. An employer who does not wish to grant a major trade union sole collective bargaining status is thus insulated from all forms of pressure.

This provision will facilitate a pattern of anti-union practices in recent years. This involves employers bolstering the position of non-union employees or minority union members at the expense of members of majority unions. In a number of cases, the industrial court has been firm in ruling that the preferential treatment of non-union members or non-strikers is unfair. It has ruled that granting non-union members, who did not engage in strike action, wage increases from an earlier date than union members, and then paying all employees an additional benefit to settle the strike, is unfair.[9]

But the new provision will increase the scope of employers to disparage majority trade unions by dealing with individuals or smaller groupings on a differential and preferential basis, and dilute the bargaining power of a majority union by forcing it to negotiate jointly with minority unions.

Damages actions

Another change to the law will make it easier for employers to hold trade unions accountable for losses suffered during strikes. The purpose of strike action is to inflict financial loss upon an employer through the withdrawal of labour. In terms of general legal principles, a trade union calling on its members to strike would be liable for the employer's economic damages suffered as a result of the loss of production and loss of business caused by the strike. But this liability would cripple trade unions and make strike action, as well as effective collective bargaining, impossible.

Accordingly, most countries' legal systems protect trade unions from liability for economic loss occasioned by strikes. In South Africa registered trade unions are protected against claims for damages and interdicts arising from legal strike action. This is a major incentive for trade unions to register and channel disputes through the statutory conciliation procedures, and is retained with some amendments in the new act.

However, the position is changed dramatically in the case of unlawful strikes or where conduct during a strike involves a criminal offence. Here the trade union has no legal immunity and is assumed to be responsible for all actions of its officials, office-bearers or members. To avoid liability for the legal consequences of these actions, the trade union will have to establish that it did not authorise the activities that may have caused the employer damage. This reverses the conventional burden of proof in court cases, and will generally be impossible to prove.

The aim of the provision is to compel trade unions, through fear of financial penalty, to police the activities of members and restrain unlawful industrial action. However, it could have the opposite effect. Trade unions, fearing possible damages action, may distance themselves from wildcat strike action rather than attempting to settle the dispute causing the strike.

In the last few months of 1988 a number of actions against trade unions were threatened amid considerable publicity. But major employers are unlikely to make use of this provision in the near future. They will not wish to risk the charges of anti-unionism that will follow the institution of a damages action with the potential to cripple a trade union. The use of the provision is, at least initially, likely to be limited to maverick employers. However, a single successful damages action could financially cripple the country's major unions.

Other provisions in the act may have more immediate financial consequences for trade unions. The powers of the industrial court to order unsuccessful litigants to pay the legal costs of the other side have been increased. This, coupled with the automatic right of appeal to the labour appeal court, will place a heavy premium on trade unions instituting or opposing legal action. With the high cost of litigation and the greater economic resources of employers, these provisions could have disastrous financial consequences for trade unions.

Conciliation procedures under the new act

The Labour Relations Act contains two channels for the resolution of industrial disputes - industrial councils, and conciliation boards which function in industries without industrial councils. Changes have been made to the conciliation procedures introducing additional technicalities. These could prevent certain disputes from being channelled through the statutory procedures.

For example, if a dispute is not referred to a conciliation board or industrial council within 90 days of it arising, no lawful industrial action can be staged. This new formality may render industrial action unlawful because of minor technical breaches of statutory provisions.

A significant amendment to the conciliation board system is the only feature of the new act welcomed by the trade unions. Previously, the minister of manpower had a discretion whether to appoint a conciliation board. He frequently refused to do so, particularly in disputes over dismissals. The effect of this was that the dispute could not proceed to the industrial court for determination as an unfair labour practice. However, legal industrial action could be staged once the board was refused.

This discretion has now been abolished and, provided the application for a conciliation board complies with the formal requirements of the act, the board must be established. This will streamline the conciliation process in industries without industrial councils where the minister has, due to the massive backlog of applications, often taken over six months to decide whether to appoint a conciliation board.

The minister has used these discretionary powers to the last. All applications for conciliation boards arising out of dismissals during the 6-8 June stayaway in protest against the changes to the Labour Relations Act were refused. In some cases the industrial court had, in applications for interim reinstatement, already ruled that certain of these dismissals were unfair. The minister adopted an overtly political stand on the stayaway which will presumably extend to any attempt by black workers to influence the content of legislation by industrial or protest action.

The private resolution of disputes

The new act is expected to increase the number of employers and trade unions who 'contract out' of the provisions of the Labour Relations Act and establish - by agreement - private dispute resolution procedures.[10] These agreements are already a significant factor with the Independent Mediation Service of South Africa providing extensive arbitration and mediation services. The complexities and ambiguities of the new act are expected to lead to the negotiation of agreements and procedures that by-pass the provisions of the Labour Relations Act. These may include negotiated unfair labour practice codes, compulsory arbitration agreements for dismissal cases and undertakings by employers not to dismiss employees for participation in procedural strikes.

The changes will also ensure that the labour arena continues to provoke a high level of legal activity. The urgent unfair labour practice jurisdiction of the industrial court, new technicalities in disputes procedures, and the ambiguities in the act will ensure that labour disputes frequently reach the courts.

The last two to three years have seen employers adopting a more active approach to the institution of legal actions. For the unions, much will depend on how the industrial court interprets its new powers and to what extent it intervenes in labour disputes. The behaviour of courts is less predictable than that of other state agencies, and the industrial court may assert its independence and remain a forum in which significant worker rights can be gained.

Notes

1 See E Cameron, H Cheadle and C Thompson, *The New Labour Relations Act*, Juta & Co, 1989, v-vi.

2 According to A Levy and J Piron, *Annual Report on Labour Relations in South Africa 1988-89*, disciplinary issues were the cause of 16,8% of strikes in the period January-November 1988. In contrast, A Levy, *Unfair Dismissal*, 1984, 63, states that in the period 1979-1983 dismissals caused between 20% and 43% of strikes in each year.

3 An indication of the increasing use made of the statutory conciliation procedures is the number of applications for conciliation boards: 1983 - 118; 1984 - 279; 1985 - 514; 1986 - 1 294; 1987 - 2 312.

4 See, for instance, E Webster, *A New Frontier of Control*, Carnegie Conference Paper 111, University of Cape Town, April 1984.

5 The terminology 'acceptable strike' has its origins in the *Report of the National Manpower Commission on Dispute Settlement, Levels of Collective Bargaining and Related Matters (RP 115/86)*, 69-82.

6 *HARWU v Karos Hotel*, unreported judgement of the industrial court, 30 September 1988.

7 See Cameron, Cheadle and Thompson, *The New Labour Relations Act*, 105-106.

8 'BTR Dunlop vs Numsa', *1989 Industrial Law Journal*, vol10, 181 and 701.

9 See *NUM v Henry Gould Chrome Mines, Industrial Law Journal*, (9), 1988, 1149 (IC).

10 For a discussion of private dispute resolution procedures see C Thompson, 'The Contractual Regulation of Strikes and Lock-outs' in P Benjamin et al, *Strikes, Lock-outs and Arbitrations in South African Labour Law*, Juta & Co, 1989, 69-82.

Recent Trends
in Industrial Action

Mark Bennett

The escalation of industrial action after the 1979 incorporation of African workers into the collective bargaining system involved a complex interplay of factors. These included:

- expansion of the black labour movement, in terms of membership, union personnel and resources. Workers saw the benefits of united action and flocked to the unions, which in turn were immediately pressured into militant actions to advance and protect members' interests;
- a sustained economic recession characterised by double-digit inflation, high unemployment and low economic growth;
- black workers' desire to eradicate the imbalance in wages and conditions of service between themselves and white workers;
- South African employers' acceptance, albeit only in principle, of black trade unionism. A legal order and industrial relations ethos has emerged which accepts that black workers organise and use strike action in pursuit of demands;
- black workers' use of their power on the factory floor, in the absence of access to political institutions, to advance wider economic, social and political interests.

Monitoring industrial action

The annual release of South African strike data is eagerly awaited by businessmen, industrial relations consultants, academics, government bureaucrats, politicians and trade unionists every year. For each interest group the statistics reveal a great deal about the country's social, economic and political order. But this collected information, which is repeatedly used in support of a variety of different claims, is often at variance with reality.[1]

Statistics available from officialdom are poor, while those available from private sector industrial relations consultants and academic sources, although in many respects more precise, are severely limited. In both cases the data is conservative and represents a substantial undercount.[2]

Department of Manpower statistics, gathered from reports which employers are legally obliged to submit in terms of the Labour Relations Act, are unreliable. Employers are often unaware of their statutory obligations to report strikes; and many believe it is either unnecessary or simply a waste of time to record short stoppages. In addition, the official definition of a strike or work stoppage excludes politically-motivated strikes, some go-slows, work-to-rules, overtime bans and sympathy strikes. And as the state believes all strikes in the public sector are illegal, Manpower figures exclude the massive strike actions that have occurred in public sector enterprises.

Statistics supplied by independent analysts from both sides of the class divide are primarily gleaned from press reports, with additional information coming from unionists and industrial relations consultants. It is physically impossible for journalists to report on every stoppage, and in addition, the commercial media are often selective in their strike coverage.

While all monitors are flawed in one way or another, a certain consistency has been evident over time, largely due to the use of the same methodology each year. The Manpower data base, for example, has shown that since 1979 there has been a progressive increase in the number of strikes and work stoppages. The same trend in the total number of employees involved, and workdays and wages lost, is reflected in further Manpower figures (see Table 1 below). Many of these trends have been confirmed, albeit in some respects on a greater scale, by data emanating from independent sources.

Independent sources estimated that about nine million workdays were lost due to industrial action by black workers in 1987. This figure, which is more than the combined figure for the previous ten years, has marked 1987 as a watershed year in South African industrial relations.[3] More than a thousand strikes and work stoppages were recorded, resulting in an enormous loss of wages, production and sales. These high levels of industrial unrest not only brought home the huge costs of industrial action to labour and capital, but also had a significant effect on industrial relations in 1988.

One important effect of the 1987 action was that all parties along the industrial divide were pushed into making greater tactical use of the collective bargaining system and other legal processes.

But the tactically 'conciliatory' approaches used by the labour movement in 1988 did not lead to any significant drop in worker militancy during that year. Indeed, both official and independent estimates show that at least 1,5-million workdays were lost as a result of industrial action; that more than a thousand individual strike and work stoppage actions occurred in commerce and industry; and that over 2,5-million workers participated on each day of a three-day work stayaway in protest against the LRA amendments.

Thus, while worker militancy in 1988 did not exceed that of 1987, it was higher than that demonstrated in 1986. More important, however, was the subtly changing character of overt industrial action.

The conciliatory shifts which occurred in 1988 included a greater sophistication on the part of both management and workers in their dealings with each other. These included a variety of strategies - short stoppages, lock-outs, sit-ins, go-slows and work-to-rules - designed to nudge each other towards settlement.[4] Differences were also resolved through the increased use of legal options offered by industrial and civil courts. Worker militancy and management aggression were beginning to transform each other, revealing a growing maturity within the collective bargaining environment.

Growth in union membership

The growth of African worker organisation continued unabated during 1986 and 1987. The substantial membership growth experienced by Cosatu through its recruitment of previously unorganised workers, the poaching of workers from other unions, and mergers with non-aligned unions, was particularly noteworthy.

When formed in December 1985, Cosatu had a paid-up membership of 450 000. By July 1987 membership had expanded by 58% to 712 000. Together with this growth Cosatu also managed, albeit with some difficulty, to streamline its structures and establish 12 industrially-based unions.

This period of rapid growth and internal reconstitution of the labour movement had a general impact upon the overall levels of industrial action. As with earlier periods in South African labour history, levels of industrial action were again synchronised with increases in the size of trade unions (see Table 1).

Recruitment

Large numbers of workers, particularly those who were previously unorganised or had belonged to moribund or ineffectual unions, joined Cosatu affiliates in 1986 and 1987, believing this would provide solutions to their problems. This led to an increase in pressure on unions to take aggressive bargaining stances in negotiating conditions of service and wages. The large number of industrial actions in 1987 and 1988 was therefore partly attributable to the growth in union membership.

During 1987, some of the newly-formed or revitalised unions operating in relatively un-unionised sectors - Potwa, Sarhwu, Samwu, Nehawu and Vgawu, for example - undertook industrial action to recruit more members or transform signed-up membership to paid-up status. Undoubtedly the strikes by post office, Sats and metal employees were primarily responsible for massive growths in these unions' memberships. But a number of unions which experienced this membership increase failed to retain their new recruits, demonstrating that recruitment of workers on the basis of militancy and highly-inflated promises is a far cry from delivering the goods.

Strikes of long duration

The many workdays lost through industrial action in 1987 reflected a number of lengthy strikes involving large amounts of workers. The duration of individual strikes has increased steadily from an average of 1,9 days in 1983, to 2,1 in 1984, 2,8 in 1985, 3,1 in 1986, and 9,9 in 1987. The 1987 losses were due mainly to a few protracted disputes: two strikes involving the post office and Potwa (one of 63 days and another of 20); the Mercedes Benz and Numsa confrontation (46 days); the Sats strike (30 days); and the OK Bazaars strike (December 1986 - March 1987).

The Chamber of Mines/NUM confrontation involving at least 250 000 black miners and lasting 21 days, ranks as South Africa's costliest strike action. AL, JP and Associates estimated that the mine and railway strikes during 1987 accounted for at least 65% of workdays lost through (non-stayaway) industrial action.

In many instances strategically planned and executed strikes developed into classic 'trials of strength and will' where each party had definite objectives, usually premised on the expectation that the other side would capitulate (eg the mining, Mercedes Benz and OK Bazaars strikes). In other actions strikes dragged on unnecessarily while both parties adopted principled bargaining positions making it difficult to reach compromise and resolution. The state's refusal to deal with or recognise unions other than entrenched and often unrepresentative in-house staff associations contributed to and exacerbated some of these strikes.

Secondary objectives of industrial action

Some industrial actions had secondary objectives aside from those which had originally sparked the disputes. For instance, the three-month-long OK Bazaars dispute involved not only an effort to secure improved wages at the retail chain, but was also an attempt by Ccawusa to establish higher wage levels and better working conditions in a sector where these had been notoriously poor. The extended duration of the strike was instrumental in the substantial wage increases workers won from other managements in the industry, notably Pick 'n Pay.[5]

Similarly the weakening effect on NUM of the 1987 mine strike may have motivated mining houses to prove that employers had the resolve and resources to combat worker militancy. And while employers in the mining industry did not wish to eradicate NUM, they did set out to curb the union's industry-wide power.

Broadly, corporate capital wished to demonstrate its power to the labour movement, which was evolving national and regional strategies in response to the extensive and excessive power of a few corporations and multinationals.

In 1988 the average duration of strikes and work stoppages was much shorter than 1987, even though the level of worker militancy was still relatively high. Major disputes occurred both in the public and private sector where Numsa, in one of the few major sectoral strikes of the year, struck against employers party to the iron and steel industrial council.

In 1987 a strike organised by Numsa had to be called off at the last moment when the now-disgraced former minister of Manpower, Pietie du Plessis, signed the industrial council agreement. This made any industrial action illegal, and was signed after Numsa had conducted strike ballots at 652 factories where 94,8% of all workers voted in favour of strike action. The 1988 Numsa strike, involving only 15 000 members and lasting for a month, was a highly successful tactical victory in that Seifsa was forced to recognise the union as representative of most black workers in the industry.

During 1988, many unions and employers channelled industrial conflict through the collective bargaining process. This partly explains why, despite the high incidence of individual strikes, the number of workdays lost remained lower than in previous years.

However, the introduction of the LRA prohibition on 'grasshopper' (suspended and resumed) strikes has raised the possibility that the average length of strikes and the number of workdays lost will be significantly increased in the future. Unions have often suspended strikes in an effort to promote further negotiation pending the consideration of a final offer. They will be reluctant to do so if unable to recommence the strike.

Union participation in the industrial council bargaining system may also result in a larger number of sectoral strike actions should deadlocks occur.

Strike triggers

Worker dissatisfaction over wages was the major cause of industrial action during 1987 and 1988. This was followed by disciplinary and grievance procedures, retrenchments, union recognition and conditions of employment. Fewer strikes - apart from some demonstration stayaways - resulted from worker protests against the state of emergency than in 1986.

Recognition agreements: Industrial action has to some extent been contained by disciplinary, grievance and retrenchment codes which have been written into recognition agreements. However, disputes over recognition could become a significant factor once industrial unions turn their attention to smaller non-unionised enterprises.

During 1987 and 1988 the tacit state recognition of independent unions in the public sector led to a number of disputes.

Wage demands: Industrial relations consultant Andrew Levy estimated that, during 1987, wage demands were responsible for 33% of all stoppages, while for 1988 he claimed the figure had risen to 40%. The effects of the severe economic recession continued to underpin union wage demands. Narrower profit margins resulted in many employers being unable to meet worker demands which in turn created fertile ground for strike action.

The living wage campaign: In mid-1987 Cosatu launched a living wage campaign, demanding a living wage for all, a 40-hour working week, a ban on all overtime, three new paid public holidays and the abolition of tax deductions. Although these demands were raised frequently in wage negotiations between employers and unions, the campaign itself was not directly responsible for the increase in industrial action, nor for the substantial wage increases some unions won for their members. Unions tended rather to see the campaign as a useful tool to consolidate organisation on the factory floor.[6]

Cosatu's leadership acknowledged that, in its initial form, the living wage campaign was a failure. But the willingness of unions to resort to strike action in support of wage demands may have resulted in many managements agreeing to the above-inflation wage increases negotiated in 1987 and the first half of 1988. This occurred despite former State President Botha's announced wage freeze in the public sector and his appeal for 'wage restraint' in the private sector. The Institute for Industrial Relations has estimated that for the first three quarters of 1988 the average minimum wage increase was 20,7%.[7]

Racism: Union and worker efforts to eradicate racist practices in the work-place continued to spark industrial action during 1987 and 1988. Many of these disputes occurred in the public sector where railway, municipal and postal employees raised the issue of preferential treatment for white employees, and general white attitudes. One cause of the April 1987 postal strike lay in policies of racial inequality, and Potwa's general secretary Vusi Khumalo argued that racial inequality lay 'at the heart of worker grievances within the public sector'.[8] He cited authorities' preferential treatment of whites; segregated canteens, toilets, training facilities; and the general day-to-day racist attitude of fellow white employees as examples of this.

The private sector was not immune to conflicts caused by racist white attitudes. In February 1987, for example, more that 5 500 AECI employees at the company's Modderfontein plant downed tools, demanding the dismissal of a supervisor who used racially abusive language. While senior management in the private sector has made firm policy guidelines outlawing any discriminatory practices, many of these codes have not filtered down to the lower echelons of white middle-management and supervisory staff.

During 1987 and 1988 sympathy industrial action increased. This form of pressure has been facilitated by the move towards industrial unionism which has been spearheaded by Cosatu; the creation of corporation-based shop-steward councils, already existing in the Anglo American and South African Breweries-owned companies; and closer co-operation between affiliates of Cosatu and Nactu.

In an effort to head off sympathy strikes, the amended LRA makes it illegal for anyone to instigate a strike or incite other workers to take part in a strike if the employer or employee is not directly involved in a dispute. This new labour law also makes it illegal for unions to engage in the organisation of product or service boycotts.

The state of emergency and the labour movement: While the on-going national state of emergency forced trade unions to take a more active political role, it has also directly impinged upon labour relations. A wave of strikes to protest the detention of union leaders wracked the retail and mining industries in 1986 - the period immediately after the imposition of the emergency - but few strikes of similar nature occurred in 1987 or 1988. Fewer union officials and active union members were detained in 1987 and 1988, and workers were generally not prepared to forego wages or risk dismissal for participation in industrial actions with little chance of success.

In some cases, community activists may have attempted to use strike actions to mobilise and politicise communities after the township-based extra-parliamentary opposition movement had been immobilised by security force action. In the OK Bazaars dispute, for example, Markham notes that the strike action had 'broader implications' than a wage increase. She asserts that the strike, with its attendant community involvement, 'provided a focus for mass mobilisation and practical involvement at a time when the state clamped down on other campaigns and there was a general lull in political activity'.[9]

Features of industrial disputes

Unions have a wide variety of strategies available to pressure management into accepting demands. In a management survey of 1987 wage settlements, 29,2% of respondents indicated that strike action was accompanied by some other form of industrial sanction.[10]

Shorter duration stoppages - both wildcat and planned - which have interfered with production included go-slows, work-to-rules, sit-ins, overtime bans, and grasshopper strikes.

Managements have a legal capacity to strike back at employees. During 1987 and 1988 management had moderate success with the legal lock-out which they used to enforce discipline among their employees and force union opponents to modify their positions. In 1987 a wage dispute at Highveld Steel resulted in the company refusing to allow about 2 100 Numsa members access to work for a month; while in another wage dispute SAPDC (South African Pharmaceutical and Drug Company) refused to allow 400 Sacwu members to attend work for over two months.

During 1988 there was an increased tendency for employers to avert or stop strike action by interdicting unions and workers through the industrial and civil courts. Employers soon discovered that courts were prepared to outlaw some strike actions even when unions had complied with all the procedures envisaged in the LRA.

The recorded number of work stayaways called during 1987 and 1988 declined significantly over previous years. While at least 21 national and regionalised stay-at-homes were observed in 1985, and 33 in 1986, only seven occurred in 1987 and six in 1988 (see Table 2). In 1987 four nation-wide stayaways were called (21 March; 21 April; 5-6 May; 16 June); locally there was one in Tembisa (15 October), one in Soweto (22-24 April), while another was confined to the Transvaal (15 April). In 1988 three stayaways had a national effect (21 March; 6-8 June; 16 June); locally there was one in Ashdown near Pietermaritzburg (1 February); and one affected White City Jabavu in Soweto (17 February).

This dramatic reduction must in many ways be attributable to the efficacy of the state of emergency. Undoubtedly many workers were also reluctant to embark upon actions which could have resulted in their losing both pay and jobs, but were still prepared to embark upon stayaway action to defend their immediate interests.

While fewer stayaways were called over the past two years, worker support for them was often spectacular. When unions initiated three days of national protest against the LRA Amendment Bill and the impending renewal of the state of emergency, the largest stayaway ever called in South Africa resulted. Cosatu estimated that at least nine million workdays were lost altogether; while AL, JP and Associates estimated that 2,5-million workers heeded the call on each day. The protest impact of some traditional stayaways - 21 March (Sharpeville Day), 1 May (May Day) and 16 June (Soweto Day) - is rapidly diminishing as many unions have negotiated paid days off for their members.

Table 1: Official strike count and union growth, 1979-1988

	Number of strikes/work stoppages	Number of employees involved	Number of workdays lost	Estimated wages lost (Rands)	Total union membership
1979	101	22 803	69 099	202 501	701 758
1980	207	61 785	174 614	1 401 516	781 727
1981	342	92 842	226 554	2 263 705	1 054 405
1982	394	141 571	365 337	4 544 630	1 226 454
1983	336	64 569	124 596	1 697 610	1 288 748
1984	469	181 942	379 712	5 174 798	1 406 302
1985	389	239 816	678 273	8 184 985	1 391 423
1986	793	424 390	1 308 958	23 166 000	1 698 157
1987	1 148	591 421	5 825 231	111 061 808	1 879 400
1988	1 025	160 000	914 388	unavailable	unavailable

Sources:
Department of Manpower, Annual Reports, 1979-1987.
National Manpower Commission, Annual Reports, 1979-1987.
Central Statistical Services, South African Labour Statistics, 1988.

Table 2: Stayaway strikes, 1984-1988*

Organisers	1984	1985	1986	1987	1988	TOTAL
Trade unions		4	4	2	1	11
National extra-parliamentary movements	1					1
Regional civic organisations		13	25	2	1	41
Alliance of groups	3	4	4	3	3	17
Total	4	21	33	7	5	70

* It is impossible to identify precisely who was responsible for organising each stayaway. 'Guesstimate' distinctions have to be made between those groupings who initiated the idea of a boycott, the actual organisers, and those groups which lent organisational support to the protest call.
* This table reflects only the numbers of individual stayaway actions observed and not worker support for stayaway calls. It must be noted, for instance, that while

only five stayaway actions were observed in 1988, more than nine million workdays were lost - perhaps more than in 1986 when 33 stayaway actions were documented.

Source: M Bennett and D Quinn (eds), *Political Conflict in South Africa*, Indicator Project, 1988.

Notes

1 A Levy, 'The Limitations of Strike Statistics', *Indicator SA*, 3(2), 1986, 10.
2 The use of industrial action data (particularly for 1988) and other statistics has been determined, in most cases, by their availability. Most of the statistics used in this paper are drawn from reports by Andrew Levy, Johan Piron and Associates (AL, JP and Associates), the Department of Manpower, and from information released in parliament by the acting-minister of Manpower.
3 AL, JP and Associates, *Annual Report on Labour Relations in South Africa, 1988-1989*, Johannesburg, 1989. The statistics supplied by this monitor do not include the number of workdays lost as a result of stayaway action in 1988, which they estimate to be a minimum of 7,5 million days.
4 AL, JP and Associates, *Annual Report 1988-1989*, 11.
5 C Markham, 'Lessons of the OK strike', *South African Labour Bulletin*, 12(3), 1987, 22.
6 I Obery, S Singh and D Niddrie, 'Disciplined mine strike a test of strength', *Work in Progress*, 49, September 1987, 34.
7 Institute for Industrial Relations, *IIR Information Sheet*, October 1988, 1.
8 C Markham, 'Racism sparks postal strike', *South African Labour Bulletin*, 12(4) 1987, 17.
9 C Markham, 'Racism sparks postal strike', 17.
10 Andrew Levy and Associates, *Annual Report on Labour Relations in South Africa, 1987-1988*, Johannesburg, 1988.

Withdrawal Pains: General Motors and Ford Disinvest from South Africa

Glenn Adler

During 1987 and 1988 General Motors (GM) and Ford dramatically ended all direct investment in South Africa, despite two decades of assurances that they were in the country to stay. Both companies were among the staunchest advocates of continued direct investment and active supporters of the sanctions-evading Sullivan Principles, themselves the brainchild of a GM company director. Neither company left South Africa altogether, but established mutually beneficial arms-length agreements with their former subsidiaries.

The withdrawals were prompted by the Comprehensive Anti-Apartheid Act in the US, but were also a response to the collapse of the South African automobile industry. The firms harmonised their disinvestments with broader programmes to restructure and restore profitability.

In recent years, workers have been unable to respond forcefully to such restructuring, and the withdrawals introduced new and more complex issues into the fight. The companies employed sophisticated strategies, developed in secret and sprung on workers at the last moment. They had vital support from the South African government and the Reagan administration, and the ultimate power to bring workers to heel by threatening plant closure and mass unemployment. Finally, the companies benefitted from a divided opposition.

Notwithstanding their general support for disinvestment, neither South African unions nor external supporters of sanctions had developed strategies to respond to actual withdrawals. Workers and their unions belatedly drew up demands to put to the departing corporations, while some in the US sanctions movement saw such calls for negotiations as setting conditions for disinvestment. While some workers were either directly or indirectly calling for continued ties between the multinational corporation and the former subsidiary, the sanctions movement was targeting such non-equity ties. Ford was able to manipulate differences over its continuing links to Samcor in late 1987 when it fended off US challenges to its sanctions-busting transfer of $61-million to the South African company. These divisions helped company efforts to weaken the Anti-Apartheid Act and made international solidarity work more difficult.

After withdrawing, the firms supplied South Africa with valuable capital investment and secured the continued transfer of technology, components, and designs. The fights over withdrawal saw workers thrown on the defensive, plant-level union structures left in disarray, and divisions sharpened between internal and external movements against apartheid.

In assessing these cases, however, it is important to distinguish between the overall impact of sanctions on the economy as a whole and their impact on a specific industry or company.[1] At the most general level, they are designed to have a cumulative effect on the political system, upping the cost to the regime for maintaining apartheid. Inevitably, comprehensive mandatory sanctions work over the long-term, in part because of the difficult political battle to enact them. In the short-run, however, the anti-apartheid movement, inside and outside South Africa, has to cope with the specific effects of sanctions, including disinvestment.

When the Anti-Apartheid Act was passed in late 1986 there were few ways for internal and external groups to debate the macro-level mechanisms of disinvestment or co-ordinate responses to the companies.

The conflicts sparked by disinvestment prompted many discussions over new paths forward. Over the last two years, unions have developed new strategies and tactics in response to withdrawals like the Ford and GM cases. At the same time, external organisations are developing techniques for increasing pressure on the South African state and capital, while supporting South African workers in their immediate fight against employers, and in their long-term movement for liberation. National and international co-ordination is not easy. But it may overcome weaknesses revealed in the GM and Ford withdrawals, while preparing the ground for more extensive challenges to capital and the South African state.

When the going gets tough, the tough change owners

Crisis and rationalisation in the motor industry

During 1983 the motor industry slumped along with the rest of South Africa's manufacturing sector. This market has always been oriented to luxury consumption, demanding the latest high-end vehicles and new technologies. With increasing automation in the international industry, local producers were forced to accept the technological requirements of their foreign-supplied designs. This involved ever-increasing investment in expensive machinery and tooling. But the small and highly competitive South African market, geared as it was to white buyers, did not allow for economies of scale: the manufacturers lacked large volumes over which to spread their investments, and had few export opportunities.[2]

The depreciating value of the rand pushed up the cost of imported machinery, tooling, components, and debt service. Along with double-digit inflation, exchange rate problems contributed to a 68% increase in operating costs between October 1983 and September 1985. New car prices almost doubled between 1983 and 1986, but manufacturers still found their profit margins shrinking.

From the historic high of 301 000 units sold in 1981, sales withered to 174 500 in 1986. Losses by the big producers mounted year by year, and exceeded R700-million in 1986 alone. Over the period manufacturers virtually stopped investing in their enterprises, and the industry slowly ground down.

The processes at work in the motor industry and the manufacturing sector as a whole formed part of a larger transformation in the international political economy, and South Africa's location in it.[3] Virtually all semi-industrialised countries shared South Africa's predicament, brought on by the massive liquidity crisis of the early 1980s after the credit-fuelled industrial expansion of the previous decade. The deep austerity measures taken in response to the debt crisis were accompanied by widespread de-industrialisation as investment was redirected to the core countries, primarily the US.

These changes occurred as the world auto industry shifted toward heavy investment in 'flexible' automation, and a manufacturing system requiring minimal inventories based on the Japanese 'just-in-time' model. This re-organisation favoured placement of component suppliers close to assembly sites and led to relocation of production to First World countries. Taken together these processes seriously undermined the historical bases of automobile production in South Africa and much of the Third World.

Watching the Fords go (good) by

'Rationalisation' was the cure government and industry prescribed for this malaise: a drastic reduction in the number of models, producers and jobs. The turning point came in late 1984 when rumours surfaced that Ford could be pulling out of South Africa. A change of heart by the American giant had immense significance. Ford was the first multinational automobile company in South Africa: it opened an assembly plant in Port Elizabeth in 1924, and for many years was the leading producer. It was also one of the most important companies in the Eastern Cape region, supporting dozens of supplier firms, and, in Ford's peak year, more than 7 000 employees depended on its fortunes.

In 1984 Ford began secret negotiations with Amcar, the Anglo American-owned car maker, over a possible merger of their operations. Local and international factors made this option particularly attractive for both parties. Each had lost money and market share over the preceding years, and had excess capacity. Amcar held the local licence to produce Mazda cars while Ford USA held a substantial share in Toyo Kogyo (Mazda) in Japan. With overlap in design between Ford and Mazda products, a merger provided room for rationalisation of models.

Both companies misled the public and their own workers, denying merger rumours virtually to the day the deal was concluded. In January 1985 Ford and Amcar announced they would become partners in the new South African Motor Corporation (Samcor). Ford would not disinvest, but would remain as the junior partner holding 42% of Samcor's equity, with Anglo controlling the remaining 58%.

Workers could do little except contest the terms of their retrenchment packages. Over the next year Ford closed its Port Elizabeth operations, except for a small engine assembly factory. Nearly 4 000 hourly-paid workers lost their jobs.

The Ford-Amcar merger was only the most dramatic example of rationalisation. In all companies, 1985 and 1986 were years of massive retrenchment. Employment in the motor industry dropped by more than 40% from a peak of nearly 50 000 jobs in 1982 to 29 000 in 1986.

What's good for General Motors?

Ford's merger with Amcar left only one wholly American-owned car maker in the country. General Motors had followed Ford to South Africa in 1926, and also built its plant in Port Elizabeth. Through the 1970s matters looked dark for GM, as it tumbled from its position as market leader. By 1985 it claimed a bare 9% share of the passenger vehicle market. As 1986 passed and the South African motor industry deteriorated, GM fell deeper into trouble, unable to hold its already low market share.

US congressional approval of the Anti-Apartheid Act in September 1986 changed the ground rules for American companies in South Africa. Among other features of the legislation, firms were barred from making new investments in their operations after November 1986. Like many other US companies, GM was busily planning its response, which remained secret until October when General Motors Corporation (GMC) Chairman Roger Smith said the corporation was reviewing its presence in South Africa.

Shop stewards from the National Automobile and Allied Workers Union (Naawu), remembering the Ford closures, promptly asked General Motors South Africa (GMSA) management to clarify Smith's remarks, but were told the chairman had been misquoted. More than a week later GMSA announced that the company would be sold, but still withheld important details.

Workers, dissatisfied with management's lack of response, called a general meeting of Naawu members. They demanded that their pension fund contributions be paid out, severance pay, and the right to appoint two worker representatives to the board of directors. Finally they asked for an urgent meeting with GMSA management to discuss the situation. Workers never asked for the factory to close nor for GM to cut all ties to South Africa. Rather, they sought to protect their own interests whatever the outcome.

In a letter leaked to the press before it was received by the union, a GMSA director rejected the union demands, ignored the request for a meeting, and threatened to hire unemployed workers to replace anyone unwilling to 'assist' the company. Two days later workers downed tools and began a sit-in, which ended after two weeks when the company called in security forces to evict strikers. Nearly 570 workers were fired. The company even allowed a police brigadier to direct his operations against workers from a strike command post on company premises.

In the course of the strike GMSA finally outlined its reorganisation scheme. Ownership would be sold to a team of GMSA executives. GMC would liquidate the massive accumulated debt of the subsidiary, estimated at R100-million, by transferring funds to GMSA ahead of the Anti-Apartheid Act's investment deadline. Finally, GMC would continue to supply components, designs and spare parts to GMSA, largely through its Opel subsidiary in Germany and its associated Isuzu Motors in Japan.

GM broke the strike when management threatened to fire workers who remained out and began hiring unemployed people to replace those fired during the sit-in. Worker support faded as hundreds of people queued daily to apply for strikers' jobs. There were repeated interventions by the police, notably an armed assault on strikers who had gathered in front of the plant, where 16 were arrested and many more injured. Within days most of the strikers went back to work, and production soon returned to normal.

With the strike broken and worker leadership decimated, the new company, renamed Delta Motors in early 1987, could restructure without organised labour opposition. Delta reportedly permitted only four shop stewards in the plant, and greatly restricted their freedom to operate on worker business. Naawu has claimed that a new company union, the Free and Independent Workers Association, appeared at the plant and agreed neither to strike nor to embrace the broad political concerns of the black community.[4]

Management used the strike to cut its hourly workforce, and drastically reduced its salaried staff in a general austerity move. With its books wiped clean after GMC had liquidated its outstanding debt, the company was no longer burdened with interest payments and, as a South African-owned firm, enjoyed a much stronger position to raise capital inside the country.

Finally, the new managers lifted GMSA's previous ban on sales to the military and police. The new managing director said the policy change secured a 'vote of confidence' from the government. After receiving 'significant' orders from two or three major departments, the executive predicted Delta would do more government business in 1987 than GM had done in the previous five years.[5]

Strengthened by such measures Delta returned to profitability in 1988.[6]

As with the Ford-Amcar merger, a combination of local and international factors led GM to reconsider its previously-held disinvestment pledge. GMSA was about to launch the Monza, on which the success of the company depended. The effort called for re-tooling and re-organisation of the workforce, as well as commitments from Opel to maintain components supply and designs for the life of the vehicle line.

Events in the US hastened this re-organisation. Steps to revive GMSA had to be taken quickly, in advance of the Anti-Apartheid Act's November deadline on new investment. In addition, GMC announced in November that it would close 11 plants in the US, affecting 29 000 workers; the South African decision was thus part of a rationalisation drive within the multinational.[7]

Lastly, under US tax law GMC would receive a substantial tax break on GMSA's losses, in terms of a loophole due to disappear by the end of 1987.

Given the exchange rate and US tax benefits, GMC could pay off GMSA's debt on the cheap, while the ownership transfer insulated the subsidiary from US sanctions without disrupting its supply links to Isuzu and Opel. At one stroke the scheme responded to the letter of the Anti-Apartheid Act in a way which comfortably served company interests.

The union response

In the course of the GM strike the unions that would soon merge to form the National Union of Metalworkers of South Africa (Numsa) issued a joint statement setting forth their demands to any company contemplating withdrawal. Naawu, Mawu and Micwu demanded that disinvesting companies give workers sufficient notice, provide full details of plans, negotiate their terms of withdrawal, and guarantee that no benefits would be prejudiced by the sale. Further, they demanded that workers' wages and benefits be protected for 12 months from the date of withdrawal.[8]

Cosatu strongly condemned the behaviour of the departing corporations, and the self-interest which defined the form of their withdrawals. The disinvestment resolution passed at the 1987 Cosatu congress criticised such moves as 'corporate camouflage which often allows these companies to increase their support for the South African regime'. Cosatu demanded that the terms of withdrawal be negotiated, but left it to individual unions to give content to that position.

Cosatu also debated the question of worker ownership in response to some withdrawing companies' offer to transfer shares to workers. While worker share ownership was an untested concept for unions, and many remained suspicious of the idea, the schemes were not rejected outright.

Instead, two general principles were accepted as guidelines for discussing such packages. First, Cosatu discouraged workers from taking positions on boards of directors, arguing that they would not gain effective control of the company but would run the risk of identifying with management rather than worker interests. Second, Cosatu criticised individual share offerings. It favoured collective ownership with shares held in trusts directed by workers, and used as financial resources for broader community purposes.

Ford has a (somewhat) better idea

The debate over shareholdings came to a head in the motor industry, and focus again fell on Naawu.[9] In June 1987 Ford announced its withdrawal from South Africa, and revealed a plan to give its workers a 24% shareholding in Samcor. The remainder of Ford's shares would be transferred to Anglo American.

Workers initially rejected this scheme, largely because Ford wanted dividends on the shares distributed on an individual basis. Instead, workers wanted the shares converted to cash and divided among Samcor employees. Numsa has criticised individual shareholding schemes 'as deliberate attempts to weaken unions and to divert workers' attention away from their more militant but justifiable and legitimate demands'.[10] This time, however, the company met with Naawu to discuss the deal, in contrast to GM's refusal to negotiate with workers before the strike, and to Ford's own failure to negotiate with its workers prior to its merger with Amcar in 1985. After months of negotiations Ford raised the stakes by threatening to withdraw from South Africa and cut all ties to Samcor unless agreement could be reached on the trust. It was widely believed that Samcor would have to close, if indeed Ford was contemplating such a step. As it became clear that Ford was not about to liquidate the shares and give the workers cash, the union devised a new plan for a community trust. In November, shop stewards put the following alternative proposals to workers at Samcor's Pretoria and Port Elizabeth plants:

1) that Ford's 24% share offer be rejected, that Ford withdraw from South Africa and cut all its links with Samcor, and that no trusts be established; or

2) that the trust be established with the 24% Ford shares and that the dividends on such shares be divided amongst the workers; or

3) that the trust controlled by Samcor workers be established with the 24% Ford shares and that the dividends be used for development projects to improve the quality of life in the communities.

According to Naawu, workers gave a mandate for the third proposal at mass meetings. In November Ford and Numsa (into which Naawu had merged) announced an agreement to establish an employee trust.

According to the plan, a 24% equity interest in Samcor would be transferred to an employee-controlled trust and the rest of Ford's 42% stake in Samcor would be held by Anglo American, the Anglo American Investment Corporation, and its associates. Dividends on the shares given to the trust would be used only for community welfare and development. Numsa also won concessions that benefits and existing employment conditions would be guaranteed, that union recognition agreements and bargaining rights would be sustained, and that the employee trust could nominate three representatives to Samcor's board of directors. Numsa viewed the trust agreement as a small step toward fulfilling Cosatu's disinvestment policy for the assets of departing corporations to be retained in South Africa and used in the interests of the people. The agreement also gave force to the federation's policy opposing individual share offerings, while winning for the workers a measure of protection against anti-worker restructuring and corporate camouflage withdrawal.

But the deal quickly ran into difficulty. In late 1987, the United States Congress learned Ford had notified the Treasury Department of its intention to transfer $61-million to Samcor, an apparent violation of the Anti-Apartheid Act's prohibition on new investment. Ford claimed the money was needed to ensure Samcor's health. This prompted an investigation by members of Congress, but the Treasury Department assured Ford that the company would not be prosecuted for violating the Anti-Apartheid Act.

Ford has been widely criticised for the way it handled the transfer. While the company dutifully informed Samcor workers, Naawu, and US anti-apartheid organisations of its general withdrawal plans, in its communications with the US anti-apartheid organisations, for example, Ford never mentioned the $61-million transfer. Congressional aides claim Ford played them off against one another, telling each that the others had agreed to the transfer.[11]

The most important factor in Ford's success was its claim that the union supported the scheme. For its part, Numsa stated that it was indeed informed about the payment, but separated itself and the trust negotiations from the company's financing plans. Nevertheless, workers not only voted for a workers' trust, but *against* cutting Ford's ties to Samcor. In the US Ford easily portrayed these positions as a demonstration of worker support for their continuing presence in South Africa, and implicitly for a transfer of funds to keep the company operating.

A far more difficult issue grew out of general discontent among workers over the terms of the plan. Before the agreement was signed many workers wanted shares converted to cash and distributed to Samcor employees. The union explained the new community trust proposal to workers and, after the November vote, believed it had general support.

But workers from Samcor's engine plant in Port Elizabeth mobilised opposition to the plan. Many of these activists were former members of Macwusa, a rival union to Naawu before the formation of Numsa. They claimed that if workers opposed the trust they could receive cash for their shares, amounting to R49 000 per employee. Exploiting workers' previous demand for a cash payout, they promised to win what the union failed to achieve. With the aid of local lawyers, these Port Elizabeth workers established an anti-trust committee to publicise their grievances and mobilise Samcor workers in Pretoria.

The committee's promises were readily accepted in the Pretoria plant. Workers rallied to the rumour of a huge lump-sum payout, while other rumours portrayed Naawu as collaborating with local community councillors and suggested the trust would be used to buy the Mamelodi Sundowns football club. Days before the agreement was to take effect in April 1988, workers at Samcor's Pretoria plant went on a wildcat strike to protest the plan. They repudiated

Numsa's mandate to negotiate with Ford over the trust and rejected the shop stewards who had supported it. A new body was designated to negotiate with Ford.

In a meeting with Samcor and Ford representatives, the anti-trust committee threatened that workers would stop building Ford products unless the company agreed to pay $100-million to its workers. Despite its threats the committee failed to move Ford and the meeting broke up without agreement. Ford ultimately transferred the shares to the employee trust and appointed company representatives to its board. Given the disruption among employees, no worker representatives were elected. When Samcor dividends for 1988 were announced the company trustees offered dividends on an individual basis to worker participants in the trust.[12]

The Ford withdrawal thus harmonised with Samcor's need for capital investment and its desire to sustain international links. Meanwhile by disinvesting in a 'respectable' manner, Ford gained a measure of protection from US sanctions legislation, especially municipal and state laws which prevent corporations doing business in South Africa from bidding on public tenders.

The departure created immense confusion among Samcor employees and their union representatives over the terms of the trust. At the end of the day Ford established a trust on the very individual shareholding basis it announced in early 1987.

Finally, by circumventing prohibitions on new investment, Ford set an important precedent weakening the Anti-Apartheid Act. The move effectively split US sanctions supporters from a group of South African workers and their union over Ford's continued presence in the country. At the moment when US groups were mounting an attack on non-equity ties between US and South African corporations, Samcor workers refused to vote themselves into what appeared to be likely unemployment by demanding that Ford cut those very ties to South Africa.

Corporate camouflage: 'you say goodbye, and I say hello'

Restructuring and withdrawal

The GM and Ford withdrawals were precipitated by the sanctions campaign, but were part of a larger process of re-organisation of an industry in deep crisis. Both companies withdrew in ways which comfortably suited the needs of the parent and the subsidiary. Both remain deeply involved in the South African car market, but now according to the arms-length model pioneered by the Japanese firms.[13] This pattern is not unique to the automobile industry. Indeed, surveys of the US corporate withdrawals show that few factories closed as a result of the new sanctions, and those that did employed very few black workers.[14] In the majority of cases, US firms either sold their operations to local management or to a South African or foreign firm.

The means used to redefine the relationship between capital and labour in these 'corporate camouflage' withdrawals were relatively new in the South African context. The recent debates in South Africa over employee share-ownership programmes and leveraged buy-outs by local managers are just two examples of local capital importing the latest US corporate management schemes. Thus these withdrawals mark a watershed of sorts in South African management strategies for labour control.

Response to corporate camouflage: South Africa

It is extremely difficult to fight corporate camouflage withdrawals and the restructuring they serve. In both the GM and Ford cases workers were largely unaware of the challenges facing them. The 1985 Ford merger and the GM buy-out were negotiated behind workers' backs, and sprung on them at the last moment. In the course of the GM strike, workers were never clear about the company's plan: was it closing down, transferring ownership, or would a new legal entity be created? The question was critical as the changing nature of ownership modified the kinds of demands workers could raise and the parties bargained with. Management easily manipulated the situation: lines of authority were kept vague as local and international managers passed responsibility from one to the other.

While Ford did negotiate the terms of the trust, workers were not party to the major withdrawal decision. Such treatment undercuts the company's claim of 'empowering' its workers. Indeed, the 24% shareholding offered by Ford fell just short of the 25% required under South African law to veto certain company decisions. Samcor refused to concede such real power to Numsa.[15]

Furthermore, the debates over trust funds involved new, complex issues for the union movement. These require extensive research and discussion of economics, finance, and political direction before clear policies can be developed. Numsa recognised the inadequacy of deciding such questions at mass meetings, as had been done at Samcor, and the need for more intense discussions and seminars at a grassroots level.

The Ford case did reveal an important difference in orientation between Samcor workers, their union representatives, and Cosatu. The workers' demand for a direct payout went against the grain of Cosatu's criticisms of individual share offerings, but nonetheless expressed the immediate needs of individuals from distressed communities. The union leadership, meanwhile, had to develop a policy with its Samcor members, respond to management, and do both in a way which contributed to the long-term goals of the workers' movement as a whole. In generating a valuable scheme inspired by the federation's policy on worker control and collective ownership, Numsa was less successful in marrying the plan to demands which would satisfy workers' immediate individual needs.

But the anti-trust committee's demand for an immediate payout completely ignored the opportunities implicit in the company's offer. While share offerings are often exercises in corporate co-option, workers can counter management intentions and develop share-ownership programmes in their own interests. Similarly, Numsa's efforts to gain representation in the board room involved a risk of co-option and confusion, but could also have gained them increasing power in the firm. The committee could gain worker support through militant propaganda on a strongly-held worker demand. The failure of the union to hold thorough discussions among workers no doubt made the committee's task easier. And the willingness of at least some workers to believe scurrilous rumours about Naawu suggests that deeper plant-level animosities were engaged which had little to do with the trust proposal. Though these could be manipulated to undermine the trust, the committee failed to deliver. In the meantime, worker demands for cash, the union's creative plan, and Cosatu's policy on employee trusts were thrown by the wayside.

It would be unwise, however, to read these conflicts as fundamental contradictions between leadership and members over sanctions policy. During the course of the GM strike and Samcor debates, the disputes were over the terms of withdrawal, not withdrawal *per se*. These issues are bound to generate friction among workers, and it is unrealistic to expect an easy accommodation of the different interests involved.

The task was all the more difficult because the issues involved were unprecedented for rank-and-file and leadership alike. Time after time, however, the union movement has demonstrated the sophistication necessary to reconcile different demands at different levels of organisation. Numsa's subsequent withdrawal deal with Mono Pumps, for example, took account of many lessons from the Samcor case. It included a direct cash transfer to workers, a company contribution to a trust fund controlled by workers, but no actual share ownership.[16]

The United States: new sanctions and solidarity

At the time of the first withdrawals in late 1986 and 1987, the international anti-apartheid movement was also overtaken by events. Supporting organisations in the United States, whether organised labour or political groups in the anti-apartheid movement, were a powerful force for the passage of sanctions, but were much less able to influence the implementation of the law they had worked to enact.[17] The compromise legislation carried weak provisions for penalising corporate violations and no effective monitoring mechanism. Indeed, the organisations which promoted sanctions quickly announced a new campaign to toughen the Anti- Apartheid Act.

There were powerful institutional forces ranged against sanctions. Though watered-down, the law was nevertheless a renegade act, passed over President Reagan's veto. The executive bureaucracy, most prominently the Treasury Department, opposed the legislation. Corporations like Ford had many friends in Washington backing up efforts to orchestrate withdrawal in their own (and ultimately the South African government's) best interests.

Conservatives were quick to express their own interpretation of the lessons of the US withdrawals.[18] In its first review of the Anti-Apartheid Act, the Reagan administration labelled sanctions a failure after doing nearly everything in its power to avoid them and, once passed, to undermine them. The GM strike and the disputes among Ford workers were often cited as indications that black workers themselves did not want US corporations to leave, and that the policy would leave blacks worse off than before.

But the enactment of sanctions in the US represented not the culmination of solidarity work, but a new phase. Most leaders of the sanctions movement never expected disinvestment, and certainly did not believe that the Anti-Apartheid Act would lead directly to liberation. Such measures were seen as one step among many in a long-term and multi-faceted movement.

Continuing efforts to tighten sanctions cut in two directions: first, extending sanctions to include South Africa's other major trading and investment partners, and second, attacking the continuing non-equity ties between US firms and their former subsidiaries.

South African car makers have become extremely worried that multilateral trade and financial sanctions will cut their access to foreign designs, components, technology and investment. The impact on the motor industry may be staggering, given its dependence on imports.[19]

The Japanese government moved to reduce its undesired status as South Africa's major trading partner, and persuaded Japanese firms to reduce exports to South Africa. Delta Motors cited such pressure when it stopped reporting detailed monthly production statistics, obscuring sales of its Isuzu trucks and thereby masking the role of Japanese involvement in its operations.[20]

GMC links to Delta through Isuzu in Japan leave GM vulnerable to the second prong of the new disinvestment strategy. The state of Michigan, for example, recently began withdrawing $5-billion in pension funds from 172 companies, including Ford and GM, because of a new state law restricting investment in companies with continuing ties to South Africa.[21] The steps were taken despite the contribution of the big Detroit car makers to the state's economy. Similar efforts are underway in many other cities and states. Pressure on Delta and other corporations will mount, as major suppliers in Japan, Germany and Detroit feel additional heat.

The South African motor industry remains extremely vulnerable to such international actions. The industry experienced a relative improvement in its fortunes in 1987 and early 1988. Government policies easing duties on imported parts and changes in consumer financing arrangements provided some relief, along with the easing of inflation and stabilisation of the rand. With costs rising more slowly, for the first time in years manufacturers contained price increases below the rise in the inflation rate.

But the re-emergence of the debt crisis led the government to rein in the consumer spending boom. These policies had an immediate impact on the industry.[22] While automobile components are not included in the new import surcharges, sales figures have declined with the rise in inflation during 1988 and the tightening of hire-purchase terms due to increases in the reserve bank prime rate.

Despite rationalisation and a temporary increase in sales, the industry is far from healthy. Its fragility is heightened by the long-term crisis in the South African manufacturing sector, with low growth rates, exchange rate volatility, a changing location in the international division of labour, external debt burdens, and the risk of tightening trade and investment sanctions.

In the meantime, workers will continue to be employed at the formerly US-owned firms, will continue to face hostile managements, and will continue to seek external support. The push for sanctions, therefore, does not pre-empt international solidarity work. The GM and Ford cases indicate the need for anti-apartheid organisations to develop new means to aid South African workers and their unions in specific conflicts as well as in the more prosaic struggles which return once the dust stirred by withdrawal settles.

While it is always difficult for external groups to assist South African workers in a strike situation, it is impossible to do so without regular institutional connections between South African and external groups and co-ordinated strategies and tactics. In the last two years activists in unions and political organisations inside and outside South Africa have been hard at work trying to overcome the problems exposed in late 1986.

But it is extremely difficult to find ways for strategies to be stretched to accommodate local, union and federation interests without cutting across efforts by the international sanctions movement. Organisations have scarce resources. Communication is distorted as activists from both countries have difficulty obtaining visas and thus cannot easily meet face-to-face. While the labour movement in South Africa faces the full wrath of government, the anti-apartheid movement in the United States has no friend in the present administration.

Nonetheless, ties are being strengthened between South African unions', leading US anti-apartheid organisations, and more progressive unions in the AFL-CIO. Many of these American unions have long-standing links with the democratic union movement. For many years, however, the solidarity efforts of US unions on work-place issues followed a separate path from US organisations aiding the liberation movements and internal political groups. But connections are growing stronger between US unions and the political, church and community organisations which have been the mainstay of the American sanctions movement.

These networks have been elaborated in many national, regional and grassroots struggles since the passage of sanctions in 1986, most prominently the campaign to free Numsa general secretary Moses Mayekiso, and the Shell boycott. Both efforts grew out of labour-based conflicts. Both engaged US trade unions - the United Auto Workers and the United Mine Workers of America - in solidarity work with their South African counterparts. Both quickly burgeoned to include active alliances with the panoply of US and South African anti-apartheid organisations. At the same time, the movement has won many local battles, resulting in stronger municipal and state sanctions laws, as witnessed in Michigan.

In New York City, activists in the Labor Committee Against Apartheid have been working to close loopholes in the municipal sanctions provisions after discussions with the Cosatu-affiliated Chemical Workers Industrial Union. In an effort which underscores the importance of such co-operation, this committee urged city council members to add clauses to existing legislation to protect workers' rights at withdrawing companies on lines inspired by Cosatu's policy on negotiated disinvestment and by the CWIU's policy, 'Conditions for acceptable disinvestment'.[23] The venture provoked debate within the anti-apartheid movement on how to combine support for South African unions facing disinvestment with the commitment to comprehensive mandatory sanctions. At an earlier point, such initiatives were seen by some in the US as setting conditions for disinvestment. Now however, similar language has been introduced into the new congressional sanctions bill, authored by representative Ron Dellums.

These developments establish a broader and deeper foundation for co-ordinating national strategies and provide a more clearly defined centre with which the South African anti-apartheid movement can work. The new structures also give US organisations greater power to help South African unions contest corporate camouflage withdrawals. They provide a base to push for tougher legislation and close loopholes in the Anti-Apartheid Act while supplying the institutional means to generate publicity, run corporate campaigns against prominent targets, and lend material aid to South African unions.

The benefits of these advances have recently been demonstrated in the fight over Mobil's disinvestment, announced in April 1989. Through contacts forged in the Shell boycott and in subsequent discussions with union and anti-apartheid activists in New York City and elsewhere, the CWIU helped to shape sanctions debates by emphasising the importance of negotiated withdrawals. Meanwhile, the CWIU became more closely acquainted with the policies, personnel, and capacities of the US sanctions movement, and could craft its responses to Mobil with these experiences in mind. Finally close communication allowed US groups to respond quickly to support the union. Anti-apartheid campaigners along with labour leaders from the United Mine Workers of America and the Oil, Chemical, and Atomic Workers pressed Mobil to negotiate. The groups publicised the dispute, especially the CWIU's strike at 15 Mobil installations, and Mobil became the target of Soweto Day demonstrations on 16 June in New York City, Chicago and Detroit.[24]

The shift during 1988 marks an important turning point in the history of US solidarity with South African unions.

New strategies for withdrawals ... and after

The General Motors and Ford withdrawals left both companies, and ultimately the South African government, stronger. Workers, their unions and the external sanctions movement were confused and effectively divided over how to respond. But in the two years since the withdrawals of late 1986 and early 1987, new strategies and tactics have developed within the South African labour movement and the international anti-apartheid movement.

Numsa and other unions have learned from the GM and Ford experiences. As witnessed in the Mono Pumps and Mobil cases, they have enjoyed greater success in accommodating worker, union, and federation interests in responding to management initiatives on disinvestment.

Internationally, recent developments contrast sharply with the earlier period, when confusion within and between movements in South Africa and the US made effective responses to capital extremely difficult. It is not surprising that conflicts arose: they are normal products of a large, diverse, long-term movement for change. However, there were no easy ways to work through the difficulties.

The stark alternatives posed by the Ford withdrawal - new investment or unemployment - were themselves produced by the absence of effective links between internal and external groups. Ford successfully split the Samcor workers and Numsa from the US sanctions movement with its threat to cut all ties to South Africa and cripple, perhaps even close Samcor. In its efforts to reach agreement with Numsa on a trust and to gain congressional acceptance of new investment, Ford played off workers' fears of unemployment against the sanctions movement's call for new prohibitions on non-equity ties.

But it was never clear why Ford would have to cut ties or why Samcor would close without the $61-million investment, or if workers failed to accept a trust. As the second-largest car maker in the country, Samcor had numerous assets: hard-earned market share; an extensive dealer network; a skilled workforce and management; and one of the most modern industrial plants in a prime location. It is hard to believe Anglo American was unable to pick up the burden of further investment. With hindsight, it is possible that Ford used the threat to force union compliance, thereby gaining a trump card to play in its dealings with the US Congress.

In the GM case, union and anti-apartheid activists in the US did not clearly understand the GM workers' demand for severance pay, as they assumed the company would not be closing. The complex and unfamiliar negotiations did not lend themselves to mobilising support in the US, especially as the recently-passed Anti-Apartheid Act was generally hailed as a major victory.

In both cases, closer contact between movements could have yielded better responses to these assaults. Over the last two years new connections between internal and external groups have allowed much closer co-ordination of strategy and tactics, and the means to harmonise differences while generating sophisticated ways to protect workers and advance the movement against apartheid.

There are three areas where further development of these processes will be crucial. First, from the point of view of South African labour, the most significant withdrawals have already occurred. There are not many large employers left among the US companies remaining in South Africa. The international sanctions campaign has now targeted non-equity ties as GM and Ford, along with many other US companies, maintain arms-length links to South Africa. The experience with Ford's $61-million investment in Samcor suggests how differences within the anti-apartheid movement can undermine common efforts against apartheid. More recent experiences, such as the mass campaigns and the Mobil fight, suggest the benefits of consultation and joint action. It is important that these procedures be employed regarding the direction of further sanctions, and that such decisions are discussed and negotiated with worker representatives from inside South Africa.

Second, trade unions and anti-apartheid groups within the US have developed means to bolster South African worker efforts to influence the terms of corporate withdrawal. The groups are working to add regulations inspired by Cosatu and CWIU policies on responsible disinvestment to existing US sanctions laws at the national, state and local levels. Similarly, US organisations have mobilised to assist unions fighting corporate camouflage. Often, however, such support comes late in the day, after a withdrawal has been announced. A stronger front can be developed against corporate camouflage if a more proactive approach is taken, strengthening ties *before* a crisis occurs.

Finally, the GM and Ford cases emphasise the need for continued international solidarity work with South African unions, whose members remain employed at the former US-owned firms. The Michigan pension fund divestment shows that US companies can still be pressured, while the Mayekiso campaign demonstrates the effectiveness of support efforts.

These are important steps in exerting increasing pressure on capital and the South African state while strengthening unions and workers in their immediate work-place battles and in the long-term movement for liberation.

Notes

I want to thank Craig Charney, Lee Cokorinos, and Bill Minter for their comments on a draft of this article.
1   I owe the points in this paragraph to suggestions by Bill Minter.
2   See Anthony Black's assessment of the relation between the manufacturing sector and the motor industry, 'Manufacturing Development and the Current Crisis: a reversion to primary production?', unpublished paper, Department of Economics, University of Cape Town, 1988.
3   For an analysis of these global trends, see Black, 'Manufacturing

Development', and D Quinn, 'Dynamic Markets and Mutating Firms: the changing organisation of production in automotive firms', paper prepared for delivery at the 1987 convention of the American Political Science Association, Chicago, 20.

4   J Kibbe and D Hauck, 'Leaving South Africa: the impact of US corporate disinvestment', Investor Responsibility Research Center, 1988, 43.

5   'Delta ready to branch out', in *The Motor Industry*, supplement to *Financial Mail*, 17 July 1987, 18.

6   'GM Pullout Benefits South African Carmaker', *Journal of Commerce*, 22 January 1989.

7   'GM Cuts Reflect a Long-Term Trend', *New York Times*, 10 November 1986.

8   Press statement from Naawu, Mawu and Micwu, 1 November 1986.

9   The following account is drawn from press clippings; a Numsa document entitled 'Update on Developments Re: Ford 24% Share Transfer to Employee Trust', Numsa automotive department, 19 May 1988; and J Matiko, 'Samcor Workers Strike Against Share Ownership', *South African Labour Bulletin*, 13(6), September 1988.

10 Numsa central committee statement on employee share ownership, 13 March 1988.

11 D Budlender, 'Assessing US Corporate Disinvestment: the Case report for the Equal Opportunity Foundation', Cape Town: Community Agency for Social Enquiry, 1989, 16.

12 'Samcor Pays 1st Worker Dividend Since Ford Withdrawal in 1987', *Automotive News*, 6 March 1989.

13 Such ties can be lucrative for the multinational as well as for the South African government. See D Katzin, 'Breaking the Ties: the call for corporate withdrawal from South Africa', *(ACAS) Bulletin* (Association of Concerned African Scholars), 26, Winter 1989.

14 See Kibbe and Hauck, 'Leaving South Africa', appendix III, 50.

15 Budlender, 'Assessing US Corporate Disinvestment', 15-16; and Kibbe and Hauck, 'Leaving South Africa', 38.

16 On the Mono Pumps case and Numsa's evolving strategies see A Bird, 'The Mono Pumps Disinvestment Story', and the speech by A Bird, 'ESOPs - part of a strategy to smash democracy', *South African Labour Bulletin*, 13(6), September 1988, 39-50. See also the contribution by J Maller in this volume.

17 KA Rodman, 'Nonstate Actors and US Economic Statecraft Toward South Africa', paper presented to the annual conference of the American Political Science Association, Washington DC, 1-4 September 1988.

18 See F de Villiers, 'The Scandal of Sanctions', *The American Spectator*, 21(3), March 1988.

19 According to Black, the motor vehicle and transport equipment sector is more import-intensive than any other manufacturing sector, with the exception of machinery. It accounted for over 9% of total merchandise imports in 1986. Black, 'Manufacturing Development', 7.

20 'Silence is Golden', *Financial Mail*, 24 June 1988.

21 'Pretoria Ties of Ford, GM', *New York Times*, 3 January 1989.

22 On the relation between the debt crisis and government efforts to stifle the boom, see A Hirsch, 'How Sanctions Work Against Apartheid: the political

implications of the South African debt crisis', paper presented to the Southern African Research Project Workshop, Yale University, October 1988, 12-13.

23 Labour Committee Against Apartheid, 'Action Alert: New Legislation Before City Council to Close the Loopholes in NYC Sanctions Against South Africa', New York, 1989.

24 'Mobil's Pull-Out: A Case in Point', *Labor Against Apartheid*, Summer 1989.

Trade Unions, Migrancy and the Struggle for Housing

William Cobbett*

In the period since the declaration of the national state of emergency in June 1986, a combination of mass detentions, bannings, restrictions, political trials, disappearances and state military power has substantially altered the terrain of resistance and struggle.

But this state offensive has not necessarily cleared the path for the untroubled implementation of the policies of the government and its allies. Many progressive organisations have succeeded in the politics of survival during this period; at the same time, their tactics and some of the major actors have changed, while most of their goals remain the same.

The focus of these organisations has moved dramatically away from issues of central state power, to more localised, issue-based and goal-oriented battles. These shifts signify a strategic reassessment, which has generally led to a longer-term time-scale and vision. This is well demonstrated when attention is focused on the issue of housing, which has become a priority on the agendas of the state, business, community organisations and trade unions.

Housing, although an issue that affects all no matter what organisational affiliation, is often seen as a concern for community organisations. Indeed, a major reason for the state clampdown was the extent to which communities had organised effective structures to articulate demands around housing and other aspects of reproduction, and attempted to defend themselves and their structures. Accordingly, the state of emergency, while all-embracing, has been felt most harshly in the townships, and by the UDF in particular.

The state and its armed agencies' attack on community structures has left a gap in the ranks of organisations capable of

*This article is a personal interpretation of aspects of the housing situation in South Africa. The primary sources, both of my understanding and my information, are a product of the collective work, discussions and debate that are a feature of working at PLANACT, a service organisation dealing with issues of housing and urban development on behalf of trade unions and community organisations.

articulating demands around housing, the built environment and unrepresentative local government structures. However, the extent to which some structures have survived, and the degree to which trade unions have been moving into the available space, suggest that the scope of competent resistance to state and business initiatives will be greater than expected.

The movement of the trade unions into the area of housing is a continuation of a trend towards engagement in issues beyond those of production. Since the formation of Cosatu in late 1985, there has been a greater degree of union willingness to extend organisational activity beyond the shop floor. However, precedents had certainly been set by Sactu in the 1950s and, indeed, by Fosatu in the late 1970s and early 1980s.[1]

This is not peculiar to South African unionism. The extent to which politics and economics are intertwined suggests that any artificial or overstated separation between the two could weaken trade unions. As has recently been argued in relation to world trade union movements as a whole, 'workerist orientations, with their exclusive focus on factory issues and organisations, will thus tend to isolate the workers and perhaps make them more vulnerable to repressive forces'.[2]

Trade union interest and activity in the field of housing mirrors, and is partly responsible for, the important national political and economic role which housing currently enjoys. With the existing backlog in housing stock, it is going to remain high on the economic and political agenda for the foreseeable future.

The crisis in housing, however, involves much more than the size of the backlog. Housing and other areas of collective consumption are intensely political issues in South Africa, mediated through a combination of race, class and region. The state is currently concerned with reducing the housing crisis to a non-political issue, one that can be solved through improved technical solutions. This points to one of the fundamental tensions yet to be played out: the increasing involvement of the democratic movement in housing precisely as the state attempts to withdraw, privatise and de-politicise the issue.

The housing crisis

The extent of the housing crisis is enormous, although there is a dispute over actual figures. According to the government, for example, there are 1 310 813 squatters in South Africa (excluding all ten bantustans), whereas other estimates put the figure at around seven million.[3] Shacks in white urban areas have increased at the rate of 60% per annum since 1985, and the increase was as high as 95% in the Witwatersrand area.[4]

Urban Foundation estimates put the national housing backlog at some 1,8-million units, rising to 4,6-million by the end of the century. An estimated 400 000 units need to be constructed each year to overcome such a backlog.[5]

These startling figures must be seen against the tension between need and affordability - only then can the true extent of the crisis be understood. The National Building Research Institute (NBRI), for example, has found that the overwhelming majority of black South Africans - 84% - cannot afford a low-cost house without some form of subsidy, while a massive 56,4% are unable to make any contribution to their form of housing.[6]

The extent of the national housing crisis in general, and the lack of housing for blacks in the urban areas in particular, are among the more visible legacies of apartheid policies. A number of laws, including the Land Act, Urban Areas Act and Group Areas Act, laid the basis for this crisis. This racist legislation, however, was only the background to the Verwoerdian offensive of the 1960s, when the current crisis was ultimately ensured. In a period of economic boom, extremely harsh measures were adopted to limit the number of Africans in 'white' urban areas; forced removals led to people being dumped in the bantustans. In the decade of the 1960s, 197 out of 308 African townships - 64% - in the 'white' areas had their populations reduced.[7]

The most crucial decisions, however, were taken in 1968. These included the abolition of the 30-year leasehold, the 'freezing' of existing townships, and the deproclamation of existing townships within 75 km of a bantustan. The state also decided to suspend any building programmes in the townships, indicating that all future housing provision must take place behind bantustan boundaries. This is reflected in state African housing expenditure patterns: in 1967 R14,4-million was spent in 'white' areas and R5-million in the bantustans. By 1975 only R7,8-million was spent in the former, with expenditure of R67-million in the latter. It was this apartheid policy decision, more than any other, which was directly responsible for the existence and extent of the current crisis.

Current state policy, instead of relieving the crisis, will exacerbate it. Under a political and economic programme increasingly guided by the imperatives of monetarism - privatisation and deregulation - the state is attempting to extricate itself from its responsibility for housing. To the majority living in poor housing, shacks and cardboard boxes, it is of little concern what particular ideological discourse is used to justify current state strategy. To those unable to afford formal housing - the majority - the effect is the same.

Current construction of urban black housing[8]

In 1986 and 1987 only 9 000 housing units were constructed by the public sector. In 1988, the private sector took up some of the burden - building 35 000 houses.

Government's current strategy is to leave the solution to this social problem to a combination of individuals, employers and the private sector. The 'free market', however, is in no position to tackle the real extent of the crisis. At the current building level, the situation continues to worsen as the combined efforts of local government and the private sector account for little more than 10% of the backlog, and these efforts often pay little attention to redressing the situation where need is greatest - at the lower end of the market.

There is more to state withdrawal from housing than the mere abdication of responsibility. Housing is also seen as a useful and material mechanism for the stratification of the working class. Implicit in the current strategy is that a hierarchy will emerge from access to housing, which is now to be governed solely by ability to pay, with need playing no role.

Confronted with a situation as grave as this, it is only theoreticians who can debate whether housing is a field for community politics or trade unions. Current union interest in the field of housing does not signify the end result of a conscious or active process to seek new terrains upon which to contest major battles. The unions, as bargaining bodies of workers, have been both pushed and pulled into the arena.

Housing, as an organisational issue, implies different things to different unions. In sectors such as mining, domestic service and parts of manufacturing, housing is an integral part of the conditions of employment and the control of the workforce. The provision and control of housing in these sectors has led to a blurring of the distinction between production and reproduction. But the issue has a different meaning in other sectors, where housing is more detached from work-place issues.

Certain unions have been engaged in the field for some time. In the mining industry, for example, housing has historically been a major issue - both for employers and employees - because of its proximity and relationship to the work process. There is, however, an overall trend in union thinking towards addressing the issue of housing. This has also been prompted by the extent of the political vacuum that the state has been able to create in the townships through its repressive actions. There are, in addition, a wide variety of forces pressuring unions to engage in housing:

- workers are putting forward the issue. While the demand for housing is not new, unionisation, in providing both organisational ability and material reward, has meant that the demand is now within the realms of the feasible. Again, the effect of the state of emergency on community organisations has led many workers to demand union assistance in the formulation of strategies in the townships;

- the unions have not found it possible to sit idly by while the state attacks existing structures within the community - attacks which affect their members. With the increasing trend towards alliance politics, political attacks on one section of the alliance necessitate responses from other partners;

- union members are affected by a whole range of issues supposedly in the 'political arena', from the continued existence of the Group Areas and Land Acts, to community responses such as the rent boycotts;

- more pertinently for the unions, the state and the business community are raising the housing issue in a sustained manner. An ever-increasing number of home-ownership schemes are being put forward by employers and the private sector. Paradoxically, unions represent the biggest obstacle to private sector initiatives, while their employed, relatively well-paid members represent the biggest market potential for home-ownership schemes.

The large-scale provision of housing offers much more than a major new area of potential capital accumulation. The accumulation aspect does, indeed, provide potentially enormous opportunities for finance houses, speculators, builders and property developers alike.

This point is demonstrated by the striking manner in which the private sector has focused on the more profitable ends of the market, governed by the imperatives of profit rather than need.

But state and business agendas also view housing as a tangible way in which to demonstrate the material benefits the system is able to provide - although for the state it may be more important to try and show the best side of the political system, while business may be more concerned to demonstrate the benefits of capitalism.

The challenge to the unions

Union power lies in organisation, democratic structures and collective muscle. However, there are potential contradictions hidden within this strength. Both the state and business see this strength as the reason to target unionised sections of the population. Given that access to housing is increasingly determined by criteria of affordability, gains made by the collective bargaining of the unions mean that they represent the most desirable sector of the working class for housing initiatives.

Widespread home ownership in the union movement, however, has far-reaching implications for the future conduct of industrial relations. The unions have a number of challenges to consider in entering the housing struggle.

There is, for example, the question of expertise and personnel, given that the unions' hard-won experience to date has been on the factory floor, representing their members' direct interests. The fact that housing is a less direct interest, albeit an important one, involves new issues and dynamics.

The packages being put forward by many employers are generous in straightforward financial terms. This makes home ownership attractive to those who can afford the terms. Management, at the same time, has a tendency to see housing in financial terms alone, whereas the unions have to consider the issue in a much wider context.

Current housing policies are based upon the premise and, indeed, the strategy, that housing is not uniformly available to union members. This division needs to be examined not only in terms of union membership itself (with different potential both within and between unions), but in terms of the working class as a whole. For when it comes to housing, the working class is potentially stratified in a number of significant ways.

The first stratum involves unionised workers. Even amongst unionised members, different skill levels, reflected in wage categories, will form a major cleavage in access to housing. This is another demonstration of the extent to which work-place issues directly affect issues of reproduction. For some unionised members, the wage levels will exclude them from most, if not all, formal housing options.

The next relates to those employed, but not unionised. In the absence of a union to bargain on their behalf, they will probably have to negotiate individual arrangements in the unlikely event of their being able to afford housing.

Then there are the urban unemployed, and finally the rural unemployed - excluded both from formal wage labour and the formal housing market. These strate are, for all practical purposes, excluded from the formal housing market entirely.

Given this stratification, the potential exists for the unions, acting in their members' interests, to unwittingly act in defence of sectional privilege. This threat of division has been understood by the unions, and their response is, with ever-increasing clarity, to reject such divisive strategies. Unions, aware of the attractions of housing for sections of their membership, have been devising ways of ensuring that the benefit of their collective bargaining power is extended to embrace other, weaker sections of the working class.

These sorts of challenges form a crucial impetus in determining trade union strategy in South Africa, as they do elsewhere in the Third World. As Petras and Engbarth put it,

> Without dissolving the basic trade union organisation into a heterogeneous movement, it is incumbent on trade unions to transcend the factory setting and help establish institutions at the community level and at the intersection of rural and urban exchanges in order to be able to muster sufficient force to establish a national presence.[9]

The unions have to respond to state and private sector initiatives with well-articulated and thoroughly researched alternatives. Yet there is little doubt that employers, and their representatives such as the Urban Foundation, have far greater resources and have been in the field for much longer. Some parastatals have been providing housing to their white employees for decades, and can usefully bring such experience to bear in the case of their black workers.

Notwithstanding the recent growth of service organisations offering support to unions and community organisations in such fields, there is still a tangible gap in the level of expertise, planning and resources. Service organisations do, however, have a significant edge over employer bodies in terms of their ability to act in the interests of the workers and their unions.

Between the trade unions, on the one hand, and the state and private sector on the other, there is a fundamental difference in approach to the question of housing. For the private sector the technical issues of financial mechanisms and delivery systems are of primary importance, with political restrictions such as the Group Areas Act also on the agenda. Unions have to consider the issue in a much wider context, including the complex questions of housing and the migrant labour system.

Migrancy and home ownership

Demands for an end to the system of migratory labour in South Africa have long been at the forefront of demands by the democratic movement. The housing provided for migrant labourers, whether by mining companies, the manufacturing industry or municipalities, has traditionally been of an extremely low standard, while facilitating a great degree of control over the workers.

At face value, it would appear that the provision of family housing close to the place of work could present an interim method of reducing the extent of the migrant labour system. This is clearly in the minds of some large companies, which have committed themselves to reducing, or even removing, migrancy as a labour form in their plants.

This is, in part, a reflection of the extent to which union organisation has challenged the tendency for migrant labour to be a structurally cheap labour form in the economy. The issue of housing and the continued existence of the migratory labour system, however, becomes increasingly complex as it is unravelled, and poses some particular challenges for unions and workers.

Some industries, historically reliant on a massive migrant labour force, seem prepared to have the workforce housed entirely in adjacent townships, in low-cost and standard family housing. Indeed, some companies have adopted this policy and are pressing ahead with its implementation, seemingly regardless of whether their plans are in accordance with the desires of their employees. The assumption that these plans will meet with worker approval needs to be treated with some caution, both by the company and relevant union.

There are a number of important factors to be considered in the provision of urban housing for migrant workers. The decision by a worker to accept the proferred urban accommodation has a bearing both in the urban sphere and in the sphere he has traditionally regarded as home - in this case, the bantustan. To understand this more fully, the issue needs to be examined through the prism of the migrant's household.

For male migrants,[10] decisions presented in respect of housing are weighted on the basis of their masculinity and thereby their role as head of household; their economic power as both salaried and unionised; and their work-place location within the mainstream of the economy. The decision on whether to 'urbanise' is usually presented to migrants as if they are single, male, metropolitan workers.

But the offer of urban housing, close to the place of work, also has effects in the migrant's rural base.[11] It is not just the migrant worker, but his entire family which is directly involved in the decision whether to accept family housing in urban townships. And going beyond the immediate family, the effects of such a decision on the base community and its social fabric have to be examined. This is particularly so when the legacy of labour zoning in recruitment of migrants is considered: the decision by a particular company to settle workers and their families close to the place of work could affect a significant number of households within a small rural community.

Any decision on whether to urbanise is not a narrow issue to be considered in terms of the union or the politics of production, but has ramifications for communities and the politics of reproduction.

In addition, caution needs to be applied in defining 'the household'. When the housing of workers in an urban settlement is considered, the model of the nuclear family is all too often assumed. This flies in the face of much of the reality of social relations in the bantustans. Decades of apartheid policy have to be considered. Specifically, in the era of Verwoerdian apartheid, the bantustans were identified as the dumping grounds for those excluded from the urban labour market - the elderly, young, sick and infirm.

Migrant workers, therefore, when considering the possibility of urban housing, are presented with the solution to only part of the problem. The remittances sent by migrants to the family home are used to support many more people than could be housed in the township dwellings of South Africa's urban areas. Workers are thus being offered the chance to urbanise with only a portion of their dependants.

In addition, many migrants work where they do in order to send remittances 'home' for the survival of the extended family network. This is as much a function of the lack of jobs in the bantustans, or the appalling rates of pay and conditions of work that generally characterise bantustan employment practices,[12] as it is a desire to break into the urban labour market on a permanent, familial basis.

Such remittances would largely disappear into bond repayments if the migrant purchased an urban dwelling. Given that the urban dwelling is extremely unlikely to house the extended family of the migrant, who had previously been supported by remittances, the choice offered to the migrant is hardly a choice at all. The removal of part of the family to the urban areas would entail the conscious abandonment of those most in need of support.

Additionally, the local rural economy the migrant worker leaves behind, then faces the burden of providing some form of welfare to those of his extended family left destitute. Again, the poorest sections of South Africa's population would be left in those areas with least resources. Such considerations, insofar as they may be generally applicable, would militate against migrant workers choosing to live with parts of their families close to their place of employment.

But there are additional features which may weigh against migrants from some bantustans opting for urban settlement. The bantustans are far from the homogeneous entities that they are so often assumed to be. A simple glance at the differing land areas and population densities will be sufficient to warn that migrants' responses are likely to vary significantly between (and even within) bantustans. By way of example, QwaQwa - which is little more than an extended impoverished urban settlement in mountainous terrain - occupies a mere 62 000 ha, whereas Transkei has 4 287 000 ha.

A pilot survey undertaken by PLANACT, where the overwhelming majority of migrants in one large plant were from the Transkei, revealed that a sizeable number of the sample had access to differing extents of land in the Transkei. These workers could relate the exact size of land and their livestock holdings with noticeably greater accuracy than they could their wages. Most envisaged retiring to their plot of land after they had completed their period of urban work, rather than moving to a township close to their place of work. This was true even if they had the possibility of having their immediate family live with them while at work.

While all of them expressed a strong desire to live with their families, concern was expressed as to who would look after the extended family, land and livestock if that was to happen. This tension was mediated by some workers suggesting that they could envisage their families joining them for limited periods in the year - school holidays for example - but returning home, especially for critical periods of the agricultural calendar.

It is an orthodoxy that the subsistence basis of the bantustan economy has long collapsed, and debate has focused on establishing the precise historical period for this decline. It is arguable that what has been overlooked is the extent - however limited - to which systems of tribal patronage and bantustan politics have been able to resuscitate the rural base of bantustans. While this may not be of national uniformity and significance, it is an area which warrants re-examination.

Economic factors apart, considerations such as family ties, historical links and quality of life may have an important bearing on migrants' evaluation of their rural base, and their attitudes to urban home ownership. Again, differences within and between bantustans warn against their treatment as homogeneous entities.

Access to such land and livestock may have an important bearing on the consciousness of workers deemed to be fully qualified members of the proletariat. This will also influence their perceptions of the relative attractions of urban housing that may be on offer.

Those Transkeian migrants in the sample with access to land were paying, on average, an annual figure of between R10 and R20 for such land from the local chief. They have to weigh this against the reality of the average township house. An urban worker from the same survey described his rented PWV dwelling as a 'small four-roomed house - no electricity - no ceiling - not painted - no hot water - yard is too small - not properly built - cracks in wall - bricks protruding - not well maintained'. This is not an atypical house. It would hardly be surprising if access to such housing were insufficient to entice the Transkeian migrant away from his rural home especially when the purchase of such an abode will remove the possibility of meaningful remittances to dependants in the bantustans. Ultimately, the smallholdings in the Transkei represent to the migrants a better form of insurance against old age than such an urban dwelling. And notwithstanding their active union membership and involvement - aspects of urbanism in South Africa - these workers often perceived the townships as unruly places, destructive of family ties and no place to raise and educate children.

While the question of migrancy and home ownership cannot be applied beyond certain sectors of the economy, it does highlight the complex nature of resolving the housing crisis. The challenge to trade unions is to force employers away from their generally narrow, financial approach to housing, and to put a wide range of options on the table.

To do this in a concerted and organised manner, the union movement needs to stop reacting to managerial and private sector initiatives, and start setting the agenda itself, working actively in the formulation of a housing policy.

Such a policy could act as a mechanism for challenging the state's withdrawal from its responsibilities. But most importantly, there is an ideological battle to be fought - to put questions of basic human rights, decency and need at the forefront of housing policies.

Notes

1 See, for example, Rob Lambert and Eddie Webster, 'The re-emergence of political unionism in contemporary South Africa', in W Cobbett and R Cohen (eds), *Popular Struggles in South Africa*, London, 1988; and Rob Lambert, *Political Unionism in South Africa: an analysis of the South African Congress of Trade Unions*, unpublished PhD thesis, University of the Witwatersrand, 1988.

2 James Petras and Dennis Engbarth, 'Third World Industrialisation and Trade Union Struggles', in Roger Southall (ed), *Trade Unions and the New Industrialisation of the Third World*, London, 1988, 100.

3 Urban Foundation, cited in *The Soweto Rent Boycott*, PLANACT, Johannesburg, March 1989, 35.

4 South African Institute of Race Relations, *Race Relations Survey 1987/88*, Johannesburg, 1988, XLVII.

5 *Race Relations Survey, 1987/88*, XLVIII.

6 *The Soweto Rent Boycott*, 39.

7 J Kane-Berman, *Soweto: Black Revolt, White Reaction*, Johannesburg, 1978, 82.

8 *The Soweto Rent Boycott*, 32.
9 Petras and Engbarth, 'Third World Industrialisation and Trade Union Struggles', 101.
10 While female migrants as a sector of the economy are often overlooked, it is those sectors dominated by male migrants that have been more extensively unionised.
11 There are important issues of terminology in the whole debate around 'urbanisation' that need to be considered. For example, many African areas in the bantustans are already 'urban' although a common reference to urbanisation implies urban settlements in 'white' areas of South Africa. Equally, all bantustans are often reduced to the simple description of 'rural' - a term that has little consistent value to anyone who has travelled in the bantustans.
12 See W Cobbett, 'Industrial decentralisation and exploitation: the case of Botshabelo', *South African Labour Bulletin,* 12(3), March-April 1987, for one example.

Accommodating Black Miners: Home-Ownership on the Mines

Jonathan Crush

The National Union of Mineworkers 'declared war' on the migrant labour system at its 1987 annual congress, demanding that migrancy and the compound system be dismantled within two years.[1]

In the weeks following the congress, black miners on several coal mines in the Eastern Transvaal openly defied management, bringing wives and girl-friends onto mine property. To pre-empt further protest, various mining houses unveiled their plans for settling more labour in family accommodation on or near mine properties. But these new housing policies were not developed in direct response to the calls of NUM and do not, in their present form, satisfy the union.

The mining houses' plans not only predated the 1987 congress; they had been formulated before NUM itself was formed. But black miners did have informal influence over the direction of policy. Mass organisation and protest on the mines had forced mine-owners to consider alternative methods of dividing and controlling the workforce.

This has set the scene for housing and accommodation issues to develop into a critical arena of struggle on the mines.

The 3% rule

During the 1970s elements within the Chamber of Mines began questioning the industry's total reliance on migrant labour. Debate soon quickened over whether mines should or could move more quickly towards reliance on a settled workforce.

This rethinking was a response to fundamental changes in the political economy of gold mining,[2] the most important of which involved rising production costs and sizeable wage increases for black miners.

As black labour became more expensive, mining capital needed it to be more productive. To this end, mine-owners began to mechanise operations with the intention of displacing unskilled labour and raising the productivity of those who remained. To be effective, mechanisation required a more stable, skilled and committed black workforce.

The proportion of skilled and semi-skilled jobs held by black workers had already grown in the 1960s and 1970s (see Table 1 below). But the question was whether the migrant labour system could deliver this labour on a regular and reliable basis. Mining capitalists concluded that it probably could not.

Some mining houses predicted the development of a 'two-tier' black workforce in which an increased number of black miners (10% to 15%) would become stable, proletarianised, industrial workers with higher wages, paid leave and pension schemes, and permanently resident on or near mine properties.[3] The remainder would continue to be migrant workers from the rural areas.

To raise the productivity of migrants, the Chamber promoted the concept of 'career' mining and adopted various strategies to encourage long-term and continuous employment on the mines. These measures proved very successful, especially given the backdrop of economic recession, rising unemployment and rural poverty.[4]

Table 1: The skilling of black miners on the gold mines (%)

Category	1960	1970	1980	(est) 1990
Skilled	2,3	4,5	6,6	8,8
Semi-skilled	21,7	26,7	31,6	36,4
Unskilled	67,8	60,4	53,1	45,7
Other	8,2	8,4	8,7	9,1

Source: P Pillay, 'Future Developments in the Demand for Labour by the South African Mining Industry', International Labour Organisation, Geneva, 1987, 7.

By contrast, progress in settling more miners in family housing was slow. Between 1975 and 1980, for example, the Anglo American Corporation increased its number of married housing units for black workers from 876 to 2 653 but built very little additional housing in the early 1980s (Table 2). By 1984, the entire gold mining industry offered only 4 902 married units - housing approximately 1,1% of the workforce. Government policy - in place since the early 1950s - prohibited mines from settling more than 3% of their workforce on mine properties. But few mines even came close to that figure.

In 1975, government gave mines permission to erect family housing for married employees who qualified for accommodation rights under section 10 of the Urban Areas Act. Individual mines could also apply to go above the 3% limit.

Highly-mechanised coal and diamond mines were allowed to exceed the 3% restriction on the basis of a complicated formula which asked what their labour needs would have been had they not mechanised. By 1984, the coal mines had 4,4% of their workforce in married accommodation, and some mines, such as Rietspruit colliery, reached a figure as high as 45%. But on the gold mines the state remained intransigent.

Even if the state had relaxed the 3% rule in the late 1970s and early 1980s, influx controls prohibited migrant miners from acquiring permanent residence status in urban areas. The mines tried and failed to recruit urban workers who had the requisite rights. Foreign workers (40% of the workforce) were automatically disqualified from settling on the mines. And bantustan independence for the Transkei and Ciskei in the late 1970s significantly raised the proportion of mineworkers defined as 'foreign' by the state.

Table 2: Married accommodation on Anglo American mines, 1960-1985

	No of Units	Total Workforce	%
1960	656	51 136	1,28
1965	755	77 421	0,98
1970	856	98 347	0,87
1975	876	86 841	1,01
1980	2 653	135 893	1,95
1985	2 844	167 209	1,70

Source: AAC

Many mine-owners balked at the prospect of providing and paying for housing rental and services for settled miners, often using the high costs involved when attempting to justify the migrant labour system. Mine representatives disingenuously quoted the cost of settling the entire workforce on the mines in attempts to justify their lack of action on the matter.

In the late 1970s, Anglo estimated a total cost of R130-million to build standardised family units for 10% of its workforce.[5] No mining house was prepared to seriously countenance this relatively modest level of investment.

In 1986, as part of a broad-ranging strategy to restructure the basis of control over black urbanisation, the South African state repealed the pass laws. Within this strategy of 'orderly' urbanisation new possibilities for labour organisation and control began to open up within the mining industry.

In late 1986, the 3% rule was quietly dropped. Since rights of urban residence now depend on access to state-approved housing and formal employment, black miners are legally free to resettle at their place of work. Black miners from the 'independent' bantustans are no longer classified as foreigners and can take advantage of the new dispensation, though workers from the Frontline states do not qualify. The granting of home-ownership rights to urban blacks and renewed state involvement in township provision and upgrading have offered mine-owners a way to finance their housing programmes.

Planning from above

Developments during the 1980s have forced mine-owners to rethink the value of migrancy and the compound system. The proportion of black miners in higher job grades continues to rise. Formal abolition of the colour bar in 1987 has not automatically guaranteed rapid black promotion on the mines. But it has opened up the possibility for greater skill acquisition by black miners, skills which management will want to retain.[6] In turn, the possibility of family accommodation may tempt skilled South African blacks to join an industry they have traditionally avoided.

Worker organisation on the mines in the 1980s has mounted a direct challenge to the personnel and structures which govern mineworkers' lives. Since the founding of the union in 1982, NUM organisers have used the organisational possibilities of the compound system to great effect, despite the long-held view in labour history circles that the physical and social environment of the compounds powerfully inhibits solidarity and organisation.[7]

The compound environment has, in fact, facilitated union

recruiting, ensuring mass attendance at union meetings and solidarity during industrial action. This was particularly evident during the 1987 NUM strike when workers seized control of the compounds on several mines.[8]

As the battle for control of the compounds has developed, mine-owners have been tempted to opt for the coercive methods of the past. In their attempts to reassert control in 1988, management increased compound security while curtailing access to outsiders. NUM responded by exposing management strategy in a high-profile publicity campaign.[9] This culminated in an acrimonious exchange between NUM and Anglo in the national media.[10]

Company industrial relations advisors in some mining houses now see the delivery of family housing as far more than a technical requirement of the changing conditions of production. More than ever, housing and home-ownership are viewed as potentially powerful social instruments in the process of industrial relations. In the short term, new housing estates will allow the mines to 'remove' black supervisors from compounds where they have been subjected to continuous attack. Eventually, the mines hope to have the majority of skilled and supervisory workers in owner-occupied family housing units, supposedly creating a 'labour aristocracy' with interests very different from those of the mass of migrant miners.

Officials quite freely admit to seeing home-ownership and family accommodation as a means of 'depopulating' the compounds and dampening the militancy of the workforce.[11]

Changes in state urban housing policy have provided the mining houses with the means to settle migrant labour at a greatly reduced price by involving both the state and workers themselves. Under Anglo's new home-ownership scheme a portion of the costs will be passed on to black miners. The corporation has secured agreements with various construction companies to provide the housing stock. House prices currently range from R20 000 to R60 000. A heavily subsidised mortgage reduces the deposit from 5% to 2,5% of the purchase price (ie R500 on a R20 000 house) and pegs interest rates at 5% (amortised over 20 years).[12]

In 1986, Anglo commissioned a sample survey of mineworkers' attitudes towards housing.[13] Almost 40% of the sampled population expressed a desire to settle on or near the mines in family accommodation within a five-year period. Amongst foreign workers the proportion was much lower (at 29%).

Anglo identified a 'target group' of 53 338 workers - married, South African miners living in single-sex compounds. Some 47% of the target group sample (or 14% of the total workforce) stated their desire to bring their families to live with them on the mines. Of these, 63% indicated that the move would be a permanent one. Almost 80% of those wishing to move to the mines wanted to own their own homes.

These survey figures provided the basis for Anglo's own planning and its negotiations with the state. In-house, the corporation calculated that up to 9 500 miners might participate in their new home-ownership scheme during its first two years. By the early 1990s, the initial target figure of around 24 000 urbanised workers would be reached.

Housing options

Geographically, the mining houses are exploring four separate options for housing delivery. The first would see new housing built in established black townships - certainly not designed to challenge urban racial segregation and the apartheid city. In the Free State, Anglo and Gencor are involved in a R700-million extension to Thabong township near Welkom. This development - the Sir Ernest Oppenheimer Park - will contain 33 000 new housing units of which 16 000 are tentatively earmarked for Anglo and Gencor employees.[14] In fact, Anglo optimists expect as many as 10 000 of their Free State employees to settle in 'Sir Ernest Oppenheimer Park', Thabong by 1991.

Township extensions are also planned for Kutlanong township near Odendaalsrus and Kanana township near Klerksdorp, the latter to service Anglo's massive Vaal Reefs mine. At Odendaalsrus, a new development at Freddies Mine includes the projected settlement of 3 000 workers in married accommodation. To date, a pilot project of 240 houses has been completed but the broader plan has run into financial difficulties. At Vaal Reefs, management accepted a consultancy plan to develop 4 500 houses in an extension to Kanana township and arranged to build 100 houses in a pilot project.[15]

The township-extension model of housing delivery is not feasible where mines are distant from existing townships. An alternative option is exemplified by developments at Carletonville where a completely new township is to be constructed around the existing mine village of Wedela. The Wedela settlement, built in the late 1970s between Western Deep and Elandsrand mines, formerly contained 360 family units which were rented out to senior black (mainly clerical) employees.

In 1986 Anglo bought up the land around Wedela and applied to have the area proclaimed as a black township. The corporation required proclamation in order to convert the housing stock into owner-occupied units and establish a black community council to run the township. By mid-1988, Wedela contained 640 completed housing units and 2 300 residents. The next phase of development (from mid-1988) includes construction of 700 lower-cost block houses and up-market rental housing for skilled black artisans. Anglo estimates that 6 600 workers at Western Deep and Elandsrand are eligible to apply for Wedela mortgages under the home-ownership scheme.

Anglo also undertook to play the role of black local authority for two years during development of the township. During that period the corporation will appoint and train black councillors (presumably ex-mine employees) to take over these functions. It has also put up an interest-free loan of R43-million to the incipient local authority.

Rand Mines has adopted a less interventionist strategy of housing delivery as a result of worker resistance to evictions from company houses in the early 1980s.[16] On its coal mines, the company has now opted for a 'wage adjustment' with which employees can either buy company houses or make their own rental or home purchase arrangements. Eventually all workers who remain in company accommodation will be charged a monthly rental. The approach is being extended to the company's gold mines.[17]

Hirschsohn reports that most miners prefer to retain their place in the compounds and visit their families who have moved into neighbouring townships or shack settlements on white farms. At ERPM in Boksburg, an estimated 50% of migrants have families in neighbouring townships. At the Dobsonville extension close to Durban Deep, some 200 employees are to be provided with sites and services and left to make their own building arrangements.[18]

The fourth option for housing delivery is to re-orient labour recruiting away from the rural areas towards existing urban townships. By building up a cadre of urban-based workers already in township housing the mines will supposedly be spared the expense of providing new accommodation. After a spectacularly unsuccessful urban recruiting campaign in the 1970s, mine management lost interest in urban labour. More recently, the Chamber of Mines Research Organisation has again embarked on a serious investigation of the current urban labour market and how it might be tapped by the mines.[19] JCI's Randfontein mine reputedly draws about 20% of its workforce from neighbouring black townships but this is an exceptional case.

TEBA recruiting figures show that mines have been more successful in obtaining labour from peri-urban bantustan resettlement areas such as Botshabelo in the QwaQwa bantustan in the Free State.[20] The growth of peri-urban labour in the mine workforce has led to a form of commuter migrancy which confers some of the benefits of permanent labour without any of the costs. Miners from these settlements (as well as those from the bantustans and Lesotho who live close enough to the mines) are increasingly engaging in a form of weekly or monthly commuting. Workers on the Free State fields are known as 'weekenders' in Lesotho.

Within Anglo there remains a considerable body of opinion wedded to short-term profit horizons which questions the need for any fundamental change in the way labour is organised. This division is also reflected within the ranks of mining capital as a whole. Of the six mining groups, Anglo has gone the furthest in planning to move away from migrancy. Gencor will share some of the space at Thabong and is still in the planning stage for its schemes. The others are more or less committed to the idea of providing additional family housing but Goldfields (employer of over 70 000 migrant miners) has expressed little interest to date in stabilising labour.[21]

Paying the cost

Affirmations of interest in the principle of home-ownership by black miners do not automatically translate into home buying, and it is unclear how many miners want, or can afford, to join management-devised home-ownership schemes.

Anglo's initial optimism about the pace of housing delivery and settlement was tempered in 1988 when the flood of workers expected to sign up for home-ownership plans did not materialise. In mid-1988, two years after the new housing policy was first implemented, only 407 of the corporation's black employees (0,24% of the total workforce) owned their own homes. March 1989 figures show that this has risen to 885 - which is still only 0,5% of the workforce. Some 2 585 (1,4%) skilled and clerical workers still rent married housing units on the mines.

For the bulk of the workforce, even a R20 000 house is out of the question. Management at Vaal Reefs admits that over 70% of the 'target' group of 4 500 is unable to afford the housing at Kanana. Tying up meagre earnings in mortgage schemes is an uninviting prospect for workers whose real wages have not risen in the 1980s. And workers who wish to build informal housing are unable to obtain finance, effectively excluding low-cost site-and-service and self-help schemes from Anglo subsidies.[22]

Ascertaining the full implications of long-term urban home-ownership is not easy for workers who have been denied such rights for 100 years and who are now subjected to a barrage of promotional material from mine management. If miners find managements' housing schemes inaccessible or incomprehensible that does not, in itself, overly concern the companies since the schemes are primarily aimed at workers in skilled and supervisory positions. The companies publicly declare that no miner is precluded from participating in their housing schemes.

But given current plans, the cost of housing will keep most workers in the compounds. Through control of the housing market, the companies can settle their skilled and supervisory 'target' workers without explicitly contradicting their declared policy of housing for all. It is precisely this contradiction which the NUM has set out to challenge.

Quite apart from the cost factor, the mining companies assume that migrant workers can and will want to cut their rural links given the opportunity. For decades the migrant labour system has forced workers to orient their working life towards building a rural base and purchasing control over rural resources. In many areas, rights of access to land and other resources depend on continued occupancy. Being asked to summarily abandon such rights is unlikely to meet with a chorus of assent.

Miners are certainly being offered the chance to move to their place of work with their immediate families. But they are also being asked to abandon the wider network of rural dwellers who depend on migrant remittances - extended families, unemployed relatives and neighbours, and mutual support organisations.[23]

There is little evidence that the mining companies have seriously investigated the rural impact of their projected policies nor what options mineworkers actually prefer. A top-down, inflexible housing policy which takes no account of rural complexity and variability is unlikely to meet with much success.

Many workers in upper job categories are not even eligible for participation in permanent mine housing. For the state has made it clear that the mining industry's 200 000 workers from Botswana, Lesotho, Swaziland, Mozambique and Malawi are to remain migrant workers. Their home governments agree. One legacy of the 1970s is that the upper job grades on most mines are still dominated by non-South African workers (Table 3).[24] Of 20 280 higher-skilled supervisory and technical positions (Job Grade 5-8) on Anglo mines, 11 827 (58,3%) are occupied by foreign workers. These workers are prohibited from settling in family housing units on the mines and must remain in the compounds.

Table 3: Foreign labour and skill levels on Anglo American gold mines, 1986

Job Grade	Foreign		South African	
	No	%	No	%
1	4 741	28,1	12 139	71,9
2	9 757	30,0	23 905	70,0
3	12 852	35,8	23 006	64,2
4	25 556	46,8	29 004	53,2
5	2 534	53,5	2 203	46,5
6	2 041	50,4	2 006	49,6
7	2 898	65,7	1 510	34,3
8	4 354	60,0	2 734	40,0

Source: Adapted from F de Vletter, 'Foreign Labour on the South African Gold Mines', *International Labour Review*, 126, 1987.

By hitching their new housing policies to the state's policy of 'orderly urbanisation' the mining houses hoped to be relieved of the immense cost of providing infrastructure and services for the new townships and township extensions. Anglo, for example, estimates that its programme of township development will require over R500-million between 1988-1992 for roads, water, electricity, sewerage, schools, clinics and recreation facilities.

At Kanana township near Vaal Reefs, workers were dissatisfied with the houses, there were conflicts with the black Kanana council and the state refused to bankroll the supply of services required for the development. A company spokesman noted that the Kanana project was 'hamstrung by bureaucracy' and the state's refusal to make sufficient resources available for the development.[25] Caught up in a severe fiscal crisis, the state prefers to privatise the delivery of services and is clearly unwilling to provide the required infrastructure itself.

Negotiations between state and capital on this question occurred throughout 1988. Anglo argued that its own expenditure on housing and services should be tax-deductible. The Receiver of Revenue disagreed. Until some form of cost-sharing arrangement is negotiated between state and capital, progress will be slow. In partial recognition of this, Anglo has considered instituting a living-out allowance for workers who choose to find their own accommodation in townships. Some 3 000 black miners receive such allowances.

Many miners will take matters into their own hands and move closer to the mines regardless of state and industry planning. And management will try to keep their employees out of the shack settlements that have sprung up on private land around some mines.

The mine-owners' new housing schemes have involved little or no consultation with NUM. The union has demanded that housing policy should be negotiable, not simply devised and implemented by management. During 1987 and 1988, NUM initiated its own research into the housing and accommodation needs of its membership. The union's basic demand is for a broad range of flexible housing options which offer all workers, rather than a select few, the possibility of opting out of the migrant labour system. NUM will also continue to demand vastly improved living conditions for workers who cannot, or who choose not to, take this route.

Home-ownership is a central component of management's housing strategy for both fiscal and ideological reasons. But the union is responsive to a more varied constituency, only part of which is interested in owning homes. Home-ownership can thus only be one of a number of options which must be made available to miners. NUM has been quick to seize upon an uncomfortable historical precedent: housing policy for white miners where the mines have traditionally provided heavily-subsidised rental housing for white workers. The union demands no less for its own membership. It remains to be seen how far mining capital will go to accommodate the real and diverse needs of the mine workforce. But initial indicators are that the gap is a wide one.

Notes

1   '1987 - The Year Mineworkers Take Control', *South African Labour Bulletin,* 12(3), March/April, 1987.
2   See J Crush, A Jeeves and D Yudelman, *The Chains of Migrancy*, Boulder, Westview Press, 1989.
3   This model was articulated in private as well as in Chamber of Mines' submissions to government commissions of enquiry such as the Du Randt, Riekert and Wiehahn commissions.
4   J Crush, 'Restructuring Migrant Labour on the Gold Mines', in G Moss and I Obery (eds), *South African Review 4*, Johannesburg, 1987.
5   M Lipton, 'Men of Two Worlds', *Optima*, 29, 1980, 114.
6   W James and J Lever, 'Industrial Relations and the Abolition of the Job Colour Bar on the South African Gold Mines', Stellenbosch, 1987.
7   J Leger and P van Niekerk, 'Organising on the Mines: The NUM Phenomenon', *South African Review 3*, Johannesburg, 1986; and J Crush, 'Migrancy and Militancy: Unionisation on the South African Gold Mines', *African Affairs*, 88, 1989.
8   C Markham and M Mothibeli, 'The 1987 Mineworkers' Strike', *South African Labour Bulletin*, 13(1), November, 1987; and W James, 'The Group with the Flag: Mine Hostels as Contested Institutions', African Studies Institute seminar paper, University of Witwatersrand, Johannesburg, 1989.
9   NUM, 'Collective bargaining at Anglo American Mines - A Model for Reform or Repression?', Johannesburg, 1988.
10 'Fear Stalks the Compound. But Who's to Blame', *Weekly Mail*, 13-19.01.89; 'Flak Flies Between Anglo and the Miners', *Sunday Times*, 15.01.89.
11 R Taylor, 'Migrant Labour in the Mining Industry: A Future Perspective', *South African Journal of Labour Relations*, 11, 1987; J Crush and W James, 'Depopulating the Compounds', Department of Sociology, University of Cape Town, 1989.
12 'Anglo Pioneers A New Housing Scheme for Workers', *Mining Sun*, 23.06.88.
13 'Family Ties', *Business Day*, 23.09.87; Market Research Africa, *Accommodation Study*, 2 volumes, Johannesburg, 1986.
14 'R700-m Plan for Housing in OFS', *Star*, 23.04.87.
15 R MacGregor, 'Urbanisation - The Challenge of the Future', in D MacArthur and A Moerdyk (eds), *The Implications of Urbanisation for the South African Mining Industry*, Johannesburg, 1988.
16 P Hirschsohn, 'Management Ideology and Environmental Turbulence', MSc thesis, Oxford University, 1988.
17 N Penstone, 'Housing the Workforce - The Rand Mines Experience', in MacArthur and Moerdyk, *The Implications of Urbanisation*.
18 'Owner-Built Houses for Initial 200 Miners', *Mining Sun,* 18.02.88.
19 W James, 'Urban and Proletarian Labour in the Gold Industry', African Studies seminar paper, University of Cape Town, 1988.

20 Crush, Jeeves and Yudelman, *Chains of Migrancy*, Chapter 6.

21 'Other Mine Groups Plan Houses Like Anglo', *Business Day*, 23.09.87; 'Bigger and Better Mine Housing', *Mining Sun*, 14.05.87.

22 Hirschsohn, 'Management Ideology', 68.

23 W Cobbett, 'Trade Unions and the Struggle for Housing' (see this collection).

24 F de Vletter, 'Foreign Labour on the South African Gold Mines', *International Labour Review*, 126, 1987.

25 R MacGregor, 'Urbanisation'.

Employee Share Ownership: Co-option or Co-operation

Judy Maller

A new style of management characterised by worker participation rather than direct control and supervision is emerging in South African industry, particularly amongst bigger, unionised firms. 'Participative management' has become a catch-all phrase which incorporates a wide range of strategies - from simple communication schemes between management and the shop floor, to worker participation in top levels of management.[1]

Employee share ownership, one form of participative management, has acquired increasing significance within industrial relations as it has come under trade union scrutiny. The debate on this issue was given particular impetus by the implementation of Anglo American's share offer to employees in early 1988. This employment share ownership scheme crystallised the debate on both sides.

Bobby Godsell of Anglo American, in attempting to justify Anglo's decision not to subject the scheme to the rigours of collective bargaining, argued that the employee share ownership scheme was 'an individual thing in which the interaction was properly between shareholders and the individual employees'.

Cyril Ramaphosa, on behalf of NUM, responded that 'workers demand that a bigger share of the profits should go towards wages and not to share ownership schemes... This scheme is a ploy to undermine unions'. Despite a union call not to participate in the share scheme, almost 70% of eligible employees, including many union members, accepted shares.

The Anglo scheme offered those employees who had two years' service or more, five Anglo shares to be held in trust for four years. During this time employees receive dividends - the first amounting to a meagre R8,13 - on their shareholding. Thus the shares remain essentially a bonus - an offer which few workers would refuse.

The real litmus test for the scheme will come at the end of the four-year trust period when employees have the option of selling their shares. If most workers sell, then the scheme will simply have been a way of earning a little extra cash, and not a demonstration of commitment to the company or free enterprise.

When the Anglo share scheme was launched it was widely contrasted with that established at Samcor on Ford's disinvestment from South Africa. Ford had entered into protracted negotiations with Numsa, and the two parties had signed an agreement in November 1987 whereby a portion of Ford's shareholding, amounting to 24% of Samcor's equity, would be held in trust and administered by workers. Future dividends would be channelled via the trust into community development.

In addition, Ford established educational trusts in Pretoria and Port Elizabeth to the value of R4-million to be administered jointly by worker and community representatives. In this way, the union successfully sidestepped the contradiction inherent in 'worker shareholders' by directing the benefits of the shareholding into the community. The agreement also made provision for workers to gain representation at board level.

The Samcor agreement was hailed by management and labour alike as the appropriate model for setting up structures of co-operation in the work place. But the scheme experienced a major setback when workers at the Pretoria plant went on strike in opposition to the deal. The motivations for the strike remain unclear but workers demanded that their shareholding be liquidated and paid out in cash to the individual workers. The existing shop-steward committee was voted out of office and an entirely new one was elected. Workers then demanded the individual remission of dividends and in mid-December 1988 'all 4 500 employees who had been in service for the full year each received a cheque for R940,00' (*BD*, January 1989) Samcor chairman Les Boyd announced that 'the trustees had changed the rules in order to give each worker the choice of accepting the money or handing it back for use in community projects' (*BD*, January 1989).

The Samcor and Anglo share ownership schemes point to the severe problems associated with employee share ownership in particular, and the project of participative management as a whole, in recent South African labour history. But it is important not to conflate worker participation with employee share ownership, although some managers have assumed they will derive the productivity advances associated with greater worker participation by simply distributing a few shares to employees.

As 'participation' schemes have mushroomed in the work place, confusion has spread as to the precise meaning of the term. Participation is based on a capacity to influence company decision-making. The ownership of less than 5% of equity does not confer this right and therefore cannot be included under the title of 'worker participation'.

Participative management in South Africa

The range of participation schemes currently being implemented by larger companies includes quality circles, working groups, briefing groups, employee share ownership and profit-sharing, worker directors, collective bargaining and full-time shop stewards. A survey of industrial, commercial and retail companies carried out in 1987 attempted to quantify the schemes and to investigate their nature and effects in the work place.[2] This survey suggested that South African workers are not given much opportunity to make decisions about their own jobs, let alone decisions about production or investment at a company level.

Management respondents identified the following structures as part of their participative management programmes:

Table 1: Management identification of participative management structures

Structures	% Respondents
Joint health and safety committees	23%
Employee trustees on pension funds	20%
Briefing groups	18%
Communication schemes	15%
Quality circles	11%
Suggestion schemes	9%

All these institutions are characterised by low levels of worker participation. The high incidence of joint health and safety committees can be attributed to legislative imperatives - the Machinery and Occupational Safety Act requires companies employing over 100 employees to establish health and safety committees on which safety representatives are obliged to sit. Employee participation in the administration of pension funds is also relatively high, and may reflect high rates of participation by white employees, especially in the commercial sector, as well as the recent push by black unions to control their own pension funds.

The presence of quality circles in 11% of respondent companies is significant and indicates the increasing popularity of this institution since the early 1980s when quality circles were first introduced into South African companies. Quality circles are problem-solving groups of workers in the same department who meet regularly to discuss solutions to productivity or quality problems experienced in their work.

Employee share ownership

The survey also investigated financial forms of participation, which turned out to be extremely limited as regards hourly paid workers.

Table 2: Financial forms of participation

Types of financial participation	% hourly-paid workers	% salaried staff	% senior management
No participation	79	58	33
Profit sharing or value-added bonus	21	38	50
Share options	-	4	-
Share trusts	-	-	-
Executive share schemes	-	-	17

This table demonstrates that where employee share ownership schemes have been introduced, they are neither widespread, nor do they affect the majority of black, unskilled or semi-skilled workers.

Despite their low incidence, it was decided to investigate ownership schemes further, given the massive publicity associated with them during 1987, when business commentators promised high productivity levels and reduced industrial relations conflict by linking workers' fortunes to those of their companies.

The characteristics of a 'typical' share ownership scheme were drawn from an analysis of 30 JSE-listed company statements. Most of these companies have created ownership schemes that are very similar in structure. The 'share incentive scheme', as it is usually called, offers two methods of purchasing shares:

- an outright purchase (or a minimum deposit with the remainder to be paid off over a period of time);
- an option to purchase, which is typically valid for ten years.

In both cases, a trust fund is usually set up to facilitate employee purchase of shares. The shares are acquired in the name of the employee, but are pledged to the trust as security against the loan of the purchasing capital. In many cases the shares are released in batches over time to employees. Typically, shares are made available to all full-time *bona fide* employees including directors, salaried staff or office-holders. In some cases, a minimum period of service is specified for eligibility, for example in one case it was stated that 'beneficial ownership (of the shares) could not be acquired until ten years of uninterrupted employment was completed' (*BD*, November 1987).

Shares are made available at middle market price, as listed on the JSE the day immediately prior to the purchase or granting of an option.

Most schemes specify a maximum limit of between one and 10% of the company's issued share capital which is made available to the scheme. All the companies who have set up these schemes have adhered to the maximum limit pegged by the JSE. The effect of this has been that even collectively, the employees' stake has never amounted to more than a minority shareholding.

The final significant feature, based on the 30 surveyed share ownership schemes, relates to trust funds. Until such time as the purchase price for the shares is fully paid up, employees effectively own shares in a specially set up trust, from which they derive the attendant dividend rights. But they cede their voting rights to trustees until they own their shares directly. Fund trustees who represent employees' interests at company AGMs are, however, all company-appointed, white and drawn either from the board of directors, senior management, company attorneys or company brokers. None of the schemes make any provision for participating employees to elect their own trustees, despite the fact that trustees vote on behalf of employees at company AGMs.

Disinvesting companies, through their own peculiar position in the South African economy, have become a separate, significant category of companies offering financial participation to employees. Some US companies have, for example, established trust funds which involve an equity stake for their South African employees.

Some of the better-known examples are:

- IBM, which disinvested in 1986, and set up a trust fund for employees 'to share some profits and an equity stake'. Black employees gained 13% of equity;
- Exxon Corporation, which sold its petroleum and chemical operations to an off-shore trust. This trust now pays Exxon's 200 local employees 2,5% of pre-tax profits;
- General Motors Corporation, which disinvested in 1986. The company sold a 50% equity stake to a trust which now shares an undisclosed proportion of profits with 'selected' employees. The remaining 50% of issued share capital was acquired by local management;
- Ford Motor Company, which placed 24% of its Samcor shareholding into a trust on behalf of 4 500 workers when it disinvested in 1987;
- Coca Cola, which placed 11% of equity in a trust when it sold out its 30% stake in Amalgamated Beverage Industries. Ten percent was allocated to black traders and 1% to previous Coke employees;
- Mono Pumps, owned by Gallagher Ltd in the United Kingdom (owned in turn by US-based American Brands Inc) which was sold to Fluid Corporation in May 1988. The sale involved the establishment of a R200 000 community-oriented trust fund and a financial settlement for each of the 350 workers amounting to approximately R575 per worker.

These initiatives must be seen as part of a current of overseas opinion that foreign companies should play a significant role in facilitating the 'economic empowerment' of black people. Some companies may be motivated by a desire to improve their public image back home by extending ownership to black people. Other companies may simply implement share ownership schemes to justify the continuation of their business in the South African market. In addition, many disinvesting companies have negotiated buy-back options, thereby providing a foothold for their possible return in the future.

Monitoring of JSE listing statements and press announcements revealed that a total of about 120 companies had introduced some form of employee scheme prior to the Anglo share offer. With disinvestment on the increase and more companies adopting measures to improve employee productivity and company loyalty, the incidence of the ownership schemes is likely to rise. What seems unlikely, however, is the possibility of existing ownership plans offering black workers any significant participation in the schemes.

The profile of employee ownership plans attached to JSE-listed companies shows that they are generally modelled on executive share schemes. Some of the newer schemes in large companies such as Pick 'n Pay have been structured so that many of their eligible un-skilled or semi-skilled employees will never directly own company shares. Instead, the schemes offer ownership of shares in a separate trust. This exclusion from direct participation must constitute a severe limitation on the motivational effects of ownership schemes especially as - at this level - the link between individual employee performance on the job and company share price becomes very tenuous.

Employee participation in share ownership, as a strategy designed to enhance worker loyalty and productivity, is fundamen-tally limited by the lack of a direct association between share price movements and company performance. Other variables, such as general economic activity, international developments, political upheavals, and the day-to-day manipulation of prices by consor-tiums buying and selling shares, often have more to do with setting share prices than employee performance.

Financial participation merely 'shares money, not power, authority or decision-making within the organisation', and is 'on its own and without any associated joint regulation of management decision-making, a very dilute form of industrial democracy'.[3] This places a question mark over employee share ownership as a motiva-tional strategy, or one that engenders employee loyalty to the firm.

Union responses

The democratic trade unions affiliated to Cosatu have publicly rejected the employee share plans and many other management-initiated forms of participation such as quality circles. They perceive these schemes as designed to obviate the need for unions by center-ing participation exclusively on the individual worker and by preventing collective bargaining. Quality circles, for example, are seen by unionists primarily as a 'con' to get workers to work harder and resist calls to strike action. The limitations of share ownership schemes are clearly outlined by a Cosatu official who stated that 'such proposals give workers a stake in ownership without giving them sufficient control to carry out policies which would be beneficial to them' (*WM*, 19.06.87).

South African managements have extended very limited opportunities for workers to exercise independent decision-making. This in turn has resulted in a union strategy of refusal to accept what are regarded as low benefits acquired at the cost of compromising union independence. Nonetheless, many workers have become involved in a range of participation schemes, particularly at shop-floor level. In a few notable cases, such as the Anglo scheme, workers have ignored union calls to refuse participation. In some cases, participation schemes have provoked real enthusiasm from the shop floor, although the long-term effects on productivity, employee turnover rates and job satisfaction have not been measured due to their recent introduction.

The union strategy of 'refusal' towards locally introduced employee share ownership schemes has been a response to the co-operative nature of participation strategies. However, this response has been inadequate. The traditional role of the union is one which aims to establish and defend a number of protective rights such as procedures for grievance resolution, dismissal and minimum wage levels. The advent of participation calls on unions to play a far more active role within companies - a role which could potentially compromise their independence.

Despite these dangers, it is important to recognise the contradictions faced by capital in the work place. Capital, while asserting its dominance, also needs to ensure a relatively co-operative relationship with labour to maximise productivity. Some structures of participation can therefore be seen in the context of opening up space for greater worker discretion and independence in an attempt to improve labour productivity. At the very least this provides workers with opportunities to derive greater job satisfaction, increased earnings from higher levels of productivity and improved working environments.

If workers are able to extend structures of participation into areas of production planning or investment decisions, they can develop their skills as well as influence strategic corporate decision-making. Strategies of worker participation can be extended even further, depending on how labour responds and moulds participation schemes to advance its own interests, and challenges accepted conventions about the nature of work and its purpose.

Some unions which have entered the terrain of worker participation on a factory level - like Numsa at the Volkswagen plant - have negotiated the terms of participation involving union office-bearers, such as full-time shop stewards. While it is clear that the union strategy of refusal has been inadequate, so too company participation schemes designed to boost worker productivity and reduce industrial conflicts have not significantly extended worker control over decision-making and are thus unlikely to succeed.

The space for independent decision-making on the part of workers is severely circumscribed. This situation requires both management and unions to take greater risks than they are currently willing to do.

Notes

1   Most of the information in this article was taken from a research report written by the author in 1988, entitled 'Esop's Fables', published by the Labour and Economic Research Centre, Johannesburg.

2   The population surveyed comprised all industrial, retail and commercial companies employing 100 or more employees. The list, supplied by the Bureau for Market Research, amounted to 5 047 companies. A random sample of 5% of this population was generated to select a sample for surveying. Questionnaires were sent out to the companies' personnel managers and an 11% response rate was received.

     A broad range of industries is represented in the sample, covering all the sectors identified by the Department of Manpower. There was also a range with respect to company size, with the average size being 300 employees.

3   M Salamon, 'Employee Participation', *Industrial Relations*, 1987, 303.

The Privatisation of Working-Class Health Care

Max Bachman, Jean Leger, Ian Macun and Max Price

Ideas of privatisation and deregulation have been promoted evangelically by government, big business and the liberal press. They have also been included in the everyday jargon of state officials formulating policy on health services and occupational health and safety.

Under the rubric of deregulation and privatisation, the state is legislating and consolidating a new health framework. It is concerned, at least formally, to deracialise health services. Indications of this concern include proposals for a non-discriminatory workers' compensation system, and the fostering of a non-racial but private health-care sector.

But, at the same time, the state is anxious to minimise its costs in this process, whether incurred through factory inspections, compensation or health services. Deracialisation of services under the existing apartheid dispensation would radically increase costs since existing services to black workers, such as medical examinations, would have to be greatly enhanced. Thus privatisation is invoked, shifting the costs of services directly onto workers and employers.

In addition, the state is attempting to contain and institutionalise conflict over safety and health provisions through the development of a system of 'self-regulation'. This should more properly be called 'managerial regulation'. By 'privatising' work-place inspection, regulation and occupational health services through new legislation, the state is transferring a number of its traditional activities to management. Moreover, with trade unions challenging management prerogatives over work-place safety, this new legislation is bolstering management prerogatives through new statutory powers.

Mindful of how health issues in western countries became mobilising points for industrial conflict in the 1960s and 1970s, the state is anxious to depoliticise this arena by withdrawing as much as possible. This is being done partly through 'deregulation' of racially-discriminatory regulations.

And finally, the state hopes to use deregulation to boost the informal sector as a potential site of economic growth. Through new statutory powers such as the Temporary Removal of Restrictions Act, deregulation can now be extended by administrative decree, potentially removing all the limited statutory protection of groups of workers regarding safety and health.

Deregulation generally refers to the removal of restrictions on economic activity, for example, through the repeal of regulations. But 'privatisation' conceals an array of agendas and strategies, making it important to unravel its uses. There are at least four distinct areas of privatisation discussed in this review:

- *inspection*, requiring employers to undertake inspection activities traditionally performed by the state. One instance of this involves mine managements conducting work-place safety inspections and accident inquiries instead of the state's mine safety inspectorate;
- *rule-making*, whereby legislative functions are transferred from the state to employers. The Minerals Bill, for example, proposes that mine managements draw up their own safety codes to which they will then be statutorily bound;
- *services* previously provided by the state may be shifted to the private sector. For example, the draft Occupational Medicine and Occupational Diseases Bill and the Compensation for Occupational Diseases Bill[1] propose that fitness examinations be provided by employers. Previously these were provided by the state, although to white miners only;
- *financial* privatisation, whereby the burden of financing is transferred from public to private sources. One instance of this involves a shift from public-funded health care to medical aid schemes funded by employers and employees.

Mine safety and health:
from racism to managerial regulation

Mining, the country's most dangerous industry, remains the sector in which safety and health issues are most vigorously contested. Having deracialised mining legislation, the state now intends to reduce its role further in safety regulation. A philosophy of bolstering management prerogative rather than resorting to state intervention underlies this new legislative trend. Changes include the ending of statutory job reservation, new regulations for safety officers and representatives, and proposals for a new mining act and compensation dispensation.

Deracialising mine safety

Safety and racial discrimination were interwoven through the Mines and Works Act, which legally reserved the undertaking of the bulk of safety measures for white (and in some cases coloured) blasting certificate holders. In practice black miners were required to perform most tasks reserved for whites, even though they did so without the necessary training, and in contravention of the law.

Pressure to end job reservation coincided with management and state thinking about restructuring the underground labour process - a view fiercely opposed by the exclusively-white Mine Workers Union (MWU). Although the Wiehahn Commission recommended the abolition of job reservation in 1981, it was only in August 1987 that an amendment to the Mines and Works Act formally removed the job bar.

In 1988 regulations were promulgated requiring applicants for blasting certificates to have passed Standard Eight, and onsetters (lift operators) to have passed Standard Seven. While the racial requirement was thus 'deregulated', these regulations are actually stricter than before: management sources have estimated that 15% of current white miners are not eligible to hold blasting certificates in terms of these new regulations.

These new educational requirements also prevent most of the estimated 50 000 potentially eligible black team leaders from applying for blasting certificates. Following representations from the NUM, the Department of Mineral and Energy Affairs has agreed to reconsider the new regulations in the light of deliberations of an advisory committee chaired by the government mining engineer, and including NUM, MWU and the Chamber of Mines.

NUM has argued that all literate team leaders with suitable mining experience should be eligible for blasting certificates. Employers are concerned that the formal education stipulations should not lead to extra training expenses. However, they do not want eligibility to be limited to team leaders who meet an experience criterion, and wish to be able to recruit outsiders for these positions, thus keeping open their options to develop new strata of supervisors within the workforce. Despite these differences in approach, the formal deracialisation of work-place supervision, and hence the immediate responsibility for the safety of the work place is progressing.

Privatising mine safety inspections

The state aims to end its direct intervention in the labour process by repealing job reservation. In addition, it intends transferring most of the role of the mines inspectorate to mine management.

New regulations providing for management-appointed safety officers and safety 'representatives' were gazetted in 1988. While further reducing state intervention in production, the regulations greatly strengthen management prerogatives. At each mine with over 300 employees, the manager must appoint a full-time safety officer who shall 'devote all his time to the execution of the safety and health requirements of regulations'. Safety officers are required to undertake inspections of work places and machinery, check on safety equipment, ensure that safety and health standards are complied with and that workers have the necessary training. Safety officers must also hold inquiries into reportable accidents and record every accident which results in more than one lost shift.

Management is further required to appoint 'safety representatives' for every work place. Representatives have to report any hazards they identify or are reported to them to the person in charge of the work place. They may also accompany safety officers when they undertake inspections in the representative's own work place and point out any hazards. Safety officers have to hold meetings with safety representatives at least once every three months.

According to Danie Steyn, minister of Economic Affairs and Technology, the total safety inspection system is gradually being transferred to the mining houses.[2] The two most important reasons for this are the state's need to improve inspection without increasing expenditure, and the need to counter growing NUM challenges in the safety arena.

The high underground fatality rate in the past has been linked to the limited size of the mine inspectorate.[3] More inspectors are clearly required on South African mines. But according to the government mining engineer, the state official in charge of mine safety:

> With the present clamour for reduced state spending and privatisation there is no prospect of having a sufficient number of inspectors to exercise direct required supervision. With this in view, the minister proposed that the safety organisations on mines be strengthened and given legal status.[4]

The duties of safety officers as prescribed in the new regulations give legal status to the management 'loss control officers' who presently work within the framework of the Chamber's 'Mine Safety Management System' (MSMS), (previously the International Mine Safety Rating System). But the MSMS has not lessened mine fatality rates, nor do mines which are highly rated by the system (four or five stars) experience fewer accidents than mines with lower ratings.

The state's new approach has left safety measures ever more dependent upon the whims of mine management than before. As the government mining engineer pointed out:

> (U)nless (the safety officer) is strongly supported by the general manager, who will hopefully be enlightened enough to see that the two roles of safety and production are complementary and not opposed, he will be ineffective, and the scheme will be unsuccessful.[5]

Safety officers, whose jobs and promotion are in the hands of management, are unlikely to present serious challenges to that same management. In the view of NUM, privatising the supervision of safety in this way is tantamount to 'giving an alcoholic the keys to the drinks cabinet'.[6]

The new regulations attempt to counter NUM's ongoing campaign for recognition of independent, worker-elected, safety stewards. Following their promulgation some managements withdrew from negotiations aimed at recognising NUM safety stewards, or refused to allow workers to elect safety stewards. It is ironic that the regulations have nevertheless forced mine managements to negotiate safety agreements with NUM: management had a deadline by which to appoint safety representatives, and workers refused to be appointed without NUM involvement.

In negotiations some managements attempted to restrict agreements to the narrow framework of the regulations, but NUM has insisted on improved safety rights. These include:

- work-place election of safety representatives;
- routine independent safety inspections of work places;
- independent inspection of accident sites by worker safety

representatives;
- the release of safety stewards from work to attend accident inspections and enquiries;
- procedures for workers to refuse dangerous work;
- unresolved safety issues to become subject to standard negotiation and dispute procedures rather than disciplinary offences which can lead to dismissal.

Four safety agreements have already been signed and many more are under discussion.

Deregulation: a new Minerals Bill

During December 1988, a new Minerals Bill was published for comment.[7] Described by George Bartlett, deputy minister of Economic Affairs, as 'promoting deregulation and privatisation within the broad principles of the free enterprise/free market system and the basic common law', the bill consolidates nine mining acts, including those concerned with safety and health.

A novel clause in the bill requires managers to draft and apply codes of practice which must be approved by the inspectorate. This may be valuable in ensuring that adequate standards are applied in special circumstances. But there is a danger that piecemeal provision of codes of practice will be used to rationalise the absence of regulations which should provide appropriate standards for the industry as a whole.

Unlike the existing Mines and Works Act, the new bill does not provide for a statutory mine safety council. It proposes only a minerals advisory council which the minister will be required to consult on proposed regulations and which will advise on mining and mineral utilisation. While the current act requires that there be three employee representatives out of a total of ten members on the mine safety council, the proposed minerals advisory council *may* have up to two employee representatives out of a committee of 15.

The proposed new council must have two state officials with knowledge of mineral beneficiation and two of environmental conservation, but none are required to have knowledge of safety and health. While the rhetoric of privatisation refers to greater participation of the 'private sector', here is an example of greater dominance by state officials and reduced representation of organised labour.

The bill also reduces the possibility of the mines inspectorate having a research capacity independent of mine management - another example of *financial privatisation*. The acts to be repealed if the bill becomes law include the **Deep Levels Research Institute Act** which provided for a research institute to study the problems of mining at great depths. Deep-level mining, particularly below 3 000 metres, has an extremely high fatality rate especially from rockburst accidents. Although an institute was never established, repealing the act means that state research into the unique hazards of South Africa's deep-level gold mines will never be undertaken - especially as practically all research conducted into mining is financed and controlled through the Chamber of Mines' research organisation.

While state initiatives in privatising mine safety are potentially far-reaching, it remains unclear what their result will be. Instead of pre-empting NUM's campaign for independent safety stewards, the safety representatives' regulations had the effect of putting pressure on management to negotiate safety agreements. This outcome was probably not anticipated when the regulations were promulgated by the state. In this era of legislative flux there are clearly fresh opportunities for unions to influence the shape of the new legislative and industrial relations framework.

Privatising occupational health services and compensation

The Department of Health, like Mineral and Energy Affairs, also intends deracialising and privatising occupational health services and compensation, both for the mines and general industry.

Early in 1988 the department circulated two draft bills for comment, the Occupational Medicine and Occupational Diseases Bill and the Compensation for Occupational Diseases Bill. The thrust of the bills is that privatised occupational health services should be developed with the state playing an even smaller part than it does at present. The first bill specifically proposes that employers in 'risk industries' should provide medical screening examinations for all workers involved in 'risk work'.

Screening examinations

Examinations performed on all workers when they begin work,

whether they feel ill or not, and at regular intervals thereafter are known as 'screening examinations'. These examinations can identify individual workers who need medical treatment or removal from further exposure to a hazardous environment.

By detecting adverse health effects in groups of workers, screening examinations can indicate where effective control measures are needed. For example, regular hearing tests in a noisy factory can detect loss of hearing before a worker becomes aware of it, suggesting that management should attempt to suppress the noise, that the worker should be transferred to a quieter area, or wear hearing protection.

At present regular screening is only compulsory for workers doing 'risk work' in mines and works as defined in terms of the Occupational Diseases in Mines and Works Act. Private mine health services have conducted screening of African mineworkers, while white and coloured mineworkers have been screened by the state's highly sophisticated Medical Bureau for Occupational Diseases.

Screening examinations are not legally required outside of the mining industry, and are not usually performed. The draft bills do not specify which industries will be required to undertake screening in the future, or what the examinations will consist of. These are to be defined in subsequent regulations.

Ideally, screening examinations should be integrated into a comprehensive occupational health system together with preventive procedures such as measurements of work-place hazards, development of control strategies, and rehabilitation and retraining of disabled workers. Where these presently exist they are not integrated and are fragmented further into services provided by employers, the state and unions.

Reliance on employers to provide screening services with minimal state involvement is a serious defect. Management-provided services are often of poor quality and lack expert support. For example, modern equipment makes auditory or lung function testing easy. However, if the tests are to be worthwhile, expertise is required in study design, instrument selection and calibration, training of those carrying out tests and interpreting the results. Screening services performed by general practitioners without specific training in occupational health could proliferate under privatisation, resulting in examinations of dubious value.

At this stage it is not clear who would ensure that employers provided the screening examinations. The bill proposes the establishment of a cadre of occupational health officers appointed by local authorities to police the regulations. They would probably be health inspectors or medical officers of health already enforcing public health regulations for local authorities, and untrained in occupational medicine. Effective enforcement is unlikely in these circumstances.

The bills, if passed, would require some employers to provide a degree of occupational health service in the form of screening examinations. But they do not provide an effective enforcement mechanism, they reduce state-provided services, and have no requirement for worker participation.

Benefit examinations

At present all mineworkers and ex-mineworkers are entitled to free 'benefit' medical examinations if they suspect they are suffering from diseases caused by their work. Benefit examinations can detect occupational diseases which carry compensation for affected workers. Examinations for white and coloured mineworkers are presently performed by the state's Medical Bureau for Occupational Diseases, while African mineworkers are examined by mine medical services or district surgeons. The bills propose decreased state responsibility, with white and coloured miners being examined at the mines. African miners would continue to be examined in the same way as before.

The bills also propose that ex-mineworkers - for whom examinations are presently free - will have to pay for their own benefit examinations. They will only be refunded if they are found to have a compensable disease. This cost will discourage workers from seeking examinations for compensation purposes, and will increase the number of unrecorded and uncompensated diseases.

It would be preferable for the state, through the National Centre for Occupational Health or the Medical Bureau for Occupational Diseases, to expand services to provide more support to workers and employers.

Occupational disease compensation

The bills have important implications regarding workers' compensation for occupational diseases. They propose a compensation system which removes the explicit racism of the current Compensation for Occupational Diseases in Mines and Works Act. Some concessions are made to black workers, but there is a loss of privileges to white workers where there should have been an extension of these rights to all workers.

The proposal that compensation be paid according to wages rather than racial category, which is the case for mineworkers at present, is a positive proposal in the bills. But the threatened withdrawal of the right to compensation for active tuberculosis and early-stage pneumoconiosis is a drawback. The bills also remove the right of white and coloured mineworkers to training bursaries for their dependants, and medical examinations at the Medical Bureau for Occupational Diseases.

The bill fails to address major problems with the compensation system: payments for disability which are too meagre to live on, the narrow range of compensable diseases and inefficient administration. Consequently the bills have been criticised by conservative white unions as well as democratic unions and health and safety service organisations.

Democratic unions have supported screening in hazardous industries but have argued for adequate state services, effective enforcement and union participation in work-place health services.

Assessing deregulation and privatisation in general industry

The Machinery and Occupational Safety Act of 1983 (Mosa) departed from traditional models of work-place regulation and was the first example of a *managerial regulation* framework in South Africa. State rhetoric has portrayed Mosa as an example of deregulation and privatisation although in practice increasing regulation of safety and health has occurred.

Managerial or self-regulation?

Relying on the approach proposed by the 1973 British (Robens)

Commission of Inquiry into Work Place Safety and Health, Mosa placed an emphasis on 'self-regulation'.

Department of Manpower reports equate the philosophy of self-regulation embodied in Mosa with the transfer of government functions to the private sector. But the concept of self-regulation contained in the act is in fact one of managerial regulation. It gives management the prerogative to appoint workers to committees rather than having worker-elected safety representatives. This absence of defined rights and duties for workers enables employers to maintain the monitoring of work-place health and safety as their prerogative. However, it is not the same as scrapping legislative restrictions, as implied by deregulation.

Employers can, in addition, block trade unions from raising health and safety grievances through normal collective bargaining procedures through insisting on dealing with issues only within Mosa structures. But where unions are firmly established and knowledgeable about Mosa, they are likely to contribute to the monitoring of working conditions.

It may be too soon in the history of Mosa to assess 'self-regulation' and managerial regulation in industry. But management has largely dominated the implementation of the act, while organised labour has largely been excluded from meaningful participation.[8] And although Mosa has made employers more aware of occupational safety and health, developments beyond this depend on the specifics of individual companies.

Deregulation: rhetoric and practice

Mosa is also an enabling piece of legislation. New regulations governing safety and health can be added or removed at the decision of the minister of Manpower. In the Department of Manpower's annual report there is a confusing view of this statutory process. The drafting of new regulations is equated with deregulation, as the following illustrates:

> Great success with deregulation has been achieved in this field and deregulation is one of the objects in drafting new regulations and in redrafting existing regulations. This approach corresponds with the principle of minimum state interference and helps to reduce the number of inspection staff required and ensures that the financial implications are much smaller than they would otherwise have been. A nucleus of inspectors ensure(s) that a watchful eye is

kept on compliance with the minimum standards prescribed in the legislation and that employers and employees make a joint effort to regulate themselves.[9]

The regulations promulgated under Mosa create a set of minimum standards to control adverse conditions. Although slow in being promulgated, regulations have appeared governing the use of asbestos, machinery, electrical installations and the like. These regulations are either similar to the old regulations under the Factories Act, or improvements on these. The asbestos regulations, for example, involve increased regulation of work-place health and safety rather than deregulation. Draft lead regulations have also been published for comment.[10] If promulgated this will be the first introduction of controls over the use of lead and mechanisms for measuring lead exposure in workers.

Some significant rights have been granted to workers under regulations such as general safety regulations.[11] In only one case - the draft of the facilities regulations - has there been a reduction of requirements on employers.[12]

Mosa contains a broad clause enabling 'the minister' to exempt employers from all or any part of the act (section 32) and this authority may be delegated to Department of Manpower divisional officers (section 34). It is therefore possible that deregulation is taking place on the ground. But the more generalised measures appear to be confined to the industrial parks and the initiatives of the Small Business Development Corporation, rather than affecting established, formal sectors of the economy where most unionised workers are employed.

The changing role of the inspectorate

Judging by declared state policy one might expect a reduction in 'state interference' in health and safety. One index of 'state involvement' is the activity of the factory inspectorate.

The effectiveness of the inspectorate has frequently been criticised, especially on account of the small size of its staff in relation to the number of factories requiring monitoring, and the questionable nature of the training available to factory inspectors. Published figures do not indicate a significant change in the extent of the factory inspectorate's monitoring of industry in recent years. As shown in Figure 1, the number of factory inspections and official

Figure 1: Factory inspectorate's monitoring of industry

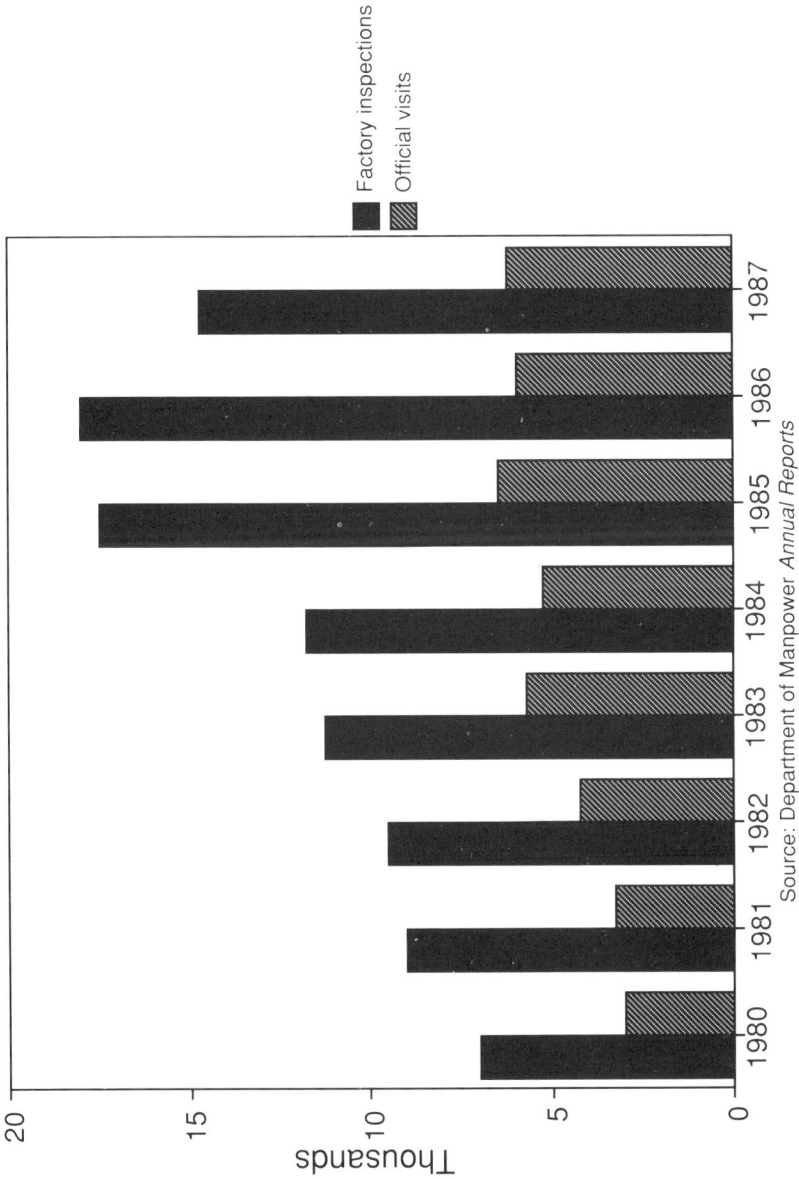

visits to factories has not declined substantially. Nor has there been a decline in the number of prosecutions recommended by the Division of Occupational Safety (see Table 2).

Table 2: Prosecutions arising out of full and formal investigations into more serious reported incidents.

Year	Prosecutions Recommended	Poesecutions instituted by Attorney-General	Successful Prosecutions
1981	169	104	94
1982	167	153	74
1983	341	308	130
1984	358	327	218
1985	348	308	175
1986	257	192	172
1987	337	204	187

General indices such as these, however, do not reflect the inspectorate's changing role. While the inspectorate claims it is targeting those employers with 'an unacceptable percentage of claims against the Accident Fund' for continued visits,[13] the key to understanding its activities lies in the way the inspectorate has re-conceptualised its role.

Rather than undertaking work-place inspections and prosecuting industry managements for contravening safety regulations, the inspectorate sees a new role for itself as 'consultant' to managements within the framework of 'self-regulation'.[14] Instead of spending time on the factory floor checking dangerous machinery, inspectors intend pursuing safety committee minutes and inspection report sheets and discussing these with factory management. The factory inspector of the past has become an 'auditor' of management compliance with the paperwork required by Mosa.

A management regulation framework has been established, but this does not necessarily mean wide-scale abdication of the state from its regulatory functions. It involves a substantial shift in emphasis from shop-floor inspection to inspection of administrative procedures and paperwork.

Trade union responses

Over the last two years both Cosatu and Nactu have formulated health and safety policies which indicate a commitment to regulate conditions faced by members. At its second national congress, Cosatu suggested unions could respond to Mosa by electing worker representatives with special responsibility for health and safety. This recognises the specific need to monitor work-place health and safety. It also raises the issue of training of work-place representatives for monitoring and tasks required by law (eg monthly inspections and accident reporting).

In August 1988 Nactu adopted a Workers' Charter on Health and Safety which sets out workers' health and safety rights which are either absent from the legal framework or ambiguous in it.

An important initiative from both Cosatu and Nactu was aimed at informing affiliates of key health and safety issues. Cosatu held a week-long residential seminar at which plans for regional activities on health and safety were devised. Nactu has established a Health and Safety Unit to assist affiliates, and convened about 20 seminars. It remains to be seen what effect these initiatives will have on individual unions, but a programmatic response to health and safety at federation level is a significant development.

Privatisation of health services for workers[15]

Health services for workers not directly related to the work place, including those used by families of workers, should ideally be the same services used by all members of the community.

However, financial privatisation of health services is creating a two-tier structure within black communities: a private 'higher' tier for those with regular, adequate incomes (usually workers employed in the formal sector) and a public, cheap, overcrowded service for the rest. The development of this two-tier system results from government privatisation policies, as well as growth in market demand[16] for private care.

State policy to privatise health services is being implemented in a number of ways. There has been a refusal to increase public expenditure on health services in line with escalating costs. These costs are pushed up by inflation, population growth, urbanisation (which results in greater health service use), increased longevity with a greater need for health care amongst the elderly, and the decline in the exchange rate which raises the costs of medication and medical equipment.

Figure 2: Membership of medical schemes by race

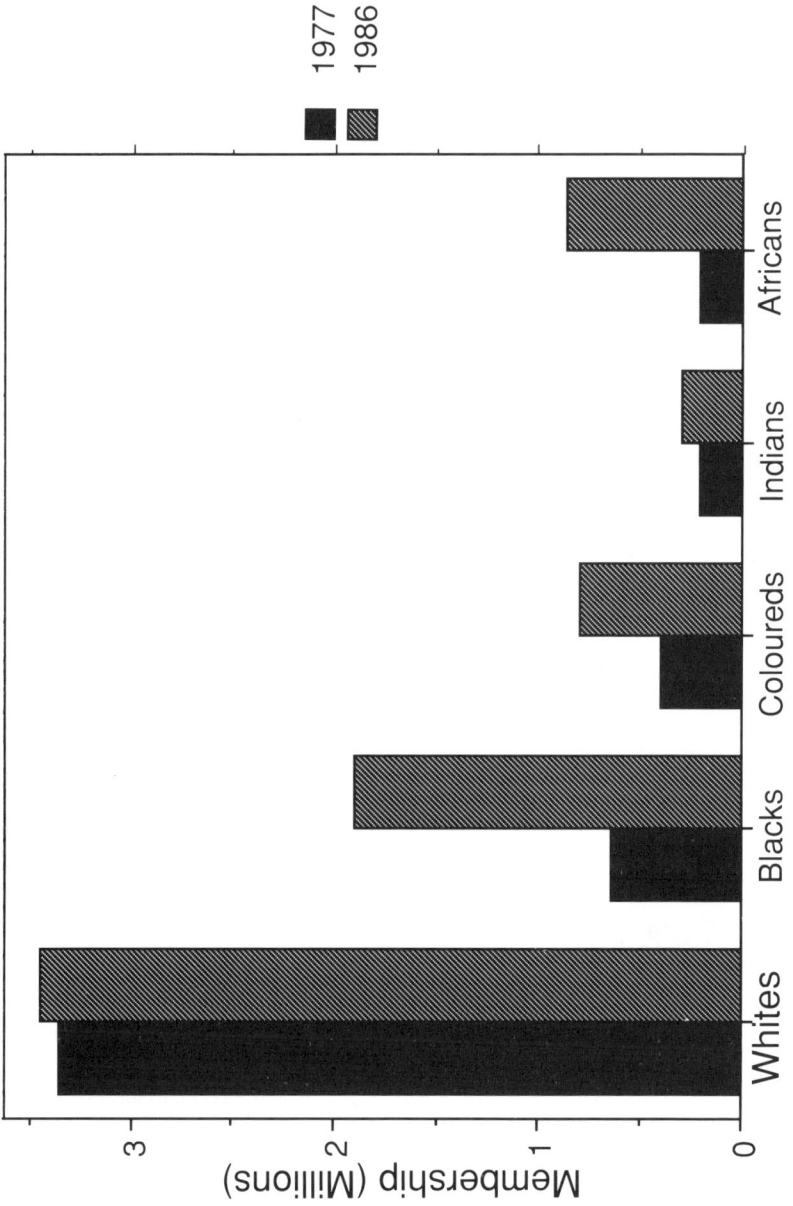

As a result of this relative decline in expenditure, public health services are more congested than ever and the quality of care has deteriorated. At the same time the population, especially the urbanised black population, has rising expectations of the health services. This has led to disaffection with public services and a move to seek private health care.

A second feature of state policy has been the tariff increases at public facilities. Almost anyone in employment earns enough to be classified into a category where he or she will have to pay a fee which may be similar to, or even exceed general practitioner fees. This encourages those with a regular income to use the private sector and join medical aid schemes.

Some employers have also encouraged workers to join medical aids. This is most commonly the case in multinational companies which are signatories to the Sullivan or other employer codes requiring the elimination of racially discriminating employment practices. Since most white employees are members of medical aids, employers encourage black employees to join as well.[17]

Finally, wages of many black workers employed in the formal sector have increased in real terms, bringing health insurance within an affordable range. In this context, a number of democratic trade unions have decided to negotiate medical scheme benefits for their members.

Medical scheme statistics shown in Figure 2 reflect these trends. For the past decade the proportion of the white population covered by medical schemes has remained roughly constant, fluctuating between 70% and 75%. It is unlikely to increase much.

However, the picture for Africans is quite different. In 1977 less than 0,9% of Africans were covered by medical schemes. In ten years, African coverage has increased four fold, to 3,6% in 1986.[18]

The consequences of this for the structure of health services in a future South Africa are far-reaching. At the moment about half the health care in South Africa is provided through the private sector and half through the public sector. Generally, only those covered by medical aid can afford private health care.

Until recently, the market demand for private care had remained static because the proportion of whites covered by medical aid had reached its maximum and the number of blacks covered had remained low. But the growth in medical scheme coverage of urbanised blacks may increase the total market demand for private health care two or three times. This will allow for a two to three-fold growth in private hospitals, general practitioners and private practice specialists.

The process of health services privatisation is therefore not simply a result of government policy. For privatisation to succeed there has to be an increase in market demand for private health care. If the number of people who can afford private health care does not increase, the new private services will collapse. The growth of medical aid coverage amongst blacks is a necessary condition for privatisation to proceed.

There are both long-term and short-term consequences flowing from this increase in the number of people using private sector, fee-for-service care. In the long term, it creates obstacles for a future transformation of the health services. The large number of independent private providers will constitute a major obstacle to redistributing health care on a more equitable basis. A significant proportion of the population will have become accustomed to private medical care. In comparing it to their *recollection* of public sector care, they will be unwilling to support a national health service (NHS). This group is likely to be relatively articulate, urbanised, employed and politically powerful.

In the short term, at least four problems will result from the growth of medical aid coverage:

- escalation of medical care costs;[19]
- aggravation of the already-severe maldistribution of health personnel between urban and rural areas, and between wealthy and poor communities;[20]
- worsening maldistribution of health personnel between private and public sectors;
- the growth of private independent providers takes the health services further away from the possibility of democratic control.

Worker demands for private care are real and urgent because public care is so poor. But providing this kind of care will lead to a two-tier health service and obstacles to the creation of an NHS in the long run. In response to this dilemma, unions and community organisations need to focus on the quality of existing health services, demand improvements and more affordable fees.

In addition, alternatives to medical aid schemes at an industry or trade union federation level require development. Such alternative schemes could incorporate the principles of an NHS - for example, in employing their own doctors and health workers and in offering worker control over the services provided.[21]

Broad-ranging deregulation and privatisation initiatives are currently underway in state policy on health services, and occupational health and safety. While these are not disparate, isolated initiatives, it is important to grasp the unevenness and contradictions in the ways these policies are applied by different departments and to different functions.

Many changes have been introduced with the promise of reforming racist policies and services. But the state has used the opportunity to reshape the terrain upon which safety and health issues are contested. In industry, and more recently in mining, new legislation has attempted to pre-empt democratic trade union moves to challenge hazardous safety and health conditions. In these moves the state has deregulated its racially interventionist role which discriminated in favour of white workers.

But the experience of Mosa since 1983 does not suggest wholesale deregulation of minimum safety standards. Indeed many of these have been, or are in the process of being reformulated. Rather, the state has reconceptualised its role and is privatising the monitoring of safety and health, health services and even the function of safety rule-making. This bolsters management prerogatives in a new framework of managerial regulation.

Simultaneously, privatisation is shifting the costs of health services more directly onto workers, as well as employers.

There is a danger of the privatisation process intensifying incipient divisions within the working class. The state's approach to health issues represents an important component of a broader strategy of differentiation. A two-tier system of work-place health regulation and health care is being created. Unionised workers in formal employment may, to a degree, be able to police safety standards and will probably have access to reasonable (private) health care through medical aid schemes. However, workers in the informal sector and the unemployed will face the prospect of practically no protection against dangerous working conditions and will only have access to poor-quality public health care. Thus material and social divisions between different sectors of the working class could be accentuated.

Just as it is necessary to unravel the different processes that go under the guise of privatisation, so there are different responses to privatisation which require critical examination. Responses to privatisation which insist that the state should maintain its traditional activities, and allocate even more resources to them, may simply bolster ineffective models and legitimise the current state. There may be more appropriate responses through which trade unions refashion the thrust of privatisation in directions more compatible with longer-term alternatives and more progressive health models.

Because of the withdrawal of state intervention, many aspects of the privatisation process may be more easily contested as they become negotiable issues between management and labour. A case in point is the appointment of safety representatives on the mines: after years of delay, some managements have been forced to negotiate safety agreements with the NUM so as to be able to appoint safety representatives to comply with the new regulations.

Similarly, unions can play a significant role in shaping workplace inspections by safety representatives and in directing occupational health services. Finally, trade unions could respond to demands to negotiate private medical aids with alternative schemes incorporating broader principles of community health care. While strictly speaking all these options would be in the private sector, they would be under the control of democratic organisation at the same time as providing models for a post-apartheid health service.

Notes

1   Department of Health, mimeos, 1988.
2   'Mining safety change', *Engineering News*, 3 June 1988.
3   HS Eisner and JP Leger, 'Safety in South African mines: aspects of supervision and inspection', *Safety and Health Congress*, Johannesburg, November 1987.
4   GP Badenhorst and D Bakker, 'Responsibility for safety and health in the production environment', *Safety and Health Congress*, Johannesburg, November 1987. This article also notes that measurement and control of underground dust levels will be shifted to management. This has major ramifications because of the increasing prevalence of pneumoconiosis.
5   Badenhorst and Bakker, *Safety and Health Congress*, 1987.
6   NUM press release, March 1989.
7   Notice 856 of 1988, Department of Mineral and Energy Affairs, Proposed Minerals Bill, 1989, Government Gazette, No 11606, 15 December 1988,

31-67.

8   I Macun, 'Yes sir! Mosa! No sir!', *Indicator SA*, 5(3), Autumn/Winter 1988.

9   Report to the Director General for the year ended 31 December 1987, Department of Manpower, RP72/1988, Government Printer, Pretoria, 103.

10 Government Gazette No 11606, 15 December 1988.

11 Technical Advice Group, 'Mosa general safety regulations', *South African Labour Bulletin*, 11(7), August, 1986, 38-41.

12 See W Hofmeyer and M Nicol, 'Deregulation: a challenge to the Labour Movement', *South African Labour Bulletin*, 12(4), May/June 1987, 79-88.

13 Report of the Director General, Department of Manpower, RP72/1988, 106.

14 Interview with M Mulder, chief director of Occupational Safety Division, Department of Manpower, 15 March 1989.

15 These views are explored more fully in M Price and P Tshazibane, 'Dilemmas posed by medical aid schemes for the labour movement', Centre for the Study of Health Policy, Paper No. 11, Johannesburg, 1988.

16 'Demand' is used here in its economic sense, ie the desire and ability to purchase services or commodities.

17 Interview with R Speedie, director of the Representative Association of Medical Schemes, 6 June 1988.

18 Department of National Health and Population Development, annual report of the Registrar of Medical Schemes, 1987, Government Printer, Pretoria.

19 M Price, 'Health care beyond apartheid: economic issues in the reorganisation of South Africa's health services', *Critical Health Dissertation Series*, 8, 1987.

20 M Price, 'The consequences of health service privatisation for equality and equity in health care in South Africa', *Social Science and Medicine*, 27(7), 1988, 703-716.

21 J Broomberg, C de Beer and M Price, 'A stepping stone to national health', *Work in Progress*, 59, June/July, 1989, 41-43.

The Level of Bargaining: New Industrial Unionism in the Food Industry

Johann Maree[1]

Following state recognition of African trade unionism in 1979, the emergent democratic unions grew rapidly, and began to win increasing recognition from capital.

In the early 1980s these unions - which had limited numbers of well-organised plants - insisted on plant-level bargaining with management. But employers were often keen to negotiate at company level, and some tried to force the democratic unions into industrial councils. Eventually, most of the democratic unions registered, allowing them to enter industrial councils. But they were generally reluctant to do so as the balance of power in the councils strongly favoured management.

The rapid growth of the democratic unions, and Cosatu's policy of consolidating one union per industry, facilitated the emergence of national industrial unions. In the second half of the 1980s the industrial relations picture began changing as the balance of power between companies and unions shifted. A key issue in collective bargaining emerged for the new industrial unions: the appropriate level and forum for bargaining.

This was the background to the Food and Allied Workers Union quest to attain national-level bargaining in the food industry. Fawu has its roots in the Food and Canning Workers Union which was founded in 1941. It was formally a registered union for coloured workers, but always worked extremely closely with the unregistered African Food and Canning Workers Union. In 1985, the two unions merged under the name of Food and Canning Workers Union. The following year the enlarged FCWU merged with the Sweet Food and Allied Workers Union and other Cosatu unions in the food industry to form Fawu - giving effect to Cosatu's policy of only one union per industry.

Fawu and the food industry

The food industry is wide-ranging in the nature of its products and production processes. Fawu, the largest union in the industry, lists 17 different food sectors in its constitution: baking, biscuits, brewing, cold storage, distribution and markets, cooldrinks, dairy, fruit and vegetables, fish, meat, milling, poultry, nuts and snacks, tea and coffee, tobacco, wine and spirits, and sugar, sweets and chocolates.

After Fawu, the Nactu-affiliated Food and Beverage Workers Union is the largest in the food industry. In 1987 Fawu had a membership in excess of 61 000 whereas FBWU had 20 000 members.[2] A breakdown of Fawu's membership and organisational presence during 1987-88 is shown in Table 1.

Table 1: Fawu membership and organisational presence, 1987- 88[3]

Sector	Members	Factories organised
Milling	11 000	82
Fruit and vegetables	11 000	45
Baking	6 700	75
Dairy	6 000	43
Brewing	5 000	25
Meat	3 900	53
Cooldrinks	3 300	12
Poultry and eggs	2 800	21
Sugar manuf/refining	2 700	10
Sweets and chocolates	2 600	12
Fish processing	2 550	25
Cold storage/distribution	2 250	47
Tobacco	1 780	4
Nuts and snacks	1 750	10
Biscuits	1 200	4
Tea and coffee	700	9
Wine and spirits	340	6
Total	65 570	483

Fawu has many more recognition agreements in the food industry than other unions. A 1987 survey of eight leading companies in the industry revealed that at least 140 of the 190 recognition agreements signed by the companies were with Fawu.[4] This is further demonstrated by the number of recognition agreements won by unions at Premier Food Holdings Limited, a subsidiary of the Premier Group. Premier Food Holdings Limited totally owns 16 of its 17 subsidiary companies. These subsidiaries process a wide variety of foods which Premier has divided into divisions ranging from milling and baking to animal feeds (Epol), edible oils (Epic Oil Mills) and others. As at May 1988, Fawu had 78 recognition agreements with the company, while another five were being negotiated. The remaining nine unions had signed a total of 21 recognition agreements, of which the FBWU had eight.[5]

The regional distribution of Fawu's signed-up membership reflects its existence as a national union.

Table 2: Signed-up membership by region, Fawu, 1987[6]

Region	Members
Western and Southern Cape	18 529
Eastern Cape	4 640
OFS and N Cape	2 041
Natal	12 521
Transvaal	24 575
Total	62 306

Fawu's number of signed recognition agreements, size and regional distribution of membership have raised the issue of national bargaining more pertinently for Fawu than any other union in the food industry. Milling, being the best organised sector of Fawu, has had to confront the issue of level and forum of bargaining most acutely. Milling, and the closely-linked baking sector, provide the best examples of national bargaining issues within Fawu.

The milling sector

Fawu has defined the milling sector as follows:
- the milling of wheat, maize, other cereals, peanuts and oil seeds;
- the manufacture and packing of wheat flour, maize flour and meal, maize rice, samp, mabella meal, rice, breakfast cereals, animal feeds, edible oils and fats, cake, margarine, peanut butter, macaroni, spaghetti, sago, tapioca, beans, peas and lentils, glucose and starches.

The milling industry is highly centralised: six groups of

companies mill 98% of the wheat produced each year while four of these groups alone mill 70% of the maize.[7]

Four companies dominate the industry:
- Tiger Oats, a subsidiary of Barlow Rand;
- Premier, a subsidiary of Anglo American;
- Tongaat-Hullett, a subsidiary of Anglo American; and
- Fedfood, a subsidiary of Federale Volksbeleggings.

There are, in addition, a number of milling co-operatives owned by farmers. The three most prominent are Sasko, which mills almost a quarter of all wheat, Senwesko, South Africa's oldest and biggest grain co-operative, and Bokomo.[8]

Apart from the centralisation of ownership, the milling industry is also very capital-intensive. Premier Milling, for instance, recently laid out R56-million to build a wheat mill in Vereeniging.[9]

Fawu's objective in the milling industry is to achieve centralised collective bargaining at a national level. The union aims to achieve parity on wage scales for all workers in the industry, and aims particularly to end the large wage differentials between urban and rural areas. In Fawu's view, national collective bargaining reduces time and resources spent on negotiations with numerous separate mills and plants. If negotiations were completely decentralised the union would have to negotiate 82 different agreements every year (see Table 1 above). Fawu also believes that centralised bargaining will create more unity between different factories and regions in the union.[10]

In dealing with companies that operate at national, and even international levels, Fawu has recognised that its bargaining strength is greater when it negotiates at national level. This is particularly so when national-level organisation is able to prevent management from switching production to functioning plants during strikes.[11]

Fawu's achievements in centralising its collective bargaining process have varied from the very successful, with the Premier Group companies and Sasko, to virtually total failure with Tiger Oats. Fawu negotiates with Premier at a national (as opposed to plant) level in certain divisions, namely the wheat and maize milling division, edible oils and derivatives (Epic Oil Mills), and the animal feed mills (Epol). Although these three divisions began negotiating separately, they have ended up together at the negotiating table, although there are still different wage levels between the divisions.

In the Barlow Rand subsidiaries bargaining is still basically at plant level, though there are a few joint bargaining situations between plants, such as the two Fatti's and Moni's outfits.

These variations in Fawu's relations with Premier and Barlow Rand subsidiaries lie essentially in differences between the two companies' philosophies, histories, managerial practices and structures.

In Barlow Rand's philosophy of decentralisation, each factory or plant is treated as a separate profit unit, with the plant manager responsible for maximising that unit's profits. This approach has been highly successful with the company which showed a R1,3billion after-tax profit for the 1987/88 financial year.[12] Listed seventy-eighth in a ranking by sales in *Fortune* magazine's 1988 directory of the biggest industrial corporations outside the United States of America, Barlow Rand has become a corporate giant employing no less than 140 000 people.[13]

This decentralisation of decision-making originated with 'Punch' Barlow, founder of the company. Now headed by Mike Rosholt, Barlow Rand still follows a policy of decentralisation which accounts for their subsidiaries' refusal to negotiate at national level and their insistence on plant-by-plant bargaining.[14]

The Premier Group's decision to negotiate nationally is a reflection of the group's structure. Companies are divisions within the group, and major decisions affecting them are taken at central level. Individual companies within the group lack the expertise to ensure that bargaining is carried out efficiently and does not conflict with Premier policy. Consistency is required between the companies, and it is more pragmatic for the group, which has negotiated a standardised recognition agreement with Fawu,[15] to negotiate nationally with the union.[16]

Signing national wage agreements with uniform wages across the country has disadvantages for Premier, in that significant competitors pay distinctly lower wages in certain regions. During 1988, for example, Premier paid R165 per week in Kroonstad, Sasko and Tiger paid approximately R145 per week, while Sentraalwes was paying its milling workers R68 per week. All these companies were organised by Fawu, but the union proved incapable of removing the differential, and was forced to settle for a 12% increase with Sentraalwes.

In trying to overcome this disadvantage, Premier made strong recommendations in 1988 to the wage board investigation of the grain and animal feeds divisions of the milling industry.

The existing minimum wage specified by the wage board at the beginning of 1988 was a meagre R66 for urban mills and R44 for rural mills. In its representations Premier argued that approximately 80% of the urban-based wheat milling employers paid R170 per week or more while approximately 90% in rural areas paid R113 per week or more. It accordingly recommended that the minimum wages in the wheat and maize milling, and animal feed industries, as prescribed in Wage Determination 447, should be increased as of 1 April 1989 to R160 per week in urban mills and R140 per week in rural mills.[17] The wage board rejected Premier's recommendations on the grounds that they would be inflationary and cause unemployment.

Premier has also proposed the establishment of an industry platform from which to negotiate wage increases. Discussions with its competitors on this matter have, however, proved fruitless. To deal with a situation where competitors are paying considerably lower wages, Premier has made concerted efforts to increase employee productivity. This has included employee involvement in some areas of decision-making, 6M training (a simulation game to explain how capitalism works and to increase productivity), and disclosure of information. At the same time the group has warned Fawu that retrenchments could follow if wages paid by competitors remain as low as they are, and productivity fails to improve.

In late 1988, Premier reversed its position on bargaining levels, and became increasingly reluctant to negotiate at divisional level or in any centralised forum. This followed Premier's decision to decentralise responsibility for profitability to divisional levels in the company, and was probably a response to major decentralised competitors being able to pay much lower wages in certain areas. However, when it attempted to introduce decentralised bargaining with Fawu, it met with very strong opposition from the union. As a result it was forced to make a concession by giving Fawu an undertaking that it would maintain bargaining at the 1988 levels.[18]

Sasko negotiates centrally with Fawu in four of its mills. Even though it has done so for the past few years, it regards national bargaining as an experiment and intends to revert to plant-level bargaining in 1989. According to the divisional personnel manager, the company began to decentralise operations in August 1988.[19]

The baking sector

The baking sector consists of two main components, namely bread and confectionery, which are manufactured in about 580 establishments employing 29 000 workers.[20] The main item produced is standard white and brown bread - which is only produced in the large factories. As the larger factories pay higher wages they have been cutting back their confectionery departments, leaving this to small outfits. By 1987, Fawu had organised 75 factories with a membership of 6 700.

The milling companies own the bakeries which produce 75% of South Africa's bread. Like the milling industry, baking is highly centralised. Six groups of companies dominate the baking industry, and between them produce 90% of standard bread.

Table 3: Ownership of the baking industry[21]

Bread	Group	Owned by	No of bakeries
Blue Ribbon	Premier	Anglo	45
Albany	Tiger Oats	Barlow Rand	41
Enterprise/Atlas	Sasko	Co-op	25 +
Boerstra/Fedbake	Fedfood	Federale	21
Duens	Bokomo	Co-op	unknown
BB Bread	Bakers	Anglovaal	unknown

The six big companies concentrate on mass bread production and leave confectionery to smaller concerns. Because standard bread is baked only in large factories, its production has now become capital-intensive and fairly mechanised.[22]

The milling and baking industries are owned and dominated by the same companies and both are capital intensive. But the level of bargaining in the two industries is quite different. Most of Fawu's bargaining in the baking sector is at plant level. The notable exception is Premier which in 1987 bargained regionally with Fawu for each of the five regions served by the company. Tiger Oats resisted regional bargaining everywhere, while Fedfood opposed it in the Transvaal.[23]

The question arises as to why the level of bargaining in the milling and baking industries should be so different when the same union and companies dominate both industries, and when milling and standard bread-baking are both capital-intensive. The reason given by the former general secretary of Fawu, Jan Theron, is the difference in the structures of the two industries. Even though six companies dominate the baking of standard bread, there are a large number of smaller bakeries and confectionery establishments. Fawu believed it would be unwieldy to attempt to represent so many establishments and for so many employers to be represented in negotiations at the national level. It thus was happier to settle for regional-level bargaining with Premier in the baking sector.[24]

But Fawu has not organised many more Premier plants in milling than in baking. In 1987 it had nine recognition agreements in milling and fourteen in baking, a number that should not have been beyond the union's ability to co-ordinate in national bargaining.

Fawu has also decided not to use industrial councils in the baking sector to try and obtain either industrial or national bargaining rights. Prior to the time Fawu organised in the baking sector, industrial councils for the baking and confectionery industry operated in many of the country's major cities. Once Fawu began organising bakery workers these industrial councils were used by employers in the baking sector to support the existing bakery unions and discredit Fawu. Consequently the industrial councils lost legitimacy in the eyes of Fawu and were dissolved.[25]

Other sectors

Fawu has attained various forms of national bargaining in other food sectors. The brewing and fishing sectors provide valuable counter-examples to the baking and milling sectors, thus preventing facile generalisations about factors influencing the level of bargaining.

In the *brewing sector*, Fawu achieved national-level bargaining in the clear beer (as opposed to the sorghum beer) industry by pressing South African Breweries into negotiating at that level. As SAB has a monopoly in the industry it could agree to national-level bargaining without fearing any consequences from competitors.

Fawu has organised eight plants in the sorghum beer division, but as most factories are situated in bantustans, it has been difficult to co-ordinate this division.

Within the *fish processing sector* the pelagic fish industry has had industrial bargaining for the past 40 to 50 years. The reasons for this, according to Fawu's former general secretary, Jan Theron, are historical and have to do with a co-operative system worked out amongst companies which all have the same marketing aim. The employers' organisations in the industry, which all operate from the offices of the SA Inshore Fishing Industry Association, find it convenient to negotiate collectively with the union. In the pelagic fish industry Fawu therefore negotiates at an industrial level.

This level of co-operation is, however, not forthcoming in other divisions of the fish processing industry. Irvin & Johnson, for example, one of the three major companies in the industry, firmly opposed company-level bargaining with Fawu in 1987 - despite enormous union pressure on the company to negotiate wages and working conditions nationally. Fawu had encouraged shop stewards from as many plants as possible to attend every meeting with I&J. As a compromise the company combined negotiation at plants where it made business sense to do so, but held out on national-level negotiations with Fawu.

Reasons for this are, in the words of the group manager (personnel and industrial relations):

> because Fawu will have the whole of I&J by the balls. Furthermore, instead of having strikes in isolated places we would have national strikes. In addition regional wage differentials would be eliminated, thus dramatically reducing the company's ability to remain competitive in certain areas. This could even lead to closure of marginally profitable businesses.[26]

Fawu has industry-wide recognition in the *fruit canning* industry where negotiations had become institutionalised by means of a conciliation board. The union has an industrial agreement which has been in existence for five years. Industrial bargaining will also take place in the *fruit packing* industry where an employers' association has been formed as a logical extension of the union's pressure for such bargaining.[27] In the *tea and coffee sector* Fawu has been putting pressure on TW Beckett, the country's largest tea and coffee supplier, to negotiate nationally.

National-level bargaining

While Fawu has made some gains in national-level bargaining, most of its negotiations are still at the local level.

The question arises whether Fawu has made greater material gains through national-level bargaining than it would have through plant-level bargaining. This has most clearly been the case in Premier Food Holdings.

In 1987 Fawu negotiated nationally with three divisions of the company, Premier Milling, Epic Oil Mills and Epol. Negotiations, which commenced separately, ended in a national two-day strike in August in which all three divisions, involving approximately 4 500 workers in 19 factories, participated. As a result of this action the union won a 33% wage increase - from R120 to R160 per week for workers in the lowest grade.[28]

Apart from strike action, this large increase could also be attributed to an immaturity on Premier's side whereby it failed to go through a rigorous mandatory process before commencing bargaining.[29] Even so, such an across-the-board gain would be hard to envisage if bargaining had not been centralised and the strike not a national one.

In two other food sectors in which national bargaining takes place, wages are also high when compared to other wages in the food industry. In brewing the wages at SAB are amongst the highest in the food and beverage industry. The minimum, which applied up to June 1988, was R180,90 per week.[30] In the pelagic fish industry the minimum wage negotiated at industry level was R103,04 per week up to July 1988. Wage Determination 348 for the fish processing industry was R40 per week at the time.[31]

But national bargaining also places demands on the union. The challenges facing Fawu have been well summarised in the union's 1988 annual report:

> Presently we negotiate nationally with some companies, but we have not had much success in persuading more employers to negotiate at this level. Employers realise that if they negotiate nationally, workers can strike nationally. We should realise that this is not a demand that we are going to win easily, and where we have national negotiations we should learn how to use them. The negotiations are far more complex and demanding than plant level negotiations. Unfortunately there are too few people with the experience and skills to cope with national negotiations. In any event national negotiations will never be successful without a high degree of national co-ordination. It would be unwise to demand national negotiations where the co-ordination is not yet established.[32]

In 1988 Fawu did not make the headway it had achieved with

national bargaining in 1987. This was mainly due to stronger opposition from both management and the state, which put the union on the defensive. However, in addition an internal dispute within Fawu over the misuse of money and leadership struggles severely disrupted the union in 1988.[33]

In her article on collective bargaining in the food industry in South Africa, Kate Jowell has argued that the level of collective bargaining is the product of four issues: history; management's perceived best interest; the bargaining partner's perceived best interest; and the relative bargaining strength that either of the two main parties can use to secure their interest at any point in their history.[34]

This article demonstrates that Jowell's list is incomplete and that the category 'perception of the best interests' is too vague. Other factors which determine the level of bargaining are company policy and structure of the industry. This is well demonstrated by the vastly different levels of bargaining attained by Fawu in the Premier Group as opposed to the Barlow Rand companies. The variation between the two companies can primarily be ascribed to their different approaches towards management.

The multiplicity of bakeries in that industry ensured that both the union and the companies were satisfied with regional rather than national bargaining, although in the milling division the same union and companies negotiate at the national level. This demonstrates the importance of the structure of the industry in determining the level of bargaining.

For negotiations to take place effectively at national level, unions have to be well organised and co-ordinated at that level. Without this, national bargaining could create more disadvantages and problems than it solves for unions.

This article is too limited to try and tease out a general trend in collective bargaining as a result of the emergence of large and powerful national industrial unions. However, it is clear that a new form of industrial unionism is emerging that is not only profoundly affecting liberation politics, but also collective bargaining in South Africa.

Notes

1 I gratefully acknowledge the competent assistance of Jane van der Riet, my research assistant, who carried out extensive research for this article.
2 Fawu Annual Report, 1987; Bendix and Paterson, 'Industrial Relations Trends in the SA Maize and Wheat Milling Industry', *Industrial Relations Journal of South Africa*, 8(3), 1988, 33.
3 Labour Research Service, *Sector Review for Food and Allied Workers Union*, all sectors (each sector is published in a separate monograph), September 1988; Fawu Annual Report, 1987.
4 K Jowell, 'Future Collective Bargaining Arrangements in the Food Industry in South Africa', *Industrial Relations Journal of South Africa*, 8(2), 1988, 60.
5 Premier Group Document, 'Premier Food Industries Unionisation Matrix as at 16.05.1988'.
6 Fawu, 'Signed up membership by branch', (mimeograph), August 1987.
7 Labour Research Service, *Sector Review*, 'Milling', 1988, 4.
8 *Sector Review*, 'Milling', 6.
9 Premier Group Annual Report, 1988, 21.
10 Premier Group Annual Report, 1988, 3-4.
11 Interview with Jan Theron, Cape Town, 20 October 1988.
12 Barlow Rand Limited, *Annual Report*, 30 September 1988.
13 *Fortune*, 1 August 1988, Directory, D9.
14 Interview with Andre Lamprecht and Helgaard Bell, 24 November 1988.
15 Premier Group Annual Report, 1988, 17.
16 Interview with Felicity Steadman, Johannesburg, 25 November 1988.
17 'Representation by Premier Food Industries in respect of Premier Milling and Epol', 24 October 1988.
18 FJ Steadman, written communication with author, 1 February 1989.
19 Discussion with Tom Duff, Cape Town, 4 January 1989.
20 *Census of Manufacturing 1985*, Report No 30-01-01.
21 Labour Research Service, *Sector Review*, 'Baking', 1988, 2, 9.
22 'Baking', 3.
23 'Baking', 7.
24 Interview with Jan Theron, Cape Town, 20 October 1988.
25 Interview with Jan Theron, Cape Town, 20 October 1988.
26 Telephonic interview, Cape Town, 2 November 1987.
27 Interview with Jan Theron, Cape Town, 26 August 1988.
28 D Cooper, 'Premier Workers - Victory for the Living Wage Campaign', *South African Labour Bulletin*, 13(1), 1987, 1.
29 Interview with F Steadman, Johannesburg, 25 November 1988.
30 Labour Research Service, *Sector Review*, 'Brewing', 1988, 4.
31 Labour Research Service, *Sector Review*, 'Fish', 1988, 5.
32 Fawu Annual Report, 1988, 19.
33 A large part of the 1988 Annual Report of Fawu is taken up with a discussion of the 'intense struggle within Fawu'.
34 K Jowell, 'Future Collective Bargaining Arrangements in the Food Industry in South Africa', *Industrial Relations Journal of South Africa*, 8(2), 1988, 64.

Section 4: Rural Politics

Introduction

Edwin Ritchken

The mid-1980s were marked by political mobilisation and resistance unparalleled in South Africa's history, and rural and bantustan areas were not excluded from this process. KwaNdebele was wracked by a bitter and violent struggle against bantustan independence. KwaZulu has experienced the most sustained violence in the country. Venda saw a week-long stayaway - the first in the region's history - against government corruption. Kangwane was affected by education-related protests. In Lebowa, a series of targets ranging from chiefs to farmers and witches were attacked. A number of communities throughout the country opposed either removal or incorporation into a bantustan.

Although the rural areas were influenced by uprisings in the townships, their processes of mobilisation were structured by the specific history and political context faced by each region. This introduction explores some of the processes that shaped political mobilisation in the rural areas during the 1980s.

Labour tenancy, migrancy and forced removals

Struggles between labour tenants and white farmers reached a head during the 1960s and 1970s. At the beginning of the 1950s one-third of black South Africans lived on white-owned farms. However, a process of mechanisation together with legislation aimed at the abolition of labour tenancy saw the rapid implementation of capitalist relations of production on the white farmlands. By the end of the 1960s the absolute number of jobs in agriculture began to fall off. Between 1960 and 1974 at least 740 000 labour tenants were forcibly removed from white farmland, and in 1980 labour tenancy as a system was legislatively abolished.

Because of influx control many of those removed were forced to establish a residence in the bantustans before they could find jobs in the cities. However, this process was regionally uneven and it was only in the late 1970s and 1980s that Ndebele labour tenants in the Eastern Transvaal left the farms to move to the newly-established KwaNdebele. Despite concerted efforts to abolish labour tenancy, it is still being maintained in areas of the South-Eastern Transvaal, Northern Natal and the Orange Free State.

Tenants were not the only people forced into the reserves. By the early 1980s the inhabitants of most of the black-owned farms in the white platteland had been removed to the bantustans. The Surplus People's Project estimates that 475 000 people were removed from 'black spots' between 1960 and 1983. There are currently fewer than ten 'black spots' left in the Transvaal, all of which are resisting removal.

This huge influx of people into the reserves stretched rural resources beyond their limits. A significant portion of the reserve population found itself without a resource base of any note in the countryside. Migrants increasingly concentrated on establishing access to the urban areas and urban support networks, rather than the maintenance of any rural resources.

During the mid-1970s and 1980s, a new form of migration occurred within the bantustans, specifically KwaZulu. Migrants moved with their families into parts of the bantustans offering easier access to cities and reasonably stable residential rights. This was probably a result of the collapse of any reliable rural subsistence base in large parts of the bantustans, as well as the establishment of more secure urban employment in the towns.

In Molweni, a settlement on the border of KwaZulu, for example, the population increased from 33 012 in 1985 to 61 896 in 1987, an increase of 87,5%.[1] Prior to 1985, nearly 75% of all migrants coming from the rural areas to Molweni did so primarily to find access to employment. However, since 1985 there has been a change in the composition of the populati n, with over 50% of new residents being long-standing migrants and their families from rural areas.

Origins of migrants moving into Molweni[2]

| | Pre-1985 | 1985-88 |
	%	%
Township residents	26,7	15,4
Long-standing migrants with families	-	53,8
from the rural areas		
New rural/urban migrants	73,3	30,8

In KwaNdebele, the population grew from 50 779 to 261 875 between 1975 and 1984. Fifty-five percent came from white farming areas to establish access to migrant networks and educational facilities. A further 29% came from Bophuthatswana - mostly Winterveld - to escape harassment by Bophuthatswana authorities.

During 1985 the government announced that the policy of forced removals had been suspended. However, despite this announcement there have been a number of forced relocations, the most significant being from Crossroads to Khayelitsha. In the rural areas the state has changed its emphasis from forced removals to forced incorporation, whereby black people and their land are transferred into bantustans through the redrawing or redefining of the bantustan's boundaries. Newton, in an update on the state of removals in this section of the *Review*, points out the advantages this process has for the state.

However, the strategy of incorporation has also backfired in a number of ways. Paradoxically, incorporation has resulted in more violence than forced removals created. This is largely a result of harassment of those resisting the incorporation by bantustan authorities. In addition, communities have won a number of notable legal victories removing the threat of incorporation in some instances.

Kenyon and Du Toit's article on the Border corridor reveals a number of paradoxes in state policy, while posing a series of challenges to democratic residents' associations. At roughly the same time that Moutse was fighting against incorporation into KwaNdebele, the state reprieved eight communities along the Border corridor from the threat of incorporation or removal. This opened up a number of challenges to community organisations as they now had to maintain cohesion while negotiating with the Department of Development Aid around upgrading proposals. Thus far, state attempts to co-opt significant portions of the community, or impose unrepresentative structures, have failed.

The politics of contemporary rural resistance have been characterised by the creation of a number of seemingly anomalous alliances between various social forces, including traditional leaders such as paramount and ordinary chiefs, top bantustan government bureaucrats, tribal authority councillors, traders and businesspeople, civil servants, landowners, migrants, women and the youth.

Chiefs, and the creation of the bantustans

The implementation of the 'Bantu Authorities' system in the 1950s and early 1960s transformed chiefs into paid bureaucrats. However, whether a chief represented the state or community remained a keen site of struggle. The migration of adult males to the cities created a fundamental change in the balance of forces on the ground, and it was left to largely uneducated and traditionally subservient women to maintain a popular discipline on the chiefs. This opened up the space for corrupt and unaccountable behaviour by chiefs and their patronage networks.

The creation of 'homeland states' added a new dimension to this disintegration of traditional structures. It also facilitated the creation of a bureaucratic elite as chiefs, councillors, elected legislative assembly members and relatively well-educated technocrats filled the posts created by the establishment of these regional administrations. Bantustan government structures opened up new networks of patronage within the rural elites, and this created further problems of accountability at the local level as chiefs and other bureaucrats became responsible to these networks.

Exclusion from these networks entailed exclusion from whatever resources they could provide, and in some bantustans, they became formalised into cultural 'liberation' movements or organisations.

At a local level, these processes resulted in the creation of a substantial political vacuum. As paid bureaucrats it did not matter if chiefs served their communities or not. Chiefs were often more interested in regional issues and opportunities than local matters. This vacuum was exacerbated by an extremely vague division of responsibilities between local and regional government, resulting in a great deal of confusion amongst the bureaucracy as to where jurisdiction of particular tasks lay. In the absence of coherent procedures of accountability within the bureaucracy itself, a space was opened for government agents (from indunas to agricultural officers) to exert extremely oppressive bureaucratic powers.

Those who could assert the power to regulate people's lives emerged to fill the vacuum created by the above processes. However, this vacuum also provided the space for the creation of alternative forms of leadership and organisation, and resistance from below remains a threat to the abuses of chiefly power in rural areas.

The paramountcy has played a particularly significant role in certain rural areas. Bantustan structures often did not recognise the special status of the paramountcy, and it was from this position of powerlessness that paramounts forged a variety of alliances across the rural social spectrum.

The Ndzundza Ndebele paramount, for example, received no formal recognition from the South African government until the mid-1960s. The paramount had maintained a following by re-establishing initiation schools on the farms, maintaining traditional structures of justice and acting as ritual head of the Ndebele 'nation'. However, with the establishment of a formal Ndebele 'homeland', the paramount was marginalised from the political process by his councillors. In this context, the paramount formed a variety of genuinely popular alliances which resulted in the removal of 'independence' from the bantustan's agenda.

Traders and traditionalists

Small traders,[3] operating in areas with small disposable incomes, are one of the most vulnerable groups in the rural regions. The trader's access to capital depends on a family's ability to accumulate funds through informal sector activities, stokvels and migrancy. To obtain a business licence involves entrance into a local patronage network, usually a chief's council. The viability of the business then depends on the ability to monopolise and maintain a consumer base. However, in the case of a corrupt chieftaincy, traders are caught between the necessity of supporting the chief and maintaining a consumer base amongst those potentially antagonistic to the chief.[4]

Aspirant accumulators often find themselves in a political vacuum. In this potentially unstable context, businesspeople will attempt to assert some political leadership or articulate with whoever has fulfilled that function. Setting up of a church often provides the space for accumulation, the establishment of a local patronage network and political/moral leadership (functions that used to be fulfilled by the chief).

The establishment of separate bantustan governments opened up three access routes to capital. These consisted of well-paying jobs in the civil service, development corporations representing a new source of capital, and access to 'white' capital by allowing large supermarket, hotel or manufacturing concerns into the area. This latter route creates a close association between the policies and politics of the 'homeland' government and the would-be capitalist. Access to capital means accepting a particular ideology and the potentially unpopular political practices this entails.

However, the link between the local authority and the 'homeland' varies between and within the bantustans. In Kwa-Ndebele, members of the legislative assembly (who were chiefs' councillors) held an absolute monopoly on the above channels. In Venda, various chiefs in top positions of government reserved this right for themselves. In KwaZulu, Inkatha formalised networks of patronage within the region. In Lebowa under Phatudi, where no such formal networks exist, this process has been affected by a strong regionalism, resulting in extremely uneven regional links between local authorities, traders and the legislative assembly.

Given the necessity of maintaining a local political stance, and being especially vulnerable to popular actions, small accumulators are forced to tread a very tenuous path between a number of political forces. In this context, ideologies of ethnicity prove extremely useful to the trader in maintaining a sense of political solidarity with both a consumer base and local and regional government networks.

However, in a polarised situation this path may become untenable, and this creates the context for vigilante groups to protect life and property. This strategy only serves to polarise the situation further, and it is not surprising to discover that in the Kwa-Ndebele uprising against independence, 70% of the bantustan's businesses were destroyed. In Inanda and KwaMashu, some of the first targets to be attacked by the rampant youth were the shops of prominent Inkatha politicians. Conversely, in Lebowa, during a crisis in 1986, businesspeople in certain districts took an active popular role in alleviating the tense situation.

Land tenure, landlords and political violence

In his contribution to this section, Aitchison provides a broad overview of events and statistics connected with the civil war in Natal.
He gives a chronology of the war and attempts to quantify its effects,
and provides insights into issues like police collusion with Inkatha
and the interest of the state in maintaining the conflict, reaching the
following conclusions about the origin and dynamic of the conflict:

- a political dynamic, particularly at its inception, has been present
  in the conflict in Natal;
- urban township strife has spread outward into peri-urban and
  semi-rural areas and even the most conservative tribally-
  controlled territories have not been exempt from the turmoil;
- the violence seems to have a strong connection with Inkatha's
  attempt to maintain or increase its influence in the region;
- political violence is becoming embedded in the social fabric.

However, Aitchison does not attempt to analyse the specifics of
each local conflict in the area. Stavrou and Crouch, for example,
have described the central role land shortage and land tenure played
in the violence in Molweni, a community in KwaZulu.[5] As
mentioned above, Molweni experienced an 87,5% increase in its
population between 1985 and 1987. The Molweni area is broadly
divided into two sections, namely Upper and Lower Molweni.
Upper Molweni is characterised by an extremely crowded residential area, made up of original inhabitants' houses, lodgers in
individual rooms, and back-yard shanties (imijondolos) housing
whole families. Given an unemployment rate of about 38%, contributions from rents form an essential part of household income.
The chief plays a relatively passive role in this process.

A greater number of Upper than Lower Molweni residents are
employed in industries where trade union activities are prevalent.
And Upper Molweni has much greater infrastructural development
in the form of a water supply, electricity and schools.

On the other hand, Lower Molweni is less developed and has
maintained more 'traditional' land tenure with both agricultural and
residential land. The chief still plays an active role in the allocation
of land. The area is noticeably more traditional and is allegedly pro-
Inkatha. There is little doubt that the huge influx of people into the
area threatened both the viability of its agricultural subsistence
base, and the 'traditional' institutions that maintained it.

In this context, a threat of a take-over of Upper Molweni by the chief of Lower Molweni would have devastating repercussions to the upper area residents. The chief from the lower area has already tried to evict shack dwellers, migrant tenants and refugees. According to a survey in Upper Molweni, 14% of household incomes come from rentals. This led a resident to comment that 'imjondolos (shacks for rent) are important, if we could not have money from rents we could not survive, now there is talk that they might throw out our lodgers - that is bad.[6] A newly arrived migrant discussed his experience of being evicted from Lower Molweni, claiming that 'they surrounded the place - there was no escape - and walked about breaking windows and attacking people in their homes, saying they wanted the people gone and the land for grazing their cattle' (*STrib*, 04.09.89). Hence it was the threat of losing accommodation and with it access to the urban areas which placed newly-settled migrants in an antagonistic position towards the chief and his followers in the lower areas.

Similarly, the threat of losing tenants with their essential rentals placed the more settled residents of Upper Molweni in an antagonistic relation to those of Lower Molweni. It was this more immediate situation, rather than the ideological labels of UDF and Inkatha, that determined the nature of the conflict. As Stavrou notes, 'of those respondents interviewed, a fraction over one-quarter of the "vigilantes" and less than 20% of the "comrades" could identify the leaders of Inkatha and the UDF respectively; even less knew what each movement stood for'.[7] This is not to deny, however, that the terms 'UDF' and 'Inkatha' had powerful and significant local meanings to the actors involved.

In the Natal context, landowners have tended to show ambiguous political affiliations. Often freehold areas are ignored by the state, which refuses to take responsibility for their infrastructural development. This has resulted in the creation of huge informal settlements as migrants find accommodation outside of state controls. Landowners discovered that accepting first farming and then residential tenants onto their land was an effective means of ensuring a steady income. Increasing the population of the land also opened up new channels for accumulation through the provision of stores and other services. However, in these areas where there is no formal local government the issue of who owns and who can rent shack land and housing is both highly confused and highly contested.

Morris and Edwards have pointed out that it is this very process

which creates massive conflicts within shack-lands as various people try to establish control over areas of shacks. Gangs of unemployed people...become transformed into private armies which can be employed to dominate a particular area and assert the power of an individual who then gains income through controlling land and housing (*STrib*, 15.10.89).

These armies of unemployed are periodically mobilised to expand areas of control. In Newtown, Inanda, shack residents have attacked and in many cases taken control of, formal housing. However, the constant influx of people into the areas also ensures there can be no peace as shack-lords are constantly having to expand their areas of control.

This process has become politicised. In Inanda Rogers Ngcobo, chair of the 'Inanda liaison committee', a landlord association, made peace with Inkatha in 1986 after an initial uneasy relationship. It was through Inkatha and the KwaZulu government that he could get access to capital and formal recognition of his position in the area. He began to build a branch structure of Inkatha by appointing a number of councillors, who now form the nucleus of 'warlords' in the area. Violence broke out in the area in 1988 when a series of meetings were called to test people's loyalty to Inkatha. Landlord-tenant tension probably fed into the conflict.

However, changes related to land tenure are not limited to the peri-urban areas where residential struggles take prominence. An increasing amount of land is being taken over by bantustan governments or chiefs for management by white-owned companies. For example, in 1979, thousands of people were forcibly removed from Bethanie, Mangwe and Berseba in Bophuthatswana when the best tribal land in the area, covering more than 3 000 hectares, was taken over by Agricor.

A number of commissions of enquiry have recommended a removal of restrictions hindering the penetration of capitalist relations of production into the bantustans. In this section, Lala Steyn gives us a detailed account of the effects of the privatisation of land on communities living in the coloured reserves of Namaqualand. Her article also provides an account of links specific to the coloured reserves between local management boards and ministers in the House of Representatives, providing a useful comparison with the dynamics of the African reserves.

Ethnicity and ethnic organisation

In recent years radical historians have attempted to produce alternative understandings of ethnic or tribal identities. Instead of accepting allegiance to a particular tribe as a 'bottom line' of 'African' identities, these historians have attempted to show how ethnic ideologies are produced, changed and shaped by specific historical processes and actors.

Ethnic ideologies in South Africa often mould a group - on the basis of allegiance to a particular region - into a coherent grouping regardless of inequalities or status distinctions within that region.

To understand the specificity of a particular ethnic ideology, all the themes mentioned thus far have to be introduced: dispossession, labour tenancy, forced removal, migrancy, chieftaincy, changes to the rural economy and household structure, traders, landowners and regional statal structures. But although ethnic ideologies attempt to unify all social strata from a particular region into a single coherent body, this does not mean that the definition of ethnicity is passively accepted by all concerned. The definition of ethnicity is a site of struggle, related to the nature of the political practices and alliances within a particular region.

Maré and Ncube, in their article in this section, show how representations within Inkatha have attempted to forge a particular type of ethnic identity, namely a Zulu nationalist identity within the context of the KwaZulu regional government.

They demonstrate how Inkatha represents itself at three different levels in often contradictory ways. Internationally and nationally, Buthelezi presents Inkatha as a 'moderate' tribal yet nationalist movement trying to forge a negotiated non-violent settlement in South Africa. Regionally, Inkatha attempts to incorporate all members of the civil service, chiefs and traders by trying to provide the political framework within which this sector's economic aspirations can be realised. However, this process is not without tensions as these social sectors are subject to conflicting forces. At the level of the community, Inkatha functions through chiefs, traders and warlords. Here, it attempts to use the Zulu nation as its mobilising tool, and it is at this level that the meanings around Inkatha become the most diffuse as each area responds to a number of different processes in a contradictory way. Maré and Ncube show that Inkatha maintains its coherence only in so far as it is able to present itself as a totality at each of these different levels.

Ethnic identity is not created by petty-bourgeois myth-makers and passively accepted by an unsuspecting working class or peasantry. There are a variety of elements forging ethnic identity, which could include a definition of the chieftaincy, 'traditional' household relationships, land tenure, and the history, language and culture of the grouping involved. In a series of case studies, Sitas has demonstrated how a definition of ethnicity (in this case 'Zuluness') is forged from below depending on the specific historical experiences lived by a particular community.[8]

Black workers in the Howick/Mpophomeni area understand themselves to be 'Zulu' as defined by linguistic and cultural characteristics. Socio-politically, though, they see themselves as part of a broader *dispossessed* African nation. This is as a result of two major experiences of land dispossession: by white colonists and settlers, and eviction as labour tenants or farmworkers. Their area had close links with the Bambatha rebellion of 1905 and the ICU in the 1920s, and their relationship with the chieftainship was broken. As wage labourers, many suffered forced removals as they were moved up to three times before settling in Mpophomeni. There, in the 1950s, they experienced both unionisation and the defiance campaign. As a result of these experiences, the land question, white domination, and worker rights have a unique resonance, while 'Zuluness' is a relatively marginal political identity. Inkatha is extremely weak in this area.

By comparison, black workers in the Richards Bay/Empangeni areas understand themselves to be 'Zulu' in the sense of a separate and distinct people:

> They are tied together by a common culture and a prior political community (state) which was destroyed by imperialism; they strongly believe that the Zulu nation needs its own territory and a government that represents them; their political unity is mediated through the chiefs and they are available to Inkatha's non-class interpellations. In fact, the majority of them are members.[9]

In contrast to the experience of dispossession described above, most people in this area still enjoy access to land. Despite overcrowding, land remains a meaningful component of household life. The chieftainship and the system of headmen still function in these areas, and the process of proletarianisation has been much slower as migrancy is only now giving way to permanent urbanisation in border areas. Although at times Inkatha is shunned for violating unwritten codes of moral behaviour as understood by residents, as a vehicle for re-establishing 'traditional' homestead life it has a strong membership in this area.

In my article on KwaNdebele in this section, I draw on Peter Delius's history of the Ndebele to demonstrate how various definitions of 'Ndebeleness' were produced by different social sectors in the area.[10] In the absence of state recognition, the Ndzundza paramountcy, following the dispersion of Ndebele onto different farms in the Eastern Transvaal, retained contact with their subjects through initiation schools, a system of headmanship on the farms and the chief's court. From a position of powerlessness, a tradition of genuine consultation was maintained. Traditional institutions provided both a model of appropriate relationships on the farms and a vision of a brighter future in the re-establishment of a relatively autonomous Ndebele state. The establishment of Kwa-Ndebele in the 1970s fundamentally changed the life of the labour force that was previously trapped on the farms. Access to the cities, previously barred to the Ndzundza, was now a reality, as was a stable form of accommodation.

However, the establishment of the KwaNdebele legislative assembly meant an end to the benign rule of the 'king'. The councillors constituting the assembly saw the creation of KwaNdebele as a means of consolidating political power which would facilitate economic accumulation. Their definition of 'Ndebele' included absolute obedience to those in authority and the re-establishment of an independent Ndebele state, with the councillors as rulers. However, this definition of Ndebele ethnicity was not acceptable to the previously trapped farm labourers:

> The first image that comes to my mind with this idea of independence is I picture myself back to my first life under the farms. When I see where independence is coming from, ie what kind of people, it makes me feel sick and depressed... I could see myself working in the (KwaNdebele cabinet) minister's kitchens, doing their washing for them, my brothers working for them on the fields... Above all no-one under the king has problems with him, they support him, love him, he is always on the side of the oppressed... They can't do away with the king. We love him...unlike the so-called cabinet-ministers... (T)hey don't help the poor, they demand levies of the poor... (W)ith the king we create our own rules, we put the law onto ourselves.

'Ndebeleness' as defined in relation to historical experience from below, in contrast to the councillor's definition, places consultation before obedience and labour mobility before political autonomy. And it was in reference to this definition of 'Ndebeleness' that a broad alliance could be mobilised against the pro-independence movement.

Notes

1 S Stavrou and A Crouch, 'Violence on the Periphery: Molweni', Centre for Social and Development Studies, University of Natal, unpublished paper, 1989, 4.
2 Drawn from Stavrou and Crouch, 'Violence on the Periphery', 5.
3 I have deliberately not used the term 'petty bourgeois' here. Conditions faced by traders, bureaucrats, teachers and chiefs within and between bantustans differ enough to question the usefulness of this concept when trying to understand these groupings' political allegiances.
4 Based on a conversation with Peter Delius.
5 Stavrou and Crouch, 'Violence on the Periphery'.
6 Stavrou and Crouch, 'Violence on the Periphery', 9.
7 Stavrou and Crouch, 'Violence on the Periphery', 10.
8 A Sitas, 'Class, Nation and Ethnicity in Natal's Black Working Class', Sociology Department, University of Natal, unpublished paper, 1989.
9 Sitas, 'Class, Nation and Ethnicity', 21.
10 P Delius, 'The Ndzundza Ndebele. Indenture and the Making of Ethnic Identity', in P Bonner, I Hofmeyr, D James and T Lodge (eds), *Holding Their Ground. Class, Locality and Culture in 19th and 20th Century South Africa*, History Workshop 4, Johannesburg, 1989.

Forced Removals in South Africa

Deborah Newton[1]

Forced removals in South Africa have grown alarmingy over the last years. Three significant political developments substantially altered the terrain of both urban and rural removals. In September 1988 Chris Heunis, then minister of Constitutional Development and Planning, announced that the government still intended to move 24 000 people at a cost of close to R500-million. With this the government abandoned all pretence that forced removals were over and brashly admitted to the policy.

The second political development, the October municipal elections, manufactured a flurry of conservative local authorities which have altered the scene of urban struggles. These authorities have extensive powers regarding removals. Finally, amendments to the Prevention of Illegal Squatting Act have set up the administrative machinery necessary to make forced removals legal, swift and far-reaching in both urban and rural areas.

Farm evictions, demolition of squatter settlements, township removals and incorporation of land and people into bantustans have continued over the past few years. But communities threatened with removal have fiercely resisted state policy - with substantial success in some cases. The government met these successes with heavy-handed tactics. Anxious to deny a policy of forced removals whilst simultaneously attempting to legalise it, the government has introduced a range of legislation to deal with squatting, farm evictions, group areas and incorporation. Much of this legislation reduces the role of the courts while increasing that of administrative authorities.

At present about five million people remain under threat of removal. Of these the most vulnerable category remains farmworkers and their dependants. The supposed justification for pursuing this policy has become increasingly economic rather than political.

State policy

The policy of forced removals has reflected the government's increasing preoccupation with making acceptable the worst faces of oppression. During the election campaign in May 1987, then State President PW Botha declared that 'forced removals have been stopped'. In January 1988, Heunis modified this. In reprieving Bossiesgif, a community at Plettenberg Bay, he said there would be no more removals for political reasons - only development reasons. Couching old policies in new language is part of many reformist measures, intended to deflect public outcry against removals, both locally and internationally.

The massive relocation of people is presented as 'voluntary'. In some cases subtle pressures are exerted to force inhabitants to move. The local or central authorities discontinue services, increase rents, and make unrealistic promises of land and resources in re-settlement areas in order to force people to move. In other instances the pressures are overtly brutal. The most blatant example of this was the forced removal of 600 squatters from Noordhoek in the Cape. Faced with armed officials, police and a bulldozer, squatters reluctantly demolished their homes in the early hours of 2 December 1987. They were then loaded onto trucks and transported to Khayelitsha 35 kilometres away. Heunis later told parliament that 'when alternative sites at Khayelitsha were offered to the squatters on 2 December 1987, they packed their belongings and demolished their structures voluntarily'.

The government has concentrated its actions on groups which live outside the control of either local authorities or bantustan governments. Even where state attempts at forcing removal or incorporation have met with disaster, the government continues with its policy. In places such as Moutse, Peelton and Braklaagte, extreme violence accompanied incorporation into bantustans. Despite court decisions and strong opposition, the state continues with its plans to remove certain communities, and in 1989 drafted legislation which overturned court decisions limiting its powers.

The Prevention of Illegal Squatting Amendment Act

Amendments to the Prevention of Illegal Squatting Act were passed in parliament during 1989. The House of Assembly passed the bill in

1988 but because it was rejected by both the House of Delegates and the House of Representatives, it was referred to the President's Council. The President's Council raised serious objections to the bill but nevertheless recommended that it be passed. It appears that both the Transvaal and Cape Provincial administrations brought great force to bear to ensure the passing of the legislation.

Technically the amendments mean that more than five million South Africans are threatened with removal. This legislation provides for dramatic social restructuring and control in both urban and rural areas. As an alternative form of influx control it legalises forced removal and excludes the courts from any jurisdiction over the process. In placing the legislation beyond the rule of the courts, legislators have extended enormous powers to local authorities and administrators. There is virtually no provision for monitoring and policing local authority actions in terms of the act, and in areas with Conservative Party local authorities the consequences could be devastating.

Provisions aimed at farmworkers are among the more alarming aspects of the act. Only those employed have a right to stay on farms while dependants and retired farmworkers are classified as squatters. This provision is all the more disturbing in the light of South African Agricultural Union guidelines on squatting, issued late in 1988.

A du Preez, SAAU's media services manager, said the union and the Manpower Committee realised that 'squatting on farms is a serious problem'. He said the 1951 Illegal Squatting Act was difficult to implement and until such time as the amended squatter bill was passed, farmers were urged to report the presence of squatters - on their own or neighbours' farms - to the police or administration.

Since the amended act was passed there have been a number of prosecutions in terms of its revised section 1. In the Western Cape the regional services council has attempted to prove that Noordhoek residents entered and remained on land without the owner's permission. In the past the onus of proof rested on the owner. Now residents have had to provide evidence of their permission to be on the land. Should Noordhoek community members be found guilty the magistrate is obliged to order their eviction - regardless of whether there is satisfactory alternative accommodation or not.

Forced incorporation

Incorporation, the process whereby African people and their land are transferred into bantustans through the redrawing of boundaries, has had violent repercussions. It is carried out in terms of two central pieces of legislation: the National States Constitution Act which incorporates land into non-independent bantustans, and the Borders of Particular States Extension Act which enables the excision of land from South Africa which is then added to the independent bantustans.

Forced incorporation, unlike forced removal, is not a tangible process. It is carried out in Pretoria at the office of the government printer. If there is any consultation over incorporation, it is usually with imposed tribal authorities. Resistance to this process is therefore extremely difficult. As chief Pupsey Sebogodi of Braklaagte enquired, 'How do you fight the drawings of a pen?'

Recent incorporations have been much more violent than most forced removals. Invariably the bantustan authorities severely harass individuals or communities who have resisted the incorporation or refuse to show allegiance to the bantustan government. Harassment takes the form of assault, arrest, detention, and demolition of houses. Braklaagte and Leeuwfontein in the Transvaal, Peelton in the Eastern Cape and Botshabelo in the Orange Free State have all been subjected to this brutality over the last year.

The fierce anger with which communities fight incorporation is born of a deep-seated hatred of the bantustan system. There is strong political objection to the apartheid assumption that Africans should exercise their political rights in bantustans. In addition people fear the material deprivation and violent repression that bantustan residence means. As a Peelton resident put it, 'The homeland governments are so cruel that they steal from the old aged, they tax the orphans and widows. They deny us and our children access to the land of our ancestors'.

Once incorporated, resistant communities are frequently subjected to reprisals from bantustan authorities. Braklaagte and Leeuwfontein in the Transvaal, Peelton in the Eastern Cape and Botshabelo in the Orange Free State have all been subjected to this harassment for refusing to give allegiance to the bantustan authorities. These communities and others such as Matjakaneng in the Transvaal and Thornhill in the Eastern Cape met during 1989 to form a national network to publicise their common problems.

Peelton East, near East London, was incorporated into the Ciskei during August 1988. Since then the Ciskei government has continuously used its police and army to harass the Nkqonkqweni village community. Numerous residents have been detained and assaulted and police have threatened to kill community leaders. More than 100 residents have been charged over the last year with alleged offences ranging from drunkenness, to allocating land without headman status, to not paying Ciskei tax. Thus far, only one conviction has resulted.

Resistance to the Ciskei has largely been passive and disciplined. People have relied extensively on legal assistance and have sought relief from the Bisho supreme court on five occasions. They have repeatedly requested the South African government to protect them as they are South African citizens.

Events in Peelton came to a head in October 1989 when the area was cordoned off by the Ciskei police. More than 50 people were injured in conflict between residents and authorities. About 700 people fled to King William's Town as their homes in Peelton had been demolished by Ciskei authorities. As elsewhere, the central government has washed its hands of responsibility for these people, saying that it cannot interfere in the internal affairs of a 'foreign country'.

Events in Peelton have mirrored those in Braklaagte and Leeuwfontein during 1989. Braklaagte, 20 km outside Zeerust on the road to Botswana, is the home of about 10 000 members of the Bahurutse ba Sebogodi tribe, having been bought by them in 1907. For decades these people have fought for their land but despite repeated protest and broken promises the farm was incorporated into Bophuthatswana in December 1988.

In March 1989 violence erupted when the Bophuthatswana police and defence force set up a police camp in the middle of the village. A bus full of schoolchildren returning from Zeerust was stopped at a roadblock and the children were individually asked if they were citizens of South Africa or Bophuthatswana. Those who said they were South Africans were beaten by soldiers with rifle butts. Within a week many residents had been arrested and severely tortured. Many returned from detention with huge welts across their backs from beatings with sjamboks. The Bophuthatswana police finally charged 65 residents with public violence, arson and malicious injury to property.

The neighbouring farm of Leeuwfontein has faced similar harassment. On two occasions, residents fled to the farm of a white South African to avoid raids and assaults by Bophuthatswana authorities. In May 1989 two residents and nine policemen were killed in a confrontation between residents and security forces. The lawyer representing detained Leeuwfontein residents was denied access to his clients and the Transvaal Rural Action Committee and the Black Sash, who had been monitoring the situation, were banned in Bophuthatswana.

The South African government through its Department of Development Aid (DDA) under Gerrit Viljoen incorporated these areas into Bophuthatswana in the full knowledge that Braklaagte and Leeuwfontein residents were opposed to the moves. When the DDA was informed of events in Braklaagte, it referred the matter to the Bophuthatswana department of Foreign Affairs, thus continuing the South African stance that it cannot interfere in the affairs of another country - despite the fact that all these residents are still South African citizens.

In the cases of Moutse and Botshabelo, legal attempts to over-turn incorporation were successful but similar efforts by Braklaagte and Leeuwfontein failed. Botshabelo, a resettlement camp 60 km east of Bloemfontein, was partially incorporated into QwaQwa in December 1987. The incorporation was agreed upon by repre-sentatives of QwaQwa, Bophuthatswana and South Africa in 1977 before Botshabelo was established. When the incorporation took place ten years later, none of the approximately 300 000 residents were consulted. They were informed by pamphlets dropped by a helicopter that they were now to be partially administered by Qwa-Qwa.

This surprised and confused residents and officials. Clerks did not know which official rubber stamp to use nor whether to hoist the QwaQwa or South African flag above the magistrate's court! The partial incorporation seemed an acknowledgement by both Qwa-Qwa and South Africa that the poorest bantustan was incapable of servicing and administering the largest black township outside Soweto.

Although the supreme court ruled the incorporation invalid in August 1988, services have continued to be handed over to Qwa-Qwa while the matter is under appeal. Harassment by police and vigilante groupings continues. Opposition to incorporation comes not only from Botshabelo residents but also from white Bloemfontein residents. *Die Volksblad*, the mouthpiece for Bloemfontein's Afrikaner community, decried the incorporation as irresponsible because QwaQwa could obviously never develop the area and Bloemfontein would be left with dangerous and impoverished neighbours.

Despite opposition and the human costs of incorporation, the government still plans to incorporate further areas into bantustans. But the process is becoming politically expensive and increasingly difficult to justify in the light of reformist statements. Many of the communities affected by incorporation have long histories in those areas. This inevitably means a bitter struggle as they fight to retain access to and control of their most precious resource: land.

The Potsdam community was successful in resisting the imposition of bantustan authorities. Originally from Blue Rock, they were removed to the Ciskei in 1983. Since then the community has been harassed and attacked by the Ciskei, some residents even being murdered. Taxes and levies were extorted, people lost jobs and many died of tuberculosis. In January 1989 a South African supreme court finally granted the Potsdam community the right to permanent residence in South Africa. However, the court did not allocate the community any land in South Africa.

The Potsdam residents' association urged the community not to take the law into their own hands by leaving the Ciskei to squat in South Africa. Instead they began a process of deputations and petitions to various South African authorities, both local and national, pressing for land in South Africa.

Initial responses to their representations were positive. Then the state announced its intention to appeal against the court decision. Nothing further was heard in response to their representations. The community took the only option left them and marched out of Potsdam for the third time. On both previous occasions South Africa had forced them back into Potsdam.

The third exodus from Potsdam began in April 1989. Residents moved onto an empty farm in South Africa owned by the DDA. In the face of this situation, the East London office of the DDA informed the press that the minister had approved that the Potsdam squatters be settled on a portion of farm 303. By the end of June some 1 500 people had settled there.

After six years of severe impoverishment in the Ciskei, the Potsdam people are trying to break the cycle of poverty. As formal employment is unlikely, they are trying to become self-sufficient. Each family is to receive 2 000 square metres of land. This is based on studies conducted in the Ciskei which have shown that a family can subsist and produce a small surplus from this amount of land.

The Potsdam community has finally forced South Africa to acknowledge their claim to a place in South Africa. Although a significant victory, the battle is not yet over. The government has tried to limit the number of residents on the new farm to 2 000, whereas there are about 11 000 community members. And the government has not withdrawn its appeal against the supreme court decision allowing permanent residence in South Africa.

For those members of the original Blue Rock community still in Potsdam, life is severe. Vigilantes who collect taxes for the upgrading of the area have put further pressure on residents, and the leader of the vigilantes reportedly killed pigs belonging to the farm community. This is a major concern as the pigs are many families' main asset and source of income.

Although legal victories give impetus to people struggling against incorporation and removal into bantustans, they come with a price. South African legislators are quick to respond with new legislation or amendments to old, thereby closing gaps exposed in existing laws. Early in 1989 the Alteration of Boundaries of Self-Governing Territories Bill was tabled which, if passed, would have overturned the Botshabelo judgement. Although passed by the House of Assembly and rejected by the other two houses it was never referred to the President's Council, possibly because of the general outcry against it and the approaching general election. It has accordingly not become law.

Farm evictions

In the last two years evictions of farmworkers and their dependants have continued. Unlike the massive state-sponsored removals of the 1960s, recent evictions have tended to represent the private actions of farmers who wish to evict families. The agricultural economy is suffering severe pressures and farmers are facing an escalating debt. In 1986 farmers' debts stood at R12 446-million (a more than threefold increase since 1981). This debt was almost the gross value of total agricultural production for 1986 (R12 676-million). In the last couple of years the government-sponsored SAAU has declared that state subsidies for agriculture will decline as more free-market policies are introduced.

One of the most important factors in evictions of farmworkers has been the consolidation of farm-ownership and the accompanying amalgamation of smaller farms into larger single units. Between 1975 and 1983 (the last year for which figures are available) the number of farms fell by almost one quarter (22,7%). As size of farms increases so economies of scale come into effect; the total number of labourers required has fallen and mechanisation has become more profitable. Both factors mean that changes of farm ownership are often accompanied by evictions.

The situation has become so serious that the deputy minister of Economic Affairs and Technology has appealed to farmers to rethink their policies on mechanisation in the light of growing unemployment in the agricultural sector.

Other reasons for farm evictions include drought, changes in type of farming conditions, wages, working hours and evictions if labourers become elderly or too sick to work. Under the labour tenant system, which persists in Natal and the Transvaal, a contract exists between the farmer and a family. Thus an extended family can live on the farm as long as an agreed number of able-bodied members of that family work for the farm. However, in some cases, because of the highly exploitative labouring conditions, the sons and daughters refuse to work for the farmer and so the whole family is evicted.

Having received eviction notices most families are determined not to leave the farms with which they have been associated all their lives. Although the abolition of influx control in 1986 has meant that evicted farm dwellers are legally entitled to move to the cities, the difficulties of securing housing in the cities, the problems of massive unemployment, their own lack of skills, and the alienating experience of urban life for rural people make urban migration an extremely unattractive option.

This situation leads people to cling tenaciously to farms. Often stopping of wages has little effect and farmers resort to other means to enforce evictions. They use state backing to prosecute families either through civil proceeding or for illegal squatting and as a result farmworkers receive heavy fines or prison sentences. However, especially in areas of labour tenancy, families sell their stock to raise cash to pay fines.

If legal means have proved ineffective or the farmer becomes impatient with the slow legal process, other forms of harassment are employed. Common methods include the impounding of labourers' stock, the destruction of their houses and frequent police raids. Tenants' crops are destroyed, often by sending their own animals to graze them.

In the Weenen district of Natal, where over 2 700 people are threatened with removal, a number of families have reported that farmers have shot their dogs and impounded their livestock. Another family said a farmer had tried to have them arrested for 'stealing' firewood and water which they used on the farm. During legal proceedings against a south-eastern Transvaal farmer for illegally impounding cattle, the local deputy sheriff remarked that impounding cattle was a messy business and that most farmers in the district preferred to place a rope around the individual huts, attach this to a tractor and then pull them down.

Despite farmworkers' isolation, vulnerability to assaults, and lack of legal protection, there have been cases of successful resistance. On a government-owned farm, Elsenburg in Kraaifontein, workers were successful in stopping the eviction of a family. After 23 years of working on this farm a worker was told to leave because he was too sick to work. Shortly after this he died. After a few months the farmer tried to evict his widow and family. A youth group organised a petition against the eviction, but management ignored it. Eventually the farmworkers marched on the farm's offices in August 1988, and to avoid further conflict, the family was allowed to stay on the farm.

Township removals

Early in 1989 community members from Lawaaikamp (George), Port Nolloth (Namaqualand), Noordhoek (Cape Town), Kleinskool (Port Elizabeth), Cathcart (Eastern Cape), and Koster and Brits (Transvaal) met to discuss their common problems. For years these communities have successfully resisted removal.

During 1989 Port Nolloth and Brits won legal cases temporarily preventing their removal. The emergency camp which had been proclaimed in Brits was declared invalid and Port Nolloth staved off tent demolitions in terms of the Prevention of Illegal Squatting Act. Since then eviction notices due to be served on Port Nolloth residents have been withheld because the deputy sheriff for the area suffered a heart attack.

Kleinskool, a township occupied by both coloured and African people, has been declared a group area for coloured occupation only. Despite numerous threats, there have been no removals.

Lawaaikamp was due to be moved on 31 May 1988. But the community has engaged in legal battle with the George municipality which accepted early in 1989 that Lawaaikamp should be upgraded. However, the municipality has not dropped legal proceedings against residents who are adamant that they want to be in control of the upgrading process.

Removals, mostly of voiceless people concealed in small urban pockets or in remote rural areas, continue. For whatever reasons - conservation in Namaqualand and Ingwavuma, development in the southern Cape, or group areas in Hillbrow - the government continues to exercise the power it has given itself to restructure South Africa as it sees fit. This is not a consistent process and departmental policies differ regionally and even within and between the upper echelons of the DDA and the Department of Constitutional Development and Planning. To some degree these bodies are being forced to recognise the human and political costs of removal and there is a growing awareness that the struggle for land has to be confronted. In urban areas, residents want to stay near places of work regardless of group area and squatter legislation. In rural areas people refuse to leave the land they have occupied for centuries. By criminalising these issues the government has incurred huge legal costs with limited long-term gains.

Although still embryonic, attempts by communities threatened with removal and incorporation to liaise nationally are significant in making the plight of these communities known. They are attempting to show that removals and incorporation are not isolated incidents reliant on the whim of local officials.

As reform policies seemingly gain increasing credibility it is important that attention be continually drawn to some of apartheid's existing and most brutal elements.

Notes

1   This article is written on behalf of the National Committee Against Removals
     which consists of six affiliates: the Association for Rural Advancement
     (Pietermaritzburg), the Grahamstown Rural Committee, the Port Elizabeth
     Anti-Removals Committee, the Southern Cape Against Removals
     Committee, the Surplus People's Project (Western Cape), and the Transvaal
     Rural Action Committee. Information for this article has been taken from
     the research and publications produced by these affiliate organisations.

Privatisation and the Dispossession of Land in Namaqualand

Lala Steyn

In December 1987 a Namaqualand farmer, Willem Beukes, appeared before a Garies magistrate charged with assault with intent to do grievous bodily harm. Beukes told the court that Gert Links, a member of the local management board, had come to impound his stock. He hit Links three times with his kierie causing Links to run away. 'If I lost my stock it would be as good as losing my life', he explained.

The dispute arose because Beukes had been grazing his stock in a camp which had recently been privatised by the management board. Previously it had been open for all residents to use as grazing.

The court found that Links acted illegally in attempting to impound the stock without following the prescribed procedures. Links conceded he had not followed these procedures but insisted he had been acting in his capacity as a member of the management board which he described as 'autonomous'.

Beukes was acquitted. The magistrate accepted that he had used reasonable force to protect his property by hitting Links with a kierie.

This incident reflects the privatisation of communal land in the coloured reserves of Namaqualand. It illustrates the role which unrepresentative management boards have played in implementing the system and the ferocity with which farmers have fought to retain their land.

Land that had been communally farmed in three of Namaqualand's coloured rural areas, Leliefontein, Steinkopf and Richtersveld, was divided into 'economic' units which were leased out to individuals by the central government. This pushed the majority of people with communal sowing and grazing rights off most of the land, forcing them to use units that had not been leased out and the small commonage around their settlements.

The privatisation of land caused widespread dissatisfaction and impoverishment amongst people who rely on access to land for a livelihood and as a refuge from starvation when unemployed.

The Namaqualand reserves

Namaqualand is a sparsely populated and semi-desert area of 47 700 square km located in the north-western corner of South Africa. It has a population of 60 234 and contains 14 small urban settlements, six coloured rural areas and vast stretches of white-owned land and mining company property.

Nearly half the population is employed in the copper and diamond mines. Many workers have their homes in the reserves and go there at weekends or month-ends. Only 9% are employed by white farmers.

Africans are not allowed to live in Namaqualand except as labourers in mine compounds. The only African settlements are at Tentedorp and Bloukamp in Port Nolloth. Since the beginning of 1987 residents there have waged a courageous battle against a municipality which wants to deport them from Namaqualand because they insist it 'is for whites and coloureds only'.

Although Namaqualand is a vast rural area, planting of agricultural crops is limited by insufficient water supply and severe droughts. Farmers generally keep small stock, goats and sheep, and plant a little wheat, barley, rye and oats.

Employment is insecure because of fluctuations in the mining industry. According to NUM officials in Springbok this situation is very serious because smaller mines are closing and 'management has given the industry five to ten years before resources are depleted'.

The six coloured rural areas, commonly known as the 'coloured reserves', comprise 25% of Namaqualand. These areas - Steinkopf, Leliefontein, Richtersveld, Komaggas, Pella and Concordia - constitute 70% of the total of 1,7-million hectares making up the 23 coloured rural areas in South Africa. The population density is low and the total population of these reserves is about 27 000, a mere 0,2% of the total South African population.

The formation of the reserves

The formation of these reserves is part of the wider story of land dispossession. As the Dutch took land, Khoi-Khoi tribes moved northwards to Namaqualand. Two Khoi-Khoi/Dutch wars in 1658-1660 and 1673-1677, and a smallpox epidemic in 1713, devastated most of these tribes. By 1761 trekboers had settled in the Kamiesberg around which the Leliefontein reserve exists today. Between 1795 and 1847 the boundary of the Cape Colony shifted in several stages to the Orange River. Indigenous people were steadily squeezed off the land.

The reserves began as mission stations, which provided the indigenous people with some refuge from the encroaching trekboers. Leliefontein is on land which was inhabited by indigenous people and acknowledged as such by the Dutch in the eighteenth century. Komaggas was founded on crown land within the Cape Colony in 1831. Steinkopf, Concordia, Pella and Richtersveld started beyond the colonial boundaries and were incorporated in 1847 when the boundary was extended to the Orange River.

Many residents of the area refer to a time when their right to the land was recognised by Queen Victoria. This was during the time when the Cape Colony granted 'tickets of occupation' to the reserves, giving residents some guarantee of permanent occupation of the areas around the mission where they were staying. This system introduced racial classification, so that the ticket for Komaggas, for example, specified that inhabitants of the reserve should be 'aborigines or bastards of aboriginal descent'. By 1874 Komaggas, Leliefontein, Steinkopf and Concordia had received these tickets.

In 1909 the Mission Station and Reserves Act was passed. The communal system was retained but space was left for changes in land tenure to take place. An annual tax was imposed on all adult males in the reserves. This forced subsistence farmers to seek work on the mines but, as elsewhere, did not insist on the mines providing housing because workers retained homes in the reserves.

These reserves should not be seen as the coloured equivalent of bantustans, as residence there is voluntary. They are a product of the specific form of colonisation that took place in Namaqualand, rather than the result of grand apartheid bantustan policy.

The history of the reserves is one of dispossession, yet without them those affected would probably have become totally landless.

The reserves have been a refuge against the harshness of apartheid and have provided residents, who have a long historical tradition on the land, with social security and a sense of communal identity.

After 1909 the reserves came under the direct control of the state in the form of the Department of Native Affairs, with local magistrates as its agents. Local boards of management were set up, and chaired by magistrates. The reserves were passed from department to department until, in 1983, they came to be administered by the Department of Local Administration, Housing and Agriculture (DLAHA) in the House of Representatives.

Origins of the privatisation scheme

Land in the reserves was privatised in the 1980s. But the historical roots of privatisation lie in the rural development scheme for the coloured reserves first legislated in 1963. This made provision for settlement schemes whereby the reserves would be divided into residential and agricultural zones, and communal land would be reserved for the exclusive use of bona fide farmers. This division aimed to increase agricultural production, enabling farmers to farm productively without subjecting the land to overgrazing and soil erosion.

Farmers required a certain amount of money or stock to hire a unit of land. Only those with non-agricultural sources of income, such as shops, were able to gain access to land. The majority of peasant farmers were forced to live in residential areas and were denied access to land.

Opposition to this scheme was most intense in Komaggas. The community removed those members of the management board who showed sympathy for the plans, and took direct protest action against the superintendent of the reserve when he tried to by-pass the newly-elected board.

The 1963 law presumed that the millions of hectares of land locked up in the reserves should be opened to coloured farmers who wanted to farm commercially. Communal farming was regarded as backward and outdated by the authorities. In the report of the 1987 commission of inquiry into land use in Leliefontein, communal farmers were labelled 'backward traditionalists' while those who want to farm commercially were viewed as 'progressives'.

The House of Representatives has inherited this mode of

thinking. Many of the officials who served in the Department of Coloured Affairs presently work for the House of Representatives. David Curry, Minister of Housing and Agriculture in the House of Representatives, has been extremely unwilling to listen to the complaints of reserve residents who support communal farming. He has only responded to their letters when they have been very persistent or when they have won court cases.

Effects of the camp system

The way in which the camp system was implemented varied from reserve to reserve, but the effects and the forms of resistance were fairly similar.

Leliefontein reserve

The Leliefontein reserve spans an area of approximately 200 000 hectares and has a population of 6 000. There are ten settlements in the reserve. Until 1985 residents used the land communally to graze their stock. Many had rights ('toekenningsbriewe' - letters of allocation) to small patches of land for wheat cultivation.

These letters of allocation, which were issued to heads of families in the 1950s, confirmed family rights to occupy and use land which they had farmed in the past. Residents had sowing plots where they practiced dry-land wheat farming. Under the system of 'economic units' (commonly called 'kampe' or camps) they have lost access to sowing plots, which now fall into privately-owned camps.

The system of communal land tenure allowed any registered occupier to graze stock anywhere in the reserve. Most residents moved seasonally with their stock to winter outposts located ten to fifteen km from the village. They lived in reed houses ('matjieshuise') which were easy to move. Residents in the settlements have more solid structures and matjieshuise are used as outhouses and kitchens. But to use the land properly and because the rainfall is sporadic, stock grazers have to move with their stock. The camp system renders this impossible.

The DLAHA divided the Leliefontein reserve into 47 camps, 30 of which were leased out to individuals or small groups. To qualify for a camp a person had to have 250 head of stock or R3 000 in assets. Rent was R300 per year. The size of camps varied from 1 700 to 4 800 hectares. Of the 30 camps leased out, 18 were hired by people with other sources of income, such as shop-owners, management board officials, teachers and pensioners.

A system of patronage determined camp allocation. Those who were allotted camps by the management board were the wealthier members of the community. The severe hardships that people suffered when this system was implemented left the community with divisions still evident today.

The majority of residents were thus forced to graze their stock on the commonage around villages. They could no longer rely on wheat as a source of food and income, and it became increasingly difficult to obtain firewood as even this could not be collected in the camps.

Johannes Brand of Nourivier told how his stock of 200 goats and 100 sheep was reduced to 18 goats and 27 sheep over two years. He stopped lambing after he lost 300 lambs during those two years. The ewes did not have milk, and often died alongside the young lambs.

Sanna Joseph, who lives at an outpost outside Nourivier, had her 40 goats impounded when they wandered through a broken fence onto privately hired land. She had to pay R172,40 to get them back from the pound in Kharakams.

Residents are very concerned about the rumour that they are all going to be moved to three villages - Kharakams, Leliefontein and Rooifontein. Residents believe the authorities want them to move so that the residential areas can be centralised, leaving the rest of the land for commercial farming.

The threat of removal has been hanging over residents' heads for many years. But the authorities do not appear serious about physically forcing people to move in the near future. They hope that those living in the seven smaller settlements will move of their own accord when they see the improved facilities in the bigger settlements. For many years there has been little state assistance with water provision or refuse removal and people have been told that they build at their own risk.

Most residents rejected the camp system. Some continued to sow wheat on sowing plots for which they have letters of allocation - even though the fields fell within a camp. Others continued to graze on land leased by individuals. Many had animals impounded or were visited by police.

Frederick Cloete from the Paulshoek village refused to move off land he had occupied for years when it became part of a camp hired by the secretary of the management board. He received two letters from the secretary telling him to vacate the land but he refused to leave his vegetable plot and fruit trees.

When police from nearby Garies visited him he told them: 'Jy sal my voor die hek doodskiet, maar trek sal ek nie. Ek het die tuin so opgebou'. (You will have to shoot me at this gate, but I will not move. I have painstakingly built up my garden.)

Towards the end of 1987 the management board demanded that people using the communal camps reapply for grazing rights. Many applications were refused. This was an attempt by the board to control the few remaining communal camps and squeeze opponents off the land.

Since the camp scheme was first mooted residents have formed themselves into various lobby groups through which they have registered their opposition. They continually appealed to the local management board with little success. In 1982 they sent a petition to the board asking for the plans to be set aside because they had not been consulted in the matter and believed the new system would be their downfall.

These grievances were never properly addressed by the board. In October 1985 the board called a meeting, attended by about 300 residents who voted overwhelmingly against the system. This did not stop the board, which continued implementing it.

In May 1987 a more structured and representative organisation was formed to resist the economic units and deal with other community problems. By this time the management board had been rejected by the majority of the people. Two representatives from each settlement were elected or nominated onto the Leliefontein Community Committee.

In December 1984 Curry, the minister responsible for the area, visited Kharakams where a crowd waited to discuss the land issue with him. They later stated that they had been chased away 'like dogs' and that Curry had only held discussions with the board and a small group of dissatisfied residents. During 1985 and early 1986 residents sent delegations to Curry and officials in his department. In September 1986, following requests from residents, Curry appointed a committee of inquiry to investigate the land question in Leliefontein.

When the committee's report eventually became public late in

1987, it recommended that the land should not be returned. While acknowledging that many farmers rejected the camp system, the report explained this away by labelling them backward and traditional. Their inability to accept the system was dismissed as a consequence of their 'inherently weak character...including their laziness and alcoholic tendencies'.

The report stated that the communal system had led to overgrazing which would result in continued poverty. But botanists argue that whether the communal system causes overgrazing has never been properly researched. Tim Huffman, a University of Cape Town botanist, for example, has suggested that 'it may well be shown that the trek system per se is not the cause of veld degradation in the region and with reduced stock numbers, veld improvement may well be superior'.

Leliefontein residents rejected the report outright, protesting that 'the camp system is illegal because the burgers did not give their permission. We were not consulted in the matter'.

In July 1987, Leliefontein residents sought a court order declaring that they had been wrongfully deprived of their land and ordering Curry to restore it to them. In April 1988 the court found in their favour, albeit on technical grounds. Although the DLAHA lost this costly court case and suffered adverse publicity, it did not shelve its privatisation policy. In reply to a letter from Leliefontein residents to Curry in May 1988 he said that 'economic units for farming is the policy of the future'. Soon after the court decision Curry told the second conference of the association of management boards of rural areas (a co-ordinating body of local authorities in the reserves), that 'the economic units have come to stay and thus a new law will have to be made'.

On 3 June 1988 new grazing regulations were passed in Leliefontein. The regulations, which are currently also being considered for Steinkopf, stipulate that the grazing fee is to increase tenfold and people must now apply for grazing rights. If the application is successful then the applicant will be given a specific plot to graze upon. This arrangement is contrary to the rotational grazing system practised by stock farmers in the area.

Stock found grazing without the requisite permission will be impounded and the owner fined R50 or three months' imprisonment. Residents have refused to comply with the new regulations, which they regard as an extension of the economic unit system.

Some of the members elected to the management board in October 1988 are sympathetic towards the communal farmers, and the board has thus been hesitant to implement the regulations and act against those who defy them. Board members who support land privatisation are reluctant to take action because of the threat of another court case.

Steinkopf reserve

Steinkopf, with a population of about 7 000 and an area of 329 301 hectares, is the third largest reserve. It lies north of Springbok and extends up to the Orange River.

In the early 1980s the 'economic unit' system was implemented there without consulting the residents. There are 43 units of which 27 have been leased out to individuals. The remaining 16 (called 'vennootskapplase' or partnership farms) are worked jointly.

When fences were erected, those who traditionally had sowing plots or outposts in a specific fenced-off area automatically became partners of that farm, which became a vennootskapplaas. Those farming in one vennootskapplaas could move to graze in another if they had permission from the farmers of that area.

Steinkopf is divided into a winter rainfall area in the west and a summer rainfall area in the east. Most of the dry-land sowing is done in the winter rainfall area and stock needs to move between both areas for grazing, depending on the season. All of the vennootskapplase were in the winter rainfall area while all but five of the individually-hired farms were in the summer rainfall area. The farmers of the vennootskapplase were not allowed to graze their stock on the individually-hired farms which meant they no longer had access to summer rainfall grazing.

The biggest problem for vennootskapplaas farmers has been overcrowding and overgrazing. This, coupled with the lack of access to the summer rainfall area, has led to severe stock losses. Vennootskapplase consist of approximately 98 000 hectares for about 640 farmers while individual farms consist of approximately 200 000 hectares used by 27 farmers. Of the 27 individuals who hired farms, 11 had other sources of income.

As in Leliefontein, Steinkopf residents opposed the economic unit system. They formed the Namaqualand Residents' Association and petitioned local, regional and central authorities. A detailed memorandum setting out their grievances was sent to the minister of Agriculture in 1986. In May 1988 they brought a successful court application for the economic unit system to be set aside.

Steinkopf residents are threatened by the same grazing regulations recently passed in Leliefontein. Although Curry instructed the management board to accept the regulations, they asked his permission to make some changes. At a meeting in 1988, residents voted overwhelmingly against the regulations, but the issue remains unresolved.

Richtersveld reserve

The Richtersveld, which is situated in the north-western corner of the Cape Province just south of the Orange River, is the biggest coloured rural area. It is the most sparsely populated with a population of about 2 700 - less than 0,5 persons per square kilometer. The area is mountainous and rugged in the north, has less rainfall than Leliefontein or Steinkopf and is used only for stock farming.

The original population of Richtersveld were Nama-speaking herders descended from the Namaqua, a Khoi people. Although most people in the settlements speak Afrikaans fluently, Nama is also still spoken there today. Over the past 80 years the original inhabitants have lost more and more of the land they traditionally used for grazing.

In 1981 the Richtersveld board, dominated by residents from the south, accepted the system of economic units without consulting the community. The southern part of Richtersveld was divided into 37 camps, most of which were leased out. The north was also divided into camps but the people from Kuboes were totally opposed to this system and no one applied to hire units.

Opposition was so strong that in December 1986 Kuboes formally broke away from Richtersveld to form the Northern Richtersveld. They now have their own board which has refused to accept the unit system. But recently, they have shown more interest in doing so and a Northern Richtersveld community association has been formed as residents became unhappy with the board's stance.

It is not only the landless who object to the scheme: unit leaseholders in Eksteenfontein complain that units are badly divided so that water and seasonal grazing are unevenly distributed. They also say it is difficult to farm economically on the units.

Residents from the Northern Richtersveld have temporarily thwarted plans to declare more than 160 000 hectares of their land a national park. The National Parks Board and the management board were due to sign a 99-year lease, in terms of which stock farmers would have had to stop using the area for grazing, but a court application halted this.

Residents are not opposed to the formation of a national park if they are allowed to continue grazing their stock in the area. They also want to retain control over the land and over how the park is run. In terms of the proposed lease, residents will receive compensation but the land allocated for this is only about a third of the size of that intended for the park. Added to this, water in the area is scarce.

Northern Richtersveld will become a new border with Namibia, and there is also growing concern that the land might be expropriated for security reasons.

Moves to privatise land have caused impoverishment and dissatisfaction amongst Namaqualand residents. In an area where employment is unstable and scarce, land has provided many with a subsistence livelihood. The extent to which people have felt robbed of what is rightfully theirs is demonstrated by their sustained resistance to the process of dispossession. Since the system of economic units was declared invalid by the courts, land has been re-occupied, and residents are grazing stock and planting crops. Despite this, they know that their court victories were won because of the DLAHA's failure to implement the economic unit scheme in accordance with set regulations, and not because the scheme was regarded as unjust. They expect the DLAHA to pursue the privatisation of land.

Residents are aware that the land has to be protected from overgrazing and that some habits of the past will have to change. But they believe these changes can occur without the destruction of their lifestyle. They are determined to farm their land and plan their own development. Consequently they are establishing a regional community structure both within and outside the reserves to strengthen the struggle for their land.

The KwaNdebele Struggle against Independence[1]

Edwin Ritchken

Between May and August 1986 KwaNdebele was wracked by a sustained and bitter struggle against independence, which was further complicated and brutalised by the actions of a vigilante group, the Mbokotho. During this conflict an estimated 160 people died, many went missing and up to 70% of the bantustan's businesses were destroyed.

There was also a two-week stayaway by the entire civil service, a three-day stayaway by the bantustan's workforce, a sustained school boycott and the creation of a number of youth organisations, village committees and civic associations which spearheaded resistance to independence plans.

Unlikely alliances were established both among those who supported and those who opposed independence. Students, youths, migrants, women, bureaucrats, teachers, peasants and some chiefs formed the anti-independence bloc, which found its leadership in the Ndzundza royal family. The pro-independence bloc was formed by a grouping of shop-owners, taxi drivers, tribal authority councillors, squatters, the Manala royal family and certain chiefs, and was led by senior cabinet ministers in the KwaNdebele legislative assembly.

Vigilantes, students and the prince

Mobilisation in KwaNdebele occurred largely as a result of the KwaNdebele government vigilante group, the Mbokotho, which forced the bantustan's populace in general, and the youth in particular, to organise a campaign of mass resistance in defence of their lives and property.

The Mbokotho were officially formed in late January 1986. Chief SS Skosana was appointed president of the organisation, with Minister of Interior Piet Ntuli as vice-president. The aims of the organisation included protection of the 'interests of the community', dealing with people who enforced boycotts, solving problems between preachers and their congregations, looking into problems affecting family life, and dealing with any 'troublemaker'. The Mbokotho could 'fetch such a person from the police and hit him'.[2]

The vigilante group had been active in various other forms before it was officially launched. As early as 1975, councillors at Pungutsha, Manala, Litho and the Ndzundza tribal authorities had attacked a political opponent within the Litho tribal authority. In 1977, a member of the royal family, Makhosana Mahlangu, was publicly assaulted and humiliated for a number of days because he refused to adopt a particular political position. During 1980, a vigilante force in Kalkfontein was used to crush resistance to a change in the local administrative structure from community to tribal authority. When conflict around the incorporation of Ekangala into KwaNdebele emerged in early 1985, prominent community leaders at Ekangala opposed to incorporation were abducted and assaulted.

The Mbokotho emerged publicly in KwaNdebele in December 1985, when they began a mass recruitment campaign to attack the 'Pedis' at Moutse. During the same period the Mbokotho started to attack real and imagined political enemies within KwaNdebele. Makhusana Mahlangu was attacked on 19 December for, amongst other things, his anti-incorporation stance on Moutse. On 25 December a group of youths from Mamelodi were assaulted and abducted while picnicking. They were accused of being Cosas members and of wanting to influence the children of KwaNdebele to revolt against their parents.

It was clashes with the youth in general, and schoolchildren in particular, which brought the matter to a head. On 28 February 1986, at Mmashadi High in Siyabuswa, the principal called in the Mbokotho and police to 'discipline' students after a meeting they had held the previous day. Students were tear-gassed and sjambokked. Five students and five teachers were abducted and assaulted in the Siyabuswa community hall.

A class boycott in mid-April 1986 at Benginhlanhla high school was launched to protest the absence of free textbooks. The Mbokotho arrived and sjambokked the students. Ten were abducted and tortured. On 18 April students at Mandlethu high school marched to the local headman to complain about the activities of

the Mbokotho. Their objections to the vigilante group revealed the extent to which the Mbokotho had offended the populace's sense of justice and hampered its ability to lead day-to-day life:
- the Mbokotho were preventing students from studying at the school at night. This was severely affecting students' studies because of a shortage of textbooks and overcrowded households, which necessitated communal studying on school premises;
- migrants who returned home late at night were being assaulted and many were robbed of their pay;
- students were being forced to join the Mbokotho against their will.

Numerous other examples of Mbokotho atrocities were publicised. For example, a youth who had been sent by his grandmother one night to inform relatives of the death of his sister, was brutally assaulted.

Executive members of the Mbokotho were present and promised to report back on 28 April. When there was no reply from Mbokotho leaders, students went on a protest march to the homes of leaders - most of whom were members of the legislative assembly. However, they were met on the road by truckloads of vigilantes who attacked them. Police, who fired tear-gas and rubber bullets at the group, had to free 30 students captured by the Mbokotho. That evening vigilantes went on the rampage in Vlaklaagte, abducting students, whom they took to a hall and tortured. The same evening the Mbokotho also abducted a parent, Jacob Skosana, who was tortured and killed. When his body was dumped at the entrance to his home, vigilantes lit a fire at the gate to ensure that no one took the body inside.

These events formed the backdrop to PW Botha's 7 May announcement that KwaNdebele would become 'independent' in December.

The royal family, who had not been consulted about the decision to opt for bantustan independence, called a meeting of their subjects and headmen for 12 May. Between 20 000 and 30 000 residents attended the meeting which drew up three main demands to be conveyed to the KwaNdebele government by the three cabinet ministers present. The demands were that Chief SS Skosana should withdraw his acceptance of independence; the Mbokotho should be dismantled; and all MPs and cabinet ministers should resign. A report-back meeting was scheduled for 14 May.

The local magistrate had refused the Ndzundza tribal authority permission to hold the report-back meeting. However, some 25 000 people, unaware of the banning, arrived at the royal kraal. Before the meeting could begin, the police and army, using casspirs and a helicopter, fired rubber bullets and tear-gas canisters into the crowd. At least three people died and many more were injured. Youths burnt down the homes and shops of a number of MPs in retaliation.

After this event, many students stopped attending school and took to the bush to avoid capture by the Mbokotho. However, they did maintain close relations with the royal family, local villagers, and parents, and were allowed by some headmen to use their tribal offices.[3]

In the following two weeks businesses and homes belonging to 41 of KwaNdebele's 72 MPs were burnt down. Successful stayaways were held on 3-4 June and 16 June. Meanwhile, violent conflict continued between the youth and South African security forces and the Mbokotho.

By 16 July, according to the Transvaal Rural Action Committee (TRAC), the tide had turned against the Mbokotho. Many vigilantes were dead or in hiding. According to one informant, this change was due to the SADF taking a less partisan stand after six people had been killed in a 12 June attack led by Piet Ntuli. The National Security Management System also allegedly intervened from the most senior levels. And by this stage police appeared to have withdrawn some of their support for the Mbokotho.

On 15 July, the entire civil service started a two-week stayaway. However, it was the death of Piet Ntuli on 29 July that marked the turning point in the violence. By this stage about 70% of the businesses in KwaNdebele had been destroyed. According to a priest who kept a diary of deaths in KwaNdebele, about 160 people died between 12 May and 25 July and well over 300 KwaNdebele residents were detained. Over 50 people had been abducted and tortured by the Mbokotho. On 12 August the legislative assembly banned the Mbokotho and cancelled the proposed independence plans.

Throughout the conflict, the royal family, and Prince James in particular, played a crucial role. Prince James was heir to the Ndzundza throne and chairperson of the regional authority: this gave him access to, and the respect of, older residents. He also had a university qualification, gaining him the respect of the youth and civil service. The royal family had access to the youth, the civil service, surrounding white farmers and central government and security force members. The royal kraal was also used as a crisis centre during the struggle.

A number of informants mention the powerful effect the actions of the Mbokotho had in mobilising people. According to a Kwa-Ndebele businessman,

> I am always differing with people who are saying that KwaNdebele people fought against independence. Yes, they fought against independence because they did not know, nobody told them about independence. They were never asked or told about it, they were just told we are going to be independent. It was an instruction from somewhere, they were not asked their opinion... All we practically saw was the Mbokotho molesting people so people revolted against that. I believe that the Mbokotho was formed to force people into independence. As simple as that. And that was what sparked the whole thing in KwaNdebele.[4]

The mobilisation in KwaNdebele created a consciousness of issues beyond the bantustan's borders. 'All of us, the black people in South Africa, we are suffering. ...(W)hen I think about it I become mad, they just shoot at us nakedly. This...is where I saw that the black people are really under heavy oppression. We do not have life here in South Africa'.[5]

At this stage a number of questions remain unanswered. Was there more to the independence issue than opposition to the Mbokotho and the people behind them? Who exactly were the Mbokotho? Why did the royal family play such a leading role in the opposition to independence? To answer these questions, it is necessary to look at the establishment of KwaNdebele and what it meant to all the different social groupings within the bantustan. To do this the specific historical experience of the Ndzundza paramountcy and their subjects needs to be examined.

The establishment of KwaNdebele

The Southern Ndebele divide into two main groups - the Manala and Ndzundza chiefdoms. Both groupings were deeply affected by the Difaqane. The Manala chiefdom barely recovered and by the 1870s its remnants were living on the Wallmansthal mission station and surrounding Boer farms. The Ndzundza, on the other hand, emerged as a significant chiefdom in the 1830s with the establishment of a series of almost impenetrable mountain fortresses. Early Boer settler attempts to extract land or labour from the chieftaincy proved unsuccessful and by the 1860s the remaining Boers had recognised the authority of the Ndzundza king over a considerable area of land.[6]

However, Ndzundza hegemony was short-lived. With the re-establishment of the Zuid Afrikaansche Republiek (ZAR) in 1881, conflicts between the Ndzundza and the Boers around land, labour and taxes re-emerged. The matter came to a head in 1882 when the Ndzundza gave shelter to Mampuru who was wanted by the ZAR for the murder of his brother, Sekhukhune. A war of attrition followed with the Boers adopting a siege strategy.

By mid-1883, widespread starvation resulted in the surrender of the Ndzundza king, Nyabela. Nyabela and 22 Ndzundza royals and subordinate chiefs were taken captive to Pretoria. Ndzundza land was distributed among the Boer commandos involved in the war. Ndzundza subjects were allocated to Boers throughout the ZAR for a five-year indenture period. In return farmers had to pay the state a tax of five pounds on behalf of each household head.

There was little change in the position of the Ndzundzas after the formal period of indenture was completed. No land was set aside for Ndebele settlement. Farmers were also reluctant to give their labourers permission to leave. Reserve areas were effectively shut off to the Ndzundza and were monitored by the state through the co-operation of certain chiefs and a network of informants.

The squatter law of 1887 prevented Ndebeles settling on unoccupied land in large numbers. However, despite the constraints, there is evidence to suggest that there was a significant movement of Ndzundzas between farms both during, and especially after, the formal period of indenture. They attempted to migrate to bigger farms where there would be fewer labour demands. Movement between farms was aimed at regrouping homesteads and the re-establishment of kinship networks. There were also attempts to return to farms in the Middelburg and Pretoria districts in order to revive wider social networks.

Compared with other 'ethnic' groupings, there were a number of factors limiting the choices available to Ndzundzas. They began their period of indenture without capital. Hence, when the period of indenture was completed they were not in a position to compete with established tenants on the larger farms. As a result they were forced to stay on the smaller, more isolated farms where there was a year-round demand for each family member's labour. This prevented them from establishing migrant networks through the migration of members of the family who were not part of a tenancy agreement.

As conditions on farms grew harsher with the development of a more capitalised agriculture, the Ndzundza did not have the option of settling in a reserve the way, for example, their Pedi counterparts could. On the reserves, the Pedi chiefdoms retained some of their coherence and power until at least the 1950s thus ensuring Pedi settlers access to networks of political authority. As the 20th century progressed, there was probably an exodus of Pedi tenants to the reserves, with their promise of some subsistence farming, grazing land and relatively easy access to migrant networks.

The Ndzundzas, on the other hand, were forced to remain on the farmlands where conditions were getting harsher and harsher:

> The system worked this way. You worked six months without pay for staying on the farm. Families would rotate the six months so people could have the six months paid. As time goes on the system of six months payment changed from money to a bag of mealie meal per month. This system was unsatisfactory so people tried to get into town, but because of influx control people found it very difficult... The worst thing is even if a brother got married and he brought his wife home, the farmer would bring a tough rule that the brother must now do the six month system. If the husband works for six months, then the wife must work the next six months.[7]

This is the context within which initiatives by the Ndzundza royalty to re-establish themselves in the late 19th century and throughout the 20th century should be understood. Nyabela's brother Matsitsi escaped from prison. He went to Kafferskraal where his family was living and managed to convince the white farmers to allow him to rule in Nyabela's place. He then called a meeting of Ndzundzas and informed them that Nyabela had given him a mandate to rule. This was accepted.

The Ndzundzas probably placed their hope of regaining access to land and some political autonomy in the royals. The royals made a number of attempts to purchase land in the early 20th century but always defaulted on payments. It was only in 1923 that they were able to secure a small independent base at Weltevrede where they are presently living. Pressure from farmers ensured that chiefly requests for land from the state were routinely rejected.

Matsitsi re-established a number of crucial chiefly institutions, the most important of which were probably the holding of male initiation and the formation of age regiments. By 1886 the first initiation school had been held.

The Ndzundzas also re-established a system of headmanship on any farm where there were a significant number of Ndzundza households. Although this move met with initial resistance from farmers, a compromise was reached whereby workers would elect a headman who had the dual role of acting as a foreman for the farmer and traditional official for the workers. Headmen were to communicate any problems their subjects were facing to royalty. They also had to officiate over 'traditional' proceedings and settle inter-household disputes. If parties were dissatisfied with a headman's judgement they could take the issue to the royal court for a final ruling.[8]

Later, in cases of severe abuse of tenants by farmers, the royals employed lawyers to represent tenants. Thus, a tradition of genuine consultation and communication with the paramount was established. In return - and in the absence of any power to do otherwise - the paramount played a relatively passive role in the taking of decisions, often 'rubber-stamping' rulings according to 'traditional' administrative and judicial procedures.

A tradition of meeting yearly at the site of the Ndzundzas' defeat in 1884 was also established. Ndzundza headmen from round the country attended the ritual where the battle was commemorated and the dream of re-establishing an Ndzundza state was rekindled.

Alternative cultural institutions and practices were blocked off to the Ndzundza, resulting in a further reinforcement of 'traditional' values and institutions as both a model of present appropriate behaviour and a vision of a brighter future. Farmers resisted any signs of 'Westernisation' in their tenants. In the early decades of the 20th century the Ndzundza wore 'skins... You were not even able to wear a pair of trousers. You dare come onto the farm wearing...clothes - you would be shot dead! A kaffir wearing clothes - No!'[9]

Migrancy was also actively discouraged by farmers, and tenants believed that farmers would go to nearby towns and tell people not to employ their workers during the six months in which they were not paying labour rent. If one was lucky enough to get a job in the town 'the farmer would say you think you are white, you think you are the boss if you get a good job in towns. He would taunt you and make your life a misery'.[10] Christianity was also out of bounds to the tenants:

> The father of the former chief magistrate of Kwaggafontein (in KwaNdebele) was sjambokked for holding a church service after Dingaan's Day. The Boer asked him 'What are you asking God for? You are confusing Him with your requests. We are asking him for power so you cannot ask him for it too'. The Boer then sjambokked him. It was very risky to hold a church service near those farms.[11]

By the end of the 1950s the Ndzundzas were faced with a situation where life on the farms had become intolerable. Cash payments were replaced with payments in mealie meal. Influx control was tightened, making access to the cities almost impossible. The youth were resisting farmwork and demanding access to education, a right that was stringently denied them on the farms. As a result there were constant desertions by youths, placing great strain on the relationship between household heads and farmers.

This was accompanied by an increase in the process of mechanisation on farms, making more and more labourers redundant. The combination of redundancy and youth resistance resulted in constant evictions. Those evicted could try to find employment on other farms, settle in already-established bantustans with no real access to administrative or political authority, or obtain access to urban markets by squatting outside cities.

It was in this context that Poni (Matsitsi's grandson resident at Nebo) accepted some land near Nebo and tribal authority status within Lebowa in 1959. This led to a split between Poni and Mapoch, the Ndzundza king at Weltevrede, who continued to insist that the only land that would be acceptable to the Ndebeles was Mapochsgronden, the site of the 1884 battle.

Members of Poni's tribal authority went from farm to farm informing people that land was available for Ndebeles in Nebo. Makhusana Mahlangu, a member of the royal family, was arrested in 1965 for encouraging people to leave the farms illegally. However, by the 1960s most of the land for agricultural use within the tribal authority's area was already allocated, hence the necessity for the more recently settled households to have migrant labourers. As a means of helping these people get access to the cities, members of the royal family arranged with a northern Ndebele chieftainess, Esther Kekana - whose tribal authority was located at Majaneng (near Hammanskraal in Bophuthatswana) - to allow people to squat on her land.

In 1967 Mapoch, the present Ndzundza king, accepted tribal authority status under Lebowa at Weltevrede. In 1969 there was a meeting of both northern and southern Ndebele chiefs to discuss the establishment of an Ndebele bantustan. Chieftainess Kekana, who hosted the meeting, was involved in constant conflicts with the Bophuthatswana authorities under whose jurisdiction the tribal authority fell.

In the early 1970s the Department of Bantu Administration and Development began responding to the demand for an Ndebele bantustan. This was probably due to the growing squatter problem the department was facing from evicted Ndzundza tenants. There was also a decrease in the demand for farm labour during this period, although farmers had previously been the most vociferous opponents to the establishment of an Ndebele bantustan. By 1974 a number of farms totalling 51 000 hectares were added to Welte-vrede. By July 1974, the Ndzundza tribal authority had been excised from Lebowa, and under the paramount chief, had received the status and functions of a regional authority. Skosana, the future chief minister, was chair of the regional authority.

The population of KwaNdebele grew from 50 779 in 1975, to 166 477 in 1980, to 261 875 in 1984. Fifty-five percent of those who arrived in KwaNdebele during the previous five years came from white farming areas, 29% from Bophuthatswana and 5% from Lebowa.[12]

For the previously trapped farmworkers, the creation of Kwa-Ndebele was popularly received. According to an ex-farmworker,

> the structure of headman was carried on but did not solve the immediate problems the farm labourers were facing on the farms (working conditions)... These grievances built up tension and the people could not take it any longer. They would take it to the headmen and cry to them. All these grievances would go from the headmen to the king. And then one day the king answered our prayers. He addressed us to say he has got a piece of land for people to stay. My brothers came here to investigate under what conditions we would be living here. Each and every married couple and mother was entitled to their own stand. People thought they would work for the king, but the king said it did not work like that. All you need to do is pay R30 and R5 registration fee. When my brothers came back and told us of the situation here the whole

family felt as if there was freedom... The king lent us two trucks and a tractor
to transport our goods... When we arrived at the village the headman asked us
where we wanted to settle. My brothers quickly built a house with thatch and
then went to pay their respects to the king who said to them 'Welcome home'.
They asked the king for permission to find work in the towns. The king said
'With pleasure'. That day was the best day of my life.[13]

Access to education was another crucial issue which attracted
people to KwaNdebele. According to the above informant: 'There
was a lot of conflict (on the farms) particularly over education, but
farmers would refuse all education. No one would put a school on
their farm. The youths are the ones who really pushed their parents
to leave the farms. Schools are built here and this made the children
wild. Many parents followed their children here'. And the change to
the labour force was dramatic: 'We were fortunate that all of us got
a job and our life improved completely. I felt free as if I had moved
from the darkness into the light. The family worked hand in hand to
help one another and their mother. Before the Mbokotho there
were no problems, no strict laws from the king, we lived in peace'.

It is also important to document the experiences of those who
moved from Bophuthatswana. By 1978, non-Tswanas were suffering
severe harassment from the Bophuthatswana police for squatting
without permits and not taking out Bophuthatswana citizenship.
There were constant threats of expulsion. Residents had difficulty
getting their pensions and work permits. Government subsidies for
schools had been suspended since 1976. By 1979 the Department of
Co-operation and Development said that 10 000 families had been
relocated in KwaNdebele from Winterveld.

However, one should be careful not to romanticise the situation
that confronted newly arrived residents. According to a Surplus
People's Project survey conducted on 533 people at Kwaggafontein,
facilities in KwaNdebele were very poor when people arrived there.
'More than 90% had no water, latrines or local authority. More than
60% had no shops, schools or clinics. No fuel was available to 59%
when they came'.

Splits within the elites

In 1977 three tribal authorities in the Hammanskraal district in
Bophuthatswana - the Litho under Lazarus Mahlangu, the Pungutsha under Isaac Mahlangu and the Manala under Alfred Mabena -

seceded from Bophuthatswana with the land and people under their jurisdiction, and joined KwaNdebele. These three tribal authorities combined to form the Mnyamana regional authority, and formed part of the South Ndebele Territorial Authority.

These developments were not without tensions between the Ndzundza royalty and councillors in their regional authority. Skosana was replaced as chair of the regional authority in 1976, as he was dominated by two central government employees, Dr Bothma, an ethnologist, and a Mr Victor who was supervising the independence process in KwaNdebele. Bothma had also been intimately involved in Bophuthatswanan independence. According to a member of the royal family, 'Skosana was accepting the separate development policy which we were against because it would destroy the paramountcy'. When Makhusana Mahlangu, a royal family and territorial authority member, queried the constitution of the territorial authority, he was brutally assaulted and publicly humiliated by a group of people led by Skosana and Ntuli. He fled to KwaZulu for the next seven years.

With the establishment of a legislative assembly in 1979, tensions in the agendas of some of the Ndzundza chiefs and their councillors began to emerge. The legislative assembly involved a 46-member body with a six-member cabinet appointed by the chief minister. All 46 members, none of whom were chiefs, were nominated by the four tribal authorities. However, once nominated, a tribal authority could not recall the MP. Only the assembly itself could remove an MP.

The chief minister also had the right to appoint or remove chiefs:

> In the beginning Skosana was very loyal to Chief Mapoch... After 1979, when KwaNdebele got legislative assembly status and Skosana got to know his rights, that he had more power than the chiefs, he started ignoring his base. Initially any decision relating to government would not be taken without the approval of Chief Mapoch; he would consult the chief and the chief would give a green light. Later, when Skosana realised his right he could take decisions without consulting the chief. He dictated some things over the chief's people... He never consulted and some careless people aired this thing: the chief minister is above everybody and the chief minister can do anything above chiefs, even dispose of them, and the strain increased.[14]

In 1982 tensions increased when Prince James became chairperson of the Ndzundza regional authority. He had a university degree and a sense of the national situation, whereas the top cabinet ministers were largely very poorly educated and were dominated by officials from Pretoria. According to a close associate of the Prince: 'The way they saw you. You are either part of the system or part of the ANC and the communist onslaught. There was no middle ground'.

By 1983 an attempt had been made by legislative assembly members to sideline the royal family in particular, and the chiefs in general, by creating a separate body for them called the Ibandla Lamakhosi. By early 1985, the split between 'traditionalist' chiefs and the legislative assembly became apparent when chief Lazarus Mahlangu of the Litho tribal authority wrote a letter in which the tribal authority stated that it wished to excise itself from Kwa-Ndebele and rejoin Bophuthatswana. Mahlangu was a Ndzundza traditionalist who had seceded from Bophuthatswana in 1977.

Reasons given were that Skosana's administration interfered in 'tribal affairs' and dictated to, rather than consulted with, the tribal authority. A symptom of this subordinate relationship was the tribal authority's desire to replace its nominated MPs with other nominees, as the present MPs were not carrying out the tribal authority's instructions. However, once nominated, MPs could only be removed by the assembly.

The tribal authority also complained that it was being ignored by the magistrate and the commissioner-general. In July 1985, Skosana withdrew recognition of Mahlangu as chief. However, in a court case in October 1985 Skosana's actions were declared illegal.

This process did not go unobserved by the citizens of Kwa-Ndebele. 'You know that thing angered not only people in the leadership, it angered the subjects. Those are the people who believe in this procedure, they were surprised to see an ordinary person, not from the royal family, doing things on their own, in their terms they feel such a person is taking over chieftainship, and they cannot allow that'.[15]

What was on the agenda of the former tribal authority councillors who were now cabinet ministers? According to one report, cabinet ministers or their families are responsible for almost 48% of all commercial activity in KwaNdebele. A 1982 report claimed that 'within a short time the visitor to KwaNdebele hears the allegations that the territory is controlled by a cabal of businessmen-politicians whose tight hold over business life keeps competition to a minimum'.[16]

There were also allegations that certain members of the assembly controlled the granting of trading licences and monopolised loans from the Corporation for Economic Development (CED).

According to the late Minister of Internal Affairs, Piet Ntuli, all but two ministers had had businesses before they became cabinet ministers.

To these people, the creation of KwaNdebele meant the consolidation of political power which would facilitate economic accumulation. Many of them were poorly educated and had previously been migrants. The establishment and maintenance of a business became dependent on the patronage of Ntuli who had control of the allocation of business licences and access to loan capital from the CED. It became common for councillors to join the assembly and shortly thereafter open a business. This power cemented the alliance between members of the assembly: 'In fact they were saying to him, if he was not joining the vigilante group, his shop is not going to be protected and he would lose his seat in parliament and then sometimes he might be detained because the old people were afraid to see themselves in jail, you see so he was on the hook as well'.[17]

It was along this axis - Ndzundza royalty and members of the legislative assembly, who were nominated tribal authority councillors - that alliances around the independence issue began to be forged.

Local struggles and the forging of alliances

From 1979 to 1986, the legislative assembly began to work closely with particular chiefs and ambitious businesspeople who were involved in local struggles for economic and political power.

In 1979 the assembly gave the Pungutsha tribal authority (PTA) control over a community in Kalkfontein. The community - previously controlled by a community authority - had purchased the land in 1922 and managed to sustain a viable agriculture. In giving the tribal authority control over the area, ownership of the land was taken out of the hands of the community and given to the tribal authority. The PTA immediately moved a number of 'squatters' onto the land creating conflict with previous landowners. However, anyone who raised opposition to this rule was brutally assaulted in their homes. This led an informant to say that 'Mbokotho was born long before it was born'.

A 'squatter' was made an MP and later became leader of the Mbokotho in the area. As the issue of independence moved further up the agenda, the cabinet was forging alliances at both tribal authority and local levels. Furthermore, in an attempt to marginalise the Ndzundza royal family, they engineered a succession dispute over the paramountcy of the Ndebele as a whole.

In the late 18th century there had been a succession dispute between two brothers in the royal family - Manala and Ndzundza. This had resulted in Ndzundza and his followers leaving the Pretoria region and moving north-east to the Steelepoort district. The Manala chiefdom purchased a farm and settled at Allamansdrift. Relations between Ndzundza and Manala were good and the Manala paramount helped Mayisha purchase land at Weltevrede. However, the Manala leaders chose to fall under Bophuthatswana when it was proclaimed.

Between 1969 and 1977 the Ndzundza persuaded the Manala to join KwaNdebele. However, when the independence issue emerged in the early 1980s, members of the cabinet promised to make the present Manala paramount - previously a taxi driver in Pretoria who had opened a number of businesses in KwaNdebele - supreme paramount of the Ndebele on the basis that the land where Kwa-Ndebele was created was historically Manala land.

In early 1986, Rhenosterkop, previously under the Ndzundza regional authority, was handed over to the Manala tribal authority as part of the new-found alliance. The Manala paramount was both a businessman and an enthusiastic member of Mbokotho.

In order to assess the local effects of this administration transfer, it is useful to delve briefly into the history of Rhenosterkop and a neighbouring community, Bloedfontein. In 1978 the administration for a number of communities moved from Bophuthatswana to the Ndzundza regional authority. Two of these communities, Bloedfontein and Rhenosterkop, were on freehold land which the communities had purchased in the 1920s. Both communities were predominantly North Sotho.

There had been a great deal of conflict between these communities and the Bakgatla Ba Mocha tribal authority (BBM) in Bophuthatswana, which had administered the communities since 1972. The BBM had

appointed headmen who have wrongfully, unlawfully and fraudulently put

people on the farm Bloedfontein and the said headman have collected money
from the new arrivals and placed them on our property. The money so collected from these people for a period is enough to buy five farms the size of
Bloedfontein but the said headman and the BBM have not accounted to the
Bloedfontein community as owner of the land nor has it accounted to the
authorities.[18]

Both communities had problems with school-building funds being
used by the BBM for developments outside of their area. They also
failed to receive subsidies due to them from the Bophuthatswana
government for school-building. Shop-owners were forced to pay
exorbitant rents, and pensioners had difficulty obtaining their
pensions.

The two communities found life substantially improved under
the Ndzundza regional authority:

There were no taxes charged by the Ndzundzas. The community made a
decision, which Mapoch rubber-stamped, to contribute R30 a year towards
the construction of schools. Mapoch would use the money from Bloedfontein
for Bloedfontein. Mapoch even built a library at the high school with money
received in subsidies from the KwaNdebele government... Mapoch works
hand in hand with the community.[19]

Many in Rhenosterkop immediately felt the change in local government from Ndzundza to Manala. The headman was forced to sign
papers agreeing to move to Manala under the 'threat of a sjambok'.
Shortly thereafter the headman and his council were deposed and
replaced by 'a group of six' which included three shop-owners, a
priest and a weekly migrant from Pretoria. This 'group of six' had
arrived at a meeting at the end of 1985 with forms for people to sign.
Young men were expected to join the Mbokotho and older men the
Manala. Treatment under the Manala tribal authority was reminis-
cent of the communities' experiences under Bophuthatswana. Taxes
of R5, R10 and then R20 a month were charged and taken to the
Manala tribal authority by members of the 'group of six'.

Alliances were also established with the chairman of the community authority in Bloedfontein, a Mr F, who was half-brother to
the headman of the community. Mr F achieved his leadership position through the vigorous role he played in opposing the
Bophuthatswana administration. In 1983 there were a number of
crimes in Bloedfontein which led to the community forming its own
(popular) vigilante group made up of unemployed residents. Under

Mr F the vigilante group changed. Group members began collecting money from the community which they shared amongst themselves. This led to conflict between the community authority and the youth. It became apparent that Mr F wanted to take over as headman. Apart from the vigilantes, Mr F also had a following of three other community figures. When he opened a shop at Ekangala in 1985 this indicated his alliance with the KwaNdebele government.

The headman of Bloedfontein said:

> When I was first elected in 1972, I introduced myself to the commissioner. (He) told me to sell the community's land to him and to go and live somewhere else, while the commissioner and I will then be in a position to open shops and make a lot of money. I refused him. While we were under Bophuthatswana we were treated as slaves. Our land and our money was taken away from us by BBM's phony headman... The main fact is that Bloedfontein is rich. A power station is nearby, there is good clay for bricks and also tin underground. With independence the relationship between the community and Mapoch will come to an end. Already during the introduction of independence we were forced to tax houses R2, R4 and R10 for independence. This money went to KwaNdebele government, not to Mapoch... With independence they will chase Mapoch away and introduce somebody else... Now the central government, with the KwaNdebele government will take Mr F to be headman and all of them will make a lot of money at the expense of the community by exploiting the community's wealth. We have already experienced it with Bop.[20]

The case studies presented here are not exhaustive in their representation of the local alliances established in KwaNdebele prior to the revolt against independence. However, they do convey some of what independence meant to particular communities as well as illustrating how actors involved in local conflicts allied themselves to the two different power blocs involved in the independence issue.

Migrants, women, the youth and independence

Rural resistance in the 1950s and 1960s often involved migrants trying to protect their rural resource bases. In KwaNdebele in 1986, migrants were struggling to protect their access to the urban areas. If KwaNdebele was to become independent, its citizens would probably lose their South African citizenship and become aliens in South Africa. Only those who had lived in South Africa for an extended period would retain their citizenship. Access to the cities would depend upon administrative agreements between the two governments.[21]

However, subtle legal points did not make up residents' collective understanding of independence in KwaNdebele. They understood independence to mean social and economic independence from South Africa. They remembered constant arrests in the cities when their 'passes' were not in order. They also remembered brutal harassment by the Bophuthatswana authorities and being 'trapped' because they could not get work permits if they failed to co-operate with the Bophuthatswanan authorities. Farmworkers too had been trapped on farms. Residents had little faith in their government's ability to create and maintain administrative agreements to their benefit.

As one farmworker said:

> The first image that comes to my mind with this idea of independence is I picture myself back to my first life under the farms. When I see where independence is coming from, ie what kind of people, it makes me feel sick and depressed... Independence was going to prevent us from getting a job, I could see myself working in the (cabinet) minister's kitchens, doing their washing for them, my brothers working for them on the fields... Above all no-one under the king has problems with him, they support him, love him, he is always on the side of the oppressed. When there was this migrant labour system people would always get stamps easily and he would wish them luck. They can't do away with the king. We love him...unlike the so-called cabinet ministers, which to me, I see them having complete apartheid, they don't help the poor, they demand levies of the poor... I wish to see these structures abolished, people want to live under the rule of the king, with the king we create our own rules, we put the law onto ourselves.[22]

The same farmworker described her position as a woman:

> If independence had occurred I would have been buried. On the farms a woman had never been given a platform to talk. I would rather die than go back to that life... Now this KwaNdebele government refuses to talk to women, it refuses to let women vote. I don't have a husband, many women's husbands are migrants, and we must have the ability to speak for ourselves. The king said each and every woman has the right to come together, form organisations and discuss their problems... Usually when we have a problem the king's wife greets us and voices our grievances but we are in a position to hear how the king responds. But in the time of the crisis, I asked the king for a platform to speak to his council directly, and he granted it to me.

Youths initially organised themselves as a form of self-defence, their aim being to disband the Mbokotho. A youth leader said that 'while they (the Mbokotho) were around no youth was safe. They had declared war on the youth'. However, tension between the youth and the political powers went deeper than that single issue. Unlike those on the farms, the youth had a number of cultural alternatives and value systems to choose from. The power of the household head, and the corresponding emphasis on youthful obedience, so powerfully reinforced on the farms, was beginning to fall apart.

A large number of urban people had established homes in Kwa-Ndebele, largely because parents had sent their children to school in KwaNdebele because of the unrest which began in September 1984. The experience of mobilisation and organisation which had occurred in the urban centres since 1984 was a subject of intense discussion in the schools. This tended to widen the gap between the political consciousness of the students and their parents. It is therefore not surprising that a number of children of prominent Kwa-Ndebele politicians (the majority of whom were very poorly educated) took leading roles in youth organisation during the rebellion.

Independence took on various meanings for different groupings. To cabinet ministers, within the context of local power struggles, independence opened up new possibilities for access to political power and the potential to accumulate wealth. For certain squatters it meant permanent access to land; to the Pungutsha tribal authority it meant control over disputed land. The Manala elite stood to gain greater political power at the expense of the Ndzundza paramountcy, which faced the loss of political power and influence. Ndzundza subjects would have an autocratic, corrupt authority in the place of their relatively democratic regional authority.

To residents of Kalkfontein, Bloedfontein and Rhenosterkop it meant the loss of their land and the wealth that went with it, as well as the replacement of local representatives with imposed corrupt headmen. Migrants from the farms feared they would be trapped there for life. The youth and those who had experienced Bophuthatswanan rule feared constant harassment, assaults and the loss of voice within government. Women too feared the loss of political rights.

However, there was a level of significance to the revolt that might have escaped popular understanding of the action at the time.

By successfully removing the threat of independence, the residents of KwaNdebele hammered the final nail into the coffin of the grand apartheid scheme.

Notes

1   I wish to thank Peter Delius, Brenda Goldblatt, Colleen McCall, Clive Glaser and Justin Maboyoane for their invaluable help. This article is dedicated to Makhosana Mahlangu for his help and courage.
2   The following narrative of the events surrounding the uprising has been taken from C McCall, *Satellite in Revolt*, SAIRR, 1987, 77.
3   TRAC, *KwaNdebele: The struggle against independence*, Johannesburg, 1986, 23.
4   Brenda Goldblatt, Interview with Guy Mthimunye, a KwaNdebele businessman.
5   Brenda Goldblatt, Interview with women's leader, KwaNdebele.
6   The following details on Ndebele history are taken from Peter Delius, 'The Ndzundza Ndebele: Indenture and the Making of Ethnic Identity', in P Bonner, I Hofmeyr, D James and T Lodge (eds), *Holding Their Ground: Class, Locality and Culture in 19th and 20th Century SA*, History Workshop 4, Johannesburg, 1989.
7   Edwin Ritchken, Interview with Jane Mahlangu (women's leader), Johannesburg, 18.03.89.
8   Edwin Ritchken, Interview with Makhusana Mahlangu (member of the Ndzundza royal family), Johannesburg, 21.03.89.
9   Quoted in P Delius, 'Ndzundza', 19.
10   Ritchken, Interview with Makhosana Mahlangu, 21.03.89.
11   Ritchken, Interview with Makhosana Mahlangu, 21.03.89.
12   McCall, *Satellite*, 14.
13   Edwin Ritchken, Interview with J Mahlangu, 21.03.89.
14   Edwin Ritchken, Interview with S Migidi (ex-general secretary, Ndzundza regional authority), Pretoria, 16.03.89.
15   Ritchken, Interview with S Migidi, 16.03.89.
16   McCall, *Satellite*, 36.
17   Brenda Goldblatt, Interview with Geelbooi Ngoma (son of MP and shop-owner), KwaNdebele.
18   TRAC, 'Bloedfontein and Geweerfontein', Newsletter 12, 1987, 2.
19   Edwin Ritchken, Interview with B Tema, headman at Bloedfontein.
20   Edwin Ritchken, Interview with B Tema, headman at Bloedfontein.
21   G Budlender, *Constitutional Issues in KwaNdebele*, Lawyers for Human Rights, 1988.
22   Ritchken, Interview with J Mahlangu, 18.03.89.

From Forced Removal to Upgrade: State Strategy and Popular Resistance in the Border Corridor

Mike Kenyon and Barry du Toit

The 'black spot' communities of the Border corridor have, through organised resistance, survived decades of forced removals and the redrawing of political boundaries resulting from the creation of the Ciskei and Transkei bantustans.

During 1986, after a long period of resistance, the state reprieved eight of these communities from the threat of incorporation or removal. Effective residents' associations with substantial popular support emerged from this period of successful resistance.

Today these communities face the new and complicated challenge of state upgrading proposals. The interaction between the popular organisations, with their history of independence and resistance to state control, and the Department of Development Aid (DDA) which administers the upgrading process, provides interesting insights into state strategies on local government, as well as community response to state reform initiatives.[1]

Situating the Border corridor

The 'white' or Border corridor is a strip of South African land lying between the Transkei and the Ciskei, and stretching some 200 km from Queenstown to East London. Its width varies from 20 km in the north-west to 70 km in the south-east. According to Pretoria, the corridor divides two antagonistic parts of the Xhosa nation, the AmaGcaleka of the Transkei and the AmaRharhabe of the Ciskei. But the boundaries of the Border corridor reflect more accurately the complex contours of modern South African political history, rather than any underlying ethnic divisions.

The eight 'black spot' communities remaining in this white corridor are Lesseyton (north of Queenstown), Goshen (west of Cathcart), Wartburg, Hekel and Mgwali (near Stutterheim), and Newlands, Kwelera and Mooiplaas (near East London). They trace their origins to the second half of the nineteenth century, when they began as settlements established for refugees from the frontier wars, as rewards to allies of the colonial government during that time, or around mission stations.

The communities vary greatly in size, from Goshen, with a population of under 5 000, to Mooiplaas which is home to some 30 000 people. Most of the inhabitants of the area have a clear historical claim to this land in that their ancestors had titles to their plots of ground. The official view, however, is that this has little force, as legislation introduced in the 1920s stipulated that only the oldest male could inherit his father's land. After the introduction of 'betterment' in the 1950s, most of the titles were converted to Certificates of Occupation. Thus, very few residents have title deeds, and a large proportion of the residents, including descendants of the original title-holders, are referred to as 'squatters' by state agencies.

As pressure on land in these areas has increased, and as farmworkers have been forced off white-owned farms, the number of 'squatters' has grown. Although most residents retain access to land, it is too little for survival. Those families that do not have members employed in the towns are in dire straits, often trying to survive on the meagre pensions of elderly family members.

The undermining of the agricultural base in these communities is so extensive that many families have another home in urban townships like Mdantsane. Although access to land is deeply valued, the tendency is for people of working age to move to the towns, leaving behind the young and old. This has serious consequences for the development of these areas.

The eight 'black spots' represent an anomaly for apartheid, being black rural communities with some historical claim to tenure outside of the bantustans. As such they have consistently presented a political challenge to apartheid's planners. While the level of organisation varies among the communities, there is a history of resistance in most, and many of the region's political leaders over the past 30 years have come from these areas. These communities are a significant testing ground both for the state in its attempts to co-opt, repress and control, and for popular strategies which aim to develop organisation and community resources.

The struggle against removal

The decade prior to the 'independence' of the Ciskei in 1981 was characterised by the most intensive programme of forced removals yet seen in the region. During that period, half of the current residents of the Ciskei were moved great distances to their 'homeland', or suddenly informed that their historical dwelling-places had become part of the Ciskei. Although the eight communities were targets for a consolidated Ciskei, they were neither incorporated nor removed into the Ciskei by the time of 'independence' in December 1981. Instead Pretoria entered into a politically expedient, if somewhat eccentric, arrangement: the Ciskei was to administer all eight areas, with the ultimate object of overseeing their removal into the consolidated Ciskei.

Most of these communities had not lived under government-appointed or approved chiefs and headmen for many generations. Goshen, for example, never had any tribal administration. It was administered by a popularly-elected board constituted in terms of a 1904 act of the Cape parliament.

From 1981 the Ciskei attempted to impose its oppressive tribal authorities. This provoked large-scale resistance, not only to tribal administration but to the Ciskei itself. The threat of removal into the Ciskei, in particular, provoked widespread opposition. In the period from 1981 to 1985, democratic residents' associations were formed to unite the people of the communities against the chiefs, the removal and the Ciskei.

In a parallel development, militancy grew, especially among the youth, who sometimes took matters into their own hands. In Mgwali, for example, a headman, who was pro-removals, was killed in a clash with the youth.[2] And in Kwelera, many residents died in strife between opponents and supporters of the Ciskei.

In May 1986 Mgwali residents, at that time probably the most cohesive of the communities, applied to the South African supreme court for an order declaring the agreement which gave the Ciskei power of administration over them unlawful. The applicants won the order and administrative responsibility for all the communities reverted to Pretoria. Shortly afterwards, Pretoria announced that all eight areas were to be reprieved from the threat of removal to the Ciskei and would remain part of South Africa.

In their victory against the Ciskei and Pretoria governments many sectors of these rural communities had united around a common rejection of chiefs, removals and the Ciskei. A foretaste of life under Ciskei administration, together with the inferior quality of many services in the bantustan, contributed to the unity of residents. Inferior services included substantially lower old-age pensions, and were compounded by a variety of taxes and 'voluntary' levies which the new tribal administration had attempted to squeeze out of residents.[3]

Progressive community leaders were able to harness and direct this broad-based opposition effectively, and democratic residents' associations were established in all communities except Lesseyton and Goshen. These drew the villages, some of which already had their own local committees, into a coherent and politically directed social movement.

The residents' associations also linked up at a regional level to co-ordinate resistance. This co-ordination took place with some assistance from the newly-formed United Democratic Front, but was essentially generated by the immediate struggles of the communities. The result was that the communities won their battle against removals, and in the process developed strong, democratic, organisational structures.

The growth of the residents' associations coincided with a period of developing political consciousness and mobilisation in black communities. Some, affected by the exuberant political climate, affiliated to the newly-formed UDF. Regardless of formal political affiliation, however, none of the residents' associations remained unaffected by resistance politics in the Border region.

While widespread popular support was a crucial feature in the successful struggle waged by the residents' associations, communities varied in their degree of unity and cohesion. For older residents especially, the re-establishment of traditional tribal structures with chiefs, councillors and headmen was appealing. This tribal ideology contributed to the formation of pro-Ciskei and pro-tribal groupings within these communities during the anti-removal struggle.

Political cohesion was complicated by the broader political

dynamics of the Border region. Debates within the trade union movement in the East London area, in particular, were carried home by trade union members. The PFP, drawn, probably unwittingly, into the internal political dynamics of both Kwelera and Mgwali, provided resources and status for opportunistic individuals and groupings within the communities. This resulted in divisions which widened subsequent to the PFP's withdrawal.

Factors which contributed to the strengths and weaknesses of the residents' associations in the period of resistance to removal were carried over to the next phase of struggle, directly against Pretoria.

State upgrading: a new challenge

In May 1987 the MP for King William's Town announced that the government was embarking on an upgrading programme in the eight communities and that R12-million had been allocated for this purpose in the budget of the Department of Development Aid, which was to administer the programme.

Like many other communities dealing with state reform initiatives, the popular organisations now had to deal with a more complex political environment. The political and organisational strategies which had succeeded in the battle against incorporation were to be tested by a very different challenge.

Although the figure of R12-million for upgrading of the corridor communities was not substantial, it could not be ignored by the organisations.[4] Being rural, the eight communities faced a critical absence of basic infrastructure and amenities, undoubtedly worsened by four years of Ciskeian administration. This was exacerbated by the six-month delay between the 1986 reprieve and South Africa's resumption of services to the area. The years of uncertainty under the Ciskei, as well as severe financial and legal constraints, had also retarded local enterprise, housing development and agriculture. In these circumstances the residents' associations could not easily refuse the offer of development. But the process they would have to enter into would be difficult, since the terrain would be largely determined by the DDA.

The risk of co-option by DDA activities proved to be very real and led to a severe test of the democratic practices, unity of purpose and political independence of the residents' associations. But it was also clear to the communities that a confrontationist approach would have risked further repression and the denial of basic services. In addition, such a decision could have resulted in less accountable, opportunistic 'leadership' arising, with whom the DDA could advance its strategy.

The challenge before the residents' associations was very different to that of the anti-incorporation phase of their struggle. In that case, resistance to South African and Ciskeian initiatives was directly in the interests of the vast majority of residents. There was a united stance in resistance: the authorities, their agents and their initiatives were uniformly undesirable and the task of the popular organisations was accordingly simplified. Democratic unity was achievable because the interests of most residents were equally threatened by incorporation, and preservable in the context of a unified rejection of all government initiatives.

The new state initiatives offered programmes potentially beneficial to residents. This confronted residents' associations with the complex task of utilising a process which involved ongoing liaison with state agents and institutions manifestly opposed to the political practices and programmes of the associations. In addition, the associations were being forced into an area in which they had no experience - local authorities and the provision of local services.

This aspect of the DDA's programme - the setting up of local authorities in accordance with government policy, and the simultaneous attempts to weaken and destroy the residents' associations - was to become the dominant item in the new political process.

DDA representatives who began the process of liaison with the communities must have been aware of this aspect of the strategy, and tailored their activities to this end. But the setting up of local authorities was at no time raised in discussion with community organisations, and the DDA spent most of 1987 constituting liaison structures within each community.

With the exceptions of Lesseyton and Goshen, residents' associations *de facto* administered the areas involved. In Lesseyton - a community with no apparent history of organisation - the remnants of a pre-incorporation tribal authority still operated, while in Goshen the existing board was a fairly democratic structure.

The DDA refused to deal directly with the residents' associations, claiming they were political and that the DDA was concerned with development, not politics. They requested two representatives from each village in each community to sit on liaison committees. With this approach, the DDA was attempting to find an alternative route into the communities.

The committees were chaired by DDA officials and meetings were attended by numerous DDA technocrats and administrators. The assessment of the residents' associations was straightforward:

- the DDA was trying to isolate a small group of influential residents and separate them from the residents' associations into independent liaison committees;
- it would then channel all development through these particular residents, granting them a status dependent on DDA resources;
- this would undermine the residents' associations, which would become increasingly irrelevant in the provision of material resources for the communities;
- through the co-option of the liaison committees, and the weakening of the residents' associations, the DDA would be able to establish conventional forms of local authority - tribal or community authorities.

The experience of the residents' associations over the next year not only confirmed this assessment but provided even more fundamental reasons for opposing the DDA plans.

Threats and promises

The R12-million initially allocated for upgrading in the corridor area decreased rapidly. In early 1988, when the DDA was pushed to begin delivering on its promises, community liaison committee representatives were told that half the money budgeted for upgrade in the corridor had been diverted to flood relief in Natal.

By the end of 1988 the DDA had delivered no more than the communities had obtained from various state departments and the divisional council before 1981. With few resources to offer the corridor communities, the DDA turned towards crude attempts at control instead. Gradually it became clear that a major, if unstated, item on its agenda was the establishment of officially-sanctioned local authorities for the communities.

The DDA faced two major obstacles in its attempts to impose these local authorities: its failure to provide the promised funds, and resistance from the residents' associations. At the end of the successful opposition to incorporation, the residents' associations commanded a high level of popular support in their communities. They functioned as partial 'local authorities', taking responsibility for the co-ordination and administration of a number of civic functions.

In some of the communities, for example, the residents' associations were responsible for the vetting of pension applications, and the DDA concluded that it could not simply ignore these structures in its community liaison. The DDA accordingly acknowledged the authority of the residents' associations in at least Kwelera and Mooiplaas. However, it consistently advanced proposals aimed at weakening their position and incorporating sections of the associations into co-opted structures.

Local authorities

The following two case studies illustrate how DDA practices are adjusted to the local context - in particular to the strength of local organisation.

Mgwali

The Mgwali Residents' Association (MRA), which had been at the forefront of the anti-Ciskei struggle, was severely affected by the national state of emergency declared in June 1986. Many executive members were detained or forced out of the area for fear of detention.

The repressive climate curtailed MRA meetings, and allowed pro-Ciskei and other opportunistic elements to make their way on to the DDA liaison committee without accountability to their communities.

The major component of the DDA plan unfolded in Mgwali, Hekel and Wartburg. Because of the proximity of the three settlements, the DDA proposed the establishment of a single liaison committee and, subsequently, a joint community authority. In late November 1988 the remnants of the residents' association learnt there was to be an election for the 'Gqolonci Community Authority' of Wartburg, Hekel and Mgwali three days later. Deputations protesting to the DDA were brushed aside.

The election was boycotted by the majority of residents, and voting for the community authority - by show of hands - involved

some 20 participants. Many of those serving on the liaison committee were elected to the local authority, and the DDA announced its
intention to establish community authorities in all the corridor
communities.

Mooiplaas and Kwelera

The state of emergency did not weaken the Mooiplaas and Kwelera
residents' associations as seriously, and it was difficult for the DDA
to ignore or circumvent them. Instead, DDA representatives
proposed the formation of a liaison committee with which they
could discuss plans for a local authority. After thorough discussion
of the merits and dangers of such a structure, the residents' associations accepted the proposal - but on condition that the liaison committees would be sub-committees of, and thus accountable to, the
residents' associations. This strategy, they hoped, would minimise
the risk of liaison committee members being co-opted by the DDA.

The importance of these safeguards has been emphasised by
subsequent DDA tactics. It has, for example, attempted to hold
meetings with community representatives in East London, rather
than within the communities themselves. On one such occasion, the
entire membership of the village committee of Mooiplaas, some 140
people, travelled to the DDA offices in East London to demonstrate
that they acted as a community, not individuals.

Another DDA proposal, first made in early 1988, was to reduce
village representation on the liaison committee from two to one per
village. At the same time the DDA offered to pay community representatives on the committee. The residents' association rejected
both proposals, seeing them as steps in the establishment of rule by
village headmen, and the creation of a tribal or community
authority.

Later in the year the DDA suggested dividing Mooiplaas into
two groups of villages, to be dealt with separately. This too was
rejected. More recently the DDA invited the liaison committee to
its offices in East London to elect an executive structure to operate
within the liaison committee. At the same time, the DDA produced
a claimant to the chieftaincy of the whole area of Mooiplaas and
Kwelera.

After the 1985 conflict, when the PFP backed the less-than-
progressive leadership of Tuba, one of the five Kwelera villages, it

remained outside of the residents' association structures. The DDA, quick to seize on the relative compliance of this village and reward it, improved roads in the village and added classrooms to the school. Tuba village, situated in full view of the national road, now serves as a showpiece for the DDA.

In the second half of 1988 the DDA sent a planner from its Pretoria head office to the communities, supposedly to tell residents 'where to place graves and what were good fields'. In the course of his travels around Kwelera the official talked to residents about electrification, water supplies and sewerage systems, mentioning that monies collected in payment for these services would go to an elected committee which would administer the services.

Kwelera falls within the Greater East London Metropolitan Area. This excludes Mooiplaas, but includes Kwelera and Newlands. The services the DDA planner spoke of in Kwelera are not those which South Africa usually provides for black rural areas. It accordingly appears that some state structures may be considering developing Kwelera into an urban township, although no attempt has been made to canvass community opinion on the issue, which would have far-reaching consequences for the current residents. Recently DDA officials have talked explicitly of Kwelera as an urban township.

Regional intrigues

The regional nature of the DDA's operations is a potentially useful mechanism for manipulation and control. For example, the DDA recently told residents of Needs Camp, another DDA-administered area but one which is unreprieved, that the reprieved areas had asked that police stations be set up because 'they wanted law and order in their areas'. Residents of the reprieved areas maintain they have requested no such thing.

Tactics such aas disinformation and playing one community against another are likely to remain part of the DDA's approach. This danger makes it imperative for community organisations to remain fully informed of the situation in other communities, and of the implications of regional developments.

The DDA's policy of approaching residents' associations in connection with its upgrade proposals is little more than a temporary and expedient measure occasioned by the strength of support which the associations command in their communities. Its intentions are to weaken the associations, and co-opt their members, clearing the way for the imposition of approved local authorities. Their tactics range from threats to promises, depending on local conditions, but the common thread is clear.

The provision of resources is linked to acquiescence in a form of local authority of Pretoria's design. But Pretoria has few resources to provide, and the DDA is accordingly not in a position to co-opt significant portions of these communities. As a result, the DDA operates in a piecemeal and *ad hoc* fashion according to local conditions, and has yet to reveal a co-ordinated strategy for winning the 'hearts and minds' of the local communities.

The DDA will continue to use a range of strategies, probing for weaknesses in the unity and strength of the residents' associations. It will continue to provide needed resources, and seek to provide them in a way which advances Pretoria's objectives in the region.

The residents' associations will have to find creative and effective strategies to continue their successful traditions of resistance against this persistent and complex adversary.

Notes

1   This report is based largely on primary material in the possession of the Grahamstown Rural Committee. Very little published research is available on this area, despite its fascinating and significant recent history.

2   In early 1986 headman Dyosi, leader of a pro-removal vigilante group, was killed in a clash with the youth. The trial of the residents allegedly responsible for the killing was still continuing in 1989.

3   Some services, for example clinics, were superior in the Ciskei.

4   In terms of the financial requirements of typical state upgrade proposals (such as the tarring of roads), the amount was relatively small. Had the money been made available, it would have been insufficient to cover even part of the engineering works envisaged in tentative DDA plans.

The Civil War in Natal

John Aitchison

Understanding the dynamics of the violent conflict in Natal and especially the Natal Midlands around Pietermaritzburg is a particularly complex task made more difficult by the assumption that there are identifiable sides and aggressors. The identity of aggressors varies, however, so that for the UDF and Cosatu it is Inkatha, for the minister of Law and Order it is 'radicals' and for Chief Buthelezi it is the ANC (external mission).

Most monitoring and human rights groups identify both Inkatha vigilantes (*otheleweni*) and UDF comrades (*amaqabane*) as involved in the violence but tend to place more blame on Inkatha. By contrast, the Inkatha Institute claims that people involved in the violence are not acting as members of any particular group but for reasons of crime or in response to poverty.

However, evidence shows that political allegiances have been crucial in deciding who should live and die. This does not necessitate a rejection of the influence of criminal activity in the violence, nor the socio-economic factors which fuel it, nor indeed the confusion in any conflict which makes the apportionment of blame a risky undertaking.

Events in Pietermaritzburg

The Natal Midlands, with Pietermaritzburg at its centre, covers a region of about 374 square kilometres. Its population is variously estimated as ranging from a quarter to three-quarters of a million people. The region includes the city of Pietermaritzburg (and the township of Sobantu); the Edendale complex (including the black-owned freehold area of Edendale); the adjoining area of Slangspruit (another freehold area); the two townships of Ashdown and Imbali; and the Vulindlela area (previously known as Zwartkop Location or Reserve, administered by KwaZulu).

Other townships in the Natal Midlands are located near Hammarsdale (Mpumalanga), Greytown (Enhlalakahle), Howick

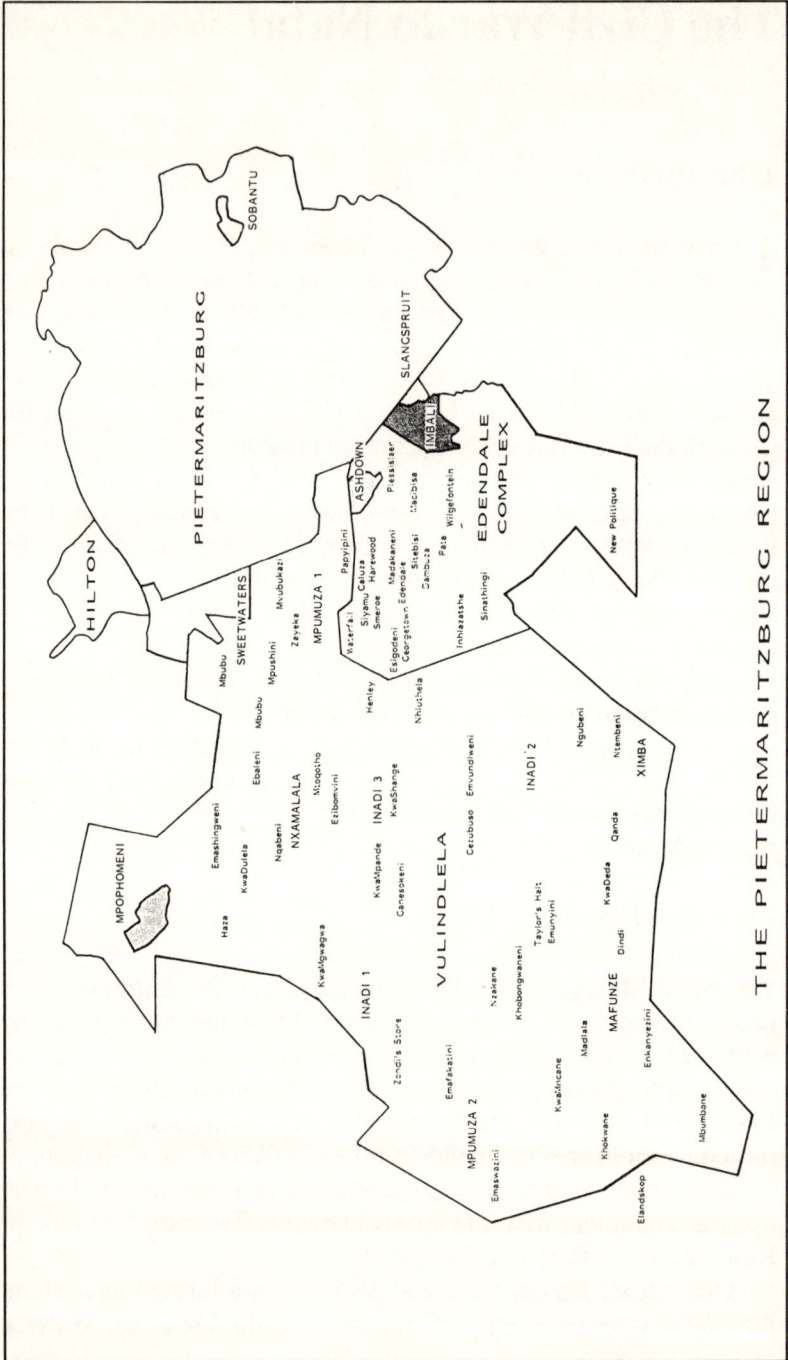

THE PIETERMARITZBURG REGION

PIETERMARITZBURG

SOBANTU

SLANGSPRUIT

IMBALI

ASHDOWN

EDENDALE COMPLEX

HILTON

SWEETWATERS

Mbubu

Mpushini

Mvubukazi

Zeyeke

MPUMUZA 1

Papyipini

Vaterkall

Siyamu

Caluza

Smeroe

Harewood

Ezigodeni

Madakeneni

Georgetown Edendale

Sitebisi

Gamba za

Pata

Wilgefontein

Machibisa

Pleasslaer

Plessislaer

Henley

Nhlazatshe

Sinathingi

Inhlazatshe

Nhluthela

New Polilque

Cetubuso

Envundlweni

INADI 2

Ngubeni

Ntembeni

XIMBA

Emashingweni

KwaDutela

Ebaleni

Mbubu

Ngabeni

NXAMALALA

Mtzogctho

Ezibomvini

INADI 3

KwaShange

KwaNgande

Ganespkeni

QandaKwaDeda

Haza

KwaNgwagwa

INADI 1

Zondi's Store

Emafakazini

Nzakane

Khobongwaneni

Taylor's Halt

Emunyini

Taylor's Halt

VULINDLELA

Dindi

MAFUNZE

Emanyezini

Emaswazini

MPUMUZA 2

KwaNcane

Madlala

Khokwane

Elandskop

Mbumbane

MPOPHOMENI

(Mpophomeni), Mooi River (Bruntville), and various rural towns and villages.

Until recently, the Natal Midlands was regarded as 'peaceful'. Events in Soweto in 1976 had little impact on the region and the response to the Vaal Triangle revolt of 1984 was also limited. In Sobantu the community council resigned and in Ashdown attempts to establish a council in 1983 collapsed when only one candidate could be found. A factor in this 'stability' was that civic associations had not developed. This was still the case in African areas during 1988.

However, in 1984 youth organisations affiliated to the UDF were founded. Key among them were Edeyo, Iyo, Ayo and Soyo - the Edendale, Imbali, Ashdown and Sobantu youth organisations. Conflict between UDF and black consciousness youth in Imbali during 1985 (a precursor to the fighting which broke out between the two groups in Sobantu in 1987) was initially defused through negotiations.

In mid-1985 the initiation of a civic association in Imbali was halted by severe intimidation when the house of its leader, Robert Duma, was petrol bombed. In August 1985, Inkatha supporters threatened the Federal Theological Seminary and demanded that it be vacated. The petrol bombing of councillor Patrick Pakkies's house in April 1985 was a demonstration of violence against Inkatha followers.

After Uwusa was formed as the trade union wing of Inkatha in 1985, conflict between Inkatha and Cosatu became inevitable. The spark for this was the strike at BTR Sarmcol in Howick following which the entire African workforce was dismissed. A stayaway on 18 July 1985 and a consumer boycott, organised on behalf of the dismissed workers, were strongly opposed by Inkatha. Conflict erupted when local youths acted coercively to enforce the stayaway. And the bussing of a large Inkatha group into Mpophomeni township in December 1986 sparked fresh violence which resulted in the deaths of three Cosatu supporters. Inquest findings in March 1988 stated that nine Inkatha members were responsible for the murders. But as yet, there have been no prosecutions related to this finding.

In 1985 and 1986 Inkatha leaders in Imbali had gathered a group of Inkatha youths and unemployed men who acted as a paramilitary group or 'impi'. As a consequence, a number of UDF and black consciousness-supporting youths, and in many cases, their families, were forced out of Imbali, and later Ashdown. Some fled to Sobantu and Mpophomeni while others retreated into Edendale and Vulindlela. This exodus undoubtedly had a politicising effect on many youths, particularly in Vulindlela where, in mid-1987, a number of UDF-affiliated youth organisations were formed.

In May and June 1987 there were reports of Inkatha recruitment drives in the region. Often these were facilitated by heavily-armed groups being bussed into an area. Deaths began to rise, reaching an average of 13 a month between March and August 1987. The UDF and Cosatu argued that these deaths were largely the result of Inkatha attacks meant to intimidate individuals and communities.

The UDF and Cosatu claimed that 90% of Pietermaritzburg workers responded to the call for a stayaway on 5 and 6 May 1987 in protest against the whites-only election. This seemed a considerable defeat for Inkatha as Chief Buthelezi had urged his supporters to campaign against a stayaway. Inkatha blamed the success of the stayaway on the Transport and General Workers Union, whose bus drivers, through their strike action, had made it difficult for workers to travel to their places of employment.

The deaths of 12 TGWU members in the following months and bus stonings by Inkatha youths may have been a direct consequence of this. Whether the call for the stayaway had been planned as a test of Inkatha strength or not - and there is no evidence that it was seen this way by the UDF and Cosatu - it demonstrated that Inkatha did not have clear dominance in the Pietermaritzburg region.

In September 1987, as devastating floods destroyed hundreds of houses in Edendale and Vulindlela, there were reports of a heavy Inkatha recruitment drive backed by threats and coercion. Reports also indicated that in Vulindlela, 4 October was the proclaimed date by which everyone had to join Inkatha.

According to the UDF, the recruitment drive and associated violence met with growing resistance. A violent conflagration ensued and a possibly idealised picture was drawn of community defence groups forming, assisted by young UDF-supporting comrades.[1]

Horrifying levels of violence occurred between September 1987 and January 1988: there were 162 deaths in January alone. Criminal groups were involved in some of the violence while revenge was also a motive. Varying estimates have been made on the extent to which poverty, unemployment and criminality fuelled the fighting.

A large number of UDF and Cosatu supporters were detained in December. Reports about key Inkatha leaders - dubbed 'warlords' by their critics - alleged that they too had engaged in large-scale acts of violence in the area without any serious state intervention against them. The only visible attempts to halt their activities involved interdicts brought against them by a Cosatu legal team.

By the beginning of 1988, Vulindlela tribal authorities were in disarray: many chiefs were no longer performing their official functions and agriculture was seriously affected. The extent of disruption and fear in the region is illustrated by the fact that at one stage there were no patients in the paediatric section of Edendale hospital because parents were too scared to leave their children there. Large numbers of people sought refuge in safer locations, or were accommodated in white areas.

The UDF and Cosatu attributed the increased violence during this period to an Inkatha counter-attack named 'Operation Doom' or 'Operation Cleanup'. Slangspruit came under heavy Inkatha pressure in January and the month ended with an Inkatha 'invasion' of Ashdown. This attack occurred after security forces had allowed a meeting of 15 000 Inkatha supporters in Mpumuza at which the crowd was allegedly incited to attack the UDF and Cosatu. Security forces, who did little to stop the resultant attack on Ashdown, are also alleged to have escorted and helped transport armed Inkatha members to the township.

The decline in deaths after January is partly accounted for by the heavy police reinforcements - including KwaZulu police and 150 special constables - who entered the area at the end of the year. A further 289 'kitskonstabels' who were deployed at the beginning of March created a new controversy, as several were Inkatha supporters with previous records of violence.

The police, stiffened by Brigadier Mellet's message of 3 February 1988 that 'the violence has to stop at all costs', seemed largely to act against UDF forces. There was growing evidence of security forces actively colluding with Inkatha, which enabled the movement to restore much of its lost control in the Vulindlela area. There were also reports of police handing captured comrades over to Inkatha or tribal authorities who then killed them, as in the well-documented case of Makhithiza Ndlovu (aged 13) who was killed on 1 January 1988. Large Inkatha meetings of armed men and youths took place without any police interference.

In the early part of 1988 a number of well-publicised interdicts were brought before the courts, requesting protection for applicants against Inkatha 'warlords'. The attorney-general's representative was petitioned in connection with the delay in bringing these cases to court, and the applications were further hampered by the assassination of a number of key applicants and witnesses, including two Mthembu brothers (Simon Mthembu on 24 January and Ernest Mthembu on 4 July) and Johannes and Phillipina Nkomo (on 13 February 1988). None of the alleged killers - named in a number of affidavits placed before the courts - have been brought to trial.

At the same time schooling was disrupted when many teachers resigned. Some scholars were refused admission to schools and others were afraid to attend after a card system was instituted at a number of Department of Education and Training schools. Many KwaZulu Department of Education and Culture schools refused to accept non-Inkatha pupils. There were reports of a drop in attendance of at least 4 000 pupils at schools near Pietermaritzburg. In Hammarsdale estimates suggested that only 25% of male pupils remained in senior classes (*S Trib*, 12.02.88). Intermittent disruption of schooling continued throughout 1988 and into 1989.

In early February 1988, about 50 weapon-wielding Inkatha youths from Harewood in Edendale swarmed off a bus near the centre of Pietermaritzburg and attacked black shoppers and pedestrians. Ten were injured, three of them seriously. Although 43 people were tried and convicted, it was on a relatively minor charge and the light sentences occasioned public comment. The trial of the six killers of Ester Molevu, a 61-year-old UDF supporter who was brutally murdered, caused further consternation when the accused received partially-suspended sentences, the heaviest jail term being three-and-a-half years.

On 21 March 1988 there was a mass round-up of youths (259 males in Ashdown and 218 in Sobantu), during which many people were assaulted. On a number of occasions delegations of women from Ashdown and Imbali appealed for police and 'kitskonstabels' to be withdrawn from the area.

Intermittent violence continued between youths supporting black consciousness and the UDF in Sobantu, and in Hammarsdale the violence continued unabated in 1988. On 13 April a confrontation between two ANC guerillas and the police resulted in the death of the guerillas and one policeman, with three others wounded.

During June 1988 there were two huge stayaways in the region: a three-day 'national peaceful protest' from 6 to 8 June, called by Cosatu and Nactu against the Labour Relations Amendment Bill and the banning of 17 organisations; and Soweto Day, 16 June. During the first stayaway 65% of the black workforce was absent. The Soweto Day stayaway was even more effective.

Peace initiatives, which first started in late 1987 and continued into 1988, including calls from the ANC, church and other community leaders, achieved little. The situation was exacerbated after key UDF negotiators were detained again in February 1988, central witnesses in a number of interdict applications were assassinated, and the UDF, Cosatu and 16 other organisations were restricted on 22 February 1988.

Various Inkatha spokesmen suggested that the peace talks were unproductive and blamed the anti-Inkatha forces and the ANC for this. Police spokesmen were similarly dismissive. The UDF compromised to the extent of agreeing to involve its national leadership in peace talks (thus implicitly allowing Inkatha's claim to be a national, rather than a regional movement). However, one hopeful move in mid-1988 was Inkatha's establishment of a special 'watchdog' group under the leadership of Inkatha general secretary Oscar Dhlomo to investigate allegations of corruption and violence against senior members of the organisation. This was the first sign of Inkatha action against some of the 'warlords' after months of violence in Natal.

At the beginning of September 1988 Inkatha and Cosatu signed an agreement to set up a complaints adjudication board headed by a retired judge. The agreement rapidly faltered when Inkatha members refused to appear before it on the grounds that criminal charges were pending against them. Cosatu withdrew support from the board in 1989 when witnesses who had testified before it were murdered.

Despite a desire by the media and white Pietermaritzburg to believe that the worst was over when killings subsided in March 1988, the death toll began to rise again. By the end of 1988 a total of 691 had been killed.

Inkatha supporters continued to 'clean up' Vulindlela during the latter part of 1988. Ashdown was frequently attacked and there was a massacre at Trust Feeds. The conflict in and around Vulindlela and Mpophomeni continued into 1989. And the stresses of two years of war became evident as strife broke out within the comrade refugee groups.

The state officially responded to a variety of anguished appeals with a curious mixture of denial and aggression. Minister of Law and Order Adriaan Vlok, who had in early 1988 rallied the police to fight and beat 'the radicals' in Pietermaritzburg, on several occasions denied that law and order had broken down. But within days of such statements he would issue warnings, such as the 'iron fist' speech of April 1989, indicating that the situation was degenerating.

Assassinations continued, notably of a complaints adjudication board witness and a relative of a prominent Cosatu shop steward. A Cosatu-led stayaway followed in Pietermaritzburg from 5 to 7 June.

The renewal of peace moves received much publicity in the first half of 1989, as did unsuccessful attempts by, amongst others, the maverick tribal chief Mhlabunzima Maphumulo to get the state to institute a judicial inquiry into the violence and the police role in it. After much negotiating, and appeals from an Anglican Church delegation, it appeared that Chief Buthelezi would be willing to engage in high-level talks with the presidents of the UDF, Cosatu and the ANC. Two initial meetings between Cosatu/UDF and Inkatha delegations took place in Durban before the end of June 1989.

Quantifying death and destruction

Reports produced by the Centre for Adult Education monitoring project at the University of Natal, Pietermaritzburg, show that there has been a devastating period of destruction, dislocation and abnormality over the last few years. But this assessment must be qualified by the fact that the region does not normally provide a good life for the bulk of its inhabitants. There are a growing number of unemployed, and violence of a less overtly political kind was already high in the region. In areas near Pietermaritzburg, for example, 298 murders were reported from the black townships in Edendale during 1985-86.

The Unrest Monitoring Project adopted a statistical, census-type approach in its reports to counterbalance 'stories' about the conflict. These stories were either anecdotal or more sustained narratives which intellectuals or para-intellectuals from the different camps -

UDF/Cosatu, Inkatha, or police/state - developed according to their positions. The statistics were tabulated to counteract the anecdotal understanding of the situation and to test the plausibility of various accounts.

The basic statistics for the Natal Midlands for the period 1987 to June 1989 are as follows:

Deaths: In 1987 there were 397 political killings. In 1988 there were 691, and in the first six months of 1989, 275 people were killed, bringing the total deaths in the political conflict over 30 months to nearly 1 400. These figures do not include deaths from the areas around Durban like Inanda and Shongweni, where many were killed in late 1988 and in 1989.

Wounded: Between January 1987 and the end of June 1988 some 408 people were wounded. This is an obvious underestimate unless an astounding seven people are killed for every four wounded. That these figures are underestimates seems evident from local hospital records which show that approximately 20% to 30% of shot or stabbed people taken to hospital, die. At Edendale Hospital the number of people admitted with violent injuries increased by 80% during December 1987.

Arson: Over 432 houses and vehicles were destroyed in the 18 months up to July 1988. It is probably safe to assume that as many as 1 000 houses have been destroyed since the conflict began in 1987.

Refugees: Estimates of the number of refugees from the conflict range from 10 000 to 60 000. On the basis of the houses which have been destroyed, it is assumed that at least 10 000 people have been refugees who have had to leave their home areas for at least some time. Other refugees may have been more transient, leaving their homes for short periods when the war was at its height in their area or, as was frequently the case, sleeping in fields at night.

Cases of young men who fled for fear of the destruction their presence might bring on themselves and their families were also common. Many of the young comrades in Edendale, Mpophomeni and Sobantu were refugees of this type. A personal estimate is that about 10 000 people have moved house permanently and about another 10 000 to 15 000 were away from home for brief periods.

Interpreting the war

Six key questions arising from the historical narrative and the

statistics point towards a plausible interpretation of current information:

*1. Is the Pietermaritzburg conflict simply part of the general revolt
against apartheid and its structures that flared up in late 1984 and
which the succession of states of emergency has suppressed elsewhere?*
An Indicator Project publication claims that at least 4 012 people
were killed in political conflict across the country, including 1 113 by
the police and army, from September 1984 until the end of December 1988.[2]

This period can be divided into five distinct phases:
- *early rumblings of revolt* from September to December 1984 in
  which 149 people died;
- *build up* from January to 20 July 1985;
- *regional emergency* from 21 July 1985 to 3 March 1986;
- *post-emergency* from 8 March 1986 to 11 June 1986;
- *country-wide emergencies* from 12 June 1986 to date.

The trend nationally is very different from the one in Pietermaritzburg, where violence appears to be escalating rather than
declining.

There are two possible interpretations for this second surge of
violence in the 1980s. One is that the violence in Natal is simply a
later occurrence of the revolt against government structures that
started elsewhere in 1984-85. The other is that the conflict is entirely
different in nature and concerns Inkatha's ability to command the
allegiance of black residents of Natal.

These two interpretations are not mutually exclusive: the revolt
against government-installed township structures was spreading in
Natal, but the conflict has been essentially about Inkatha's desire to
maintain its support (or at least to maintain its ability to claim such
support without contradiction) among blacks in Natal.

In the national context the motivation for this was clear. Orkin
has shown that between 1977 and 1988 Inkatha's support in the industrial heartland of South Africa (the Pretoria/Witwatersrand/Vaal
area) declined from over 30% (more or less equal to what the ANC
had in 1977) to less than 5% in 1988. By that time the ANC's
support had risen to nearly 50%.[3]

Inkatha faced major problems as a national force and its need to
stop a reduction of support in Natal was evident. There was also a
danger of revolt against township structures, a number of which

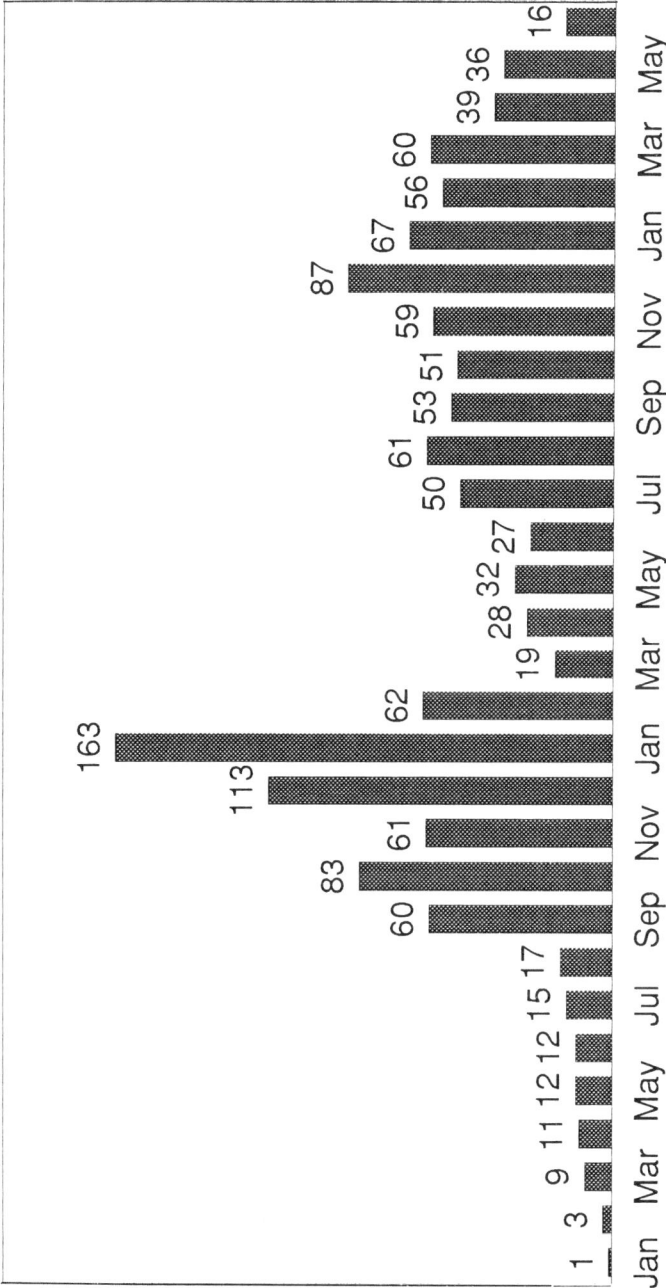

DEATHS : 1987 - June 1989
Pietermaritzburg and Natal Midlands

Source: Centre for Adult Education

were administered by Inkatha in Natal. Available evidence suggests that it was Inkatha's aggressive attempt to recruit membership both for its broader structures and its trade union, Uwusa, which laid the ground for the dramatic escalation of violence in 1987 and 1988.

Another factor differentiating the Natal conflict from the national one is the inability or unwillingness of state forces to crush the violence raging in the region.

2. What is the origin and dynamic of the conflict?

Who fired the first shot is a traditional preoccupation of war historians. In the case of the Natal violence this question could lead to an endless regression. Was it the murder of four Cosatu members in Mpophomeni by Inkatha in December 1986? Was it the coercion by consumer boycott supporters prior to that? Was it Inkatha supporters in Imbali township who chased out radical youth? Was it some youth who threw a petrol bomb at a town councillor's house?

Despite speculation, it is common knowledge in Pietermaritzburg that the 'big war' began in September 1987 when comrades and residents resisted forced recruitment. Based on data from the Natal Midlands,[4] the following conclusions can be drawn:

- a political dynamic, particularly at its inception, has been present in the conflict in Natal;
- urban township strife has spread outward into peri-urban and semi-rural areas and even the most conservative tribally controlled territory is not exempt from turmoil;
- the violence seems to have a strong connection to Inkatha's attempt to maintain or increase its influence in the region;
- political violence is becoming embedded in the social fabric.

Evidence from the Pietermaritzburg area reveals that the 1987 violence started in an urban township and was therefore likely to have been of a political nature rather than a result of floods, unemployment and starvation on the margins of the region.

Eye-witnesses confirm that the violence flowed outwards from the townships adjacent to Pietermaritzburg, engorging itself with deaths.[5] The ripeness for conflict in the peri-urban and rural areas of Natal was severely underestimated. This undoubtedly had much to do with the influx of people who had been thrown off white farms and faced the pressure of having to seek work in urban areas; the battle for land and fuel; and the decay and corruption of the tribal system of leadership.

3. What role have Inkatha and the UDF played?

The UDF was never a political party, but constituted as a loose alliance of small and large organisations with different agendas. Any impressions it gave of political and organisational coherence in Natal were largely attributed to it by intellectuals and para-intellectuals in the organisation's leadership. It is also important to note that since the 1986 emergency declaration the UDF has been a restricted organisation.

Inkatha, by contrast, is a political party with a clear hierarchical structure and chain of command. The state allowed it to hold regular public meetings and rallies at which it engaged in paramilitary activities. Many of its leadership carry licensed firearms.

Statistics from January 1987 to June 1988, show that Inkatha supporters and vigilantes killed more UDF supporters and comrades than they themselves lost in the violence. However, the political affiliations of a large number of those killed during this time were not known.

One explanation for this is the superior firepower of Inkatha supporters. In an analysis of 88 incidents in 1987 in which Inkatha was alleged to be the initiator and in which people were killed, 35% of incidents involved guns, with 30% involving knives or hand weapons and 35% unknown. By contrast, in the 29 incidents alleged to have been initiated by the UDF or Cosatu in the same period only 21% of the incidents involved guns, with 48% involving hand weapons and 31% unknown. Of the 40 incidents in which people were killed by Inkatha supporters in the first three months of 1989, 45% involved guns. In seven comrade-initiated incidents only 14% involved guns.

Reliance on police statistics is limited, not only because of their tendentiousness but also because it is often hard to determine what period or area they refer to. However, a statement by Brigadier Jac Buchner in Pietermaritzburg on 1 March 1989 is instructive. In defending police impartiality he said that of 261 violent unrest-related cases brought to court, ranging from murder to assault, arson and rape since April 1988, some 252 Inkatha and 156 UDF supporters were involved (*NW*, 02.03.89).

4. Has the state sided or colluded with Inkatha?

It may be fruitless to speculate on police understanding of the conflict or their agendas in relation to it. At first, when the number of the killings rose dramatically in September and October 1987, there appears, at least for public consumption, to have been a denial of the problem. At the time, Brigadier J Kotze, the SAP's divisional commander said: 'There is no reason to worry, entry in and out of the townships is no problem and under control. There have been no stepped-up patrols and the police can cope with the situation... I repeat, the situation is normal and under control' (*S Trib*, 18.10.89).

The police under-reported killings as did virtually every other source during 1987. This may be due to a deliberate unwillingness to publicise what was in effect a civil war which had erupted in spite of the state of emergency. The Minister of Law and Order's refusal to provide statistics for deaths and injuries in 1987 is indicative of this attitude (*NW*, 09.04.88).

The state response is further illuminated by correspondence between the state president and the Anglican Bishop of Natal during 1988 and 1989. In mid-1988 PW Botha said that senior police officers had investigated the region and declared that calm had returned. Either Botha was misled by his subordinates or he had a dubious conception of calm. In 1989 Brigadier Buchner made a similar announcement, seemingly on the basis that huge mobs were no longer 'spontaneously' killing people. However, premeditated assassinations continued.

Allegations of more active collusion between the state and Inkatha arose in relation to the Inkatha attack on Ashdown township on 31 January 1988, Inkatha attempts to regain control in Vulindlela during 1988 and 1989, killings in Imbali at the beginning of 1989 (the subject of a Cosatu dossier and press conference), and attacks on Mpophomeni (the subject of a successful interdict against the police).

The deployment of 'kitskonstabels' in the region gave rise to many complaints because of Inkatha recruits in their ranks. The response of a police spokesman that 'while every special constable is screened as far as their previous criminal records are concerned, we do not question their affiliation to any political organisation. If we enlist men into the police force we do not take into account whether they are members of Inkatha or the UDF', was unconvincing (*NW*, 10.04.88).

This collusion was in effect officially blessed when Minister of Law and Order Adriaan Vlok said at a police ceremony in Pieter-maritzburg that 'the police intend to face the future with moderates and fight against radical groups... Radicals, who are trying to destroy South Africa, will not be tolerated. We will fight them. We have put our foot in that direction, and we will eventually win the Pieter-maritzburg area' (*NW*, 27.02.88). Vlok reiterated this approach with his 'iron fist' speech in parliament during April 1989.

The way the police played their role on the ground is as important as their leaders' public pronouncements. In 1988 and 1989 the police appeared to many observers to be supportive of Inkatha. This need not be a matter of conspiracy but may be the tendency of an apartheid-nurtured police force to side with the forces of conservative tribalism. Colonel JJA Fourie's statement after killings in KwaMashu that an Inkatha crowd might have appeared to 'outsiders' as though it was armed, but it was not in fact an 'armed group' because 'traditionally Zulu men carry arms', explains much.

A similar conclusion must be drawn about the use of emergency detention as a means of halting the bloodletting. The rationale for the declaration of the 1986 and 1987 states of emergency was that unrest and killings had required the use of extraordinary measures such as emergency detention. The detention of over 734 UDF/Cosatu supporters and comrades in 1987 is understandable in terms of this rationale as some UDF supporters had undeniably been involved in killing people (at least 62, according to statistics for 1987).

But the fact that no Inkatha members or supporters were detained in 1987 was inconsistent with the emergency rationale, especially as Inkatha supporters had been responsible for 125 deaths.

In 1988 a similar situation prevailed with at least 460 anti--Inkatha individuals detained from January to June whilst only 21 Inkatha supporters were detained - most of them very briefly.

Complaints that state forces had colluded with Inkatha through lack of action against alleged killers were made by Chief Mhlabun-zima Maphumulo, a member of the KwaZulu Legislative Assembly, who argued that police should have detained all known figures who had killed and maimed people (*NW Echo*, 07.01.88).

In large-scale Inkatha attacks on Slangspruit on 25 and 27 June 1988, victims stated that attackers included Inkatha members who had bussed in from Mpumuza (Vulindlela) and Hammarsdale. One person was killed, several injured and a number of houses destroyed. On the day before the first attack, 30 Slangspruit women had gone to a police station in Pietermaritzburg to request police protection against Inkatha attacks.

On 2 December 1988 Pierre Cronje, MP for Greytown, warned the police of an impending attack on Trust Feeds. That night 11 people were massacred in a single house.

The extent of passive collusion between the security forces and Inkatha is further illustrated by the police and army's tolerance towards Inkatha meetings and rallies, which were clearly illegal in terms of the emergency regulations.

5. Why have the peace initiatives not succeeded so far?
It is not in the perceived interests of one or more of the parties to have peace.

Whilst a variety of factors complicate the peace process, more substantial progress would have been made if a genuine desire for peace had been present amongst all parties.

Some of these complicating factors are undoubtedly the business sector's tendency to favour Inkatha and particularly Chief Buthelezi, and to discount negative reports about them, as well as its cowardice in refusing to confront the security establishment. Other problems relate to communicating with and disciplining UDF-supporting comrades, given that the organisation is banned and structurally not organised as a political party.

Whilst there may be some truth to the short-lived triumphalism experienced by the UDF at having beaten off Inkatha in 1987, the reality of 1988 and 1989 made it clear that peace was in their interests. There was no way that they could take on both Inkatha and the state.

Cosatu had, from the start, a vested interest in peace and appears to have consistently worked for it. The main stumbling block appears to have been the state and, to some extent, Inkatha.

6. In whose interest is the continuation of the conflict?
Continuation of the conflict is not in the interests of the UDF, ANC or any other groups arguing for a democratic, non-racial and unitary state. It is not in the interests of Cosatu. Neither is it in the interests of Inkatha. Apart from destroying the basis of its *realpolitik* moderateness - its non-violence and associated willingness to enter into acceptable compromises with the existing white order, particularly in the economic sphere - it is fast losing international

credibility and has fashioned a long-term burden of revengeful hatred for itself in Natal. The conflict makes its chances of governing Natal in some compromise settlement fraught with dangers. Worst of all, the conflict makes Inkatha increasingly dependent on the South African state.

By contrast, the Natal conflict has been in the short-term interests of government for the following reasons:

- it is cheap. The state has not poured material resources into Natal to 'mop up the oilspots' as it has done in other troubled localities;
- it keeps Inkatha, and hence also KwaZulu, occupied so that it does not have time or energy to address the long-term cost-benefits of its current relationship to the South African state;
- it keeps pressure off the state. It was a great accomplishment that comrades in Natal saw Inkatha as 'the main enemy' rather than the apartheid government.

The South African government has built up a formidable body of expertise in setting up and co-opting traditional elements in collapsing tribal societies. It has decades of experience with bantustan systems which have led to remarkable stability. It has growing experience with destabilisation, often of a violent kind.

In these terms the Natal Midlands conflict could be seen as one of apartheid society's greatest achievements.

Notes

1   For example in the articles in the *South African Labour Bulletin*, 13(2), February 1988.
2   M Bennett and D Quinn, *An Overview of Political Conflict in South Africa: data trends 1984-1988*, Indicator Project, 1988.
3   M Orkin, 'Politics, social change and black attitudes to sanctions', in M Orkin (ed), *Sanctions against apartheid*, Cape Town, 1989.
4   JJW Aitchison, *Numbering the Dead: Patterns in the Midlands Violence*, Centre for Adult Education, University of Natal, Pietermaritzburg, 1988; and JJW Aitchison, *Last Blanket in Pietermaritzburg*, Centre for Adult Education, University of Natal, Pietermaritzburg, 1989.
5   T Smith, 'Experiencing the violence', unpublished paper given at the Seminar on Political Violence organised by the Centre for Adult Education of the University of Natal, Pietermaritzburg on 18.04.88.

Inkatha: Marching from Natal to Pretoria?

Gerhard Maré and Muntu Ncube

The scale and nature of political violence in Natal, together with the ambiguous role of KwaZulu Chief Minister Buthelezi as a political figure, continue to raise questions about Inkatha's role in struggles to abolish apartheid. This has been emphasised by the organisation's involvement in the destruction of democratic anti-apartheid structures in various Natal townships.[1]

While a civil war, which has cost more than 1 000 lives in the past two years, rages on in Natal, Buthelezi and Inkatha have continued to take tentative steps towards a role in national politics. Since the split with the ANC in 1979, and the consequent loss of credibility that had rubbed off through contact with the banned movement, Inkatha leaders have consolidated an ideological and political base in the Natal/KwaZulu region, aiming to use the region as a stepping stone into national politics.

The move into national politics will probably bring Inkatha leaders into negotiations with the ANC, Cosatu national, and the mass democratic movement, thus enhancing its status and moving negotiations away from the embarrassment of the behaviour of its local-level supporters. However, Buthelezi cannot simply abandon the region, his mobilising arena, in favour of wider ambitions.

Since 1980 Inkatha's major public interventions - the Buthelezi Commission and the Indaba - have strengthened regional alliances and advanced regional constitutional blueprints. The Indaba constitution, if implemented, would make South Africa a federation, providing a constitutional framework favouring specifically regional movements like Inkatha. But at the national level Inkatha would have to operate without its regional support structures, administration and control, and the central state would set the agenda and the pace of change - hence the importance of a national base for Buthelezi.

Inkatha is moving into uncharted waters. Some of its staunchest supporters have accepted that even before the period of 'Zulu' consolidation, Inkatha was largely restricted to Natal and KwaZulu, and more specifically, to rural areas. Inkatha leaders claim 40% non-Zulu membership, but this is based on a misreading of a German study of the 1970s. The figure has gained a measure of credibility through repetition, but Inkatha leaders and advisors are probably wise enough not to believe their own propaganda.

Even in Natal and KwaZulu, Inkatha has not had it all its own way. Attacks by its members or supporters on progressive organisations and activists have led to an increasingly sophisticated, if uneven, response from 'amaqabane' (the 'comrades') in the violent conflict with the 'otheleweni' (the 'ones who throw over the cliff').

At one stage Pietermaritzburg businessmen, traditionally Buthelezi's strongest allies, questioned the ability of Inkatha leaders to control their members, particularly within the movement's stated policy of 'non-violence'. Inkatha, here operating as regional government, has refined its own strategy - for example, in calling for the extension of 'Zulu' police control over the affected townships.[2]

Many progressive activists describe Inkatha as a sell-out organisation. This is often reinforced by the argument that since the bantustans are creations of the apartheid state, the organisations and institutions there are simple extensions of the state. But this formulation prevents the development of a detailed understanding of the tensions within the bantustans. It also obscures the fact that only small sections of the bantustan populations ever benefit from apartheid policy.

While the 'sell-out' epithet is not entirely incorrect, it is an inappropriate analytical starting point. For Inkatha, unlike political parties in the other bantustans, has a significant support base and has come closest to forming a popular alliance of various social classes and groups within the region.

KwaZulu, Inkatha and the bantustans

There has been a tendency to push the exceptional character of Inkatha and Natal politics to an extreme, thereby preventing an analysis of similarities between the various bantustans. The conflict that Inkatha has to confront is a reflection of the same fragile hold that the bureaucratic petty bourgeoisie[3] has over civil society in all appendages of the central state.

It is necessary to specify the 'core' of the African petty bourgeoisie, otherwise the erroneous impression is created that the 'petty bourgeoisie' as a whole is the overall beneficiary of the bantustan programme. In fact, the crisis in these bantustans has produced some remarkably strange alliances. For instance, in Kwa-Ndebele an anti-independence alliance of comrades, chiefs, the royal family, and sections of the civil service and trading class has arisen. Such alliances are not limited to KwaNdebele.

In almost all the other bantustans there has been some conflict between the 'core' and the rest of the petty bourgeoisie, as well as serious tensions within the bureaucratic petty bourgeoisie itself. For instance, in Kangwane the bureaucratic petty bourgeoisie has made significant overtures to the MDM; in QwaQwa and Venda teachers have resisted the ruling clique; and there have been internal disputes reflected in the Transkei coup, the Ciskei, and the attempted coup in Bophuthatswana. The same is becoming true for KwaZulu.

At the same time there are the specific and unique features of Buthelezi's leadership style, of KwaZulu, and the region as a whole. It is important to account for the contradiction between Inkatha's regional and national operations and profile, and to be aware of the consequences of Inkatha's attempts to resolve this contradiction.

The basis of the crisis within the bantustans is the sharpening of contradictions facing the apartheid regime itself. The central state's fiscal crisis has exposed the extent of dependence of bantustan ruling cliques on the apartheid state. The increasing reluctance of the central state to finance corruption in the bantustans has severely weakened patronage networks and halted their extension - at least for the moment. This has left dissatisfied clients of the old order.

In KwaZulu, patronage at the level of the bantustan parliament is being curtailed, while local-level patronage which depends on individualised extortion and redistribution is increasing. The degree of social collapse has made possible the growth of 'warlordism', a form of patronage at 'community' level.

It would be absurd to argue that Inkatha has been involved in all violence in the region. But as much as Inkatha, as governing party, takes credit for the services it provides, it also shares in the blame associated with neglect in the region. The immensity of social deprivation is something Inkatha cannot deal with as a 'government' within a 'liberated zone' - KwaZulu. Thus it is increasingly left with the task of controlling the effects of this deprivation.

The regional context

Over the 18 months from July 1987 to December 1988 more than 1 000 people died in political violence around Pietermaritzburg and the Natal Midlands. At least 3 000 houses were burnt down, leaving some 30 000 people homeless. These deaths do not include those incurred in 'faction fights' - a term frequently serving as a cloak for a range of incidents; nor do they include the high and growing number of people killed in violence in the rest of Natal.

Explanations for the violence have varied. The Inkatha Institute advanced a 'social deprivation' theory,[4] while National Party MP Schutte 'explained' the violence to the BBC as a glorified faction fight between the 'Zulu' Inkatha and the 'Xhosa' ANC/UDF. Others perceive the conflict in terms of a structural imbalance between a state-supported conservative organisation on a drive for regional control (Inkatha) and the 'mass democratic movement', suffering under both the state of emergency and the actions of the regional governing party.[5]

While the Inkatha movement is unique in the context of apartheid politics, it too has had to deal with social upheaval similar to that experienced in all the other bantustans. In 1980, after the break with the ANC, Inkatha was, for example, confronted with school boycotts in KwaMashu. Its robust crushing of the boycotts set the tone for regional conflict during the 1980s and lost Inkatha support among pupils and parents.

The formation of Cosatu in 1985 confirmed that organised labour was also lost to Inkatha. In 1986 it formed its own labour grouping - Uwusa. Responses in the labour arena paralleled those of the school boycotts and the Ngoye massacre, with Inkatha members murdering Mawu trade unionists in Mpophomeni during December 1986. Inquest findings into the incident identified nine Inkatha members, including national youth organiser, Joseph Mabaso, as being involved. Those who died in the Mpophomeni incident had worked at BTR-Sarmcol, where Uwusa later organised scab workers and entered into a recognition agreement with management.

The tension and violent conflict between Inkatha/Uwusa and Cosatu unions was referred to in mitigation in the trial and conviction of Pietermaritzburg TGWU regional chairperson and national vice-president, Alfred Ndlovu, early in 1989. He was sentenced to an effective five years' imprisonment for aiding an ANC gunman who had fired on a group of Inkatha youth during August 1987, wounding 14. Ndlovu himself had been ambushed and attacked, and five of his fellow-drivers killed in the violent period leading up to the shooting (*DN*, 24.02.89). These were not isolated incidents in the struggle for allegiance of the working class, but examples of a pattern that has existed for years.

In September 1988 the conflict in and around Pietermaritzburg took a new turn with the signing of a 'peace accord' between Inkatha and Cosatu. In terms of the agreement Cosatu agreed not to pursue a number of cases arising out of interdicts against Inkatha members (including some of the 'warlords'); a board, headed by a retired judge, was to hear complaints against members of both organisations, and make recommendations on action. Both signatories also agreed to discipline their members, publicise the accord, and allow free political action within the area.

The accord falls within the general Cosatu strategy of attempting to contain the violence within acceptable channels. But there were serious limitations to its success, if measured in terms of incidents and deaths. Many of the participants in or victims of the violence were unemployed youths, not members of Cosatu unions. Other factors that detracted from the potential effect of the accord were the non-involvement of the UDF (through its proscription and the detention of many of its officials), and the geographical limitation of the accord to Pietermaritzburg.

The adherence of participants is central to the success or failure of any accord. The complaints adjudication board (CAB) issued one report in which it reprimanded Inkatha for its lack of action against members who breached the accord. The CAB withheld its report for two months after Inkatha said it needed more time to respond to the recommendations. When no action was forthcoming the report was released, stating that there was strong evidence to suggest that two Inkatha leaders, Shayabantu Zondi and Lawrence Zuma, had broken the agreement through threats of violence and intimidation, and that violence and threats had been used in recruitment of members (*NM*, 26.01.89; *WM*, 27.01.89). Inkatha has taken no action against the two men.

Buthelezi did, however, take the near-unprecedented step of publicly warning certain Inkatha members that they were embarrassing the movement (*NW*, 20.02.89). This step might just as well have been taken against former Inkatha central committee member and Lindelani warlord Mandla Shabalala, as against Midlands Inkatha leaders. Two of Shabalala's officers were sentenced to death for the murder of seven schoolboys (*NM*, 08.02.89). Another five members of the gang ('guards' employed by Shabalala) received prison sentences for kidnapping the murdered children.

Classes and levels of operation

Inkatha operates on at least three different levels. But these do not necessarily indicate constitutionally-determined hierarchies of authority.

The first is the national and international context in which the movement is almost exclusively represented by Buthelezi. This personalised politics is partly because Inkatha is not formally a national party competing in a national parliament. Lacking the structures to have anything other than a personalised presence at this level, Inkatha is presented as a national extra-parliamentary liberation movement.

At the second level, Inkatha functions within the KwaZulu legislative assembly as a bantustan regional government and political organisation.

At the third level, that of community, individuals, rather than corporate entities or 'sectors' of society, are recruited into, or act in the name of, the movement.

Inkatha is essentially a petty-bourgeois organisation. But it is a 'core' of that class which Inkatha represents, rather than the interests of the African petty bourgeoisie in Natal as a whole. Furthermore, identifying Inkatha as petty bourgeois is not done on the basis of class membership or the origins of its leaders alone. This is based on an analysis of the politics and ideology of the movement. These are distinctly petty bourgeois in that they advance economic interests aiming for full bourgeois status.

The various layers of the petty bourgeoisie which might benefit from such a course include, first and foremost, a stratum of civil servants and other professionals, including teachers and nurses, who are employed by the bantustan branch of the state or other semi-public institutions. This group largely reproduces the dominant patterns within civil society, and has increased rapidly, given the multiplication of state departments and self-conscious formation under apartheid.

The second stratum consists of the collaborative bureaucracy both in the bantustans (chiefs, top bureaucrats, central legislative assembly members, etc), and in the urban areas (community councillors). In the Natal/KwaZulu case, it is increasingly defined both politically and through its insertion into private and parastatal economic structures. This stratum, particularly in its bantustan version, has evolved over a long time and directly out of colonial and pre-colonial African societies. But its present form makes it a new class stratum with characteristics deriving from the apartheid era.

Lastly, there is the traditional petty bourgeoisie, consisting of traders (the biggest component part), small farmers and landowners. This is the second largest grouping within the African petty bourgeoisie, albeit small in absolute terms.[6]

Formal class representations within Inkatha, or in its functional levels, are not those of the working class. Recruitment into Inkatha takes place at the lowest level, where it recruits 'the people' as *individuals* having no necessary class effect within the movement.

This examination of classes and social groups within Inkatha enables us to draw out the contradictions and fault-lines which characterise this populist movement. Recent events can also be better understood and explained in terms of the contradictions they illustrate, both at the level of the central state and economy and within regional branches.

Level one: 'the anvil on which apartheid ultimately faltered'[7]

The 'anvil', referred to by Anglo American's Gavin Relly, is not Inkatha, but that marketable political commodity, Chief Minister Buthelezi. Buthelezi has collected together a multiplicity of symbols, with the result that he can speak on behalf of a range of classes and other social interests. The working class, for example, is present only indirectly, not through representatives of its own choosing, but mediated through claims made on its behalf by Buthelezi. These claims are centred around the issue of support for the Inkatha leadership's rejection of sanctions; its belief in the positive effect that direct foreign investment has in the South African economy and politics; and its belief in a just equilibrium between workers and capitalists.

However, Buthelezi does not simply operate as champion of the working class. He is also an hereditary leader of 'Zulus', a democratic spokesperson for the oppressed, and repository of the values of 'the founding fathers of the African National Congress'. He features internationally as one of the most familiar black politicians, along with Nelson Mandela and exiled ANC leaders. Through him Inkatha has an image of being both indispensable and desirable in a settlement moving away from apartheid. Buthelezi's pronouncements and actions have an importance at this national and international level largely because they dominate in image formation, and Buthelezi manages national and international personal contact himself. A recent 'official' biography is aimed at this 'market'.[8]

At this first level, Buthelezi has established himself as a serious participant in shaping a future South Africa. He has done this through highly publicised meetings with world leaders to whom he speaks on behalf of 'black South Africa'.

Many alliances are formally cemented at this first level, through projects like the Indaba. The Pietermaritzburg violence, subsequent court interdicts, murder convictions, and the generally unfavourable publicity for Inkatha did not deter the Indaba and its supporters. It is astonishing to note how infrequently Inkatha has been criticised by the 'liberal' opposition in South Africa. The only examples that come readily to mind involve the PFP when 'oaths of allegiance' were required of civil servants, and the late Alan Paton, who reprimanded Buthelezi for anti-Indian threats.

The Indaba team has shifted its activities to a lower-profile campaign of winning over opinion-makers in the state and capital. Offices were opened in both Cape Town and Johannesburg for this purpose, while the ex-director of the Indaba, Dawid van Wyk, continues his attempts at obtaining state approval for the project in Pretoria. Early in 1989 meetings were held in Cape Town between the Indaba secretary, KwaZulu minister of education and culture, and Inkatha secretary-general, Oscar Dhlomo, and Minister Chris Heunis, on the Indaba proposals. These talks coincided with the first meeting of the joint committee to examine obstacles in the way of negotiations between Inkatha and Cosatu/UDF.

At this first level, Inkatha benefits from an image of negotiation and non-violence largely uncluttered by what occurs at the other levels. Events at the level of regional government and 'the

community' feature little in the local press and even less so internationally. Debates in the KwaZulu Legislative Assembly are a
mystery to all but listeners to Radio Zulu and readers of the
Inkatha-owned *Ilanga* newspaper.

Because of its 'uncluttered' image in some national quarters,
Inkatha has been presented as an essential partner in moves to
restructure white opposition politics. It had vocal support from the
Independent Party, the PFP and some members of the NDM prior
to their merger into the DP - despite the misgivings of those with
greater local-level knowledge and experience.

At this first level of operations, Inkatha projects an image of representing the true wishes of a very large percentage, if not the
majority, of Africans in South Africa. This is important both in shaping current policy and the development of a position with regard to
future national negotiations within the country.

Level two: 'there will always be a region to govern'

Inkatha has a direct presence within the structures of administration
and legislation. Models of European parliamentary politics, with an
apparent distinction between governing party and state, do not
apply. This is partly due to the subordinate status of the bantustans
as branches of the central state, and the lack of pressure to
reproduce the 'state' through multi-party democracy. In KwaZulu,
public funds are used for activities which enhance the party because
it is not clear where the party ends and the 'state' begins.

At this level there is an intricate interaction between Inkatha's
status as a 'liberation' movement and its position as a complaining,
if central, element of the apartheid system. It is at this level that
'state' apparatuses are available to Inkatha, through the KwaZulu
branch of the central state; through its dealings with a civil service;
and through its mobilising call to a 'Zulu nation'. This is also the
level of operations where some of the interests of the petty bourgeoisie can be met.

Here Buthelezi is the chief minister of KwaZulu, a chief in his
own right, historically destined to be 'prime minister' to the Zulu
royal family, and a member of the royal house.

Chiefs, as the numerically dominant group in the KwaZulu
bureaucratic petty bourgeoisie, constitute Inkatha's immediate
power base in the KLA. The co-operation of chiefs in KwaZulu and

their position within Inkatha have been maintained mainly through two mechanisms:

- as traditional leaders of their respective tribes;
- through the KwaZulu government's system of patronage which pays them and which has treated chiefs more sympathetically than civil servants with regard to salary increases.

Buthelezi's stature as chief and member of the Zulu royal house has also conferred a particular status on the chiefs.

These factors contribute to the chiefs' loyalty to Inkatha. Additionally chiefs have access to a political organisation which is committed to their longer-term political and administrative role. However, potential for conflict between their membership of Inkatha and their roles as 'fathers of the nation' at grassroots level, cannot be ruled out.

An indication of this tension is the role that chief Mhlabunzima Maphumulo of Maqongqo has played. In October 1988 he called a meeting of Inkatha officials, his own indunas and Cosatu officers to celebrate his 15-year office during which time there had been no political violence. At this gathering, Maphumulo explained that

> I am an Inkatha man, but that does not mean I have to discriminate. I have to accommodate every member of my tribe irrespective of their political allegiance, be it UDF, Cosatu, Inkatha or Azapo. I will not tolerate people who go house to house forcing others to join their organisation (NW, 10.10.88).

Chief Maphumulo believes that this stance has not endeared him to other chiefs, or to Ulundi (NW Echo, 27.10.88). There have been at least two other cases involving similar 'rebellious' incidents.

Town councils in KwaZulu are also beginning to manifest signs of similar discontent. The mayor of Umlazi, Prince Patrick Zulu, has been critical of the direct involvement of Ulundi in local matters, and accusations of unfair practices with regard to the allocation of land have been made. Even cabinet members, like Interior Minister Dennis Madide and Education Minister and Inkatha Secretary-General Oscar Dhlomo, have been critical on issues like this (STrib, 12.03.89; NM, 10.09.88).

The most strongly represented petty-bourgeois stratum within Inkatha consists of traders. They, unlike the chiefs, exist within capitalism, and have economic aspirations demanding extra-economic assistance for their realisation. The traders stand to gain most from Inkatha's control of the KwaZulu Legislative Assembly and of the regional government (licensing, business sites, access to finance from the KwaZulu Finance and Investment Corporation and monopoly capital, contracts, etc).

The incorporation of traders has, however, not been without conflict as divisions surface around economic policy, the distribution of sorghum beer within KwaZulu, and participation in or support for boycotts of white-owned shops. These conflicts came to a head with a split from Inyanda (the African chamber of commerce for the region) and the formation of KwaNacoci (the KwaZulu/Natal Chamber of Commerce and Industry).

The Inyanda split expresses the choices available to the regional trading petty bourgeoisie - between a longer-term vision with a role in a liberated South Africa, and the immediate benefits of a close relationship with a regional government prepared to protect their interests on a racial and ethnic basis in the foreseeable future. The former is dependent on establishing a relationship with the ANC and MDM, and a commitment to national politics. The latter depends on the establishment of links with monopoly capital which will benefit a select few.

After Nafcoc's 1986 visit to the ANC and conditional support for sanctions, Inyanda chairman PG Gumede said that while his region supported the pro-sanctions motion at the Nafcoc summit, 'We (in Natal and KwaZulu) trade where there is strong opposition to sanctions and Chief Buthelezi - who leads Inkatha, of which we are an affiliate - has been campaigning vigorously against it'.[9] The same point was reiterated by the Inyanda management committee (*CP*, 06.03.88).

As a consequence of the tension which arose out of this position, Gumede resigned his regional post and relinquished his national role in 1988. Two other Inyanda and Nafcoc officials, Roger Sishi and Reuben Tshabalala, followed suit (*STrib*, 21.02.88). Mdu Lembede, writing in the *Sunday Tribune* (21.02.88), commented: 'Inyanda's very survival depends on the strong membership and support it enjoys among rural KwaZulu traders, most of whom are staunch royalists and would not dream of questioning Inkatha policies'. The power of the numerically strong rural traders was used to bring the rebels into line.

The KwaZulu/Natal Chamber of Commerce and Industry (KwaNacoci) was formed at an August 1988 meeting in Ulundi, attended by 800 businessmen. New president, SJ Mhlungu, had the right credentials for the position: he had served as managing director of Khulani Holdings (the Inkatha investment arm), is a KFC director and KLA member.

The struggle for the allegiance of the traditional petty bourgeoisie is not over yet. The depleted Inyanda will continue to function. But the split in the organisation reflects tensions within Inkatha, which is unable to represent traders adequately at the national level (level one) while it holds patronage reins at lower levels (through licensing, access to finance, and joint ventures with large-scale capital). If communities take action here (levels one and two), it puts the traders in an untenable position, especially those located in the more politicised urban areas.

The civil servants are probably the element of the petty bourgeoisie least adequately incorporated into Inkatha. The teachers' body, Natu, is affiliated to Inkatha but this seems a reluctant contact, and teachers appear to have little influence over educational policy.[10] Civil servants recently staged a mini-revolt over salaries, after an announcement by Buthelezi in February 1988 that all lower-level civil servants would achieve pay parity with their white counterparts. This was the group most clearly discriminated against within the bantustan system, deprived of the high salaries of the bureaucratic petty bourgeoisie.

Civil servants were promised that the increases would only come into effect in May or June, but would be back-dated to March. By October, when nothing had been forthcoming, the KwaZulu Staff Association blamed the KwaZulu government for the delay in paying out the amount allocated by the central government the previous year. Strike action was threatened from 1 December. The KwaZulu Public Service Commission responded by blaming the delay on Pretoria. Attention was drawn to a Buthelezi statement saying that the increase would be paid no later than January 1989! The Department of Development Aid admitted that they had not provided the bantustans with 'the necessary detailed information to implement the parity measures' (*NM*, 08.11.88).

In response, the KwaZulu Staff Association (Kwasa) held a protest meeting in Durban rather than Umlazi, because they feared an attack by Inkatha. It was claimed that about 500 of the 5 000 affected civil servants attended the meeting at which the KwaZulu Public Service Commission was rejected. Some of those present expressed resentment over Radio Zulu announcements that they would lose their jobs if they stopped work. KwaZulu was blamed for the delay in payouts. And at a meeting he had called in Umlazi, Kwasa chairman Jetro Sokhela was asked to resign. Sokhela subsequently suspended three members for 'involving trade unions in the issue', and for making press statements. Trade unions are forbidden for KwaZulu employees.

Kwasa won the showdown with the southern region of the association, who backed down after a promise that increases would be paid on 23 December. However, the flurry of activity confirmed widespread dissatisfaction with the KwaZulu government. The protest also established links between protesters dissatisfied with their staff association, and Cosatu's National Education, Health and Allied Workers Union (Nehawu).[11]

Dissatisfaction also boiled over early in 1989 when teachers held boycotts and meetings in Mpumalanga and KwaMashu to protest the issue of salaries. With teachers making overtures to Nehawu (*DN*, 06.03.89), the fields of education and employment are set to be sites of conflict in the future.

The issue of 'pledges of allegiance' to the KwaZulu government and its chief minister (and hence to the Inkatha movement) has also caused tension. Since 1984 pledges have been demanded of all KwaZulu employees. At the beginning of 1987 teachers were asked to 'solemnly declare' not to 'vilify, denigrate, or speak in contempt' of Buthelezi, his cabinet or the KwaZulu government. By mid-1987 this pledge was made mandatory with Buthelezi warning that failure to sign would amount to 'misconduct'.

The pledge, which had previously been 'unofficial' for doctors, was now formalised. Frank Mdlalose, KwaZulu Minister of Health, said he was 'disgusted' when members of his department had refused to sign. Namda president Diliza Mji blamed the shortage of doctors in KwaZulu largely on the pledge. Mdlalose denied this (*STrib*, 24.04.88 and 25.05.88). Whatever the case, in October 1988 a KwaZulu/Natal delegation went on an overseas recruiting drive for doctors (*NW*, 25.10.88) as only 36% of rural posts for doctors were filled in KwaZulu.

Level three: Inkatha and constituency politics

Inkatha leaders have always made much of the movement's large membership, its involvement in 'constituency' politics (referring to a mode of organisation and democratic representation), and its policy of 'non-violence' (qualified by an equally strong commitment to revenge politics). However, over the years it has become clear that the top structures of the movement are in no position to ensure disciplined behaviour at lower levels. Social tensions, endemic to the bantustan system, have created contradictions (such as unemployment) which neither Inkatha nor the central state can control. For example, Inkatha (as government in KwaZulu) has never been able to solve 'faction fights', despite exaggerated claims of what it intends doing, a policy of unity of 'Zulus', and police powers of detention specifically for 'faction fighters'.

At the 'community' level, Inkatha functions through tribal authorities and chiefs who provide one of the few links between the second and the third levels. The role of the chiefs and their loyalty assumes great importance. In the 1970s there was, therefore, great sensitivity to insignificant dissent from certain chiefs at the direction taken by Buthelezi, and to their initial reticence in accepting that Inkatha was advancing their interests.

However, at that stage the only strong alternative to Inkatha was the central state. The rebellion was, therefore, soon discredited. At present chiefs are subjected to contradictory pressures - from the top where Inkatha makes decisions relating to national politics and international events; and from below, where there is day-to-day interaction with 'subjects' who need jobs, education, law and order, pensions, etc, or who are caught in other anti-apartheid political allegiances to trade unions or the MDM.

This is the level where the 'warlords' function. As a fairly recent phenomenon they lack the formal powers and 'legitimacy' which chiefs have. However, warlords have the ability to operate outside of formal processes, with brute force and intimidation. Because their power is not bestowed, and their patronage is local, the warlords also pose a threat to Inkatha.

For Inkatha to respond by putting warlords and their followers in the uniforms of police reservists or 'kitskonstabels - as has happened and been suggested[12] - would remove them from their present methods of concentrating and using power. It would also reflect badly on Inkatha at other levels, as did the co-option of Shabalala onto the central committee, and as the election of 'warlords' to the KLA might still do.[13] If Inkatha's various levels of operation are brought together through the KLA and the Inkatha central committee, this could unsettle the image so meticulously created and maintained at the higher levels.

At the 'community' level of operation, Inkatha's presence takes many forms, as might be expected from a movement with a sophisticated image at the top but very little hold over the day-to-day actions of those who are associated with it. The violence that has devastated the townships and 'informal settlements' in Natal, especially over the last 18 months, is found at this level.

To analyse the specifics of local class composition, formation and representation, it is necessary to make sense of what Inkatha is, and how it operates.

Inkatha faces a number of major problems in the way it operates at present: how local-level Inkatha leaders and supporters separate the justification for revenge ('an eye for an eye') from the first-level strategy of 'non-violence'; how they deal with the demands for rapidly increasing membership (a demand generated by Inkatha's ridiculous claims of more than 1,5-million paid-up members), and the reticence of many individuals to be associated with Inkatha or the bantustan system. These are contradictions intrinsic to the operation of the whole organisation.

The central state is not simply handing out tickets to the 'Great Indaba'. Inkatha has to prove itself worthy in its march to Pretoria. Contrary to the fear that Shepherd Smith claims exists in the Transvaal of 'an army of rampaging Inkatha warriors', there is probably a great deal of acclaim for such a force.[14] Inkatha is, in fact, taking control of more, rather than fewer, of the central state's control functions. This amounts to an attempt to eliminate one of the tensions between the levels at which Inkatha operates.

If level two has no real authority over the local level (operating as warlords and chiefs), then formal structures of the 'state' must be extended down to this level. This serves two purposes: first, it allows control through the uniformed presence of the Inkatha government rather than through undisciplined members or sympathisers; and, second, it gives an alternative presence when chiefs or councillors show too much 'neutrality' because of conflicting pressures. However, it creates another tension through Inkatha's integration into *formal* structures of the state.

This is the central issue for Inkatha on its road to national politics: how does it avoid going into negotiations as an extension of the state (Inkatha as government), and continue to benefit from those powers? And how does it convince its actual and potential followers that it continues to be a 'liberation' movement shaping and winning a larger 'Indaba' region that no longer contains a bantustan?

Notes

1 Not only have these organisations been very hard hit through state action (such as the banning of organisations and detentions of activists - with a recent trend being the release of detainees but under restrictions), but Inkatha leaders have specifically included the UDF, Cosatu and the ANC in the Inkatha list of enemies.

2 See Gerhard Maré and Georgina Hamilton, 'Policing "Liberation Politics"', paper presented at Centre for Adult Education Seminar on Violence, University of Natal, Pietermaritzburg, April 1988. By 1982 seven police stations had been transferred to KwaZulu. By 1986 Buthelezi was threatening court action if more stations were not transferred: 'It undermines us if we are not seen to be responsible for law and order in our townships'. In 1987 KwaZulu took control of the townships around Durban, and in early 1989 of Mpumalanga and KwaNdengezi.

3 By the 'bureaucratic petty bourgeoisie' we refer to the top echelons within the bantustan branches of the central state; those individuals who collectively ensure that the bantustans reproduce themselves as administrative, repressive and ideological fragments in apartheid society.

4 (Inkatha Institute Director) Gavin Woods, 'Identifying all the players in a typical South African township conflict', paper presented at National Conference on Negotiation and Mediation in Community and Political Conflict, University of Durban-Westville, 1988; also *NW*, 21.02.89.

5 Unrest Monitoring Project, 'The tightening noose: violence against anti-apartheid organisations in Natal/KwaZulu during the 1980s'; and Paulus Zulu, 'Crisis in the Politics of Opposition: A Look into Natal, 1981-1987', unpublished paper, Race Relations Unit, University of Natal, Durban, 1988.

6 By 1982 there were 1 742 African-owned general dealers in KwaZulu; also 231 butcheries, 973 registered hawkers, 310 fresh produce/tearoom owners, and several other categories of trade. There were also about 26 000 African teachers (1987), and 13 600 people employed in the department of health (6 300 being nurses). Figures drawn from KwaZulu Government Departmental Reports, various years.

7 Gavin Relly, quoted in Jack Shepherd Smith, *Buthelezi: the biography*, Johannesburg, 1988, 105.

8 Shepherd Smith, *Buthelezi*.

9 South African Institute of Race Relations, *Survey of Race Relations, 1987/88*, Johannesburg, 373-74.

10 See Praisley Mdluli, 'Ubuntu-Botho: Inkatha's people's education', *Transformation*, 5, 1987, and Gerhard Maré, 'Education in a "liberated zone": Inkatha and education in KwaZulu', in *Critical Arts*, 4(4), 1988, and 5(1),

1989, for mention of lack of enthusiasm in implementing policy.

11 Buthelezi attacked 'UDF and Cosatu elements' in early December 1988 for wanting 'confrontation with me at any cost' and 'using and abusing' the civil servants in the matter. Buthelezi speech, Department of the Chief Minister and Department of Police, End of Year Function, Ulundi, 02.12.88.

12 Maré and Hamilton, *An Appetite for Power: Buthelezi's Inkatha and the Politics of 'Loyal Resistance'*, Johannesburg, 1987, 215.

13 In the February 1988 KLA elections only three of 26 seats were contested. Two alleged 'warlords', Chief Shayabantu Zondi and David Ntombela, became members of the KLA (*WM*, 24.02.89).

14 Shepherd Smith, *Buthelezi*, 232.